Digital Logic and **Microprocessor Design** with **Interfacing**

Digital Logic and Microprocessor Design with Interfacing

2nd Edition

Enoch O. Hwang
La Sierra University
Riverside, California, USA

Australia • Brazil • Mexico • Singapore • United Kingdom • United States

***Digital Logic and Microprocessor Design with Interfacing,* Second Edition**
Enoch O. Hwang

Product Director, Global Engineering: Timothy L. Anderson

Associate Media Content Developer: Ashley Kaupert

Product Assistant: Alexander Sham

Marketing Manager: Kristin Stine

Director, Higher Education Production: Sharon L. Smith

Senior Content Project Manager: Kim Kusnerak

Production Service: SPi Global

Senior Art Director: Michelle Kunkler

Cover/Internal Designer: Red Hangar Design, LLC

Cover/Internal Image: Jonathan Y. Hwang

Intellectual Property

Analyst: Christine Myaskovsky

Project Manager: Sarah Shainwald

Text and Image Permissions Researcher: Kristiina Paul

Manufacturing Planner: Doug Wilke

Library of Congress Control Number: 2016952181

ISBN: 978-1-305-85945-6

Cengage Learning
20 Channel Center Street
Boston, MA 02210
USA

Cengage Learning is a leading provider of customized learning solutions with employees residing in nearly 40 different countries and sales in more than 125 countries around the world. Find your local representative at **www.cengage.com**.

Cengage Learning products are represented in Canada by Nelson Education Ltd.

To learn more about Cengage Learning Solutions, visit **www.cengage.com/engineering**.

Purchase any of our products at your local college store or at our preferred online store **www.cengagebrain.com**.

Printed in the United States of America
Print Number: 01 Print Year: 2016

To my wife Windy,
the love of my life,
for her endless love and support.

CONTENTS

CHAPTER 3
Combinational Circuits 65

CHAPTER 4
Standard Combinational Components 112

CHAPTER 5
Sequential Circuits 157

CHAPTER 6
Finite-State Machines 215

CHAPTER 7
Dedicated Microprocessors 283

CHAPTER 8
General-Purpose Microprocessors 363

CHAPTER 9
Interfacing Microprocessors 415

APPENDIX A
Xilinx Development Tutorial 486

APPENDIX B
Altera Development Tutorial 512

APPENDIX C
Verilog Summary 533

APPENDIX D
VHDL Summary 553

PREFACE

This book is about the digital logic design of microprocessors, and is intended to provide both an understanding of the basic principles of digital logic design, and how these fundamental principles are applied in the building of complex microprocessor circuits using current technologies. Although the basic principles of digital logic design have not changed, the design process and the implementation of the circuits have. With the advances in fully integrated modern hardware computer-aided design (CAD) tools for logic synthesis, simulation, and the implementation of digital circuits in field-programmable gate arrays (FPGAs), it is now possible to design and implement complex digital circuits very easily and quickly.

Many excellent books on digital logic design have followed the traditional approach of introducing the basic principles and theories of digital logic design and the building of separate standard combinational and sequential components. However, students are left to wonder about the purpose of these individual components and how they are used in the building of more complex digital circuits, such as microcontrollers and microprocessors that are used in controlling real-world electronic devices. The primary goal of this book is to fill in this gap by going beyond the logic principles and the building of basic standard components. The book discusses in detail how the basic components are combined together to form datapaths, how control units are designed, and how these two main components (datapath and control unit) are connected together to produce actual dedicated custom microprocessors and general-purpose microprocessors. The book ends with an entire chapter containing many examples on how microprocessors are interfaced with real-world devices.

Many texts on digital logic design and implementation techniques mainly focus on the logic gate level. At this low level, it is difficult to discuss larger and more complex circuits that are beyond the standard combinational and sequential circuits. However, with the introduction of the register-transfer technique for designing datapaths and the concept of a finite-state machine for control units, we can easily design a dedicated microprocessor for any arbitrary algorithm and then implement it on a FPGA chip to execute that algorithm. The book uses an easy-to-understand ground-up approach with complete circuit diagrams, and both Verilog and VHDL codes, starting with the building of basic digital components. These components are then used in the building of more complex components, and finally the building of the complete dedicated microprocessor circuit. The construction of a general-purpose microprocessor then comes naturally as a generalization of a dedicated microprocessor. At the end, students will have a complete understanding of how to design, construct, and implement fully working custom microprocessors.

Design of Circuits using Verilog and VHDL

Although this book provides coverage on both Verilog and VHDL for all of the circuits, this information can be omitted entirely while gaining an understanding of digital circuits and their design. For an introductory course in digital logic design, learning the basic principles is more important than learning how to use a hardware description language (HDL). In fact, instructors may find that students can get lost in learning the principles while trying to learn the language at the same time. With this in mind, the Verilog and VHDL code in the text is totally independent of the presentation of each topic and may be skipped without any loss of continuity.

On the other hand, by studying the HDL codes, the student can not only learn the use of a hardware description language but also learn how digital circuits can be designed automatically using a synthesizer. This book provides an introduction to both Verilog and VHDL and uses the "learn-by-examples" approach. In writing either Verilog or VHDL code at the dataflow and behavioral levels, the student will see the power and usefulness of a state-of-the-art hardware CAD synthesis tool.

New to This Edition

In this newly revised second edition, a new chapter on interfacing microprocessors with external devices has been added. Just knowing how to design and implement a microprocessor is not sufficient. The main purpose and usage of a microprocessor is to control external devices. This entire chapter contains many real-world examples on interfacing microprocessors with external devices. Students can use these examples to help them in doing their final projects.

Throughout the book, many new examples have been added and old examples updated. This new edition also covers the usage of both Verilog and VHDL, the two industry standard hardware description languages for describing digital circuits. All circuit examples, in addition to having schematic diagrams, also include codes written in both VHDL and Verilog.

In addition to the Altera FPGA development software, a new section in the Appendix is added for using the Xilinx FPGA development software. Using either the Altera or the Xilinx FPGA development software and their respective FPGA hardware development boards, students can actually implement these microprocessor circuits and see them execute, both in software simulation and in hardware. The book contains many interesting examples with complete schematic diagrams and Verilog and VHDL codes for implementing them in hardware. With the hands-on exercises, students will learn not only the principles of digital logic design but, also in practice, how circuits are implemented using current technologies.

To actually see your own microprocessor come to life in real hardware and being able to control real-world external devices is an exciting experience. Hopefully, this will help students to not only remember what they have learned but will also get them interested in the world of microprocessor controllers and digital circuit design.

Using This Book

This book can be used in either an introductory or a more advanced course in digital logic design. For an introductory course with no previous background in digital logic, Chapters 1 and 2 are intended to provide the fundamental basic concepts in digital logic design, while Chapters 3 and 4 cover the design of combinational circuits and standard combinational components. Chapter 5 on the design of sequential circuits can be introduced and lightly covered.

An advanced digital logic design course will start with sequential circuits in Chapter 5, and the design of finite-state machines in Chapter 6. Chapters 7 and 8 cover the design of datapaths and control units, and the building of dedicated and general-purpose microprocessors. Finally, Chapter 9 concludes with the interfacing of microprocessors with the external world.

It is strongly recommended that a lab component be fully integrated with the lecture. With an integrated lab, students can have a hands-on learning experience alongside the theoretical concepts that they have learned in class. In fact, many teachers find that too often not enough hours are given to the lab. As we probably know, it is often easier to understand the theory, but to actually implement a circuit and to get it to work requires much more detail and time. Ready-to-use labs that complement the lecture are available for download from the teachers' resource website at https://login.cengage.com.

Chapter 1—Introduction to Microprocessor Design gives an overview of the various components of a microprocessor circuit and the different abstraction levels in which digital circuits can be designed.

Chapter 2—Fundamentals of Digital Circuits provides the basic principles and theories for designing digital logic circuits by introducing binary numbers, the use of truth tables, Boolean algebra, and how the theories get translated into logic gates and circuit diagrams. Also a brief introduction to Verilog and VHDL is given.

Chapter 3—Combinational Circuits shows how combinational circuits are analyzed, synthesized, and optimized.

Chapter 4—Standard Combinational Components discusses the standard combinational components that are used as building blocks for larger digital circuits. These components include the adder, subtractor, arithmetic logic unit, decoder, multiplexer, tri-state buffer, comparator, shifter, and multiplier. In a hierarchical design, these components will be used in the building of the datapath used in the microprocessor.

Chapter 5—Sequential Circuits introduces latches and flip-flops as basic storage elements and then continues with larger storage components such as registers, register files, and memories. Special sequential components such as shift registers and counters are also covered.

Chapter 6—Finite-State Machines shows how finite-state machines are analyzed, synthesized, and optimized.

Chapter 7—Dedicated Microprocessors first introduces the need for a datapath, and then explains how a control unit, in the form of a finite-state machine, is used to control the datapath. The chapter expands further showing how dedicated microprocessors are constructed by connecting the datapath and the control unit together as one coherent circuit.

Chapter 8—General-Purpose Microprocessors continues on from Chapter 7 to suggest that a general-purpose microprocessor is really a dedicated microprocessor that is dedicated to only read, decode, and execute instructions. The chapter discusses the complete design and construction of two simple general-purpose microprocessors with their own custom instruction set, and how programs written in machine language are executed on them. The highlight of this chapter and this book is that these two fully-working general-purpose microprocessors can be implemented in hardware and have programs executed by them.

Chapter 9—Interfacing Microprocessors provides several complete examples on how to interface microprocessors with real-world external devices. Examples include interfacing with a real-time clock IC using the I^2C protocol, Bluetooth communication using RS-232, and drawing graphics on a VGA monitor.

The **Appendixes** provide tutorials on using both the Altera and Xilinx software development tools, and summaries on the Verilog and VHDL hardware description languages.

Supplements

Resources for the book can be found at https://login.cengage.com/. The instructor site is password protected and requires a verified instructor login to access the site.

Student Resources

- Chapter on Implementation Technologies
- Labs for each chapter
- All of the example codes from the book in VHDL and Verilog
- Altera FPGA development software download
- Xilinx FPGA development software download

Instructor Resources

- Chapter on Implementation Technologies
- Labs for each chapter
- PowerPoint lecture slides
- Solutions to problems at the end of each chapter
- All of the example codes from the book in VHDL and Verilog
- Altera FPGA development software download
- Xilinx FPGA development software download

Acknowledgments

I want to thank Professor Zhiguo Shi, Ph.D., and many of his graduate students from Zhejiang University, Hangzhou, China, for translating this book into Chinese. In the process, we have become lasting friends.

I also want to thank the following reviewers for their constructive feedback:

Christopher Doss, North Carolina A&T State University

Eric Durant, Milwaukee School of Engineering

Rajiv J. Kapadia, Minnesota State University, Mankato

Emma Regentova, University of Nevada, Las Vegas

Darrin Rothe, Milwaukee School of Engineering

I wish to acknowledge and thank the Global Engineering team at Cengage Learning for their dedication to this new book:

Timothy Anderson, Product Director; Ashley Kaupert, Associate Media Content Developer; Kim Kusnerak, Senior Content Project Manager; Kristin Stine, Marketing Manager; Elizabeth Brown, Learning Solutions Specialist; and Alexander Sham, Product Assistant. They have skillfully guided every aspect of this text's development and production to successful completion.

I also want to thank the College of Information Science and Electronic Engineering at Zhejiang University for inviting me as a visiting professor to teach their Digital Systems Design course (in English) using the contents of this book. During this time, I was able to gather many valuable ideas and feedbacks from the bright and enthusiastic students on how to make the book better. As a result, numerous changes have been made. This book truly is field-tested.

I also want to thank my school, La Sierra University in sunny California, for giving me the time off to be at Zhejiang University and to work on this book. It would have been extremely difficult without this extra time.

Finally, I want to thank my wife, Windy, for her support and giving me the time to focus and to finish this book.

Enoch O. Hwang, Ph.D.
Riverside, California

ABOUT THE AUTHOR

Enoch Hwang has a Ph.D. in Computer Science from the University of California, Riverside. He is currently a professor of computer science at La Sierra University in Southern California teaching digital logic and microprocessor design. In 2015, he was invited as a visiting professor to Zhejiang University in Hangzhou, China, where he taught their Digital Systems Design course. Many new ideas from that class have been incorporated into this edition of the book.

Even from his childhood days, he has been fascinated with electronic circuits. In one of his first experiments, he attempted to connect a microphone to the speaker inside a portable radio through the earphone plug. Instead of hearing sound from the microphone through the speaker, smoke was seen coming out of the radio. Thus ended that experiment and his family's only radio. He now continues his interest in digital circuits with research in embedded microprocessor systems, controller automation, power optimization, and robotics.

CHAPTER 1

Introduction to Microprocessor Design

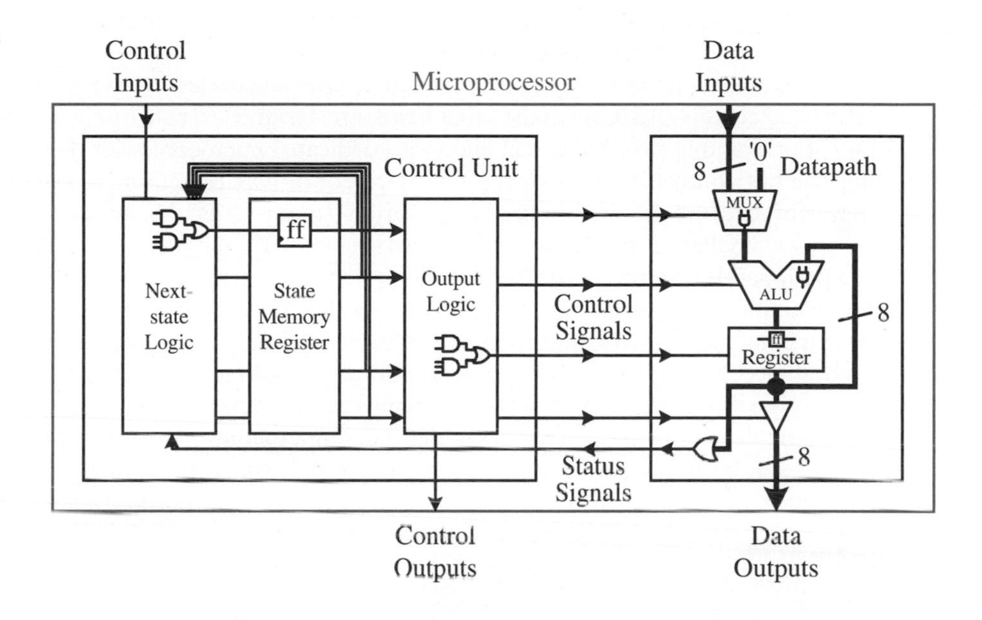

Electronic devices are an integral part of our lives. Every day and everywhere we see and use electronic devices, from cellular telephones to electronic billboards, cars, toys, TVs, elevators, musical greeting cards, personal computers, traffic lights, and many more. Inside each and every one of them, there is a microprocessor that controls their operations. Microprocessors are at the heart of all of these "smart" devices. Their smartness is a direct result of the work of the microprocessor, without which none of these electronic devices would be able to operate as they do.

There are generally two types of microprocessors: **general-purpose microprocessors** and **dedicated microprocessors**. General-purpose microprocessors, such as the Intel Core™ i7 CPU shown in Figure 1.1(a) can perform different tasks under the control of different software programs. General-purpose microprocessors typically are much more powerful in terms of processing power and speed. However, they usually require external components for their memory and supporting input/output (I/O) peripherals. They are used in all personal computers.

Dedicated microprocessors, also known as **microcontrollers** or **application-specific integrated circuits** (ASICs), on the other hand, are designed to perform just one specific task. For example, inside your cell phone is a dedicated microcontroller that does nothing else but control its entire operation. Microcontrollers therefore are usually not as powerful (because they do not need to perform so many tasks) as a microprocessor and are much smaller in size. However, they usually will have the memory and supporting I/O peripherals included inside the chip, hence the entire system can be on a single chip. For example, the Atmel ATtiny13A microcontroller shown in Figure 1.1(b) has built-in flash memory, electrically erasable programmable read-only memory (EEPROM), static random-access memory (SRAM), general-purpose I/Os, timers, serial interface, and analog-to-digital converters (ADC). Dedicated microcontrollers are used in almost all smart electronic devices. Although the small dedicated microcontrollers are not as powerful and are slower in speed as compared to general-purpose microprocessors, they are being sold much more and are used in a lot more places than general-purpose microprocessors.

(a)

(b)

FIGURE 1.1 Microprocessors: (a) General-purpose Intel Core™ i7 CPU; (b) Dedicated Atmel ATtiny13A microcontroller.

In this book, I will show you in detail how to design, implement, and interface a microprocessor. At the end, you will be able to design your own custom microprocessor and use it to control your own electronic device. I will use a hands-on approach to guide you step-by-step through the entire design process with complete circuits that you actually can implement in hardware. The exciting part is that at the end, you can actually, very easily and inexpensively, implement your own custom microprocessor in a real integrated circuit (IC) and see that it really can execute software programs, make lights flash, or do whatever you have designed it to do.

We will start with the fundamentals of digital logic circuit design in Chapter 2, which will provide you with a good foundation and basic building blocks for creating larger and more complex digital circuits. Chapters 3 and 4 will discuss the design of simple digital circuits and common circuits that are used as building blocks for larger circuits. Chapter 5 talks about the design of memory circuits. Typically, an introduction to digital logic design course will cover the materials from Chapters 1 to 5 only. Moving on to more advanced digital logic design, Chapter 6 talks about control unit circuits. Chapter 7 talks about the datapath and how to connect it with the control unit to produce a dedicated microcontroller. Chapter 8 extends the dedicated microcontroller from Chapter 7 to produce a general-purpose microprocessor. Finally, Chapter 9 concludes with examples of how to interface these microprocessors and microcontrollers in the real world.

1.1 Overview of Microprocessor Design

The microprocessor or microcontroller is an electronic digital logic circuit that is implemented inside an IC chip. Any digital electronic circuit at the lowest physical level understands only whether there is electricity or no electricity, which is typically represented by the use of a 1 or a 0. The question is how do we design a microprocessor so that it can understand the 1s and 0s, and then do something meaningful with that understanding? To design a microprocessor is to design its logic circuit to do whatever it is intended to do. To implement the microprocessor is to put the logic circuit of the microprocessor onto an IC chip.

Previously, making an IC chip with any circuit was a long and expensive process. With the advance of large-capacity field-programmable gate array (FPGA) chips, digital circuits of almost any size can be implemented in a chip easily and quickly. Moreover, because FPGA chips are erasable, you can use the same FPGA chip over and over again to implement different circuits. If you put an adder circuit in the FPGA chip, that chip will be an adder, and if you put a traffic light controller circuit in the FPGA chip, that chip will be a traffic light controller. So implementing any digital circuit in a FPGA chip is quite simple. The challenge now is how to design the circuit; how do we design the adder circuit or the traffic light controller circuit?

A block diagram of a microprocessor circuit is shown in Figure 1.2. As you can see, it is divided into two main parts: the **control unit** and the **datapath**. The datapath is responsible for the execution of all of the microprocessor's data operations, such as the addition of two numbers inside the arithmetic logic unit (ALU). The datapath also includes registers for the temporary storage of data and comparators for testing data values. These and many other functional units are connected together

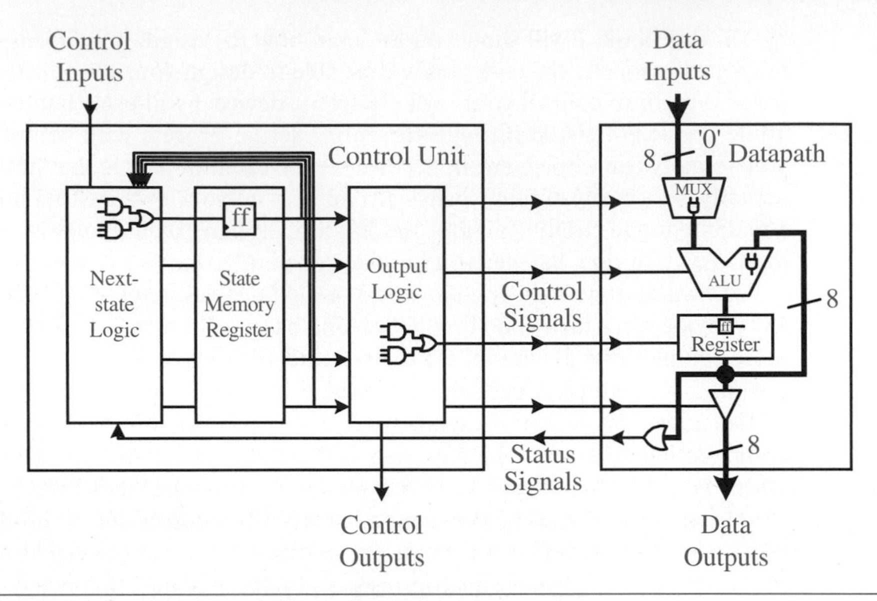

FIGURE 1.2 Internal parts of a microprocessor.

with multiplexers and data signal lines. The data signal lines are for transferring data between two functional units. Sometimes, several data signal lines are grouped together to form a **bus**. The width of the bus (i.e., the number of data signal lines in the group) is annotated next to the bus line. In the figure, the bus lines are thicker and are 8-bit wide. Multiplexers, also known as MUXs, are for selecting data from two or more sources to go to one destination. In the figure, a 2-to-1 multiplexer is used to select between the input data and the constant "0" to go to the left operand of the ALU. The output of the ALU is connected to the input of the register. The output of the register is connected to three different destinations: (1) the right operand of the ALU; (2) an OR gate used as a comparator for the test "not equal to 0"; and (3) a tri-state buffer, which is used to control the output of the data from the register.

Even though the datapath is capable of performing all of the microprocessor's data operations, it cannot, however, do it on its own. In order for the datapath to execute the operations automatically and correctly, a control unit is required. The control unit, also known as the **controller**, controls all of the operations of the datapath and therefore, the operations of the entire microprocessor. The control unit is also called a **finite-state machine** (FSM) because it is a machine that executes by going from one state to another, and there are only a finite number of states for the machine to go to. The control unit is made up of three parts: (1) the **next-state logic**; (2) the **state memory**; and (3) the **output logic**. The purpose of the state memory is to remember the current state that the FSM is in. The next-state logic is the circuit that determines what the machine's next state should be. The output logic is the circuit that generates the actual control signals for controlling the datapath and/or external devices.

Every digital logic circuit, regardless of whether it is part of the control unit or the datapath, is categorized as either a **combinational circuit** or a **sequential circuit**.

A combinational circuit is one where the output of the circuit is dependent only on the current inputs to the circuit, and therefore has no memory about what has happened before. For example, an adder is a combinational circuit because it will produce a sum when given any two input numbers.

A sequential circuit, on the other hand, is dependent not only on the current inputs, but also on all of the previous inputs. In other words, a sequential circuit has to remember its past history. For example, a register is a sequential circuit because it can remember a value indefinitely. Because sequential circuits are dependent on the history, they must contain memory elements to remember that history. Combinational circuits, on the other hand, do not need to remember the history, and so they do not have memory elements.

An analogy of the difference between a combinational circuit and a sequential circuit is the combination lock that we are familiar with. There are actually two different types of combination locks as shown in Figure 1.3. For the lock in Figure 1.3(a), you just turn the three number dials in any order you like to the correct number and the lock will open. For the lock in Figure 1.3(b), you also have three numbers that you need to turn to, but you need to turn to these three numbers in the correct sequence. If you turn to these three numbers in the wrong sequence the lock will not open even if you have the numbers correct. The lock in Figure 1.3(a) is like a combinational circuit where the order in which the inputs are entered into the circuit does not matter, whereas, a sequential circuit is like the lock in Figure 1.3(b) where the sequence of the inputs does matter.

Examples of combinational circuits inside the microprocessor include the ALU, multiplexers, tri-state buffers, and comparators in the datapath, and the next-state logic and output logic circuits in the control unit. Examples of sequential circuits include the register for the state memory in the control unit and the registers in the datapath.

FIGURE 1.3 Two types of combination locks: (a) the order in which you enter the numbers does not matter; (b) the order in which you enter the numbers does matter.

All digital logic circuits, whether they are combinational or sequential, are made up of the three basic logic gates: AND, OR, and NOT gates. From these three basic gates, the most powerful computer can be made. Furthermore, these basic gates are built using transistors—the fundamental building blocks for all digital logic circuits. Transistors are simply electronic binary switches that can be turned on and off. The 1s and 0s that we, as computer scientists, often talk about are used to represent the on and off states of a transistor.

To summarize, transistors, as the lowest-level building blocks, are used to build the basic logic gates. Logic gates are connected together to form either combinational circuits or sequential circuits. The difference between these two types of circuits is only in the way the logic gates are connected together. Certain combinational circuits and sequential circuits are used as standard building blocks for larger circuits and so are kept in standard libraries. These standard combinational and sequential components are connected together to form either the datapath or the control unit. Finally, combining the datapath and the control unit together will produce the circuit for either a dedicated or a general-purpose microprocessor.

1.2 Design Abstraction Levels

Digital circuits can be designed at any one of several abstraction levels. When designing a circuit at the **transistor level**, which is the lowest level, you are dealing with discrete transistors and connecting them together to form the circuit. The next level up in the abstraction is the **gate level**. At this level, you are working with logic gates to build the circuit. In using logic gates, a designer usually creates standard combinational and sequential components for building larger circuits. In this way, a very large circuit, such as a microprocessor, can be built in a hierarchical fashion. Design methodologies have shown that solving a problem hierarchically is always easier than trying to solve the entire problem as a whole from the ground up. These combinational and sequential components are used at the **register-transfer level** to build the datapath and the control unit in the microprocessor. At the register-transfer level, we are concerned with how the data is transferred between the various registers and functional units to realize or solve the problem at hand. Finally, at the highest level, called the **behavioral level**, we can describe the behavior or operation of the circuit using a high-level hardware description language, and we can use a synthesizer, which is equivalent to a compiler, to automatically generate the logic circuit for it. Designing at this level does not require knowledge of the underlying logic gates and circuits because the synthesizer will automatically create the logic circuit for you. This is very similar to writing a computer program using a high-level programming language, and then using the compiler to automatically translate the program into machine language that the computer can execute.

An important point to realize is that there are many different ways to create the same functional circuit. Although they are all functionally equivalent, they are different in other respects, such as the actual circuit (how the transistors or gates are connected together), size (how big the circuit is or how many transistors or gates it uses), speed (how long it takes for the output result to be valid), cost (how much it costs to manufacture), and power usage (how much power it uses). Hence, when designing a circuit, in addition to being functionally correct, we also should consider the economic versus performance tradeoffs. In this book, we will focus mainly on how to design a functionally correct circuit with some discussion about how to optimize the circuit size.

1.3 Examples of a 2-to-1 Multiplexer

As an introduction example, let us look at the design of the 2-to-1 multiplexer from different abstraction levels. At this point, don't worry too much if you don't understand the details of how all of these circuits are built. This example is intended just to give you an idea of what the circuit looks like at the different abstraction levels. We will get to the details in the rest of the book.

The multiplexer is a component that is used a lot in the datapath. An analogy for the operation of the 2-to-1 multiplexer is similar in principle to a railroad switch in which two railroad tracks are to be merged onto one track. The switch controls which one of the two trains on the two separate tracks will move onto the one track. Similarly, the 2-to-1 multiplexer has two data inputs, d_1 and d_0, and a select input, s. The select input determines which data from the two data inputs will pass to the output, y.

Figure 1.4 shows the graphical symbol, also referred to as the **logic symbol**, for the 2-to-1 multiplexer. From looking at the logic symbol, you can tell how many signal lines the 2-to-1 multiplexer has, and the name or function designated for each line. For the 2-to-1 multiplexer, there are two data input signals, d_1 and d_0, a select input signal, s, and an output signal, y.

FIGURE 1.4 Logic symbol for the 2-to-1 multiplexer.

1.3.1 Behavioral Level

We can describe the operation of the 2-to-1 multiplexer simply (using the same names as in the logic symbol) by saying that

if $s = 0$ then d_0 passes to y,

otherwise

d_1 passes to y

Or more precisely, the value that is at d_0 passes to y if $s = 0$, and the value that is at d_1 passes to y if $s = 1$.

We use a hardware description language (HDL), which is quite similar to many high-level computer programming languages, to describe the circuit at the **behavioral** level. When describing a circuit at this level, you would write basically the same thing as in the description, except that you have to use the correct syntax required by the hardware description language. Figure 1.5 shows the description of the 2-to-1 multiplexer using the hardware description language called **Verilog**, and Figure 1.6 shows the description of the same 2-to-1 multiplexer using another hardware description language called **VHDL**, which stands for VHSIC Hardware Description Language (VHSIC, in turn, stands for Very High Speed Integrated Circuit). Verilog and VHDL are two standard hardware description languages used for digital logic design.

```
module multiplexer (
  input s,
  input d0,
  input d1,
  output reg y
);

  always @ (s or d0 or d1) begin
   if (s == 0) begin
    y = d0;
   end else begin
    y = d1;
   end
  end

endmodule
```

FIGURE 1.5 Behavioral Verilog code for a 2-to-1 multiplexer.

In the Verilog code shown in Figure 1.5, the declaration of the component begins with the keyword `module` followed by the name of the component, which in the example, is the user identifier `multiplexer`. All of the words used in Verilog are case sensitive. The input and output interface signals are listed next using the keywords `input` and `output`. For ease of reference, the user-defined names used for these signals match those shown earlier in the logic symbol. The `always` block is followed by its sensitivity list of signals inside the parentheses. The `always` block is executed each and every time when any one of the signals in the sensitivity list changes value. The statements inside the `always` block (bracketed by the `begin` and `end` keywords) are executed sequentially. In the example, there is only one `if-then-else` statement inside the block. Like any `if` statement in other programming languages, the assignment statement `y = d0` is executed when the condition "s equals to 0" is true, otherwise the assignment statement `y = d1` is executed. For the two assignment statements, the value for the expression on the right side of the equal sign is assigned to the signal on the left side of the equal sign. Notice that the output signal `y` on the left side of the equal sign is declared as a `reg` because assignment statements inside the `always` block cannot drive a `wire` data type, but can only drive a register or an integer data type. Finally, the module is terminated with the keyword `endmodule`. A summary of the Verilog language can be found in Appendix C.

In the VHDL code shown in Figure 1.6, the LIBRARY and USE statements are similar to the "#include" and "using namespace" preprocessor commands in C++. None of the words used in VHDL is case sensitive, however, in the examples, the keywords are shown in upper case. The IEEE library contains the definition for the STD_LOGIC type used in the declaration of signals. The ENTITY section declares the interface for the circuit by specifying the input and output signals of the circuit. In this example, there are three input signals of type STD_LOGIC and one output signal also of type STD_LOGIC. Again, the names used for these signals match those shown earlier in the logic symbol. The ARCHITECTURE section defines the actual operation of the circuit.

```
LIBRARY IEEE;
USE IEEE.STD_LOGIC_1164.ALL;

ENTITY multiplexer IS PORT (
  s, d0, d1: IN STD_LOGIC;
  y: OUT STD_LOGIC);
END multiplexer;

ARCHITECTURE Behavioral OF multiplexer IS
BEGIN
  PROCESS(s, d0, d1)
  BEGIN
   IF (s = '0') THEN
    y <= d0;
   ELSE
    y <= d1;
   END IF;
  END PROCESS;
END Behavioral;
```

FIGURE 1.6 Behavioral VHDL code for a 2-to-1 multiplexer.

The ARCHITECTURE keyword is followed by a user identifier name and the entity that it is for. The PROCESS block with its sensitivity list is like the `always` block in Verilog. The operation of the multiplexer is defined in the conditional IF-THEN-ELSE statement. The two signal assignment statements, which use the symbol <= to denote the signal assignment, in conjunction with the IF-THEN-ELSE statement, says that the signal y gets the value of d_0 if s is equal to 0; otherwise, y gets the value of d_1. The PROCESS block is terminated by the END PROCESS statement, and the ARCHITECTURE block is terminated by the END keyword followed by the name of this architecture. A summary of the VHDL language can be found in Appendix D.

Having written the behavioral code, either in Verilog or VHDL, we will use a synthesizer to automatically construct the netlist (which is the circuit connections) that will operate according to the description of the code. As you can see, when designing circuits at the behavioral level, we do not need to know what logic gates are needed or how they are connected together. We only need to know their interface and functional operation, and then describe it using an HDL.

1.3.2 **Gate Level**

At the gate level, you can draw a **schematic diagram** or **circuit diagram**, which shows how the logic gates are connected together. Two different schematic diagrams of a 2-to-1 multiplexer circuit are shown in Figures 1.7(a) and (b). In Figure 1.7(a), the circuit uses three NOT gates (-▷o-), four 3-input AND gates (≡D-), and one 4-input OR gate (≣▷-). In Figure 1.7(b), only one NOT gate, two 2-input AND gates, and one 2-input OR gate are needed. Although one circuit is larger (in terms of the number of gates needed) than the other, both of these circuits realize the same 2-to-1 multiplexer function. Therefore, when we want to actually implement a 2-to-1 multiplexer circuit, we will want to use the second, smaller circuit rather than the first.

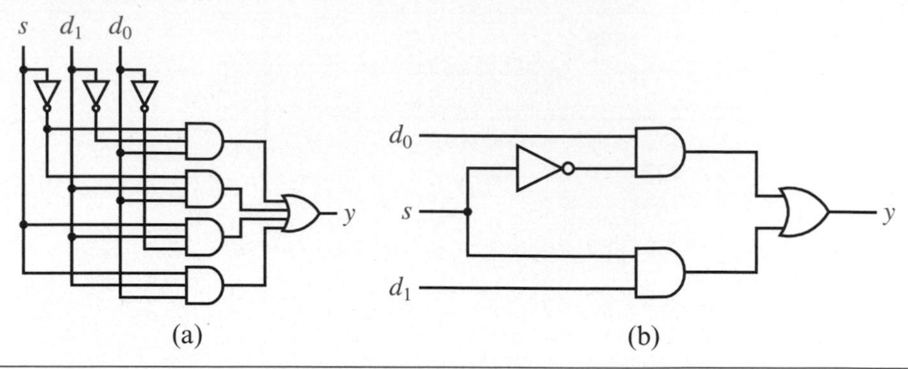

FIGURE 1.7 Gate level circuit diagram for the 2-to-1 multiplexer: (a) circuit using eight gates; (b) circuit using four gates.

At the gate level, you also can describe the 2-to-1 multiplexer using a **truth table** or with a **Boolean equation** as shown in Figures 1.8(a) and (b), respectively. For the truth table, we list all possible combinations of the binary values for the three inputs, s, d_0, and d_1, and then determine what the output value y should be based on the functional description of the circuit. We see that for the first four rows of the table when $s = 0$, y has the same values as d_0; whereas, in the last four rows when $s = 1$, y has the same values as d_1.

The Boolean equation in Figure 1.8(b) can be derived from either the schematic diagram or the truth table. The first equality in Figure 1.8(b) matches the truth table in Figure 1.8(a) and also the schematic diagram in Figure 1.7(a). The second equality in Figure 1.8(b) matches the schematic diagram in Figure 1.7(b). To derive the first equality equation from the truth table, we look at all of the rows where the output y is a 1. Each of these rows results in a term in the equation. For each term, the variable is primed ($'$) when the value of the variable is a 0, and unprimed when the value of the variable is a 1. For example, the first term $s'd_1'd_0$ in the equation is obtained from the first row in the truth table where y is a 1, since s is a 0, d_1 is a 0 and d_0 is a 1.

s	d_1	d_0	y
0	0	0	0
0	0	1	1
0	1	0	0
0	1	1	1
1	0	0	0
1	0	1	0
1	1	0	1
1	1	1	1

(a)

$$
\begin{aligned}
y &= s'd_1'd_0 + s'd_1d_0 + sd_1d_0' + sd_1d_0 \\
&= s'd_0 + sd_1
\end{aligned}
$$

(b)

FIGURE 1.8 Gate level description of the 2-to-1 multiplexer: (a) using a truth table; (b) using a Boolean equation.

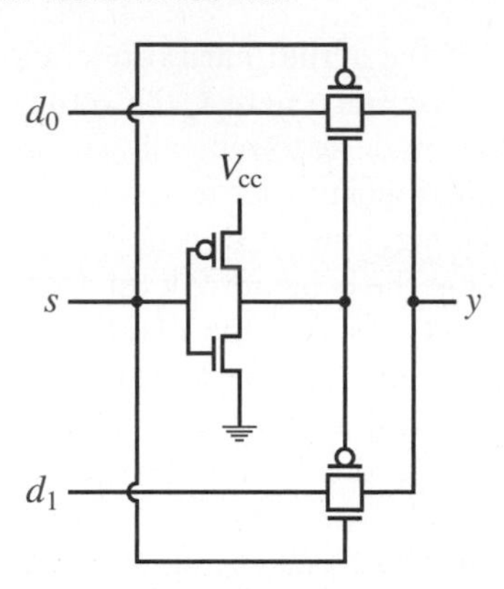

FIGURE 1.9 Transistor circuit for the 2-to-1 multiplexer.

1.3.3 Transistor Level

The 2-to-1 multiplexer circuit at the transistor level is shown in Figure 1.9. It contains six transistors, three of which are P-type metal-oxide semiconductor (PMOS) transistors ( ), and three are N-type metal-oxide semiconductor (NMOS) transistors ( ). The pair of transistors on the left forms a NOT gate for the signal s, while the two pairs of transistors on the right form two transmission gates. The transmission gate allows or disallows the data signal d_0 or d_1 to pass through, depending on the control signal s. The top transmission gate is turned on when s is a 0, and the bottom transmission gate is turned on when s is a 1. Hence, when s is 0, the value at d_0 is passed to y, and when s is 1, the value at d_1 is passed to y.

A more detailed discussion about how to design digital circuits at the transistor level can be found in the online chapter on Implementation Technologies.

1.4 Introduction to Hardware Description Language

The popularity of using hardware description languages (HDL) to design digital circuits began in the mid-1990s when commercial synthesis tools became available. Two popular HDLs used by many engineers today are VHDL and Verilog. VHDL was sponsored and developed jointly by the U.S. Department of Defense and the Institute of Electrical and Electronic Engineers (IEEE) in the mid-1980s. It was standardized by the IEEE in 1987 (VHDL-87), and later extended in 1993 (VHDL-93). Verilog, on the other hand, was first introduced in 1984, and again later in 1988, as a proprietary hardware description language by two companies: Synopsys and Cadence Design Systems.

Both Verilog and VHDL, in many respects, are similar to a regular computer programming language, such as C. For example, it has constructs for variable assignments, conditional statements, loops, and functions (just to name a few). In a computer programming language, a compiler is used to translate the high-level source code to machine code. In HDL, however, a synthesizer is used to translate the source code to a description of the actual hardware circuit that implements the code. From this description, which we call a **netlist**, the actual, physical digital device that realizes the source code can be made automatically. Accurate functional and timing simulation of the code is also possible in order to test the correctness of the circuit.

We saw in the previous section how we used Verilog and VHDL to describe the 2-to-1 multiplexer at the behavioral level. HDL also can be used to describe a circuit at other levels. Figure 1.10 shows the VHDL code for the multiplexer written at the **dataflow** level. The main difference between the behavioral VHDL code shown in Figure 1.6 and the dataflow VHDL code is that, in the behavioral code, there is a PROCESS block statement; whereas, in the dataflow code, there is no PROCESS statement. Statements within a PROCESS block are executed sequentially as in a computer program, while statements outside a PROCESS block (including the PROCESS block itself) are executed concurrently or in parallel. The signal assignment statement, using the symbol <=, is derived directly from the Boolean equation for the multiplexer, as shown in Figure 1.8(b), using the built-in VHDL operators: AND, OR, and NOT.

The corresponding Verilog version of the 2-to-1 multiplexer written at the dataflow level is shown in Figure 1.11. For Verilog, the `assign` keyword is used for the signal assignment, and the symbols &, |, and ~ are used for the logical operators AND, OR, and NOT, respectively.

In addition to the behavioral and dataflow levels, we also can write HDL code at the **structural** level. Figure 1.13 shows the Verilog code for the multiplexer written at the structural level and Figure 1.14 shows the VHDL code. The code is based on the circuit shown in Figure 1.7(b) and duplicated here in Figure 1.12.

```
LIBRARY IEEE;
USE IEEE.STD_LOGIC_1164.ALL;

ENTITY multiplexer IS PORT(
  s, d0, d1: IN STD_LOGIC;
  y: OUT STD_LOGIC);
END multiplexer;

ARCHITECTURE Dataflow OF multiplexer IS
BEGIN
  y <= ((NOT s) AND d0) OR (s AND d1);
END Dataflow;
```

FIGURE 1.10 Dataflow level VHDL description of the 2-to-1 multiplexer.

```
module multiplexer (
  input s, d0, d1,
  output y
);

  assign y = ((~s) & d0) | (s & d1);

endmodule
```

FIGURE 1.11 Dataflow level Verilog description of the 2-to-1 multiplexer.

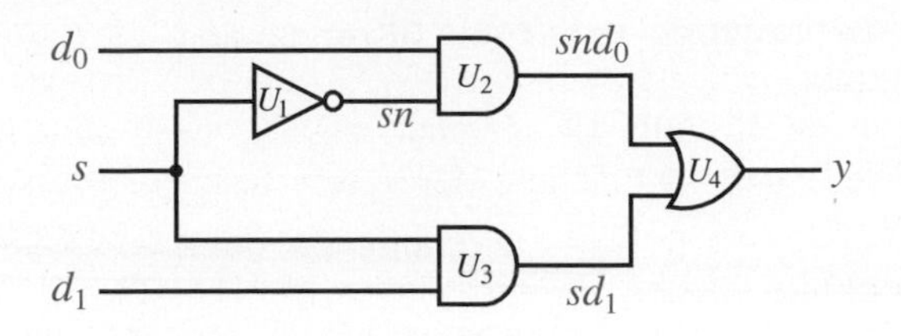

FIGURE 1.12 2-to-1 multiplexer circuit.

```
module multiplexer (
  input s, d0, d1,
  output y
);

  wire sn,snd0,sd1;

  // first parameter is the output;
  // remaining parameters are the inputs
  not U1(sn,s);
  and U2(snd0,sn,d0);
  and U3(sd1,s,d1);
  or U4(y,snd0,sd1);

endmodule
```

FIGURE 1.13 Structural Verilog code for the 2-to-1 multiplexer.

```
----------------- NOT gate -----------------------
LIBRARY IEEE;
USE IEEE.STD_LOGIC_1164.ALL;
ENTITY notgate IS PORT(
  i: IN STD_LOGIC;
  o: OUT STD_LOGIC);
END notgate;

ARCHITECTURE Dataflow OF notgate IS
BEGIN
  o <= NOT i;
END Dataflow;

----------------- 2-input AND gate ---------------
LIBRARY IEEE;
USE IEEE.STD_LOGIC_1164.ALL;
ENTITY and2gate IS PORT(
  i1, i2: IN STD_LOGIC;
  o: OUT STD_LOGIC);
END and2gate;
```

FIGURE 1.14 Structural VHDL code for the 2-to-1 multiplexer.
(continued on next page)

```
ARCHITECTURE Dataflow OF and2gate IS
BEGIN
  o <= i1 AND i2;
END Dataflow;

----------------- 2-input OR gate ----------------
LIBRARY IEEE;
USE IEEE.STD_LOGIC_1164.ALL;
ENTITY or2gate IS PORT(
  i1, i2: IN STD_LOGIC;
  o: OUT STD_LOGIC);
END or2gate;

ARCHITECTURE Dataflow OF or2gate IS
BEGIN
  o <= i1 OR i2;
END Dataflow;

----------------- 2-to-1 multiplexer ------------
LIBRARY IEEE;
USE IEEE.STD_LOGIC_1164.ALL;
ENTITY multiplexer IS PORT(
  s, d0, d1: IN STD_LOGIC;
  y: OUT STD_LOGIC);
END multiplexer;

ARCHITECTURE Structural OF multiplexer IS
  COMPONENT notgate PORT(
    i: IN STD_LOGIC;
    o: OUT STD_LOGIC);
  END COMPONENT;
  COMPONENT and2gate PORT(
    i1, i2: IN STD_LOGIC;
    o: OUT STD_LOGIC);
  END COMPONENT;
  COMPONENT or2gate PORT(
    i1, i2: IN STD_LOGIC;
    o: OUT STD_LOGIC);
  END COMPONENT;

  SIGNAL sn, snd0, sd1: STD_LOGIC;

BEGIN
  U1: notgate PORT MAP(s, sn);
  U2: and2gate PORT MAP(d0, sn, snd0);
  U3: and2gate PORT MAP(d1, s, sd1);
  U4: or2gate PORT MAP(snd0, sd1, y);
END Structural;
```

FIGURE 1.14 Structural VHDL code for the 2-to-1 multiplexer.

In the structural Verilog code shown in Figure 1.13, the keywords (`not`, `and`, `or`) for the various gates are used. The first parameter for these gate statements is the output from the gate. The remaining parameters in the statements are the inputs to the gate. The `wire` keyword defines user identifiers using the wire data type for connecting the gates together based on the schematic diagram. For example, looking at the schematic diagram, the NOT gate labeled U_1 has *s* as the input and *sn* as the output. Hence in the corresponding code, we have the statement `not U1(sn, s)`. The user-identifier name U1 in the statement is optional, and is added only for easy identification of the gate in the circuit.

The structural VHDL code shown in Figure 1.14 looks more complicated than the Verilog code, but it actually does the same thing. The three different gates (*notgate*, *and2gate*, and *or2gate*) used in the circuit are declared and defined first using the ENTITY and ARCHITECTURE statements, respectively. After this, the multiplexer is declared (also with the ENTITY statement). The actual, structural definition of the multiplexer is in the ARCHITECTURE section for *multiplexer*. First of all, the COMPONENT statements specify what components are used in the circuit. The SIGNAL statement declares the three internal signals (*sn*, *snd0*, and *sd1*) that will be used in the connection of the circuit. Finally, the PORT MAP statements declare the instances of the gates used in the circuit and specify how they are connected using the external and internal signals. So if you ignore all of the preliminary declaration stuff, the last few statements in the VHDL code matches the statements in the Verilog code.

The focus of this book is not to teach the details of how to write Verilog or VHDL codes. Because these two hardware description languages are quite similar to high-level computer languages such as C, which you already should be familiar with, we will take the approach of learning by examples. Throughout the book, there are Verilog and VHDL codes for all of the circuits discussed, and in the appendix there is a syntax summary of the two languages for your reference. By looking at the examples and the summary references, you should be able to write codes using these languages to describe your digital circuits.

1.5 Synthesis

Given a gate-level circuit diagram, such as the one shown in Figure 1.7, you actually can get some discrete logic gates and manually connect them together with wires on a breadboard. Traditionally, this is how electronic engineers actually designed and implemented digital logic circuits. However, this is not how electronic engineers design circuits anymore. They write programs, such as the one in Figure 1.6, just like what computer programmers do. The question is how does the program that describes the operation of the circuit actually get converted to the physical circuit?

The problem here is similar to translating a computer program written in a high-level language to machine language for a particular computer to execute. For a computer program, we use a compiler to do the translation. For translating a digital logic circuit, we use a **synthesizer**. Instead of using a high-level computer language to describe a computer program, we use a hardware description language (HDL) to describe the operations of a digital logic circuit. Writing a description of a digital logic circuit is similar to writing a computer program except that a different language is used. A synthesizer then is used to translate the HDL program into the circuit **netlist**. A netlist is a description of how a circuit actually is realized or connected using basic gates. This translation process from an HDL description of a circuit to its netlist is referred to as **synthesis.**

The netlist from the output of the synthesizer can be used directly to implement the actual circuit in a FPGA IC chip. With this final step, the creation of a digital circuit that is implemented fully in an IC chip can be done easily. The appendices give a tutorial of the complete process: from writing the Verilog or VHDL code to synthesizing the circuit and uploading the netlist to a FPGA chip using an FPGA development board.

1.6 Going Forward

We will now embark upon a journey that will take you from a simple transistor to the construction of a microprocessor. Figure 1.2 will serve as our guide and map. If you get lost on the way, and do not know where a particular component fits in the overall picture, just refer to this map. At the end, you will be able to design your own custom microprocessor. The exciting part is that this is not just talk and theories. You will be able to implement and try out all of the circuits on a real FPGA chip. You will be able to make lights flash, motors run, and execute your own software program on your own custom microprocessor.

Figure 1.15 is an actual picture of the circuitry inside an Intel Pentium 4 CPU. When you reach the end of this book, you still may not be able to design the circuit for the P4 because of lack of resources, but you certainly will know how it is designed, because you actually will have designed and implemented a real working microprocessor yourself.

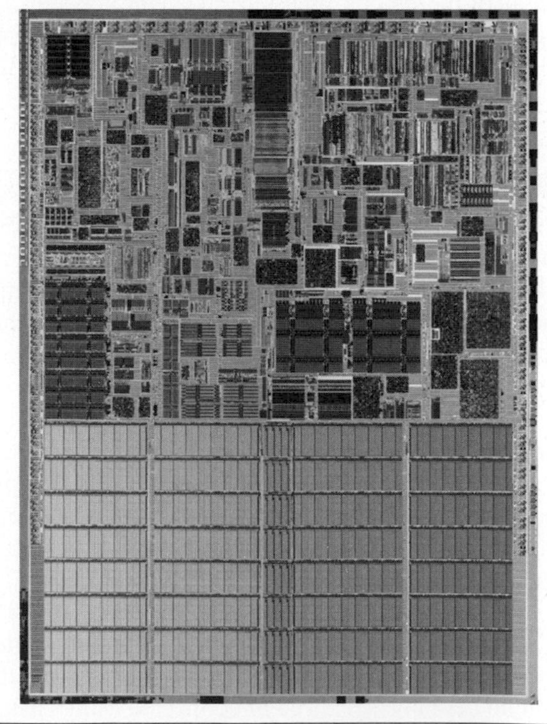

FIGURE 1.15 The internal circuitry of the Intel Pentium 4 CPU.

1.7 PROBLEMS

1.1. Find out the approximate number of general-purpose microprocessors sold in your country in the most recent year versus the number of dedicated microprocessors sold.

1.2. Compile a list of devices controlled by a microprocessor that you use during one regular day.

1.3. Describe what your regular daily routine would be like if no electrical power (including battery power) were available.

1.4. What are the inputs and outputs for the following systems?

a) Traffic light

b) Heart pacemaker

c) Microwave oven

d) Musical greeting card

e) Hard disk drive (not the entire personal computer)

1.5. The speed of a microprocessor is often measured by its clock frequency. What is the clock frequency of the fastest general-purpose microprocessor available today?

1.6. Compare some typical clock speeds between general-purpose microprocessors and dedicated microprocessors.

1.7. Summarize the mainstream generations of the Intel general-purpose microprocessors used in personal computers, starting with the 8086 CPU. List the year introduced, the clock speed, and the number of transistors in each.

1.8. The first-generation PC uses the Intel 8088 with approximately 29,000 transistors. Approximately how many transistors are in the Intel Core i7 (Quad) CPU? Approximately how many transistors are in the Xilinx Virtex-7 FPGA? Approximately how many transistors are in the Altera Stratix V FPGA?

1.9. Using Figure 1.11 as a template, write the dataflow Verilog code for the 2-to-1 multiplexer circuit shown in Figure 1.7(a).

1.10. Using Figure 1.13 as a template, write the structural Verilog code for the 2-to-1 multiplexer circuit shown in Figure 1.7(a).

1.11. Using Figure 1.10 as a template, write the dataflow VHDL code for the 2-to-1 multiplexer circuit shown in Figure 1.7(a).

1.12. Using Figure 1.14 as a template, write the structural VHDL code for the 2-to-1 multiplexer circuit shown in Figure 1.7(a).

1.13. Do the Xilinx ISE tutorial in Appendix A.

1.14. Do the Altera Quartus tutorial in Appendix B.

CHAPTER 2

Fundamentals of Digital Circuits

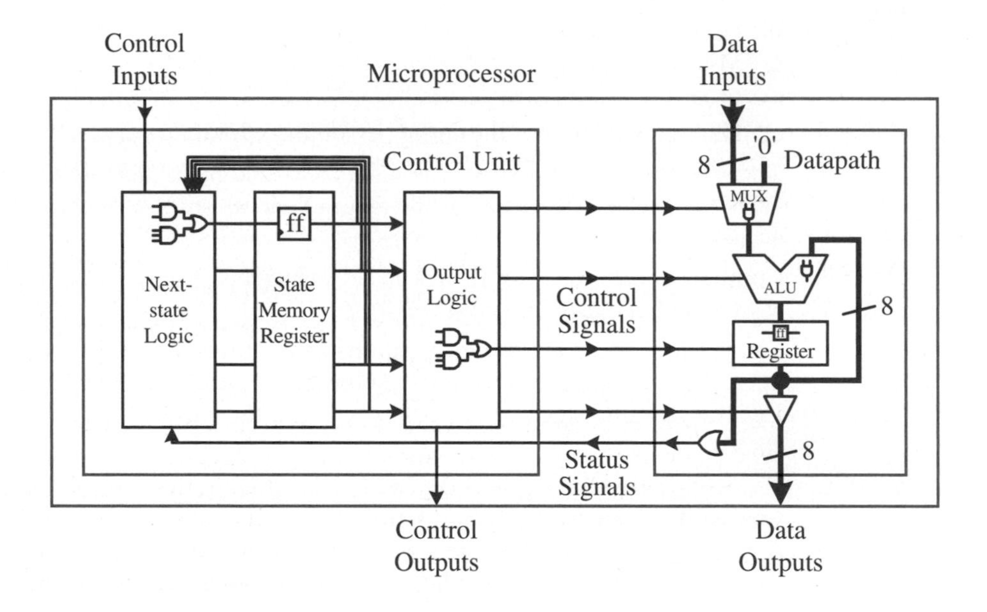

Our world is an analog world. Measurements that we make of the physical objects around us are never in discrete units, but rather in a continuous range. We talk about physical constants such as 2.718281828. . . or 3.141592. . . . To build analog devices that can process these values accurately is next to impossible. Even building a simple analog radio requires very accurate adjustments of frequencies, voltages, and currents at each part of the circuit. If we were to use voltages to represent the constant 3.14, we would have to build a component that will give us exactly 3.14 volts every time. This is again impossible because the manufacturing process is imperfect, and each component produced will be slightly different from the others. Even if the manufacturing process could be made as perfect as we can get, we still would not be able to get 3.14 volts from this component every time we use it. The reason is that the physical elements used in producing the component behave differently in different environments, such as temperature, pressure, and gravitational force, just to name a few. Therefore, even if the manufacturing process is perfect, using this component in different environments will not give us exactly 3.14 volts every time.

To make things simpler, we work with a digital abstraction of our analog world. Instead of working with an infinite continuous range of values, we use just two values, 1 and 0, to represent the two states, on and off, in a digital electronic circuit. It certainly is much easier to control and work with two values rather than an infinite range of values. We call these two values binary values because there are only two of them. A single 0 or a single 1 is a **binary digit** or **bit** for short (where "bi" comes from the first two letters of the word "binary," and "t" comes from the last letter of the word "digit"). This sounds great, but we have to remember that the underlying building block for our digital circuits is still based on an analog world.

This chapter provides the theoretical foundations for building digital logic circuits using logic gates, the basic building blocks for all digital circuits. We start with an introduction on working with binary numbers and performing simple arithmetic in binary. We then will introduce the basic logic gates for building digital circuits. Next, we cover the basic theory of Boolean algebra, Boolean functions, and how to use and manipulate them. Many students may find these theories to be boring and difficult to understand, but let me encourage you to grind through it patiently. The good news is that I will try to keep them as short and simple as possible. You also will find that many of the Boolean theorems are very familiar, because they are similar to the algebra theorems that you already learned in your high school math class. Finally, we conclude with an example of building a simple digital circuit.

2.1 Binary Numbers

Because digital circuits deal with binary values, we will begin with an introduction to **binary numbers**. A bit, having a value of either 0 or 1, can represent only two things or two pieces of information. It is, therefore, necessary to group together many bits to represent more pieces of information. A string of n bits can represent 2^n different pieces of information. For example, a string of two bits can represent four different things using the four unique combinations: 00, 01, 10, and 11. By using different encoding techniques, a group of bits can be used to represent different information, such as a number, a letter of the alphabet, a character symbol, or a command for the microprocessor to execute.

Instead of working with the more familiar decimal numbers that consist of decimal digits, we work with binary numbers, which consist of a string of bits. However, the use of binary numbers is just a convenient form of representation for a string of bits, because we can just as well use octal, decimal, or hexadecimal numbers to represent this group of bits. In fact, you will find that hexadecimal numbers often are used as a shorthand notation for binary numbers.

2.1.1 Counting in Binary

The decimal number system that we all are familiar with is a positional system. In other words, the value of the digit is dependent on the position of the digit within the number. For example, in the decimal number 48, the decimal digit 4 has a greater value than the decimal digit 8 because it is in the tens position, whereas the digit 8 is in the unit position.

The binary number system is also a positional system, with the only difference between the two systems being that the binary system is a base-2 system, using only two digits, 0 and 1, instead of a base-10 system, using ten digits, 0 to 9. So for the binary number 101, the leftmost 1 has a larger value than the rightmost 1.

Counting in binary is just like counting in decimal. When we count in decimal, we count from 0 to 9. After reaching 9 (the last digit) we go back to 0, and have a carry of a 1 by incrementing the next digit to the left by 1, giving us 10. Doing the same thing in binary, we count from 0 to 1 (which is the last digit in the binary system). After reaching 1, we go back to 0, and increment the next bit to the left by 1, giving us 10. Although we use the number 10 in both systems, they are different because one is in base ten and the other is in base two, and so they have different values. To avoid confusion, we will use a subscript to denote what base the number is in, for example, 10_2 is the number 10 in base 2, and 10_{10} is the same number, but in base 10. Continuing with the count in binary, the next few numbers in sequence are 11, 100, 101, 110, 111, and 1000.

The binary numbers from 0 to 15 (decimal) are shown in Figure 2.1. The range from 0 to 15 has 16 different combinations. We need a 4-bit binary number (i.e., a string of four bits) to represent this range because $2^4 = 16$. The count from 0 to 15 in binary is shown in the second column. The corresponding octal (base 8) and hexadecimal (base 16) numbers also are shown in the figure. We will discuss them further in later sections.

2.1.2 Converting between Binary and Decimal

The decimal value of a binary number can be found just like that for a decimal number, except that we raise the base number 2 to a power rather than the base number 10 to a power. The rightmost digit or bit is raised to the power 0, the next digit or bit to the left is raised to the power 1, then the next one to the power 2, and so on. For example, the value for the decimal number 658 is

Decimal	Binary	Octal	Hexadecimal
0	0000	0	0
1	0001	1	1
2	0010	2	2
3	0011	3	3
4	0100	4	4
5	0101	5	5
6	0110	6	6
7	0111	7	7
8	1000	10	8
9	1001	11	9
10	1010	12	A
11	1011	13	B
12	1100	14	C
13	1101	15	D
14	1110	16	E
15	1111	17	F

FIGURE 2.1 Numbers from 0 to 15 in decimal, binary, octal, and hexadecimal number systems.

$$658_{10} = (6\times 10^2) + (5\times 10^1) + (8\times 10^0) = 600 + 50 + 8 = 658_{10}$$

Similarly, the decimal value for the binary number 1011011_2 is

$$1011011_2 = (1\times 2^6)+(0\times 2^5)+(1\times 2^4)+(1\times 2^3)+(0\times 2^2)+(1\times 2^1)+(1\times 2^0)$$

$$= 64 + 0 + 16 + 8 + 0 + 2 + 1 = 91_{10}$$

To get the decimal value of a binary number (which is the same as converting a binary number to its equivalent decimal number), the least significant bit (in this case, the rightmost 1) is multiplied with 2^0. The next bit to the left is multiplied with 2^1, and so on. Finally, they are all added together to give the value 91_{10}. This calculation is very simple because you are always multiplying with either a 0 or a 1. All the 0 terms can be ignored because they will give a 0, and all the 1 terms will just be the number 2 raised to the power.

Notice also the subscript 10 in the decimal number 658_{10}, and the subscript 2 in the binary number 1011011_2. This subscript is used to denote the base of the number whenever there might be confusion as to what base the number is in.

Converting a decimal number to its binary equivalent can be done by successively dividing the decimal number by 2 and keeping track of the remainder at each step.

Combining the remainders together (starting with the last one) forms the equivalent binary number. For example, using the decimal number 91, we divide it by 2 to get 45 with a remainder of 1. Then we divide 45 by 2 to get 22 with a remainder of 1. We continue in this fashion until the end as shown next

$$
\begin{array}{r|ll}
2 & 91 & 1 \quad \uparrow \text{ Least significant bit} \\
2 & 45 & 1 \\
2 & 22 & 0 \\
2 & 11 & 1 \qquad\qquad = 1011011 \\
2 & 5 & 1 \\
2 & 2 & 0 \\
 & 1 & \quad \text{Most significant bit}
\end{array}
$$

Concatenating the remainders together, starting with the last one (most significant bit) results in the binary number 1011011_2.

EXAMPLE 2.1

Convert the binary number 100101 to its decimal equivalent

To convert the binary number 100101 to its decimal equivalent, we perform the following calculation

$$
\begin{aligned}
100101_2 &= (1 \times 2^5) + (0 \times 2^4) + (0 \times 2^3) + (1 \times 2^2) + (0 \times 2^1) + (1 \times 2^0) \\
&= 32 + 0 + 0 + 4 + 0 + 1 = 37_{10}
\end{aligned}
$$

The equivalent decimal number of 100101_2, therefore, is 37.

EXAMPLE 2.2

Convert the decimal number 58 to its binary equivalent

To convert the decimal number 58 to its binary equivalent, we perform the following calculation

$$
\begin{array}{r|ll}
2 & 58 & 0 \quad \uparrow \text{ Least significant bit} \\
2 & 29 & 1 \\
2 & 14 & 0 \qquad\qquad = 111010 \\
2 & 7 & 1 \\
2 & 3 & 1 \\
 & 1 & \quad \text{Most significant bit}
\end{array}
$$

The equivalent binary number of 58, therefore, is 111010_2.

2.1.3 Octal and Hexadecimal Notations

Binary numbers usually consist of a long string of bits, and to write them out as individual bits is sometimes tedious. A shorthand notation for writing out a lengthy string of bits is to use either the octal or hexadecimal number systems. Because the octal system is base-8 and the hexadecimal system is base-16 (both of which are a power of 2), a binary number can be converted easily to an octal or hexadecimal number, and vice versa.

Octal numbers only use the digits from 0 to 7 for the eight different combinations. When counting in octal, we count from 0 to 7, and then the next number is 10_8 as shown in the third column in Figure 2.1.

To convert a binary number to octal, we simply group the bits into groups of threes, starting from the right (least significant bit). The reason for this is because $8 = 2^3$. For each group of three bits, we write the equivalent octal digit for it using the table in Figure 2.1. For example, the conversion of the binary number 1110011_2 to the octal number 163_8 is shown next.

$$\frac{001}{1} \quad \frac{110}{6} \quad \frac{011}{3}$$

Because the original binary number has seven bits, we need to extend it with two leading 0s to get three bits for the leftmost group. Note that when we are dealing with negative numbers, we may require extending the number with leading 1s instead of 0s. This is discussed in more detail in Section 2.2.2, when we talk about sign extending negative numbers.

Converting an octal number to its binary equivalent is just as easy. For each octal number, we write down the equivalent three bits using the table in Figure 2.1. These groups of three bits are concatenated together to form the final binary number. For example, the conversion of the octal number 5724_8 to the binary number 101111010100_2 is shown next.

$$\begin{matrix} 5 & 7 & 2 & 4 \\ 101 & 111 & 010 & 100 \end{matrix}$$

The decimal value of an octal number can be found just like that for a binary or decimal number, except that we raise the base number 8 to a power instead of the base number 2 or 10 to a power. For example, the octal number 5724_8 has the value 3028_{10} as shown in the following calculation.

$$\begin{aligned} 5724_8 &= (5 \times 8^3) + (7 \times 8^2) + (2 \times 8^1) + (4 \times 8^0) \\ &= 2560 + 448 + 16 + 4 \\ &= 3028_{10} \end{aligned}$$

Hexadecimal numbers are treated basically the same way as octal numbers except with the appropriate changes to the base. Hexadecimal (or **hex** for short) numbers use base-16, and thus require 16 different symbols. In addition to the ten symbols from 0 to 9, we additionally use the first six letters of the alphabet, A to F. When counting in hex, we count from 0 to 9, from 9 we go to A, then B, and so on up to F. After F we then go to 10_{16}. The count from 0 to 15_{10} in hex is shown in the last column in Figure 2.1.

Converting binary numbers to hexadecimal numbers involves grouping the bits into groups of four because $16 = 2^4$. For each group of four bits, we write the equivalent hex digit for it using the table in Figure 2.1. For example, the conversion of the binary number 11011011011_2 to the hexadecimal number $6DB_{16}$ is shown next. Again, we need to extend it with a leading 0 to get four bits for the leftmost group.

$$\frac{0110}{6} \quad \frac{1101}{D} \quad \frac{1011}{B}$$

To convert a hex number to a binary number, we write down the equivalent four bits for each hex digit using the table in Figure 2.1, and then concatenate them together to form the final binary number. For example, the conversion of the hexadecimal number $5C4A_{16}$ to the binary number 0101110001001010_2 is shown next.

$$\begin{matrix} 5 & C & 4 & A \\ 0101 & 1100 & 0100 & 1010 \end{matrix}$$

The decimal value of a hexadecimal number can be found just like that for a binary or decimal number, except that we raise the base number 16 to a power instead of the base number 2 or 10 to a power. For example, the hex number $5CF6_{16}$ has the value 23798_{10} as shown in the following calculation.

$$\begin{aligned} 5CF6_{16} &= (5 \times 16^3) + (C \times 16^2) + (F \times 16^1) + (6 \times 16^0) \\ &= (5 \times 16^3) + (12 \times 16^2) + (15 \times 16^1) + (6 \times 16^0) \\ &= 20480 + 3072 + 240 + 6 \\ &= 23798_{10} \end{aligned}$$

Notice in the calculation that the hex digits C and F are first converted to their decimal values 12 and 15, respectively, before performing the multiplication.

Converting a decimal number to hexadecimal is similar to converting a decimal number to binary. The only difference is that we successively divide the decimal number by 16 instead of by 2, and keep track of the remainder at each step. The remainder can be any number between 0 and F (i.e., 15_{10}). Example 2.4 shows this process.

EXAMPLE 2.3

Convert the hexadecimal number $C4A_{16}$ to its decimal equivalent

To convert a hexadecimal number to decimal, we use the base 16 and raise it to the appropriate power.

$$\begin{aligned} C4A_{16} &= (C \times 16^2) + (4 \times 16^1) + (A \times 16^0) \\ &= (12 \times 16^2) + (4 \times 16^1) + (10 \times 16^0) \\ &= 3072 + 64 + 10 \\ &= 3146_{10} \end{aligned}$$

The equivalent decimal number is 3146_{10}.

EXAMPLE 2.4

Convert the decimal number 7689_{10} to hexadecimal

To convert the decimal number 7689_{10} to its hexadecimal equivalent, we perform the following calculation

Division	Remainder	
16 ⌊7689	9	↑ Least significant bit
16 ⌊480	0	
16 ⌊30	E	
1		Most significant bit

= 1E09

Note in the calculation that when we divide 30 by 16, the remainder is 11_{10}, but it is written as E in hex. The equivalent hexadecimal number for 7689_{10} is $1E09_{16}$.

2.1.4 Binary Number Arithmetic

Now that you know about binary numbers and how to convert back and forth between the different bases, you need to learn how to perform simple additions and subtractions with binary numbers. Adding binary numbers is just like adding decimal numbers. We just need to keep in mind that when we add, we carry over a 1 when the sum is a two or a three.

For example, consider the following addition of the two 4-bit binary numbers, 1001 and 0011.

$$\begin{array}{r} 1\ 0\ 0\ 1 \\ +0\ 0_{1}\ 1_{1}\ 1 \\ \hline 1\ 1\ 0\ 0 \end{array}$$

The result of the addition is 1100. The addition is performed just like that for decimal numbers, except that there is a carry whenever the sum is either a 2 or a 3 in decimal, because 2 is 10 in binary and 3 is 11. The most significant bit in the 10 or the 11 is the carry-over bit.

Starting with the rightmost least significant bit, we have $1 + 1 = 2$. Instead of writing the sum as 2, we write it as 10_2 with the leading 1 bit being the carry-over bit for the next bit position and the sum bit 0. Next, when we add the $0 + 1$, we also need to add the carry-over 1 bit from before, resulting in $0 + 1 + 1 = 2$. Again we write the 2 as 10 with a carry-over bit of 1 and a sum of 0. Next, we add $0 + 0 + 1 = 1$. The result is a sum of 1 with no carry. Finally, we add $1 + 0 + 0 = 1$. The result of the addition is 1100_2. You might want to verify that the result 1100_2 is correct by converting the two operands and final sum to decimal and doing the calculation in decimal.

EXAMPLE 2.5

Calculate $1101_2 + 1011_2$

$$\begin{array}{rrrrrcr} & 1 & 1 & 0 & 1 & = & 13 \\ + & 1 & 0 & 1 & 1 & = & +11 \\ \hline 1 & 1 & 0 & 0 & 0 & = & 24 \end{array}$$

For subtracting binary numbers, when we need to borrow, we borrow a two instead of a ten as shown in the following subtraction of the two 4-bit binary numbers, 1001 and 0011.

$$\begin{array}{rrrrcr} 1 & 0 & 0 & 1 & = & 9 \\ -0 & 0 & 1 & 1 & = & -3 \\ \hline 0 & 1 & 1 & 0 & = & 6 \end{array}$$

Starting with the rightmost least significant bit, $1 - 1 = 0$. For the next bit position, $0 - 1$ is not enough, so you need to borrow a 2. You now do a $2 - 1 = 1$. For the next bit, you also need to borrow a 2 because you have given away a 1, leaving you with a 1; so you end up with doing a $1 - 0 = 1$. Finally, for the leftmost bit, because the original 1 has been given away, you are left with a 0, to do a $0 - 0 = 0$. The result of the subtraction is 0110_2.

EXAMPLE 2.6

Calculate $1101_2 - 1011_2$

$$\begin{array}{rrrrcr} 1 & 1 & 0 & 1 & = & 13 \\ -1 & 0 & 1 & 1 & = & -11 \\ \hline 0 & 0 & 1 & 0 & = & 2 \end{array}$$

2.2 Negative Numbers

So far we have been working only with positive numbers. To work with negative numbers, we need to know how negative numbers are represented in binary. Binary encoding of numbers can be interpreted as either signed or unsigned. **Unsigned numbers** include only positive numbers and zero, whereas **signed numbers** include positive, negative, and zero. For decimal numbers, we use the minus sign ($-$) in front of the number to denote that it is negative. However, this is not the case for binary numbers. Regardless of whether a binary number is positive or negative, it is represented as a string of bits such as 101010_2. So given this number 101010_2, how do we know whether it is a positive or a negative number? Well, we don't know unless someone tells us whether to interpret it as a signed or unsigned number.

2.2.1 Two's Complement Representation

If we are told that the binary number is an unsigned number, then we simply evaluate all of the bits together to form its value as described in the previous section.

Negative or signed binary numbers, on the other hand, are encoded using the **two's complement** representation. So if we are told that the binary number is a signed number, then we need to use the two's complement method to evaluate its value. Note that we use the two terms "signed number" and "two's complement number" interchangeably to mean the same thing. In the two's complement representation, the most significant bit (MSB) tells whether the number is positive or negative. If the most significant bit is a 1, then the number is negative; otherwise, the number is positive. With this definition, the decimal number 0 also is considered a positive number because the leading most significant bit for the binary number 0 is a 0. The value of a positive signed number is obtained the same way as for unsigned numbers as discussed in the previous section.

However, to determine the value for a negative signed number, we need to perform a two-step process:

1. Flip all the 1 bits to 0s and all the 0 bits to 1s.
2. Add a 1 to the result obtained from Step 1.

The number obtained from Step 2 is evaluated as an unsigned number for its value, and the negative of this resulting number is the value of the original negative signed number.

Figure 2.2 shows the two's complement numbers for four bits. The range goes from -8 to 7. In general, for an n-bit two's complement number, the range is from -2^{n-1} to $2^{n-1} - 1$.

4-bit Binary	Two's Complement
0000	0
0001	1
0010	2
0011	3
0100	4
0101	5
0110	6
0111	7
1000	−8
1001	−7
1010	−6
1011	−5
1100	−4
1101	−3
1110	−2
1111	−1

FIGURE 2.2 4-bit two's complement numbers.

We will now illustrate with several examples.

EXAMPLE 2.7

Find the value for the unsigned number 1001_2

The value of the unsigned number 1001_2 is evaluated simply as

$$(1 \times 2^3) + (1 \times 2^0) = 8 + 1 = 9_{10}.$$

EXAMPLE 2.8

Find the value for the signed number 01001_2

We know that this signed number 01001_2 is positive because the leading most significant bit is a 0. Thus, to get its value we simply evaluate its value just like an unsigned number.

$$(1 \times 2^3) + (1 \times 2^0) = 8 + 1 = 9_{10}.$$

EXAMPLE 2.9

Find the value for the signed number 1001_2

We know that this signed number 1001_2 is negative because the leading most significant bit is a 1. To find the value of this negative signed number, we perform the two-step process as follows.

	1001	(original number)
(1)	0110	(flip all the bits)
(2)	0111	(add a 1 to the previous number)

The value for the unsigned number 0111_2 is $(1 \times 2^2) + (1 \times 2^1) + (1 \times 2^0) = 7$, therefore, the value for the original number 1001_2 is negative 7 (-7).

EXAMPLE 2.10

Find the value for the signed number 1000_2

The signed number 1000_2 is a negative number because of the leading 1 bit. We apply the two-step process to the number as follows.

	1000	(original number)
(1)	0111	(flip all the bits)
(2)	1000	(add a 1 to the previous number)

Note that the resulting number 1000_2 in Step 2 is exactly the same as the original number. This, however, should not confuse you if you follow the instructions for the conversion process exactly. We need to interpret the resulting number from Step 2 as an unsigned number to determine its value. Interpreting the resulting number 1000_2 as an unsigned number gives us the value $1 \times 2^3 = 8$. Therefore, the original number, which is also 1000_2, is negative 8 (-8).

To convert a negative decimal number to its equivalent two's complement binary number, we start with the unsigned binary number for the positive decimal number. Then we perform the same two-step process shown above to convert the positive binary number to the negative number. This is illustrated in the following example.

EXAMPLE 2.11

Find the two's complement binary number for the decimal number –58

We start with the positive decimal number 58. Its unsigned binary number equivalent is 111010, as calculated next.

Division	Remainder	
2 ⌊58	0	↑ Least significant bit
2 ⌊29	1	
2 ⌊14	0	= 111010
2 ⌊7	1	
2 ⌊3	1	
1		Most significant bit

We need to make this unsigned positive number 111010_2 into a signed positive number because to work with negative numbers, we must interpret all numbers as signed numbers; therefore, we need to add a leading 0 to make it positive. Otherwise, interpreting 111010_2 as a signed number will make it negative because of the leading 1. Hence, we start with the signed positive number 0111010_2.

Next, we perform the same two-step process to convert 0111010_2 to the negative number -58.

	0111010	(original number)
(1)	1000101	(flip all the bits)
(2)	1000110	(add a 1 to the previous number)

Therefore, 1000110_2 is the two's complement (or signed) binary number for -58.

As an exercise, convince yourself that the signed number 1000110_2 from the last example is indeed -58_{10} by converting it back to its decimal number.

2.2.2 Sign Extension

There are times when we need to add more bits to a number but without changing its value. For unsigned numbers, we simply add as many zeros as necessary to the front of the number. Just like for decimal numbers, adding leading zeros does not change the value of the number. However, for signed numbers, we cannot just add leading zeros because if the original number is negative (i.e., the most significant bit is a 1), then by adding a leading zero, you will make the number positive.

For example, if the original signed number is 1001_2 which is -7, and you add a leading 0 to make it 01001, then you will have made this new number into $+9$. On the other hand, if you add a leading 1 to make it 11001, the value for this new number is still -7 as shown next.

	11001	(original number)
(1)	00110	(flip all the bits)
(2)	00111	(add a 1 to the previous number)

The resulting number in Step 2 is +7, therefore, the original number is −7. So if a number is negative, (i.e., has a leading 1) then no matter how many leading 1 bits you add to it, it still will have the same value.

The conclusion is that to extend unsigned numbers, you always add leading 0s. But to extend signed numbers, you need to add leading 0s or 1s, depending on whether the original most significant bit is a 0 or a 1. If the most significant bit is a 0, you sign extend the number by adding leading 0s. If the most significant bit is a 1, you sign extend the number by adding leading 1s. In other words, to extend signed numbers, you always add whatever the most significant bit is. By performing this sign extension, the value of the number is not changed, as shown in the next example.

EXAMPLE 2.12

Performing sign extensions

Given the two signed numbers, 10010_2 and 0101_2, sign extend them to 8 bits. For the number 10010, because the most significant bit is a 1, we need to add three leading 1s to make the number 8 bits long. The resulting number is 11110010. For the number 0101, because the most significant bit is a 0, we need to add four leading 0s to make the number 8 bits long. The resulting number is 00000101. The following calculations show that the two resulting numbers have the same value as the two original numbers. The first number is negative (because of the leading 1 bit), so we need to perform the two-step process to evaluate its value. The second number is positive, so we can evaluate its value directly.

	Original Number	Sign Extended	Original Number	Sign Extended
	10010	11110010	0101	00000101
Flip bits	01101	00001101		
Add 1	01110	00001110		
Value	−14	−14	5	5

2.2.3 Signed Number Arithmetic

Performing arithmetic with signed numbers is the same as for unsigned numbers. And as expected, the result will be correct for both additions and subtractions when we use two's complement to encode the negative numbers.

We will now illustrate with several examples. All of the binary numbers used in these examples are to be interpreted as signed numbers using the two's complement representation.

EXAMPLE 2.13

Calculate the addition of the two signed numbers 0011 + 1001

$$\begin{array}{rcr} 0\ 0\ 1\ 1 & = & 3 \\ +\ 1\ 0_1\ 0_1\ 1 & = & +(-7) \\ \hline 1\ 1\ 0\ 0 & = & -4 \end{array}$$

EXAMPLE 2.14

Calculate the subtraction of the two signed numbers 0011 − 1110

$$\begin{array}{rcr} \overset{2}{0}\ \overset{2}{0}\ 1\ 1 & = & 3 \\ -\ 1\ 1\ 1\ 0 & = & -(-2) \\ \hline 0\ 1\ 0\ 1 & = & 5 \end{array}$$

Note that when we subtract the most significant bit 0 − 1, we can always borrow a 2 from the left even though there are no more bits to the left in the 4-bit operand. We borrowed a 2, but because we already have given away a 1, we are left with 1 − 1 = 0.

EXAMPLE 2.15

Calculate the addition 3 + (−3) using 4-bit signed binary numbers

$$\begin{array}{rcr} 3 & = & 0\ 0\ 1\ 1 \\ +(-3) & = & +\ 1_1\ 1_1\ 0_1\ 1 \\ \hline 0 & = & \not{1}\ 0\ 0\ 0\ 0 \end{array}$$

The result, 10000, has five bits. But because we are using 4-bit arithmetic (i.e., the two operands are 4 bits wide) the result also must be in 4 bits. The leading 1 in the result is, therefore, an overflow bit and we need to ignore it. By dropping the overflow bit, the remaining result 0000 is the correct answer for the problem. Although this addition resulted in an overflow bit, we obtained the correct answer by dropping the extra bit.

EXAMPLE 2.16

Calculate the addition 6 + 3 using 4-bit signed binary numbers

$$
\begin{array}{rcl}
6 & = & 0\ 1\ 1\ 0 \\
+\,3 & = & +\ 0_{1}\ 0_{1}\ 1\ 1 \\
\hline
9 & \neq & 1\ 0\ 0\ 1
\end{array}
$$

The result 1001 is a 9 if we interpret it as an unsigned number. However, because we are performing signed number arithmetic, we need to be consistent and interpret all numbers, including the result, as signed numbers. Interpreting 1001 as a signed number gives −7, which, of course, is incorrect. The problem here is that the range for a 4-bit signed number is from −8 to +7, and +9 is outside of this range.

Although the addition in Example 2.16 did not result in an overflow bit, the answer, however, caused an overflow error. In order to correct this problem, we need to add (at least) one extra bit by sign extending the number to prevent the overflow error. The range for a 5-bit signed number is from −16 to +15, and +9 is inside of this range. The corrected calculation is shown in the next example.

EXAMPLE 2.17

Calculate the addition 6 + 3 using 5-bit signed binary numbers

$$
\begin{array}{rcl}
6 & = & 0\ 0\ 1\ 1\ 0 \\
+\,3 & = & +\ 0\ 0_{1}\ 0_{1}\ 1\ 1 \\
\hline
9 & = & 0\ 1\ 0\ 0\ 1
\end{array}
$$

The result 01001, when interpreted as a signed number, is 9.

2.3 Binary Switch

Besides the fact that we are working only with binary values, digital circuits are easy to understand because they are based on one simple idea, and that is to turn a switch on or off to obtain either one of the two binary values. Because the switch can be in either one of two states (on or off), we call it a **binary switch**, or just a **switch** for short. The switch has three connections: an input, an output, and a control for turning the switch on or off, as shown in Figure 2.3. When the switch is opened, as in Figure 2.3(a), it is turned off, and nothing gets through from the input to the output. When the switch is closed, as in Figure 2.3(b), it is turned on, and whatever is presented at the input is allowed to pass through to the output.

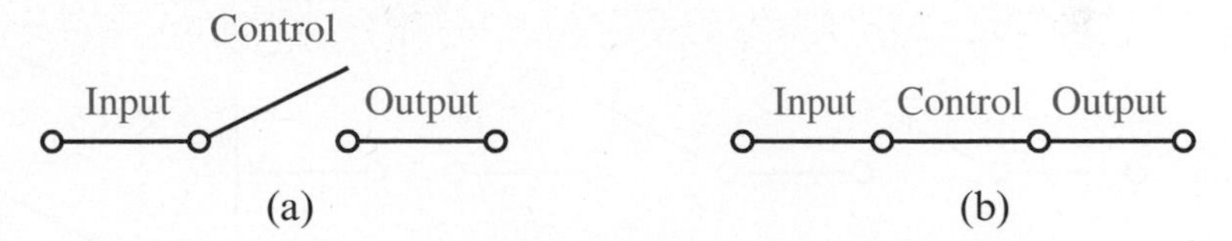

FIGURE 2.3 **Binary switch: (a) opened or off; (b) closed or on.**

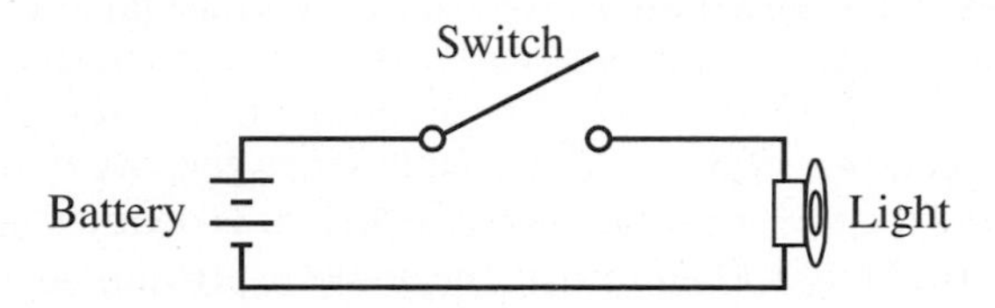

FIGURE 2.4 **A light controlled by a switch.**

Uses of the binary switch idea can be found in many real-world devices. For example, the switch can be an electrical switch with the input connected to a power source and the output connected to a light L, as shown in Figure 2.4.

When the switch is closed, the light turns on, and when the switch is opened, the light turns off. The usual convention is to use a 1 to mean "on" and a 0 to mean "off." Therefore, when the switch is closed, the output is a 1, and the light turns on. We also can use a variable, x, to denote the state of the switch. We can let $x = 1$ to mean the switch is closed, and $x = 0$ to mean the switch is opened. Using this convention, we can describe the state of the light L in terms of the state of the switch x using a simple logic expression. Because $L = 1$ if $x = 1$, and $L = 0$ if $x = 0$, we can write

$$L = x$$

This logic expression describes the output L in terms of the input variable x; in other words, L is on when x is on.

2.4 Basic Logic Operators and Logic Expressions

Two binary switches can be connected together either in series or in parallel, as shown in Figure 2.5. The left side in both of the diagrams is the source input, which is equivalent to the battery in Figure 2.4. For discussion purposes, we will assume that power (logic 1) is always available at the source. The right side in both diagrams, labeled F, is the output, which is equivalent to the light in Figure 2.4.

If two switches are connected in series, as in Figure 2.5(a), then both switches have to be on in order for the output F to be 1. In other words, $F = 1$ if both $x = 1$ and $y = 1$. If either x or y is off, or both are off, then $F = 0$. Translating this into a logic expression, we get

$$F = x \text{ AND } y$$

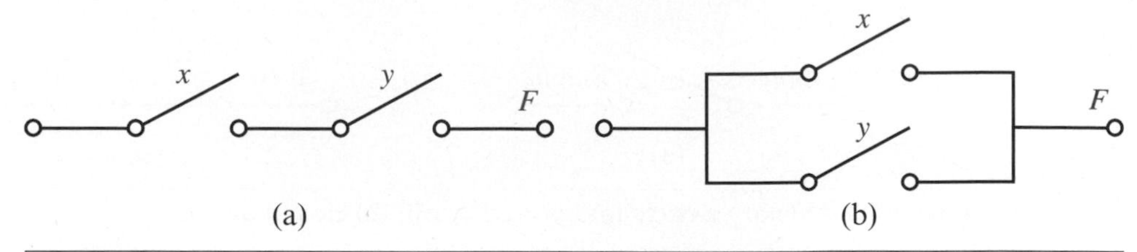

FIGURE 2.5 Connection of two binary switches: (a) in series; (b) in parallel.

Hence, two switches connected in series give rise to the logical AND operator. When used in a logic expression, the AND operator is denoted either with a dot (·) or no symbol at all. Thus, we can rewrite the above expression as

$$F = x \cdot y$$

or simply

$$F = xy$$

when it is clear that x and y are two individual variables.

If we connect two switches in parallel, as in Figure 2.5(b), then only one switch needs to be on in order for the output F to be 1. In other words, $F = 1$ if either $x = 1$, or $y = 1$, or both x and y are 1. This means that $F = 0$ only if both x and y are 0. Translating this into a logic expression, we get

$$F = x \text{ OR } y$$

and this gives rise to the logical **OR** operator. When used in a logic expression, the OR operator is denoted with a plus symbol (+). Thus, we can rewrite the above expression as

$$F = x + y$$

In addition to the AND and OR operators, there is another basic logic operator—the NOT operator, also known as the **INVERTER**. Whereas the AND and OR operators have multiple inputs, the NOT operator has only one input and one output. The NOT operator simply inverts its input, so a 0 input will produce a 1 output, and a 1 becomes a 0. In a logic expression, the NOT operator is denoted with either an apostrophe symbol (′) or a bar on top (¯) as in

$$F = x'$$

or

$$F = \overline{x}$$

When several operators are used in the same logic expression, the precedence given to the operators are (from highest to lowest) NOT, AND, and OR. The order of evaluation can be changed by means of using parentheses. For example, the expression

$$F = xy + z'$$

means (x AND y) OR (NOT z), and the expression

$$F = x(y + z)'$$

means x AND (NOT (y OR z)).

2.5 Logic Gates

Logic gates are the actual physical devices that implement the logical operators discussed in the previous section. **Transistors**, acting as tiny electronic binary switches, are connected together to form these gates. Thus, we have the AND gate, the OR gate, and the NOT gate (also called the INVERTER) for the corresponding AND, OR, and NOT logical operators. These gates form the basic building blocks for all digital logic circuits. The name "gate" comes from the fact that these devices operate like a door or gate to let or not to let things (in our case, current) through. Refer to the online chapter Implementation Technologies for a detail discussion on transistors and how logic gates are constructed using transistors.

In drawing digital circuit diagrams (also called **schematic diagrams** or just **schematics**), we use special **logic symbols** to denote these gates, as shown in Figure 2.6. The **AND gate** (specifically, the 2-input AND GATE) in Figure 2.6(a) has two input connections coming in from the left and one output connection going out on the right. Similarly, the 2-input **OR gate** in Figure 2.6(b) has two input connections and one output connection. The **NOT gate** in Figure 2.6(c) has one input coming from the left and one output going to the right. The outputs from these gates, of course, are dependent on their inputs and are defined by their logical operations.

Somtimes, an AND gate or an OR gate with more than two inputs is needed. So, in addition to the 2-input AND and OR gates, there are 3-input, 4-input, or as many inputs as are needed, of the AND and OR gates. In practice, however, the number of inputs is limited to a small number, such as five. The logic symbols for some of these gates are shown in Figures 2.7(a) through (d).

There are several other gates that are variants of the three basic gates that also are used often in digital circuits. They are the **NAND gate**, the **NOR gate**, the **XOR gate**, and the **XNOR gate**. The NAND gate is derived from the AND gate and the NOT gate connected in series, so that the output of the AND gate is inverted. The name NAND comes from the words "NOT AND." Similarly, the NOR gate is the OR gate with its output inverted, and the name comes from the words "NOT OR" The XOR or eXclusive OR gate is like the OR gate except that when both inputs are 1, the output is a 0 instead. The XNOR (or eXclusive NOR) gate is just the inverse of the XOR gate for when there are an even number of inputs

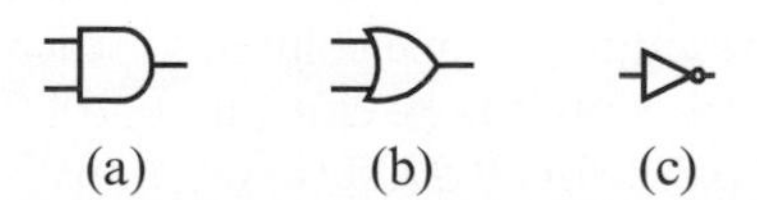

FIGURE 2.6 Logic symbols for the three basic logic gates: (a) 2-input AND; (b) 2-input OR; (c) NOT.

(a) (b) (c) (d) (e)

(f) (g) (h) (i) (j)

FIGURE 2.7 Logic symbols for: (a) 3-input AND; (b) 4-input AND; (c) 3-input OR; (d) 4-input OR; (e) 2-input NAND; (f) 2-input NOR; (g) 3-input NAND; (h) 3-input NOR; (i) 2-input XOR; (j) 2-input XNOR.

(like two inputs). However, when there are an odd number of inputs (like three inputs), the XOR is the same as the XNOR. The logic symbols for some of these gates are shown in Figures 2.7(e) through (j).

In Figures 2.7(e) through (j), notice the use of the little circle or bubble at the output of some of the logic symbols. This bubble is used to denote the inverted value of a signal. For example, the NAND gate is the inverse of the AND gate. Thus, the NAND gate logic symbol is the same as the AND gate logic symbol, except that it has the extra bubble at the output.

The notations used for the NAND, NOR, XOR, and XNOR gates in a logic expression are: $(xy)'$ for the 2-input NAND gate, $(x + y)'$ for the 2-input NOR gate, $x \oplus y$ for the 2-input XOR gate, and $x \odot y$ for the 2-input XNOR gate. We will shortly see that the 2-input XOR gate is equivalent to the following logic expression

$$x \oplus y = x'y + xy'$$

and the 2-input XNOR gate is equivalent to

$$x \odot y = x'y' + xy$$

For the 3-input gates, they are $(xyz)'$ for the 3-input NAND gate, $(x + y + z)'$ for the 3-input NOR gate, $x \oplus y \oplus z$ for the 3-input XOR gate, and $x \odot y \odot z$ for the 3-input XNOR gate. It was noted earlier that the 2-input XOR and XNOR gates are inverses of each other, so $(x \oplus y) = (x \odot y)'$, but the 3-input XOR and XNOR gates are the same, so $(x \oplus y \oplus z) = (x \odot y \odot z)$.

2.6 Truth Tables

The operation of the AND, OR, and NOT logic operators can be described formally by using a **truth table**, as shown in Figure 2.8. A truth table is a two-dimensional array with one column for each input and one column for each output. Because we are dealing with binary values, each input can be either a 0 or a 1. We simply enumerate all possible combinations of 0s and 1s for all of the inputs. Usually, we want to write these input values in the normal binary counting order. With two inputs, there are 2^2 combinations giving us the four rows in the table. The values in the output column are determined from applying

x	y	F
0	0	0
0	1	0
1	0	0
1	1	1

(a)

x	y	F
0	0	0
0	1	1
1	0	1
1	1	1

(b)

x	F
0	1
1	0

(c)

FIGURE 2.8 Truth tables for the three basic logic operators: (a) AND; (b) OR; (c) NOT.

x	y	2-NAND $(x \cdot y)'$	2-NOR $(x + y)'$	2-XOR $x \oplus y$	2-XNOR $x \odot y$
0	0	1	1	0	1
0	1	1	0	1	0
1	0	1	0	1	0
1	1	0	0	0	1

x	y	z	3-AND $(x \cdot y \cdot z)$	3-OR $(x + y + z)$	3-NAND $(x \cdot y \cdot z)'$	3-NOR $(x + y + z)'$	3-XOR $x \oplus y \oplus z$	3-XNOR $x \odot y \odot z$
0	0	0	0	0	1	1	0	0
0	0	1	0	1	1	0	1	1
0	1	0	0	1	1	0	1	1
0	1	1	0	1	1	0	0	0
1	0	0	0	1	1	0	1	1
1	0	1	0	1	1	0	0	0
1	1	0	0	1	1	0	0	0
1	1	1	1	1	0	0	1	1

FIGURE 2.9 Truth tables for: 2-input NAND; 2-input NOR; 2-input XOR; 2-input XNOR; 3-input AND; 3-input OR; 3-input NAND; 3-input NOR; 3-input XOR; 3-input XNOR.

the corresponding input values to the functional operator. For the AND truth table shown in Figure 2.8(a), we know from the previous discussion that $F = 1$ only when x and y are both 1, otherwise, $F = 0$. Hence, in the F column of the truth table, the first three rows are 0, and only the last row is a 1. For the OR truth table shown in Figure 2.8(b), $F = 1$ when either x or y is 1, or both are 1, otherwise $F = 0$. Hence, in the F column of the truth table, the first row is a 0, and the remaining three rows are 1. For the NOT truth table shown in Figure 2.8(c), the output F is just the inverted value of the input x.

The truth tables for the other 2-input and 3-input gates are shown in Figure 2.9. Note again that the 2-input XOR gate and the 2-input XNOR gates are inverses of each other, whereas, the 3-input XOR and XNOR gates are the same. For larger inputs, XOR is the inverse of XNOR when they have an even number of inputs, and they are the same when they have an odd number of inputs.

Using a truth table is one method to formally describe the operation of a circuit or function. The truth table for any given logic expression (no matter how complex it is) can always be derived. The use of truth tables to formally describe the operation of digital circuits is discussed further in Chapter 3.

2.7 Boolean Algebra and Boolean Equations

Another method to formally describe the operation of a digital circuit is by using Boolean equations.

2.7.1 Boolean Algebra

George Boole, in 1854, developed a system of mathematical logic, which we now call **Boolean algebra**. Based on Boole's idea, Claude Shannon, in 1938, showed that circuits built with binary switches can be described easily using Boolean algebra. The abstraction from switches being off and on to the use of Boolean algebra is as follows. Let $B = \{0, 1\}$ be the Boolean algebra whose elements are one of the two values, 0 and 1. We define the operations AND ($\cdot$), OR ($+$), and NOT ($'$) for the elements of B by the axioms in Figure 2.10(a). These axioms are simply the definitions as previously given in the truth tables for the AND, OR, and NOT operators.

(a)

1a.	$0 \cdot 0 = 0$	1b.	$1 + 1 = 1$
2a.	$1 \cdot 1 = 1$	2b.	$0 + 0 = 0$
3a.	$0 \cdot 1 = 1 \cdot 0 = 0$	3b.	$1 + 0 = 0 + 1 = 1$
4a.	$0' = 1$	4b.	$1' = 0$

(b)

5a.	$x \cdot 0 = 0$	5b.	$x + 1 = 1$	Null Element
6a.	$x \cdot 1 = 1 \cdot x = x$	6b.	$x + 0 = 0 + x = x$	Identity
7a.	$x \cdot x = x$	7b.	$x + x = x$	Idempotent
8a.	$(x')' = x$			Double Complement
9a.	$x \cdot x' = 0$	9b.	$x + x' = 1$	Inverse

(c)

10a.	$x \cdot y = y \cdot x$	10b.	$x + y = y + x$	Commutative
11a.	$(x \cdot y) \cdot z = x \cdot (y \cdot z)$	11b.	$(x + y) + z = x + (y + z)$	Associative
12a.	$(x \cdot y) + (x \cdot z) = x \cdot (y + z)$	12b.	$(x + y) \cdot (x + z) = x + (y \cdot z)$	Distributive
13a.	$(x \cdot y)' = x' + y'$	13b.	$(x + y)' = x' \cdot y'$	DeMorgan's

FIGURE 2.10 Boolean algebra axioms and theorems: (a) axioms; (b) single-variable theorems; (c) two- and three-variable theorems.

A variable x is called a **Boolean variable** if x takes on only values in B (i.e., either 0 or 1). Consequently, we obtain the theorems in Figure 2.10(b) for single variable and Figure 2.10(c) for two and three variables.

It is not the intent of this book to dwell on the theoretical aspects of proving theorems. Only because digital circuits can formally be described using Boolean equations, and sometimes it is easier to write an equation rather than drawing out the circuit diagram, therefore, as a digital circuit designer, we need to have a basic understanding of how to use and manipulate simple Boolean equations based on these theorems. Furthermore, we can use the Boolean theorems to help us to reduce the size of the circuit. Therefore, for our purposes, we simply assume that all of the theorems are true, and we will use them to simplify circuits.

The single-variable Theorems 5 to 9 can be shown to be true from the truth tables for the AND, OR, and NOT gates by substituting either a 0 or a 1 into the variable x. The two- and three-variable Theorems 10 to 12 are similar to the commutative, associative and distributive properties of basic mathematics.

Theorem 13 is called **DeMorgan's Theorem**. Notice in Theorem 13a that the left side of the equation contains only the AND and NOT operators, whereas the right side of the equation contains only the OR and NOT operators. Because of this, DeMorgan's Theorem is often used in digital circuit design to convert an AND gate circuit to an OR gate circuit, and vice versa. At first glance, it is not obvious that this equality is true. To understand this theorem let us create a circuit from the following truth table having two inputs, x and y, and one output F.

x	y	F
0	0	1
0	1	1
1	0	1
1	1	0

This is the truth table for the 2-input NAND gate, which is an AND gate followed by a NOT gate. We saw earlier that the equation for a 2-input NAND gate is $F = (x \cdot y)'$. Alternatively, F is a 1 when either x or y is a 0, and this translates to $F = (x' + y')$. From this, we get the equality equation $F = (x \cdot y)' = (x' + y')$ for DeMorgan's Theorem 13a.

The next example will show that DeMorgan's Theorem is true by using a truth table.

EXAMPLE 2.18

Proof of DeMorgan's Theorem using a truth table

DeMorgan's Theorem 13a states that $(x \cdot y)' = (x' + y')$. To prove the equation is true using a truth table, we need to show that for every combination of values for the two variables x and y, the left side of the equation is equal to the right side.

We start with the first two columns labeled x and y, and we enumerate all possible combinations of values for these two variables giving us the four rows as shown next.

x	y	$(x \cdot y)$	$(x \cdot y)'$	x'	y'	$x' + y'$
0	0	0	1	1	1	1
0	1	0	1	1	0	1
1	0	0	1	0	1	1
1	1	1	0	0	0	0

For each combination (row), we evaluate first the expression $(x \cdot y)$, and then $(x \cdot y)'$ for the left side of the equation as shown in the next two columns in the table. Then we do the same thing for the right side of the equation; first x', then y', and finally $x' + y'$. Finally, we note that all the values under the two columns, $(x \cdot y)'$ for the left side of the equation, and $x' + y'$ for the right side of the equation, are identical for every combination of x and y; therefore, we conclude that the theorem is true.

It turns out that DeMorgan's Theorem also can be applied to more than two variables. Thus, the following two equations for three variables are also true

$$(x \cdot y \cdot z)' = x' + y' + z'$$

and

$$(x + y + z)' = x' \cdot y' \cdot z'$$

Examples 2.19 and 2.20 show how Boolean expressions can be simplified using the Boolean Theorems.

EXAMPLE 2.19

Using Boolean algebra to reduce an expression

Use Boolean algebra to reduce the expression $x + (x \cdot y)$ as much as possible.

$$\begin{aligned} x + (x \cdot y) &= (x \cdot 1) + (x \cdot y) && \text{by Theorem 6a} \\ &= x \cdot (1 + y) && \text{by Theorem 12a} \\ &= x \cdot (1) && \text{by Theorem 5b} \\ &= x && \text{by Theorem 6a} \end{aligned}$$

Because the expression $x + (x \cdot y)$ reduces to just x, so when creating a circuit we want to implement the circuit based on the latter expression rather than the former because the circuit size for the latter is much smaller.

EXAMPLE 2.20

Using Boolean algebra to reduce the equation $F = (x'yz) + (xy'z) + (xyz') + (xyz)$ as much as possible

We will use the Distributive Theorem 12a to combine terms by factoring out the same variables. For example, both $(x'yz)$ and (xyz) have yz in common, therefore, we can factor out the two variables yz to give $yz(x' + x)$. This further reduces to just yz because

$(x' + x) = 1$ (by Theorem 9b), and $w \cdot 1 = w$ (by Theorem 6a). Continuing in a similar manner, $(xy'z) + (xyz)$ reduces to just xz, and $(xyz') + (xyz)$ reduces to just xy. The complete detail steps for the reduction is shown next.

$F = (x'yz) + (xy'z) + (xyz') + (xyz)$	
$= (x'yz) + (xy'z) + (xyz') + (xyz) + (xyz) + (xyz)$	by Theorem 7b
$= (x'yz) + (xyz) + (xy'z) + (xyz) + (xyz') + (xyz)$	by Theorem 10b
$= [(x'yz) + (xyz)] + [(xy'z) + (xyz)] + [(xyz') + (xyz)]$	by Theorem 11b
$= yz(x' + x) + xz(y' + y) + xy(z' + z)$	by Theorem 12a
$= yz(1) + xz(1) + xy(1)$	by Theorem 9b
$= yz + xz + xy$	by Theorem 6a
$= z(y + x) + xy$	by Theorem 12a

2.7.2 Duality Principle

Notice in Figure 2.10 that we have listed the axioms and theorems in pairs. Specifically, we define the **dual** of a logic expression as one that is obtained by changing all the + operators with the $\cdot$ operators, and vice versa, and by changing all the 0s with 1s, and vice versa. For example, the dual of the logic expression

$$(x \cdot y' \cdot z) + (x \cdot y \cdot z') + (y \cdot z) + 0$$

is

$$(x + y' + z) \cdot (x + y + z') \cdot (y + z) \cdot 1$$

The **duality principle** states that if a Boolean expression is true, then its dual is also true. The duality principle does not say that a Boolean expression is equivalent to its dual. For example, Theorem 5a in Figure 2.10(b) says that $x \cdot 0 = 0$ is true, thus by the duality principle, its dual, $x + 1 = 1$ is also true. However, $x \cdot 0 = 0$ is not equal to $x + 1 = 1$, because 0 definitely is not equal to 1.

We will see in Section 2.7.3 that the inverse of a Boolean expression can be obtained easily by first taking the dual of that expression and then complementing each Boolean variable in the resulting dual expression. In this respect, the duality principle often is used in digital logic design. Whereas an expression might be complex to implement, its inverse might be simpler, thus resulting in a smaller circuit; inverting the final output of this circuit will produce the same result as from the original expression.

2.7.3 Boolean Functions and Their Inverses

Any digital circuit can be described by a logic equation, also known as a **Boolean function**. Any Boolean function can be formed from binary variables and the Boolean operators $\cdot$, +, and ′ (for AND, OR, and NOT, respectively). For example, the following Boolean function uses the three variables x, y, and z. It has three AND **terms** (also referred to as **product terms**), and these AND terms are ORed (summed) together. The first two AND terms each contain all three variables, while the last AND term contains only two variables. By definition, an AND (or product) term is either a single variable or two or more variables ANDed together. Quite often, we refer to functions that are in this format as a **sum-of-products** (or **or-of-ands**).

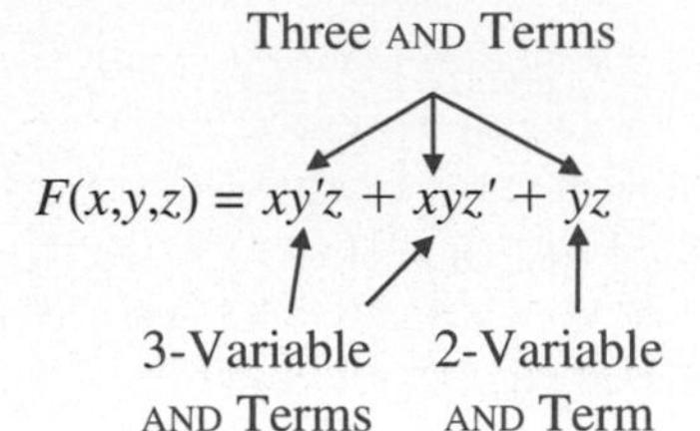

The value of a function evaluates to either 0 or 1, depending on the given set of values for the variables. For example, the above Boolean function evaluates to 1 when any one of the three AND terms evaluate to a 1, because 1 OR x is 1 by Theorem 5b. The first AND term, $xy'z$, equals 1 if

$$x = 1, y = 0, \text{ and } z = 1$$

because if we substitute the values 1, 0, and 1 for the variables x, y, and z, respectively, into the first AND term $xy'z$, we get a 1. Similarly, the second AND term, xyz', equals 1 if

$$x = 1, y = 1, \text{ and } z = 0.$$

The last AND term, yz, has only two variables. What this means is that the value of this term is not dependent on the missing variable x. In other words, x can be either 0 or 1, but as long as $y = 1$ and $z = 1$, this term will be equal to 1.

Thus, we can summarize by saying that F evaluates to 1 if

$$x = 1, y = 0, \text{ and } z = 1$$

or

$$x = 1, y = 1, \text{ and } z = 0$$

or

$$x = 0, y = 1, \text{ and } z = 1$$

or

$$x = 1, y = 1, \text{ and } z = 1.$$

Otherwise, F evaluates to 0.

It is often more convenient to summarize this verbal description of a function with a truth table, as shown in Figure 2.11 under the column labeled F. Notice that the four rows in the table where $F = 1$ match the four cases in the description above.

x	y	z	F	F'
0	0	0	0	1
0	0	1	0	1
0	1	0	0	1
0	1	1	1	0
1	0	0	0	1
1	0	1	1	0
1	1	0	1	0
1	1	1	1	0

FIGURE 2.11 Truth table for the function $F = xy'z + xyz' + yz$, and F'

The inverse of a function, denoted by F', can be obtained easily from the truth table for F by simply changing all the 0s to 1s and 1s to 0s, as shown in the truth table in Figure 2.11 under the column labeled F'. Therefore, we can write the Boolean function for F' in the sum-of-products format, where the AND terms are obtained from those rows where $F' = 1$. Thus, we get

$$F' = x'y'z' + x'y'z + x'yz' + xy'z'$$

To deduce F' algebraically from F requires the use of DeMorgan's Theorem twice (Theorems 13a and 13b). For example, using the same function

$$F = xy'z + xyz' + yz$$

we obtain F' as follows

$$\begin{aligned} F' &= (xy'z + xyz' + yz)' \\ &= (xy'z)' \cdot (xyz')' \cdot (yz)' && \text{by Theorem 13b} \\ &= (x' + y + z') \cdot (x' + y' + z) \cdot (y' + z') && \text{by Theorem 13a} \end{aligned}$$

There are three things to notice about this equation for F'. First, F' is just the dual of F (as defined in Section 2.7.2) but with all of the variables inverted. We call this the **inverted dual**. For example, the first term $xy'z$ becomes $x' + y + z'$. Second, instead of being in a sum-of-products format, it is in a **product-of-sums** (or **and-of-ors**) format where three **OR terms** (also referred to as **sum terms**) are ANDed together. Third, from the same original function F, we obtained two different equations for F'. From the truth table in Figure 2.11, we obtain

$$F' = x'y'z' + x'y'z + x'yz' + xy'z'$$

and from applying DeMorgan's Theorem to F, we obtain

$$F' = (x' + y + z') \cdot (x' + y' + z) \cdot (y' + z')$$

Hence, we must conclude that these two equations for F', where one is in the sum-of-products format and the other is in the product-of-sums format, are equivalent. In general, all functions can be expressed in either the sum-of-products format or the product-of-sums format.

Thus, we also should be able to express the original function, $F = xy'z + xyz' + yz$, in the product-of-sums format. We can derive it using one of two methods. For method one, we can start with F' and apply DeMorgan's Theorem to it just like how we obtained F' from F.

$$\begin{aligned}
F &= (F')' && \text{by Theorem 8a} \\
&= (x'y'z' + x'y'z + x'yz' + xy'z')' \\
&= (x'y'z')' \cdot (x'y'z)' \cdot (x'yz')' \cdot (xy'z')' && \text{by Theorem 13b} \\
&= (x + y + z) \cdot (x + y + z') \cdot (x + y' + z) \cdot (x' + y + z) && \text{by Theorem 13a}
\end{aligned}$$

For the second method, we start with the original F and convert it to the product-of-sums format using the Boolean theorems from Figure 2.10.

$$\begin{aligned}
F &= xy'z + xyz' + yz \\
&= (x{+}x{+}y) \cdot (x{+}x{+}z) \cdot (x{+}y{+}y) \cdot (x{+}y{+}z) \cdot (x{+}z'{+}y) \cdot && \text{(Step 1)} \\
&\quad \cancel{(x{+}z'{+}z)} \cdot \cancel{(y'{+}x{+}y)} \cdot (y'{+}x{+}z) \cdot \cancel{(y'{+}y{+}y)} \cdot \cancel{(y'{+}y{+}z)} \cdot \\
&\quad \cancel{(y'{+}z'{+}y)} \cdot \cancel{(y'{+}z'{+}z)} \cdot (z{+}x{+}y) \cdot (z{+}x{+}z) \cdot \\
&\quad (z{+}y{+}y) \cdot (z{+}y{+}z) \cdot \cancel{(z{+}z'{+}y)} \cdot \cancel{(z{+}z'{+}z)} \\
&= (x{+}y) \cdot (x{+}z) \cdot \cancel{(x{+}y)} \cdot (x{+}y{+}z) \cdot (x{+}z'{+}y) \cdot && \text{(Step 2)} \\
&\quad (y'{+}x{+}z) \cdot \cancel{(z{+}x{+}y)} \cdot \cancel{(z{+}x)} \cdot (z{+}y) \cdot \cancel{(z{+}y)} \\
&= (x{+}y) \cdot (x{+}z) \cdot (x{+}y{+}z) \cdot (x{+}y{+}z') \cdot (x{+}y'{+}z) \cdot (y{+}z) && \text{(Step 3)} \\
&= (x{+}y{+}zz') \cdot (x{+}yy'{+}z) \cdot (x{+}y{+}z) \cdot (x{+}y{+}z') \cdot && \text{(Step 4)} \\
&\quad (x{+}y'{+}z) \cdot (xx'{+}y{+}z) \\
&= (x{+}y{+}z) \cdot (x{+}y{+}z') \cdot \cancel{(x{+}y{+}z)} \cdot (x{+}y'{+}z) \cdot \cancel{(x{+}y{+}z)} \cdot && \text{(Step 5)} \\
&\quad \cancel{(x{+}y{+}z')} \cdot \cancel{(x{+}y'{+}z)} \cdot \cancel{(x{+}y{+}z)} \cdot (x'{+}y{+}z) \\
&= (x + y + z) \cdot (x + y + z') \cdot (x + y' + z) \cdot (x' + y + z)
\end{aligned}$$

In Step 1, we apply Theorem 12b (Distributive) to get every possible combination of the sum terms. For example, the first sum term $(x + x + y)$ is obtained from getting the first x from $xy'z$, the second x from xyz', and the y from yz. The second sum term $(x + x + z)$ is obtained from getting the first x from $xy'z$, the second x from xyz', and the z from yz. This is repeated for all combinations. In this step, the sum terms, such as $(x + z' + z)$, where it contains variables of the form $v + v'$, can be eliminated, because $v + v' = 1$ by Theorem 9b, and $1 \cdot x = x$ by Theorem 6a.

In Steps 2 and 3, duplicate variables and terms are eliminated. For example, the term $(x + x + y)$ is equal to $(x + y + y)$, which is just $(x + y)$ by Theorem 7b.

In Step 4, every sum term with a missing variable will have that variable added back in by using Theorems 6b and 9a, which says that $x + 0 = x$ and $yy' = 0$, therefore, $x + yy' = x$.

Step 5 uses the Distributive Theorem, and the resulting duplicate terms are again eliminated to give us the final format that we want.

Functions that are in the product-of-sums format (such as the one shown below) are more difficult to deduce when they evaluate to 1. For example, using

$$F' = (x' + y + z') \cdot (x' + y' + z) \cdot (y' + z')$$

F' evaluates to 1 only when all three terms evaluate to 1. For the first term to evaluate to 1, x can be 0, y can be 1, or z can be 0. For the second term to evaluate to 1, x can be 0, y can be 0, or z can be 1. Finally, for the last term, y can be 0, z can be 0, or x can be either 0 or 1. As a result, we end up with many more combinations to consider, even though many of the combinations are duplicates.

However, it is easier to determine when a product-of-sums format expression evaluates to a 0. For example, using the same expression:

$$F' = (x' + y + z') \cdot (x' + y' + z) \cdot (y' + z')$$

F' evaluates to 0 when any one of the three OR terms is 0, because 0 AND x is 0; and this happens when

$$x = 1, y = 0, \text{ and } z = 1 \text{ for the first OR term,}$$

or

$$x = 1, y = 1, \text{ and } z = 0 \text{ for the second OR term,}$$

or

$$x = 0, y = 1, z = 1 \text{ for the last OR term,}$$

or

$$x = 1, y = 1, z = 1 \text{ also for the last OR term.}$$

So, for example, if we have $x = 1$, $y = 0$, and $z = 1$, the first OR term will be 0, which in turn will make the entire expression 0. These four conditions in which F' evaluates to 0 match exactly the four rows in the truth table shown in Figure 2.11, where $F' = 0$.

For a sum-of-products format expression, it is just the opposite, in that it is easier to evaluate when it is a 1, but more difficult to evaluate when it is a 0.

	Sum-of-Products		Product-of-Sums
F	$x'yz + xy'z + xyz' + xyz$	← Equal →	$(x + y + z) \cdot (x + y + z') \cdot (x + y' + z) \cdot (x' + y + z)$
↑ Inverse ↓		Inverted Dual ╳ Inverted Dual	
F'	$x'y'z' + x'y'z + x'yz' + xy'z'$	← Equal →	$(x + y' + z') \cdot (x' + y + z') \cdot (x' + y' + z) \cdot (x' + y' + z')$

FIGURE 2.12 Relationships between the function $F = xy'z + xyz' + yz$ and its inverse F', and between the sum-of-products and product-of-sums formats. The label "Inverted Dual" means applying the duality principle and then inverting the variables.

From this discussion, we can conclude that, in general, the unique algebraic expression for any Boolean function can be specified by either (1) selecting the rows from the truth table where the function is a 1 and using the sum-of-products format, or (2) selecting the rows from the truth table where the function is a 0 and using the product-of-sums format. Whichever format we decide to use, the one thing to remember is that when we create a circuit based on the function, we are always interested in when the function (or its inverse) is equal to a 1.

Figure 2.12 summarizes these two formats for the function $F = xy'z + xyz' + yz$ and its inverse F'. Notice that the sum-of-products format for F is the dual but with its variables inverted from the product-of-sums format for F'. Similarly, the product-of-sums format for F is the **inverted dual** from the sum-of-products format for F'.

2.8 Minterms and Maxterms

As you recall, a product term is a term with either a single variable or two or more variables ANDed together, whereas, a sum term is a term with either a single variable or two or more variables ORed together. To differentiate between a term that contains any number of variables with a term that contains *all* of the variables used in the function, we use the words minterm and maxterm. We are not introducing new ideas here; rather, we are just introducing two new words and notations for defining what we already have learned.

2.8.1 Minterms

A **minterm** is a product term that contains all the variables used in a function. For a function with n variables, the notation m_i, where $0 \leq i < 2^n$, is used to denote the minterm whose index i is the binary value of the n variables, such that the variable is complemented if the value assigned to it is a 0 and uncomplemented if it is a 1.

This definition sounds much more complicated than it actually is. Figure 2.13(a) shows the eight minterms and their notations for $n = 3$ using the three variables x, y,

x	y	z	Minterm	Notation
0	0	0	$x'y'z'$	m_0
0	0	1	$x'y'z$	m_1
0	1	0	$x'yz'$	m_2
0	1	1	$x'yz$	m_3
1	0	0	$xy'z'$	m_4
1	0	1	$xy'z$	m_5
1	1	0	xyz'	m_6
1	1	1	xyz	m_7

(a)

x	y	z	Maxterm	Notation
0	0	0	$x + y + z$	M_0
0	0	1	$x + y + z'$	M_1
0	1	0	$x + y' + z$	M_2
0	1	1	$x + y' + z'$	M_3
1	0	0	$x' + y + z$	M_4
1	0	1	$x' + y + z'$	M_5
1	1	0	$x' + y' + z$	M_6
1	1	1	$x' + y' + z'$	M_7

(b)

FIGURE 2.13 (a) Minterms for three variables; (b) Maxterms for three variables.

and z. For example, the minterm notation m_3 is used to represent the term in which the values for the variables xyz are 011 (for the subscript 3). For minterms, we want to complement the variable whose value is a 0 and uncomplement it if it is a 1, therefore, the notation m_3 is for the minterm $x'yz$.

When specifying a function, we usually start with product terms that contain all of the variables used in the function. In other words, we want the **sum-of-minterms**, and more specifically, the sum of the one-minterms (i.e., the minterms for which the function is a 1) as opposed to the zero-minterms (i.e., the minterms for which the function is a 0). We use the notation **1-minterm** to denote one-minterm, and **0-minterm** to denote zero-minterm.

Consider the function from the previous section:

$$
\begin{aligned}
F &= xy'z + xyz' + yz \\
&= x'yz + xy'z + xyz' + xyz
\end{aligned}
$$

and repeated in the following truth table has the 1-minterms: m_3, m_5, m_6, and m_7.

x	y	z	F	Minterm	Notation
0	0	0	0	$x'y'z'$	m_0
0	0	1	0	$x'y'z$	m_1
0	1	0	0	$x'yz'$	m_2
0	1	1	1	$x'yz$	m_3
1	0	0	0	$xy'z'$	m_4
1	0	1	1	$xy'z$	m_5
1	1	0	1	xyz'	m_6
1	1	1	1	xyz	m_7

Thus, a shorthand notation for the function is

$$F(x, y, z) = m_3 + m_5 + m_6 + m_7$$

By using just the minterm notations, we do not know how many variables are in the original function. Consequently, we need to specify explicitly the variables used by the function, as in $F(x, y, z)$.

We can further simplify the notation by using the standard algebraic symbol Σ for summation and listing out the minterm index numbers. Therefore, we can rewrite the equation as follows

$$F(x, y, z) = \Sigma(3, 5, 6, 7)$$

These are just different ways of expressing the same function.

Because a function is obtained from the sum of the 1-minterms, the inverse of the function, therefore, must be the sum of the 0-minterms. This can be obtained easily by replacing the set of indices with those that were excluded from the original set. Therefore, the inverse of the above function is

$$F'(x, y, z) = \Sigma(0, 1, 2, 4)$$

EXAMPLE 2.21

Converting a function to the sum-of-minterms format using Boolean algebra

Given the Boolean function $F(x, y, z) = y + x'z$, use Boolean algebra to convert the function to the sum-of-minterms format.

This function has three variables. In a sum-of-minterms format, all product terms must have all variables. To change a product term to a minterm, we need to expand each product term by ANDing it with $(v + v')$ for every missing variable v in that term. Because $(v + v') = 1$, therefore, ANDing a product term with $(v + v')$ does not change the value of the term.

$$\begin{aligned} F &= y + x'z \\ &= y(x + x')(z + z') + x'z(y + y') && \text{expand 1st term by ANDing it with } (x + x')(z + z')\text{, and 2nd term with } (y + y') \\ &= xyz + xyz' + x'yz + x'yz' + \cancel{x'yz} + x'y'z && \text{by Theorem 12a} \\ &= m_7 + m_6 + m_3 + m_2 + m_1 \\ &= \Sigma(1, 2, 3, 6, 7) && \text{sum of 1-minterms} \end{aligned}$$

EXAMPLE 2.22

Converting the inverse of a function to the sum-of-minterms format

Given the Boolean function $F(x, y, z) = y + x'z$, use Boolean algebra to convert the inverse of the function to the sum-of-minterms format.

$F' = (y + x'z)'$	inverse
$= y' \cdot (x'z)'$	use DeMorgan's Theorem
$= y' \cdot (x + z')$	use DeMorgan's Theorem
$= y'x + y'z'$	use Distributive Theorem to change to sum-of-products format
$= y'x(z + z') + y'z'(x + x')$	expand 1st term by ANDing it with $(z+z')$, and 2nd term with $(x+x')$
$= xy'z + xy'z' +$ ~~$xy'z'$~~ $+ x'y'z'$	
$= m_5 + m_4 + m_0$	
$= \Sigma(0, 4, 5)$	sum of 0-minterms

2.8.2 Maxterms

Analogous to a minterm, a **maxterm** is a sum term that contains all of the variables used in the function. For a function with n variables, the notation M_i, where $0 \leq i < 2^n$, is used to denote the maxterm whose index i is the binary value of the n variables, such that the variable is complemented if the value assigned to it is a 1 and uncomplemented if it is a 0.

Figure 2.13(b) shows the eight maxterms and their notations for $n = 3$ using the three variables x, y, and z. For example, the maxterm notation M_3 is used to represent the term in which the values for the variables xyz are 011 (for the subscript 3). For maxterms, we want to complement the variable whose value is a 1 and uncomplement it if it is a 0, therefore, the notation M_3 is for the maxterm $x + y' + z'$.

We have seen that a function can also be specified as a **product-of-maxterms**, or more specifically, a product of the zero-maxterms (i.e., the maxterms for which the function is a 0). Just like the minterms, we use the notation **1-maxterm** to denote one-maxterm, and **0-maxterm** to denote zero-maxterm. Thus, the function:

$$F(x, y, z) = xy'z + xyz' + yz$$
$$= (x + y + z) \cdot (x + y + z') \cdot (x + y' + z) \cdot (x' + y + z)$$

which is shown in the following table

x	y	z	F	Maxterm	Notation
0	0	0	0	$x + y + z$	M_0
0	0	1	0	$x + y + z'$	M_1
0	1	0	0	$x + y' + z$	M_2
0	1	1	1	$x + y' + z'$	M_3
1	0	0	0	$x' + y + z$	M_4
1	0	1	1	$x' + y + z'$	M_5
1	1	0	1	$x' + y' + z$	M_6
1	1	1	1	$x' + y' + z'$	M_7

can be specified as the product of the 0-maxterms M_0, M_1, M_2, and M_4. The shorthand notation for the function is

$$F(x, y, z) = M_0 \cdot M_1 \cdot M_2 \cdot M_4$$

By using the standard algebraic symbol Π for product and listing out the maxterm index numbers, the notation is further simplified to

$$F(x, y, z) = \Pi(0, 1, 2, 4)$$

The following summarizes these relationships for the function $F = xy'z + xyz' + yz$ and its inverse. Comparing these equations with those in Figure 2.12, we see that they are identical.

$$\begin{aligned} F(x, y, z) &= x'yz + xy'z + xyz' + xyz \\ &= m_3 + m_5 + m_6 + m_7 \\ &= \Sigma(3, 5, 6, 7) \\ &= (x+y+z)\,(x+y+z')\,(x+y'+z)\,(x'+y+z) \\ &= M_0 \cdot M_1 \cdot M_2 \cdot M_4 \\ &= \Pi(0, 1, 2, 4) \end{aligned}$$

$$\begin{aligned} F'(x, y, z) &= x'y'z' + x'y'z + x'yz' + xy'z' \\ &= m_0 + m_1 + m_2 + m_4 \\ &= \Sigma(0, 1, 2, 4) \\ &= (x+y'+z')\,(x'+y+z')\,(x'+y'+z)\,(x'+y'+z') \\ &= M_3 \cdot M_5 \cdot M_6 \cdot M_7 \\ &= \Pi(3, 5, 6, 7) \end{aligned}$$

Σ 1-minterms
Π 0-maxterms
Σ 0-minterms
Π 1-maxterms
Inverted Duals
Equivalent
Equivalent
Inverse

Notice that it is always the Σ of minterms and Π of maxterms; you never have Σ of maxterms or Π of minterms.

EXAMPLE 2.23

Converting a function to the product-of-maxterms format using Boolean algebra

Given the Boolean function $F(x, y, z) = y + x'z$, use Boolean algebra to convert the function to the product-of-maxterms format.

In a product-of-maxterms format, all sum terms must have all variables. To change a sum term to a maxterm, we need to expand each sum term by ORing it with (vv') for every missing variable v in that term. Because $(vv') = 0$, therefore, ORing a sum term with (vv') does not change the value of the term.

$F = y + x'z$

$= y + (x'z)$

$= (y + x')(y + z)$ — use Distributive Theorem to change to product-of-sums format

$= (y + x' + zz')(y + z + xx')$ — expand 1st term by ORing it with zz', and 2nd term with xx'

$= (x' + y + z)(x' + y + z')(x + y + z)\cancel{(x' + y + z)}$

$= M_4 \cdot M_5 \cdot M_0$

$= \Pi(0, 4, 5)$ — product of 0-maxterms

EXAMPLE 2.24

Converting the inverse of a function to the product-of-maxterms format

Given the Boolean function $F(x, y, z) = y + x'z$, use Boolean algebra to convert the inverse of the function to the product-of-maxterms format.

$F' = (y + x'z)'$ — inverse

$= y' \cdot (x'z)'$ — use DeMorgan's Theorem

$= y' \cdot (x + z')$ — use DeMorgan's Theorem

$= (y' + xx' + zz') \cdot (x + z' + yy')$ — expand 1st term by ORing it with $(xx' + zz')$ and 2nd term with yy'

$= (x + y' + z)\ (x + y' + z')\ (x' + y' + z)\ (x' + y' + z')\ (x + y + z')\ \cancel{(x + y' + z')}$

$= M_2 \cdot M_3 \cdot M_6 \cdot M_7 \cdot M_1$

$= \Pi(1, 2, 3, 6, 7)$ — product of 1-maxterms

In conclusion, given a function in either the sum-of-minterms or product-of-maxterms format, we can easily obtain its inverse by either changing the operator Σ to Π or vice versa, or changing the set of indices to those that were excluded from the original set, regardless of whether it is for minterms or maxterms. Using this fact, we can easily reduce the size of a circuit as shown in the next example.

EXAMPLE 2.25

Reducing a function by using the inverse of the function, and the sum-of-minterms and product-of-maxterms transformation

Given the function $F(x, y, z) = \Sigma(1, 2, 5, 6, 7)$, reduce its size by finding its inverse and using the sum-of-minterms and product-of-maxterms transformation.

To get the inverse of a function, we simply either change the operator Σ to Π or vice versa, or change the set of indices to the other set. Hence, the inverse of the function $F(x, y, z) = \Sigma(1, 2, 5, 6, 7)$ is $F'(x, y, z) = \Sigma(0, 3, 4)$ by changing the set of indices. To get the original function back, we invert F' by changing the operator to get $F = \Pi(0, 3, 4)$.

Hence, the function $\Sigma(1, 2, 5, 6, 7)$ is equivalent to the function $\Pi(0, 3, 4)$ in terms of functional operation, but in terms of size, the function $\Pi(0, 3, 4)$ is smaller. The size of an AND gate versus an OR gate is insignificant. Therefore, the size of the circuit is determined mainly by the number of terms. Because the latter function contains fewer terms, it is, therefore, smaller.

2.9 Canonical, Standard, and Non-Standard Forms

Any Boolean function that is expressed as a sum-of-minterms, or as a product-of-maxterms is said to be in its **canonical form**. For example, the following two equations are in their canonical forms:

$$F = x'yz + xy'z + xyz' + xyz$$

$$F' = (x + y' + z') \cdot (x' + y + z') \cdot (x' + y' + z) \cdot (x' + y' + z')$$

As noted from the previous section, to convert a Boolean function from one canonical form to its other equivalent canonical form, simply interchange the symbols Σ with Π and list the index numbers that were excluded from the original set. For example, the following two equations are functionally equivalent.

$$F_1(x, y, z) = \Sigma(0, 1, 2, 3, 4, 5)$$

$$F_2(x, y, z) = \Pi(6, 7)$$

However, in terms of size, F_2 is much smaller.

To convert a Boolean function from one canonical form to its inverse, simply interchange the symbols Σ with Π and list the same index numbers from the original set. For example, the following two equations are inverses.

$$F_1(x, y, z) = \Sigma(3, 5, 6)$$
$$F_2(x, y, z) = \Pi(3, 5, 6)$$

A Boolean function is said to be in a **standard form** if a sum-of-products expression or a product-of-sums expression has at least one term that is not a minterm or a maxterm, respectively. In other words, at least one term in the expression is missing at least one variable. For example, the following equation is in a standard form because the last term is missing the variable x.

$$F = xy'z + xyz' + yz$$

To further reduce the size of an equation, common variables in a standard form expression can be factored out using the Distributive Theorem. The resulting expression is no longer in a sum-of-products or product-of-sums format. These expressions are in a **non-standard form**. For example, starting with the previous equation, if we factor out the common variable x from the first two terms, we get the following equation, which is in a non-standard form.

$$\begin{aligned} F &= xy'z + xyz' + yz \\ &= x(y'z + yz') + yz \end{aligned}$$

2.10 Digital Circuits

A **digital circuit** is a connection of two or more logic gates together. Many gates can be interconnected to form large and complex circuits called **networks**. These networks can be described either graphically using **schematic circuit diagrams**, or with Boolean equations, or with truth tables.

For example, the following is a schematic diagram of a digital circuit having three NOT gates, five 3-input AND gates and one 5-input OR gate.

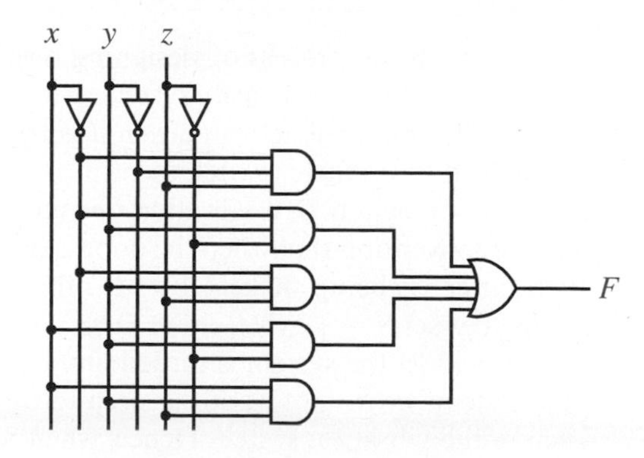

This circuit also can be described formally with the Boolean equation

$$F(x, y, z) = x'y'z + x'yz' + x'yz + xyz' + xyz$$

Notice the correspondence between the schematic diagram and the equation. The five AND gates correspond to the five AND terms in the equation; the three inputs for each AND gate correspond to the three variables in each AND term; and the one 5-input OR gate corresponds to the ORing of the five AND terms together. The inputs to the AND gates come directly from the three variables x, y, and z (or their inverted values). Notice that in the equation, there are six inverted variables. However, in the circuit, we do not need six NOT gates. Rather, only three NOT gates are used; one for each variable.

The circuit also can be described formally using a truth table as shown next.

x	y	z	F
0	0	0	0
0	0	1	1
0	1	0	1
0	1	1	1
1	0	0	0
1	0	1	0
1	1	0	1
1	1	1	1

First, all possible values of the three input variables are enumerated giving us the eight rows in the table. Then for each AND gate in the circuit there is a corresponding row in the truth table where the F value is a 1. For example, for the first AND gate where the three inputs are x', y', and z, therefore the first row where $F = 1$ is when $x = 0, y = 0$, and $z = 1$.

The process to convert back and forth between the schematic diagram, Boolean equation and truth table is discussed in detail in Chapter 3.

2.11 Designing a Car Security System

We will now go through the process of designing a small digital control system that we are all familiar with. In a car security system, we usually want to connect the siren in such a way that the siren will activate when it is triggered by one or more sensors. In addition, there will be a master switch to turn the system on or off. Let us assume that there is a car door switch D, a vibration detector switch V, and a master switch M. We will use the convention that when the door is opened, $D = 1$, otherwise, $D = 0$. Similarly, when the car is being shaken, $V = 1$, otherwise, $V = 0$. Thus, we want the siren S to turn on (i.e., set $S = 1$) when either $D = 1$ or $V = 1$ or when both $D = 1$ and $V = 1$, but only for when the system is turned on (i.e., when $M = 1$). However, when we turn off the system, we do not want the siren to turn on regardless of the state of the door switch or the vibration switch. Hence, when $M = 0$, it does not matter what values D and V have, the siren should remain off (i.e., set $S = 0$).

Given the above description of a car security system, we can build a digital circuit that meets our specifications. We start by constructing a truth table, which is basically a precise way of stating the operations for the device. The table will have three input columns M, D, and V, and an output column S, as shown in Figure 2.14(a).

Under the three input columns, we enumerate all possible binary values for the three inputs. The values under the S column are obtained from interpreting the description of when we want the siren to turn on or off. When $M = 0$, we don't want the siren to come on, regardless of what the values for D and V are. When $M = 1$, we want the siren to come on when either D or V is a 1, or both D and V are 1.

The truth table in Figure 2.14(a) can be described formally with a logic equation written in words as

$$S = (M \text{ AND } (\text{NOT } D) \text{ AND } V) \text{ OR } (M \text{ AND } D \text{ AND } (\text{NOT } V)) \text{ OR } (M \text{ AND } D \text{ AND } V)$$

or preferably, using the simpler notation of a Boolean function:

$$S = (MD'V) + (MDV') + (MDV)$$

Again, what this equation is saying is that we want the siren to activate (i.e., set $S = 1$) when:

- the master switch is on and the door is not opened and the vibration switch is on, or
- the master switch is on and the door is opened and the vibration switch is not on, or
- the master switch is on and the door is opened and the vibration switch is on.

Notice that we are interested only in the situations when $S = 1$. We ignore the rows when $S = 0$. When we construct circuits from truth tables, we always use only the rows where the output is a 1.

M	D	V	S
0	0	0	0
0	0	1	0
0	1	0	0
0	1	1	0
1	0	0	0
1	0	1	1
1	1	0	1
1	1	1	1

(a)

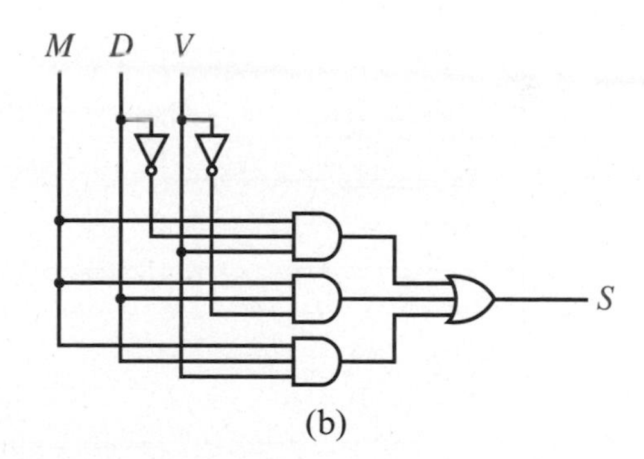

(b)

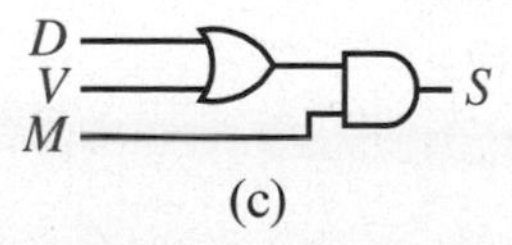

(c)

FIGURE 2.14 Car security system: (a) truth table; (b) circuit diagram derived from the truth table; (c) simplified circuit diagram.

Finally, we can translate this equation into a circuit diagram. The translation is a simple one-to-one mapping of changing the AND operator into the AND gate, the OR operator into the OR gate, and the NOT operator into the NOT gate. Thus, we get the circuit diagram shown in Figure 2.14(b) for our car security system.

A careful reader might notice that the Boolean equation shown above for specifying when the siren is to be turned on can be simplified to

$$S = M(D + V)$$

This simplified equation says that the siren is to be turned on only when the master switch is on and either the door switch or vibration switch is on. The corresponding simplified circuit diagram is shown in Figure 2.14(c). Just by using simple reasoning, we can see that this simplified circuit will do exactly what the circuit in Figure 2.14(b) does. In other words, the two circuits are functionally equivalent.

More formally, we can use the Boolean theorems from Section 2.7.1 to show that these two equations (and therefore, the two circuits) are indeed functionally equivalent as follows:

$$\begin{aligned}
S &= (MD'V) + (MDV') + (MDV) & \\
&= M(D'V + DV' + DV) & \text{by Distributive Theorem 12a} \\
&= M(D'V + DV' + DV + DV) & \text{by Idempotent Theorem 7b} \\
&= M(D(V' + V) + V(D' + D)) & \text{by Distributive Theorem 12a} \\
&= M(D(1) + V(1)) & \text{by Inverse Theorem 9b} \\
&= M(D + V) & \text{by Identity Theorem 6a}
\end{aligned}$$

Figure 2.15(a) shows a sample simulation trace of the car security system circuit. Between times 0 and 200 ns, the master switch M is a 0, so regardless of the values of

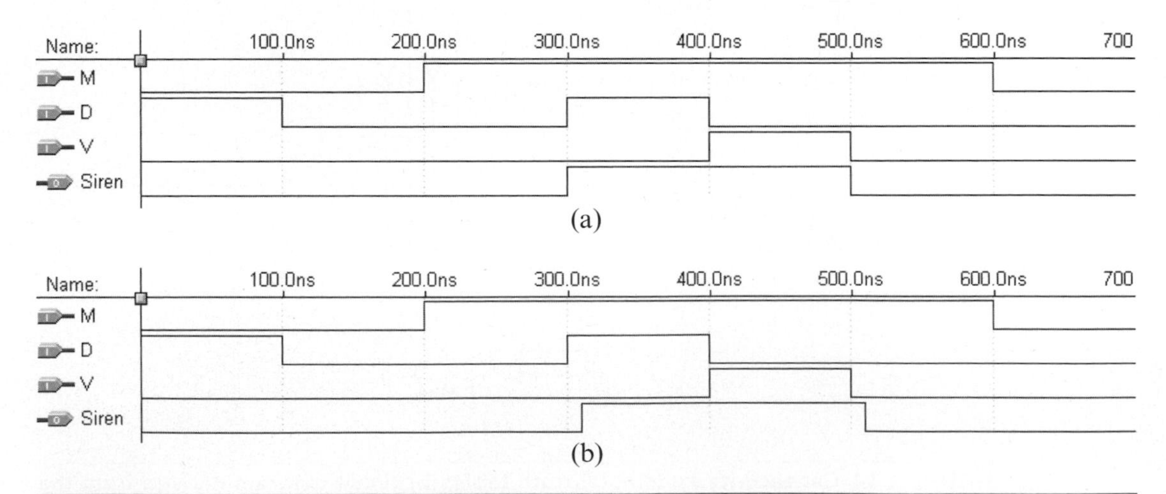

FIGURE 2.15 Sample simulation trace of the car security system circuit: (a) functional trace; (b) timing trace.

D and V, the siren is off ($Siren = 0$). Between times 200 ns and 600 ns, $M = 1$. During this time, whenever either $D = 1$ or $V = 1$, the siren is on.

Figure 2.15(a) is a **functional** trace of the circuit, so all the signal edges line up exactly, that is, the output signal edge changes at exactly the same time (with no delay) as the input edge that caused it to change. For a **timing** trace, on the other hand, the output signal edge will be delayed slightly after the causing input edge, as shown in Figure 2.15(b).

When building circuits, in addition to having a functionally correct circuit, we also want to optimize it in terms of its size, speed, heat dissipation, and power consumption. We will see in later sections how circuits are optimized.

2.12 Verilog and VHDL Code for Digital Circuits

A digital circuit that is described with a Boolean function can be converted easily to either Verilog or VHDL code using the **dataflow model**. At the dataflow level, a circuit is defined using built-in logic operators such as AND, OR, and NOT. These operators are applied to wires or signals using concurrent signal assignment statements.

2.12.1 Verilog Code for a Boolean Function

Figure 2.16 shows the Verilog code for the car security system circuit discussed in Section 2.11. The function implemented is $S = (MD'V) + (MDV') + (MDV)$. This also will serve as a basic template for Verilog dataflow codes. Note that Verilog is case sensitive. Lines starting with two slashes are comments.

Every component defined in Verilog starts with the keyword `module` followed by the name of the module, which in this example the name *Siren* is used. Next, a list of input and output signals is specified between the open and close parentheses. This parameter list is similar to a function declaration in C++ and serves as the interface between the component and the outside world. It specifies the data to be passed in and out of the component. In this example, there are three input signals called M, D, and V and an output signal called S.

```
// this is a Verilog dataflow model of the car security system

module Siren (
  input M,
  input D,
  input V,
  output S
);

  wire term1, term2, term3;

  assign term1 = (M & ~D & V);
  assign term2 = (M & D & ~V);
  assign term3 = (M & D & V);
  assign S = term1 | term2 | term3;

endmodule
```

FIGURE 2.16 Dataflow Verilog code for the car security system circuit of Section 2.11.

The `wire` keyword declares wire signals for connecting to other signals. The next four `assign` statements are the code that realizes the operation of the function. This Verilog code is written at the dataflow level because it uses the concurrent `assign` statements with the built-in Verilog operators & (for AND), | (for OR), and ~ (for NOT) to realize the operation of the circuit. Unlike statements in C++ that are executed in sequential order, concurrent `assign` statements in the module body are executed in parallel. Thus, the ordering of these statements is irrelevant. The symbol = is used for assignment. The expression on the right side of the = symbol is evaluated when the values stored in the variables change (either from 0 to 1 or from 1 to 0), and the result is assigned to the signal on the left side. The three wire variables, *term1*, *term2*, and *term3*, although unnecessary, are used to store the intermediate results of the three AND terms. The values in the three terms are ORed together and assigned to the output signal *S*. The `endmodule` keyword is used to terminate the definition of the module.

2.12.2 VHDL Code for a Boolean Function

Figure 2.17 shows the VHDL code for the car security system circuit discussed in Section 2.11. The function implemented is $S = (MD'V) + (MDV') + (MDV)$. This also will serve as a basic template for VHDL dataflow codes. Note that VHDL is not case sensitive. Lines starting with two hyphens are comments.

The LIBRARY and USE statements specify that the IEEE library is needed and that all of the components in that library package can be used. These two statements are equivalent to the "#include" and "using namespace" preprocessor lines in C++. Every component defined in VHDL, whether it is a simple gate or a complex microprocessor, has two parts: an ENTITY section and an ARCHITECTURE section. The ENTITY section is similar to a function declaration in C++ and serves as the interface between the component and the outside. It declares all of the input and output signals for a circuit. Every entity must have a unique name; in this example, the name *Siren* is used. The entity contains

```
-- this is a VHDL dataflow model of the car security system

LIBRARY IEEE;
USE IEEE.STD_LOGIC_1164.all;

ENTITY Siren IS PORT (
  M: IN STD_LOGIC;
  D: IN STD_LOGIC;
  V: IN STD_LOGIC;
  S: OUT STD_LOGIC);
END Siren;

ARCHITECTURE Dataflow OF Siren IS
   SIGNAL term1, term2, term3: STD_LOGIC;
BEGIN
   term1 <= M AND (NOT D) AND V;
   term2 <= M AND D AND (NOT V);
   term3 <= M AND D AND V;
   S <= term1 OR term2 OR term3;
END Dataflow;
```

FIGURE 2.17 Dataflow VHDL code for the car security system circuit of Section 2.11.

a PORT list, which, like a parameter list, specifies the data to be passed in and out of the component. In this example, there are three input signals called *M*, *D*, and *V* of type STD_LOGIC and an output signal called *S* of the same type. The STD_LOGIC type is like the BIT type, except that it contains the additional values Z and U besides just 0s and 1s.

The ARCHITECTURE section defines the operation of the entity; it contains the code that realizes the operation of the component. For every architecture, you need to specify its name and which entity it is for. In this example, the name is *Dataflow*, and it is for the entity *Siren*. This VHDL code is written at the dataflow level not because of the name "Dataflow" in the architecture section. Dataflow-level coding uses logic equations to describe a circuit, and this is done by using the built-in VHDL operators such as AND, OR, and NOT in concurrent signal assignment statements. These concurrent statements are written inside the body of the architecture. Unlike statements in C++ that are executed in sequential order, concurrent statements in the architecture body (between the BEGIN and END keywords) are executed in parallel. Thus, the ordering of these statements is irrelevant. The symbol <= is used for the signal assignment statement. The expression on the right side of the <= symbol is evaluated when the values stored in the variables change (either from 0 to 1 or from 1 to 0), and the result is assigned to the signal on the left side. The three signal variables *term1*, *term2*, and *term3*, although unnecessary, are used to store the intermediate results of the three AND terms.

2.13 PROBLEMS

2.1. Convert the following decimal numbers to binary numbers.

a) 66
b) 49
c) 513
d) 864
e) 1897
f) 2015

2.2. Convert the following unsigned binary numbers to decimal, octal, and hexadecimal numbers.

a) 11110
b) 11010
c) 100100011
d) 1011011
e) 1101101110
f) 10111101010

2.3. Convert the following hexadecimal numbers to unsigned binary numbers.

a) 66
b) E3
c) 2FE8
d) 7C2
e) 5A2D
f) E08B

2.4. Convert the following numbers to 12-bit signed binary numbers using two's complement representation.

a) 234_{10}
b) -234_{10}
c) 234_8
d) $BC4_{16}$
e) -472_{10}

2.5. Convert the following two's complement binary numbers to decimal, octal, and hexadecimal formats.

a) 1001011
b) 011110
c) 101101
d) 1101011001
e) 0110101100

2.6. Perform the following 8-bit unsigned binary calculations, and specify the resulting decimal number.

a) 10101010 + 00111011
b) 00111101 + 01110100
c) 11100011 + 11110011
d) 11000111 + 10010110

2.7. Perform the following 8-bit signed binary calculations, and specify the resulting decimal number.

a) 10101010 + 00111011
b) 00111101 + 01110100
c) 11100011 + 11110011
d) 11000111 + 10010110

2.8. Perform the following 8-bit unsigned binary calculations, and specify the resulting decimal number.

a) 10101010 − 00111011
b) 00111101 − 01110100
c) 11100011 − 11110011
d) 01000111 − 10010110

2.9. Perform the following 8-bit signed binary calculations, and specify the resulting decimal number.

a) 10101010 − 00111011
b) 00111101 − 01110100
c) 11100011 − 11110011
d) 01000111 − 10010110

2.10. Perform the following 4-bit binary calculations. (The first one has been done for you.) Then:

a) Specify whether there is an overflow in the binary calculation.
b) Specify whether there is an overflow error if the binary numbers are interpreted as unsigned decimal numbers.

c) Specify whether there is an overflow error if the binary numbers are interpreted as signed decimal numbers.

Binary calculations	Unsigned decimal calculations	Signed decimal calculations
1001 + 0011 = 1100 No overflow	9 + 3 = 12 No overflow error	−7 + 3 = −4 No overflow error
0110 + 1011 =		
0101 + 0110 =		
0101 − 0110 =		
1011 − 0101 =		

2.11. Derive the truth table for the following Boolean functions.

a) $F(x,y,z) = x'y'z' + x'yz + xy'z' + xyz$

b) $F(x,y,z) = xy'z + x'yz' + xyz + xyz'$

c) $F(w,x,y,z) = w'xy'z + w'xyz + wxy'z + wxyz$

d) $F(w,x,y,z) = wxy'z + w'yz' + wxz + xyz'$

e) $F(x,y,z) = xy' + x'y'z + xyz'$

f) $F(w,x,y,z) = w'z' + w'xy + wx'z + wxyz$

g) $F(x,y,z) = [(x + y')(yz)'](xy' + x'y)$

h) $F(N_3,N_2,N_1,N_0) = N_3'N_2'N_1N_0' + N_3'N_2'N_1N_0 + N_3N_2'N_1N_0' + N_3N_2'N_1N_0 + N_3N_2N_1'N_0' + N_3N_2N_1N_0$

2.12. Derive the Boolean function for the following truth tables.

a)

a	*b*	*c*	*F*
0	0	0	0
0	0	1	0
0	1	0	1
0	1	1	1
1	0	0	0
1	0	1	0
1	1	0	1
1	1	1	0

b)

w	*x*	*y*	*z*	*F*
0	0	0	0	0
0	0	0	1	0
0	0	1	0	1
0	0	1	1	0
0	1	0	0	1
0	1	0	1	1
0	1	1	0	0
0	1	1	1	1
1	0	0	0	0
1	0	0	1	1
1	0	1	0	1
1	0	1	1	0
1	1	0	0	1
1	1	0	1	1
1	1	1	0	0
1	1	1	1	1

c)

w	x	y	z	F_1	F_2
0	0	0	0	1	1
0	0	0	1	0	1
0	0	1	0	0	1
0	0	1	1	1	1
0	1	0	0	0	0
0	1	0	1	1	1
0	1	1	0	1	0
0	1	1	1	0	0
1	0	0	0	0	1
1	0	0	1	1	1
1	0	1	0	1	0
1	0	1	1	0	0
1	1	0	0	1	1
1	1	0	1	0	1
1	1	1	0	0	1
1	1	1	1	1	1

d)

N_3	N_2	N_1	N_0	F
0	0	0	0	0
0	0	0	1	0
0	0	1	0	1
0	0	1	1	1
0	1	0	0	0
0	1	0	1	0
0	1	1	0	1
0	1	1	1	0
1	0	0	0	0
1	0	0	1	0
1	0	1	0	1
1	0	1	1	1
1	1	0	0	1
1	1	0	1	0
1	1	1	0	0
1	1	1	1	1

2.13. Use a truth table to show that the following variations of the DeMorgan's Theorem are true.

a) $(x + y)' = x' \cdot y'$
b) $(x + y + z)' = x' \cdot y' \cdot z'$
c) $(x \cdot y \cdot z)' = x' + y' + z'$
d) $(w \cdot x \cdot y \cdot z)' = w' + x' + y' + z'$

2.14. Use a truth table to show that the following equations are true.

a) $w'z' + w'xy + wx'z + wxyz = w'z' + xyz + wx'y'z + wyz$
b) $z + y' + yz' = 1$
c) $xy'z' + x' + xyz' = x' + z'$
d) $xy + x'z + yz = xy + x'z$
e) $w'x'yz' + w'x'yz + wx'yz' + wx'yz + wxyz = y(x' + wz)$
f) $w'xy'z + w'xyz + wxy'z + wxyz = xz$
g) $x_iy_i + c_i(x_i + y_i) = x_iy_ic_i + x_iy_ic_i' + x_iy_i'c_i + x_i'y_ic_i$
h) $x_iy_i + c_i(x_i + y_i) = x_iy_i + c_i(x_i \oplus y_i)$

2.15. Use Boolean algebra to show that $x \cdot (x + y) = x$ is true.

2.16. Use Boolean algebra to show that $x + (x \cdot y) = x$ is true.

2.17. Use Boolean algebra to show that $(x \cdot y) + (x \cdot y') = x$ is true.

2.18. Use Boolean algebra to show that $(x + y) \cdot (x + y') = x$ is true.

2.19. Use Boolean algebra to show that the equations in Problem 2.14 are true.

2.20. Use Boolean algebra to reduce the functions in Problem 2.11 as much as possible.

2.21. Use Boolean algebra to reduce the equation $F(x, y, z) = (x' + y' + x'y' + xy)(x' + yz)$ as much as possible.

2.22. Any function can be implemented directly either as specified or as its inverted form with a NOT gate added at the final output. Assume that the circuit size is proportional to only the number of AND gates and OR gates (i.e., ignore the number of NOT gates in determining the circuit size). Determine which form of the function (the inverted or non-inverted) will result in a smaller circuit size for the following function. Give your reason, and specify how many AND and OR gates are needed to implement the smaller circuit.

$$F(x,y,z) = x'y'z' + x'y'z + xy'z + xy'z' + xyz$$

2.23. Derive the truth table for the following logic gates.

a) A 4-input AND gate.
b) A 4-input NAND gate.
c) A 4-input NOR gate.
d) A 4-input XOR gate.
e) A 4-input XNOR gate.
f) A 5-input XOR gate.
g) A 5-input XNOR gate.

2.24. Derive the truth table for the following Boolean functions.

a) $F(w,x,y,z) = [(x \odot y)' + (xyz)'] \, (w' + x + z)$
b) $F(x,y,z) = x \oplus y \oplus z$
c) $F(w,x,y,z) = [w'xy'z + w'z(y \oplus x)]'$

2.25. Use Boolean algebra to convert the functions in Problem 2.24 to:

a) The sum-of-minterms format
b) The product-of-maxterms format

2.26. Use Boolean algebra to reduce the functions in Problem 2.24 as much as possible.

2.27. Use a truth table to show that the following equations are true.

a) $(x \oplus y) = (x \odot y)'$
b) $x \oplus y' = x \odot y$
c) $(w \oplus x) \odot (y \oplus z) = (w \odot x) \odot (y \odot z)$
$= (((w \odot x) \odot y) \odot z)$
d) $[((xy)'x)' \, ((xy)'y)']' = x \oplus y$

2.28. Use Boolean algebra to show that the equations in Problem 2.27 are true.

2.29. Use Boolean algebra to show that
$x \oplus y \oplus z = x'y'z + x'yz' + xy'z' + xyz$.

2.30. Use Boolean algebra to show that XOR = XNOR for three inputs.

2.31. Express the Boolean functions in Problem 2.11 using:

a) The Σ notation

b) The Π notation

2.32. Write the following equations as a Boolean function in the canonical form.

a) $F(x, y, z) = \Sigma(1, 3, 7)$

b) $F(w, x, y, z) = \Sigma(1, 3, 7)$

c) $F(x, y, z) = \Pi(1, 3, 7)$

d) $F(w, x, y, z) = \Pi(1, 3, 7)$

e) $F'(x, y, z) = \Sigma(1, 3, 7)$

f) $F'(x, y, z) = \Pi(1, 3, 7)$

2.33. Given $F'(x, y, z) = \Sigma(1, 3, 7)$, express the function F using a truth table.

2.34. Use Boolean algebra to convert the function $F(x, y, z) = \Sigma(3, 4, 5)$ to its equivalent product-of-sums canonical form.

2.35. Given $F = xy'z' + xy'z + xyz' + xyz$, write the equation for F' using:

a) The product-of-sums format

b) The sum-of-products format

2.36. Use Boolean algebra to convert the equation $F = w \odot x \odot y \odot z$ to:

a) The sum-of-minterms format

b) The product-of-maxterms format

2.37. Write the complete dataflow Verilog code for the Boolean functions in Problem 2.24.

2.38. Write the complete dataflow VHDL code for the Boolean functions in Problem 2.24.

2.39. Write the complete behavioral Verilog code for the car security system circuit discussed in Section 2.11.

2.40. Write the complete behavioral VHDL code for the car security system circuit discussed in Section 2.11.

CHAPTER 3

Combinational Circuits

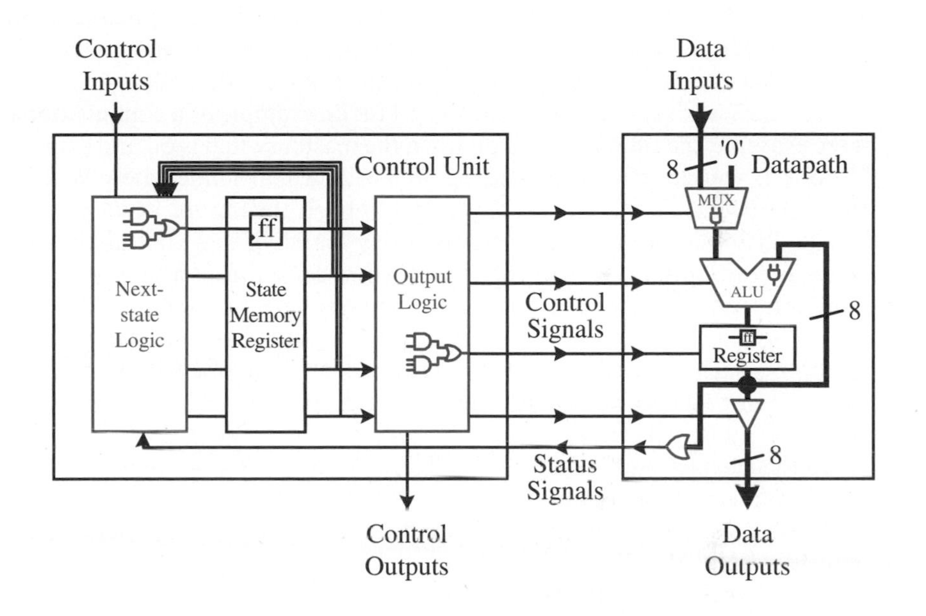

Digital circuits, regardless of whether they are part of the control unit or the datapath, are classified as either one of two types: combinational or sequential. **Combinational circuits** are digital circuits where the outputs of the circuit are dependent only on the current inputs. In other words, a combinational circuit is able to produce an output simply from knowing what the current input values are. **Sequential circuits**, on the other hand, are circuits whose outputs are dependent on not only the current inputs, but also on all of the past inputs. Therefore, in order for a sequential circuit to produce an output, it must know the current input and all past inputs. Because of their dependency on past inputs, sequential circuits must contain memory elements in order to remember the history of past input values. Combinational circuits do not need to know about past inputs, and therefore, do not require any memory elements to remember its history. A "large" digital circuit may contain both combinational circuits and sequential circuits. However, since both combinational circuits and sequential circuits are digital circuits, therefore, they use the same basic building blocks—the AND, OR, and NOT gates. What makes them different is in the way the gates are connected.

The car security system from Section 2.11 is an example of a combinational circuit. In the example, the siren is turned on when the master switch is on and someone opens the door. If you close the door, then the siren will turn off immediately. With this setup, the output, which is the siren, is dependent only on the inputs, which are the master, door, and vibration switches. For the security system to be more useful, the siren should remain on even after closing the door or when vibration stops after it is first triggered. In order to add this new feature to the security system, we need to modify it so that the output is dependent not only on the master, door, and vibration switches, but also on past inputs such as whether the door has previously been opened. A memory element is needed in order to remember whether the door previously was opened, and this requires a sequential circuit.

In this chapter and the next, we will look at the design of combinational circuits. In this chapter, we will look at the analysis and synthesis of any combinational circuits. Chapter 4 will look at the design of specific combinational components that are used frequently in building larger digital circuits. The design of sequential circuits will be discussed in subsequent chapters.

In addition to being able to design a functionally correct circuit, we also would like to be able to optimize the circuit in terms of size, speed, and power consumption. Usually, reducing the circuit size will increase the speed and reduce the power usage. In this chapter, we will look only at reducing the circuit size. Optimizing the circuit for speed and power usage is beyond the scope of this book.

3.1 Analysis of Combinational Circuits

When given a digital logic circuit, one of the first things that we often would like to know is its operation. The **analysis of combinational circuits** is the process in which we are given a combinational circuit, and we want to derive a precise description of the operation of the circuit. In general, a combinational circuit can be described precisely either with a truth table or with a Boolean function.

3.1.1 Using a Truth Table

For example, given the combinational circuit of Figure 3.1, we want to derive the truth table that describes the circuit.

We create the truth table by first listing all of the inputs found in the circuit, one input per column, followed by all of the outputs found in the circuit, again one output per column. For our sample circuit, we start with a table with four columns: three columns for the inputs x, y, and z, and one column for the output f, as shown in Figure 3.2(a).

The next step is to enumerate all possible combinations of 0s and 1s for all of the input variables. In general, for a circuit with n inputs, there are 2^n combinations, from 0 to $2^n - 1$. Continuing on with the example, the table in Figure 3.2(b) lists the eight combinations for the three variables in order.

Now, for each row in the table (i.e., for each combination of input values), we need to determine what the output value is. This is done by substituting the values for the input variables and tracing through the circuit to the output. For example, using $xyz = 000$, the output of the top AND gate is a 0 because one of its inputs is connected to z and since z is 0, the output of this AND gate is 0 (because 0 AND anything is 0). Continuing likewise for the remaining AND gates, we find that the outputs for all of the AND gates are also 0. The outputs from all of the AND gates are connected to the inputs of the OR gate, and ORing all the zeros gives a zero. Therefore, $f = 0$ for this set of values for x, y, and z. This is shown in the annotated circuit in Figure 3.2(c), and in the output column f in the truth table shown in Figure 3.2(e), the value 0 is written in the first row where $xyz = 000$.

For the second row in the table where $xyz = 001$, the output of the top AND gate gives a 1 as shown in the annotated circuit in Figure 3.2(d). Since 1 ORed with anything gives a 1, it is unnecessary to deduce the remaining AND gate outputs in order to know that $f = 1$. Hence, for this second row where $xyz = 001$, we write a 1 under the f column in the table shown in Figure 3.2(e).

Continuing in this fashion for all of the remaining input combinations, we can complete the final truth table for the circuit, as shown in Figure 3.2(e).

A different and sometimes faster way of evaluating the values for the output signals is to work backward, that is, to trace the circuit from the output back to the inputs. You

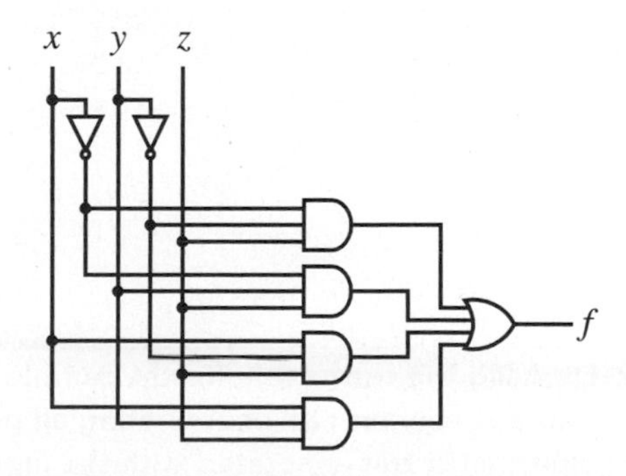

FIGURE 3.1 Sample combinational circuit.

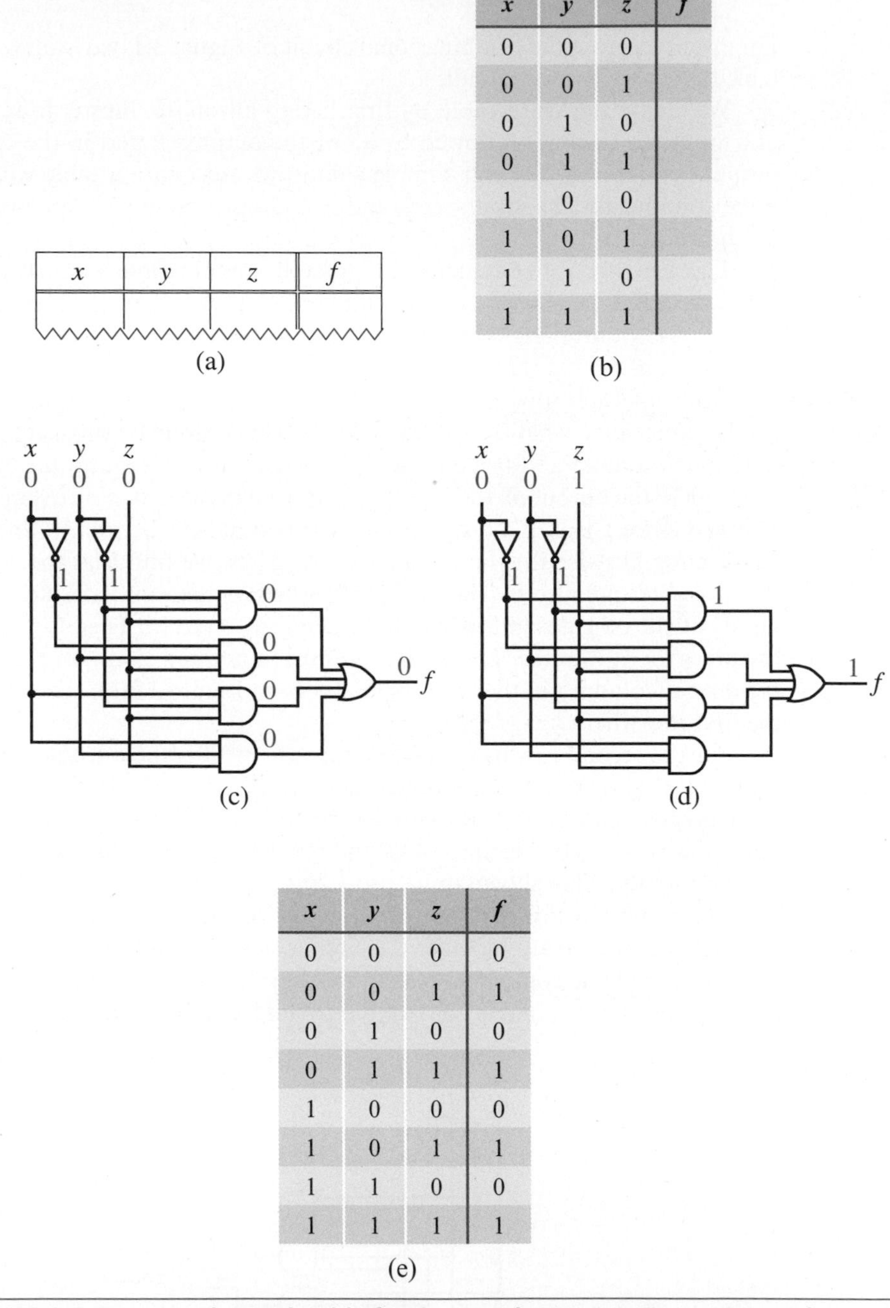

(a)

x	y	z	f

(b)

x	y	z	f
0	0	0	
0	0	1	
0	1	0	
0	1	1	
1	0	0	
1	0	1	
1	1	0	
1	1	1	

(e)

x	y	z	f
0	0	0	0
0	0	1	1
0	1	0	0
0	1	1	1
1	0	0	0
1	0	1	1
1	1	0	0
1	1	1	1

FIGURE 3.2 **Deriving the truth table for the sample circuit in Figure 3.1: (a) listing the input and output columns; (b) enumerating all possible combinations of the three input values; (c) circuit annotated with the input values $xyz = 000$; (d) circuit annotated with the input values $xyz = 001$; (e) complete truth table for the circuit.**

want to ask the question: When is the output a 1? Then trace back to the inputs to see what the input values ought to be in order to get the 1 output. Sometimes, it is easier to figure out when the output is a 0 instead. You can use whatever method is the easiest. For example, using the same circuit in Figure 3.1, f is a 1 when any one of the four OR-gate inputs is a 1. For the first input of the OR gate to be a 1, the output of the top AND gate must be a 1, which means that the inputs to the top AND gate must be all 1s. This means that the values for x, y, and z must be 0, 0, and 1 respectively, since x and y are inverted but z is not. So we have a 1 for the second row in the table where $xyz = 001$. Repeat this analysis for the remaining three inputs to the OR gate, and what you end up with are the four input combinations for which f is a 1 as shown in the truth table in Figure 3.2(e). The remaining input combinations, of course, will all be 0s for f, and so need not be worked out manually.

EXAMPLE 3.1

Deriving a truth table from a circuit diagram

Derive the truth table for the following circuit with three inputs A, B, and C, and two outputs P and Q.

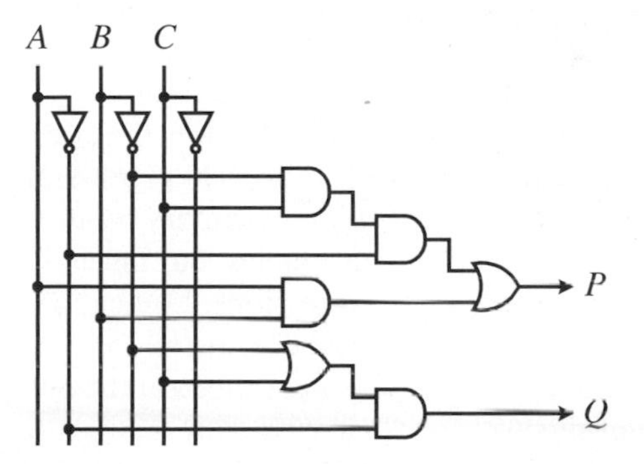

The truth table will have three columns for the three inputs and two columns for the two outputs. Enumerating all possible combinations of the three input values gives eight rows in the table. For each combination of input values, we need to evaluate the output values for both P and Q. Tracing backward from the output, P is connected to the output of an OR gate. So for P to be a 1, either of the OR-gate inputs must be a 1. The first input to this OR gate is connected to two 2-input AND gates making it a 3-input AND gate. The output of the AND gate is a 1 only if $ABC = 001$. Both A and B must be 0 because they both are connected through an inverter. C is a 1 because it is connected directly.

The second input to the OR gate is connected to the output of a 2-input AND gate. The output of this AND gate is a 1 if $AB = 11$ because both inputs are connected directly to A and B. Since C is not specified in this case, it means that C can be either a 0 or a 1. Hence, we get the three input combinations ($ABC = 001$, 110, and 111) for which P is a 1, as shown in the following truth table under the P column. The rest of the input combinations will produce a 0 for P.

Tracing backward from the output for Q to be a 1, both inputs to the AND gate must be a 1. Hence, A must be a 0, and either B is a 0 or C is a 1. This gives the three input combinations (ABC = 000, 001, and 011) for which Q is a 1, as shown in the truth table under the Q column.

A	B	C	P	Q
0	0	0	0	1
0	0	1	1	1
0	1	0	0	0
0	1	1	0	1
1	0	0	0	0
1	0	1	0	0
1	1	0	1	0
1	1	1	1	0

3.1.2 **Using a Boolean Function**

The second way to describe a combinational circuit precisely is to use a Boolean function. To derive a Boolean function that describes a combinational circuit, we simply write down the Boolean logical expression at the output of each gate as we trace through the circuit from the primary input to the primary output. This actually is similar to deriving the truth tables, but instead of substituting values of 0s and 1s, we write down the logical expressions. You can refer back to Sections 2.4 and 2.5 for the notations to use in a logical expression for the various logic gates.

To help keep track of the expressions at the output of each logic gate, we can annotate the outputs of each logic gate with the resulting intermediate logical expression. Using the sample combinational circuit shown in Figure 3.3, we note that the logical

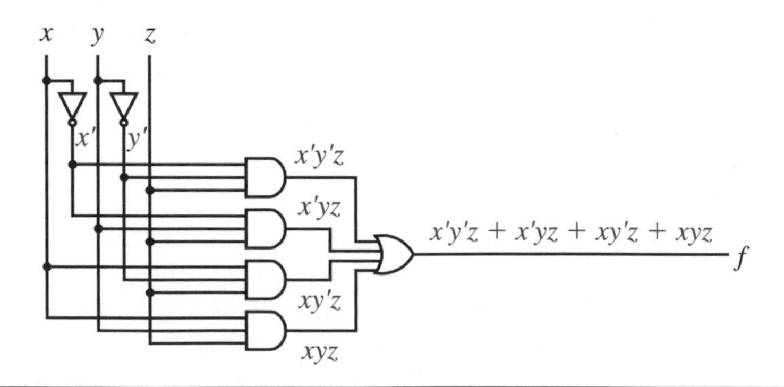

FIGURE 3.3 Sample combinational circuit.

expression for the output of the top AND gate is $x'y'z$ since both the first and second inputs come from the inverted values of x and y, and the last input comes directly from z. The logical expressions for the remaining AND gates are $x'yz$, $xy'z$, and xyz, respectively. Finally, the outputs from these AND gates are all ORed together giving us the final Boolean equation $f = x'y'z + x'yz + xy'z + xyz$.

If we substitute all possible combinations of values for all of the variables in the final equation, we should obtain the same truth table as shown in Figure 3.2(e).

If a circuit has two or more outputs, then there must be one equation for each of the outputs. The equations are derived totally independent of each other.

EXAMPLE 3.2

Deriving Boolean functions from a circuit diagram

Derive the Boolean functions for the following circuit with three inputs x, y, and z, and two outputs f and g.

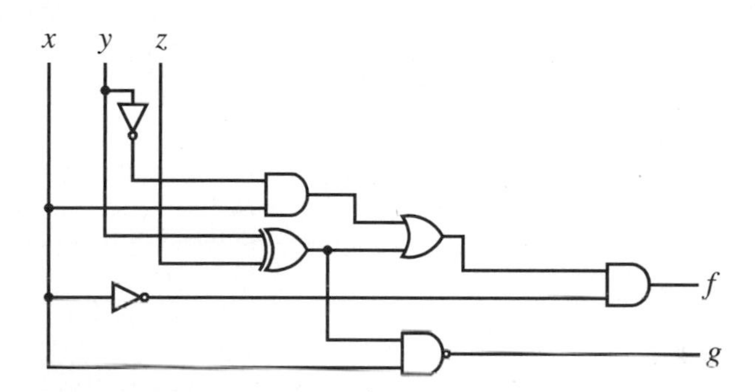

Starting from the primary inputs x, y, and z, we annotate the outputs of each logic gate with the resulting logical expression. Hence, we obtain the annotated circuit shown next.

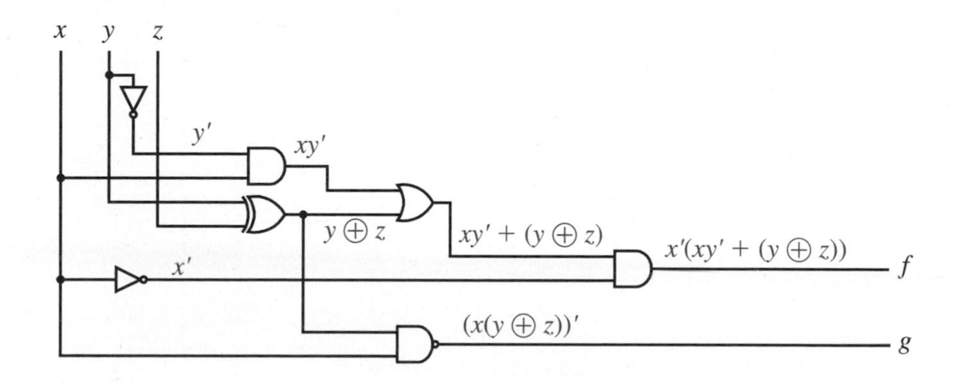

Since there are two outputs f and g in the circuit, there will be two corresponding Boolean functions for the circuit. These are

$$f = x'(xy' + (y \oplus z))$$

and

$$g = (x(y \oplus z))'$$

3.2 Synthesis of Combinational Circuits

Synthesis of combinational circuits is just the reverse procedure of the analysis of combinational circuits. In synthesis, we are given a description of the operation of a circuit, and we want to derive a circuit that realizes it. From the given description, we first derive either a truth table or a Boolean function that precisely describes the circuit's operation. In a real situation, you might not be able to come up with the precise truth table because the description might be ambiguous. If this happens, then you will have to clarify with the person who gave you the description until there are no more ambiguities. After we have either the truth table or the Boolean function, we easily can translate that into a circuit diagram.

For example, let us construct a 3-bit unsigned comparator circuit for the test "greater than or equal to five." This circuit outputs a 1 if the number is greater than or equal to 5, otherwise it outputs a 0. Since we are using a 3-bit unsigned number, the input range is from 0 to 7, so the circuit will output a 0 if the input is a number between 0 and 4 (inclusive) and outputs a 1 if the input is a number between 5 and 7 (inclusive). We use the three bits, x_2, x_1, and x_0, to represent the 3-bit input value to the comparator. From the description, we obtain the following truth table.

Decimal number	Binary number			Output
	x_2	x_1	x_0	f
0	0	0	0	0
1	0	0	1	0
2	0	1	0	0
3	0	1	1	0
4	1	0	0	0
5	1	0	1	1
6	1	1	0	1
7	1	1	1	1

It is straightforward to derive the Boolean function and then the circuit once we have the truth table. In constructing the circuit, we are interested only in when the output is a 1 (i.e., when the function f is a 1). Thus, we need to consider only the rows in the truth table where the output function $f = 1$. In other words, the Boolean function is simply the sum of the 1-minterms that we discussed in Chapter 2.

From the previous truth table, we see that there are three rows where $f = 1$. For each row where $f = 1$, we will AND all of the inputs together. The variables in the AND terms are such that it is inverted if its value is a 0, and not inverted if its value is a 1. In the case of the first AND term, we want $f = 1$ when $x_2 = 1$ and $x_1 = 0$ and $x_0 = 1$; and this is satisfied in the expression $x_2x_1'x_0$. Similarly, the second and third AND terms are satisfied in the expressions $x_2x_1x_0'$ and $x_2x_1x_0$, respectively. Finally, we want $f = 1$ when any one of these three AND terms is equal to 1. So we ORed the three AND terms together, giving us our final expression

$$f = x_2x_1'x_0 + x_2x_1x_0' + x_2x_1x_0 \tag{3.1}$$

We also can write the equation using the shorthand sum-of-minterms notation as $f(x_2, x_1, x_0) = \Sigma(5, 6, 7)$.

To draw the schematic diagram from the Boolean function, we simply convert the AND operators to AND gates, OR operators to OR gates, and primes to NOT gates. The equation is in the sum-of-products format, meaning that it is summing (ORing) the product (AND) terms. A sum-of-products equation always translates to a two-level circuit with the first level being made up of AND gates and the second level made up of one OR gate. Each of the three AND terms contains three variables, so we use a 3-input AND gate for each of the three AND terms. The three AND terms are ORed together, so we use a 3-input OR gate to connect the output of the three AND gates. For each inverted variable, we need an inverter. Note that the maximum number of inverters needed will always be the number of variables in the function. The schematic diagram derived from Equation 3.1 is shown next.

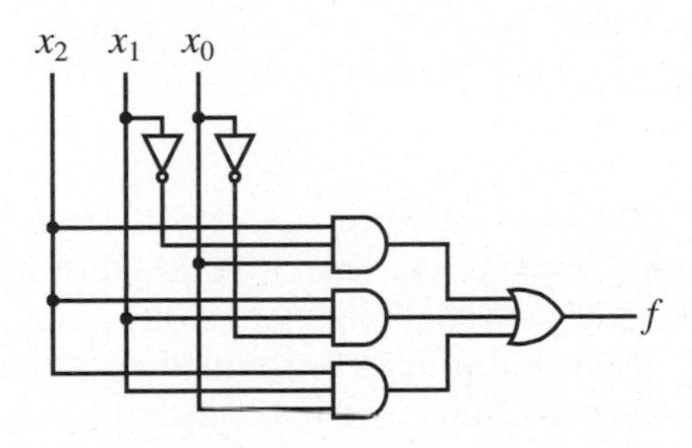

From this discussion, we see that any combinational circuit can be constructed using only AND, OR, and NOT gates from either a truth table or a Boolean equation. If we construct the circuit based directly on the truth table, then it will be in the canonical form. Circuits in this form usually can be simplified and reduced to a smaller size circuit, as we will see in the next section.

EXAMPLE 3.3

Synthesizing a combinational circuit from a truth table

Synthesize a combinational circuit from the following truth table. The three variables *a*, *b*, and *c* are input signals, and the two variables *x* and *y* are output signals.

a	*b*	*c*	*x*	*y*
0	0	0	1	0
0	0	1	0	0
0	1	0	1	1
0	1	1	1	0
1	0	0	0	1
1	0	1	1	1
1	1	0	1	0
1	1	1	0	0

We either can derive the Boolean equation from the truth table and then derive the circuit from the equation, or we can derive the circuit directly from the truth table. For this example, we first will derive the Boolean equation. Since there are two output signals, there will be two equations—one for each output signal.

From Section 2.8, we saw that a function is formed by summing its 1-minterms. For output *x*, there are five 1-minterms: m_0, m_2, m_3, m_5, and m_6. These five 1-minterms represent the five AND terms, $a'b'c'$, $a'bc'$, $a'bc$, $ab'c$, and abc'. Hence, the equation for *x* is

$$x = a'b'c' + a'bc' + a'bc + ab'c + abc'$$

Similarly, the output signal *y* has three 1-minterms, and they are $a'bc'$, $ab'c'$, and $ab'c$. Hence, the equation for *y* is

$$y = a'bc' + ab'c' + ab'c$$

The combinational circuit constructed from these two equations is shown in Figure 3.4(a). Each 3-variable AND term is replaced by a 3-input AND gate. The three inputs to these AND gates are connected to the three input variables *a*, *b*, and *c*, either directly if the variable is not primed or through a NOT gate if the variable is primed. For output *x*, a 5-input OR gate is used to connect the outputs of the five AND gates for the corresponding five AND terms. For output *y*, a 3-input OR gate is used to connect the outputs of the three AND gates.

Notice that the two AND terms, $a'bc'$ and $ab'c$, appear in both the *x* and the *y* equations. As a result, we do not need two separate AND gates to generate these two signals twice. We can reduce the size of the circuit simply by not duplicating these two AND gates, as shown in Figure 3.4(b).

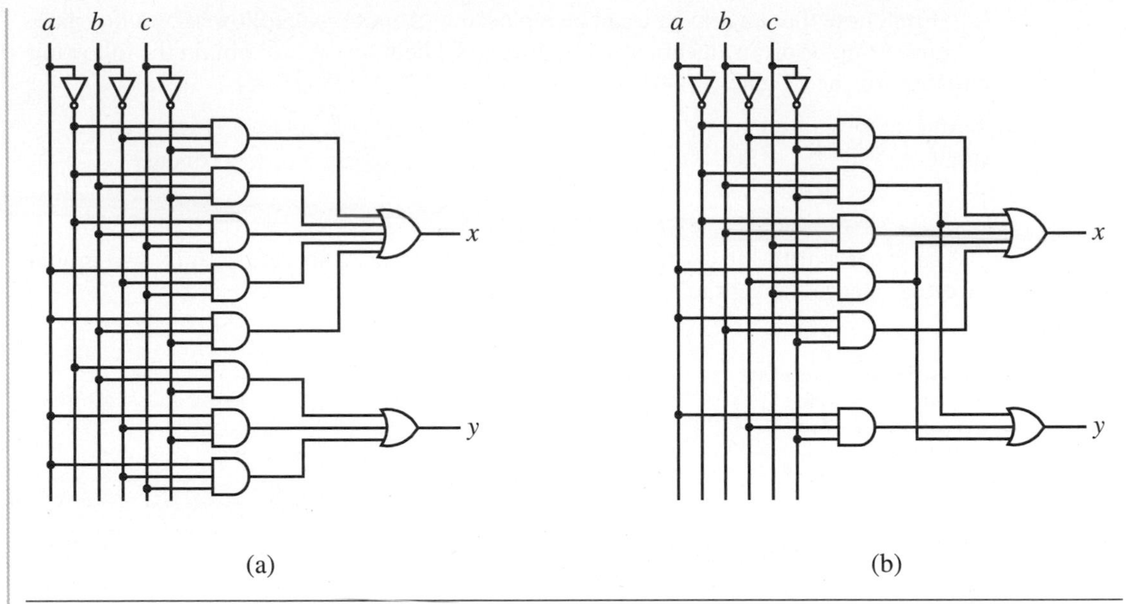

FIGURE 3.4 Combinational circuit for Example 3.3: (a) no reduction; (b) with reduction.

3.2.1 **Using Only NAND Gates**

It turns out that for all combinational circuits in the sum-of-products format, instead of using one level of AND gates and the second level using one OR gate, all of the AND and OR gates can be replaced with NAND gates, as shown in the next example. In fact, all digital circuits can be built using only NAND gates. In practice, using NAND gates instead of AND and OR gates reduces the transistor count in a circuit because a 2-input AND or OR gate uses six transistors each, whereas a 2-input NAND gate uses only four transistors. Refer to the online chapter on Implementation Technologies, for a detailed discussion on using transistors to implement basic gates.

EXAMPLE 3.4

Converting a 2-level sum-of-products circuit to use only NAND gates

Convert the following 2-level sum-of-products circuit to use only NAND gates.

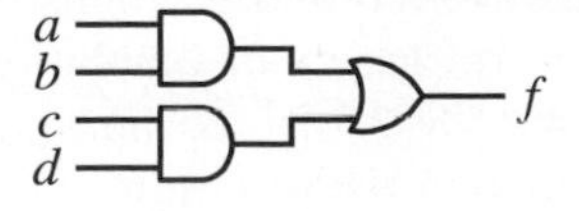

First, note that an AND gate can be replaced by a NAND gate followed by a NOT gate. Second, using Boolean algebra and DeMorgan's Theorem, we can obtain the following equality for the OR gate.

$$x + y = (x + y)''$$
$$= (x'y')'$$

In other words, the OR gate can be replaced by the NAND gate with its two inputs inverted. Hence, we get the following equivalent circuit.

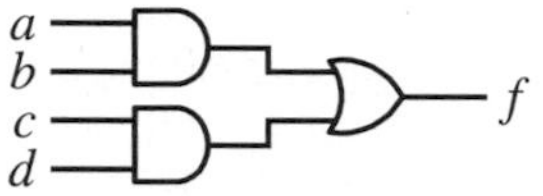

a b c d f

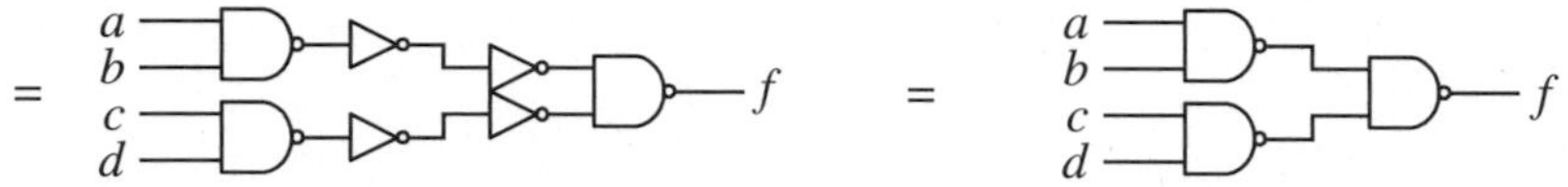

= a b c d f = a b c d f

3.3 Minimization of Combinational Circuits

When constructing digital circuits, in addition to obtaining a functionally correct circuit, we like to optimize them in terms of circuit size, speed, and power consumption. In this section, we will focus on reducing the circuit size, which usually will increase the speed and lower the power consumption. We have seen in the previous sections that any combinational circuit can be represented using a Boolean function. The size of the circuit is directly proportional to the size or complexity of the functional expression. In fact, it is a one-to-one correspondence between the functional expression and the circuit size. In Section 2.7, we saw how we can transform a Boolean function to another equivalent function by using the Boolean algebra theorems. If the resulting function is simpler than the original, then we want to implement the circuit based on the simpler function, since that will give us a smaller circuit size.

Using Boolean algebra to transform a function to one that is simpler is not a straightforward task, especially for the computer. There is no formula that says which is the next Boolean theorem to use. Luckily, there are easier methods for reducing Boolean functions. The **Karnaugh map** method is an easy way for reducing an equation manually and is discussed in Section 3.3.2. This method, however, is good only for a few variables. The **Quine-McCluskey** or **tabulation** method for reducing an equation discussed in Section 3.3.4 is ideal for programming the computer and has no limit to the number of variables used in the equation.

3.3.1 Boolean Algebra

In synthesizing a circuit from a truth table, we usually start with a Boolean equation in the canonical sum-of-minterms format that is derived directly from the truth table. To minimize a Boolean equation in the sum-of-minterms format, we need to reduce the number of product terms, which usually is done by factoring out the common variable(s). In so doing, we also will have reduced the number of variables used in the product terms. For example, given the following 3-variable function with two product terms

$$F = xy'z' + xyz'$$

Both product terms in the function have the two common variables x and z', so we can factor them out and reduce the function as follows:

$$\begin{aligned} F &= xy'z' + xyz' \\ &= xz'(y' + y) \\ &= xz'1 \\ &= xz' \end{aligned}$$

In other words, two product terms that differ by only one variable, whose value is a 0 (primed) in one term and a 1 (unprimed) in the other term, can be combined together to form just one term with that variable omitted. Thus, we have reduced the number of product terms, and the resulting product term has one less variable. By reducing the number of product terms, we not only reduce the number of AND gates needed, but also reduce the number of OR operators in the expression, which translates to reducing the number of inputs to the OR gate. By reducing the number of variables in a product term, we reduce the number of AND operators in the expression, which translates to reducing the number of inputs to the AND gate.

Sometimes, it may be advantageous to duplicate a product term one or more times in an equation. This is because a product term in the same equation can be reused as many times as needed so that more product terms can be combined and reduced as shown in the next example.

EXAMPLE 3.5

Use Boolean algebra to reduce a Boolean equation

Use Boolean algebra to reduce the 3-variable Boolean equation $F = x'yz' + x'yz + xyz'$.

We note that in the equation, the first and second product terms have two variables x' and y that are the same. Furthermore, the first and third product terms also have two variables y and z' that are the same. Since the first product term is needed twice for the reduction, once with the second term and once with the third term, we can duplicate the first term without changing the equation. After that we can factor and reduce the equation as follows:

$F = x'yz' + x'yz + xyz'$	
$= x'yz' + x'yz + xyz' + x'yz'$	duplicate 1st term by Idempotent Theorem 7b
$= (x'yz' + x'yz) + (x'yz' + xyz')$	by Commutative Theorem 10b
$= x'y(z' + z) + yz'(x' + x)$	by Distributive Theorem 12b
$= x'y\,1 + yz'1$	by Theorem 9b
$= x'y + yz'$	by Theorem 6a
$= y(x' + z')$	by Distributive Theorem 12b
$= y(xz)'$	by DeMorgan's Theorem 13a

The circuit for the original equation requires three 3-input AND gates, one 3-input OR gate, and two NOT gates. The final reduced equation only requires one 2-input AND gate and one 2-input NAND gate.

3.3.2 Karnaugh Maps

In the previous section, we saw how to reduce a Boolean equation by combining similar variables that occur in two or more terms. However, just by looking at a logic function's truth table or equation, sometimes it is difficult to see how the product terms can be combined and minimized. A **Karnaugh map** (or **K-map**) provides a simple and straightforward procedure for combining these product terms. A K-map is just a graphical representation of a logic function's truth table, where the minterms are grouped in such a way that it allows one to easily see which of the minterms can be combined. The K-map is a two-dimensional array of squares, each representing one minterm in the Boolean function. Thus, the K-map for an n-variable function is an array with 2^n squares.

Figure 3.5 shows the K-maps for functions with 2, 3, 4, and 5 variables. Notice the labeling of the columns and rows are such that any two adjacent columns or rows differ in only one bit change. This condition is required because we want minterms in adjacent squares to differ in the value of only one variable or one bit, and so these minterms can be combined together. This is why the labeling for the third and fourth columns, and for the third and fourth rows, are always interchanged. When we read K-maps, we need to visualize them as such that the two end columns or rows wrap around, so that the first and last columns and the first and last rows are really adjacent to each other, because they also differ in only one bit.

In Figure 3.5, the K-map squares are annotated with their minterms and minterm numbers for easy reference only. For example, in Figure 3.5(a), for a 2-variable K-map, the entry in the first row and second column highlighted in blue is labeled $x'y$ and annotated with the number 1. This is because the first row is when the variable x is a 0, and the second column is when the variable y is a 1. Since, for minterms, we need to prime a variable whose value is a 0 and not prime it if its value is a 1, this entry represents the minterm $x'y$, which is minterm number 1. It is important to note that if we label the rows and columns differently, the minterms and the minterm numbers will be in different locations. When we use K-maps to minimize an equation, we will not write these in the squares. Instead, we will be putting 0s and 1s in the squares.

For a 5-variable K-map, as shown in Figure 3.5(d), we need to visualize the right half of the array (where $v = 1$) to be on top of the left half (where $v = 0$). In other words, we need to view the map as three dimensional. Hence, although the squares for minterms 2 and 16 are located next to each other, they are not considered to be adjacent to each other. On the other hand, minterms 0 and 16 are adjacent to each other, because one is on top of the other.

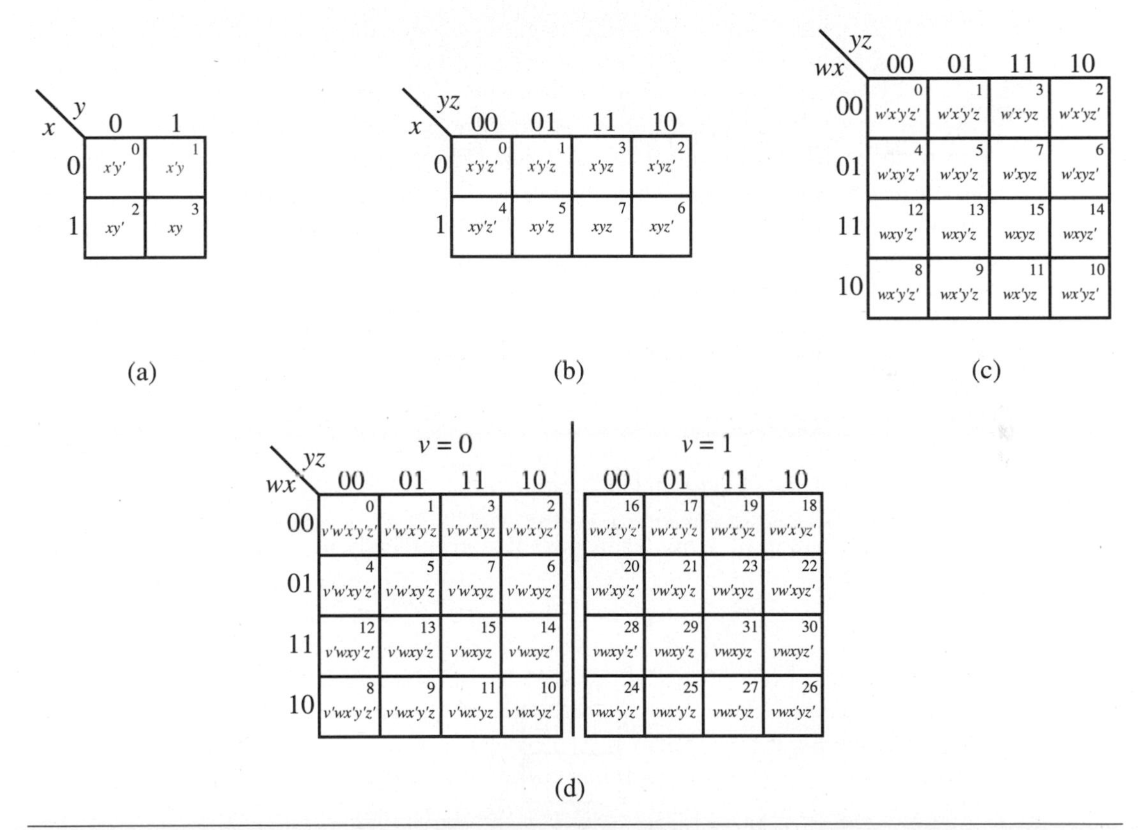

FIGURE 3.5 Karnaugh maps for: (a) 2 variables; (b) 3 variables; (c) 4 variables; (d) 5 variables.

Given a Boolean function, we set the value for each K-map square to either a 0 or a 1, depending on whether that minterm for the function is a 0-minterm or a 1-minterm, respectively. However, since we are interested only in using the 1-minterms for a function, it is unnecessary to write the 0s in the 0-minterm squares.

For example, the K-map for the 2-variable function

$$F = x'y' + x'y + xy$$

is

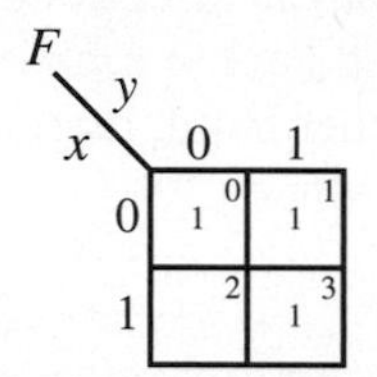

There are three 1-minterms in the function, m_0, m_1, and m_3, which correspond to the three squares in the K-map with the 1 labeled in them. Notice that the two 1-minterms, m_0 $(x'y')$ and m_1 $(x'y)$, are in adjacent squares, which means that they differ in the value of only one variable. In this case, x is 0 for both minterms, but for y, it is a 0 for one minterm and a 1 for the other minterm. Thus, variable y can be dropped, and the two terms are combined together giving just x'. The prime in x' is because x is 0 for both minterms. This reasoning corresponds to the following equality.

$$x'y' + x'y = x'(y' + y) = x'(1) = x'$$

Similarly, the two 1-minterms, m_1 $(x'y)$ and m_3 (xy), are also adjacent to each other, and y is the variable having the same value for both minterms, and so they can be combined to give

$$x'y + xy = (x' + x)\,y = (1)\,y = y$$

This combining of 1-minterms is shown graphically in the following annotated K-map.

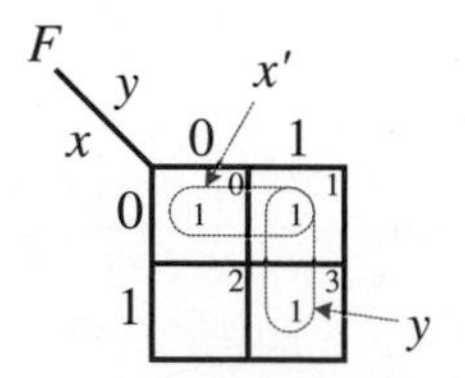

We use the term **subcube** to refer to a rectangle that encloses or covers one or more adjacent 1-minterms. In the above 2-variable K-map, there are two subcubes: one labeled with x' and the second labeled with y. Two conditions must be satisfied in order for a subcube to be valid:

1. It must be rectangular in shape.
2. The number of 1-minterms that it encloses must be a power of two.

Formally, for an n-variable K-map, an *m-subcube* is defined as that set of 2^m minterms in which $n - m$ of the variables will have the same value in every minterm, while the remaining variables will take on the 2^m possible combinations of 0s and 1s. Thus, a 1-minterm all by itself is called a 0-subcube, two adjacent 1-minterms combined is called a 1-subcube, four adjacent 1-minterms combined is called a 2-subcube, and so on. In the previous 2-variable K-map, the two subcubes are 1-subcubes.

A 2-subcube will have four adjacent 1-minterms and can be in the shape of any one of those shown in Figures 3.6(a) through (e). Notice that Figures 3.6(d) and (e) also form 2-subcubes, even though the four 1-minterms are not physically adjacent to each other. They are considered to be adjacent because the first and last rows and the first and last columns wrap around in a K-map. In Figure 3.6(f), the four 1-minterms cannot form a 2-subcube, because even though they are physically adjacent to each other, they do not form a rectangle. However, they can be separated to form the three 1-subcubes—$y'z$, $x'y'$, and $x'z$.

We say that a subcube is *characterized* by the variables having the same values for all of the 1-minterms in that subcube. In general, an m-subcube for an n-variable K-map will be characterized by $n - m$ variables. If the value that is similar for all of the variables is a 1, that variable is unprimed; whereas, if the value that is similar for all of the variables is a 0, that variable is primed. In a Boolean expression, this is equivalent to the resulting smaller product term when the minterms are combined together by factoring.

For example, the 2-subcube in Figure 3.6(d) is characterized by z', since the value of z is 0 for all of the 1-minterms, whereas the values for x and y are not all the same for all of the 1-minterms. Similarly, the 2-subcube in Figure 3.6(e) is characterized by $x'z'$. In Figures 3.6(d) and (e), although they both have four 1-minterms forming a 2-subcube, the subcube in Figure 3.6(d) is characterized with only one variable, z', whereas the subcube in Figure 3.6(e) is characterized with two variables, $x'z'$. This is because the function in the first K-map has only three variables, whereas the function in the second K-map has four variables.

For a 5-variable K-map, as shown in Figure 3.7, we need to visualize the right half of the array (where $v = 1$) to be on top of the left half (where $v = 0$). Thus, for example, minterm 20 is adjacent to minterm 4 since one is on top of the other, and they form the 1-subcube $w'xy'z'$. Even though minterm 6 is physically adjacent to minterm 20 on the map, they cannot be combined together, because when you visualize the right half as being on top of the left half, then they really are not on top of each other. Instead, minterm 6 is adjacent to minterm 4 because the columns wrap around, and they form the subcube $v'w'xz'$. Minterms 9, 11, 13, 15, 25, 27, 29, and 31 are all adjacent, and together they form the 4-subcube wz. Now that we are viewing this 5-variable K-map

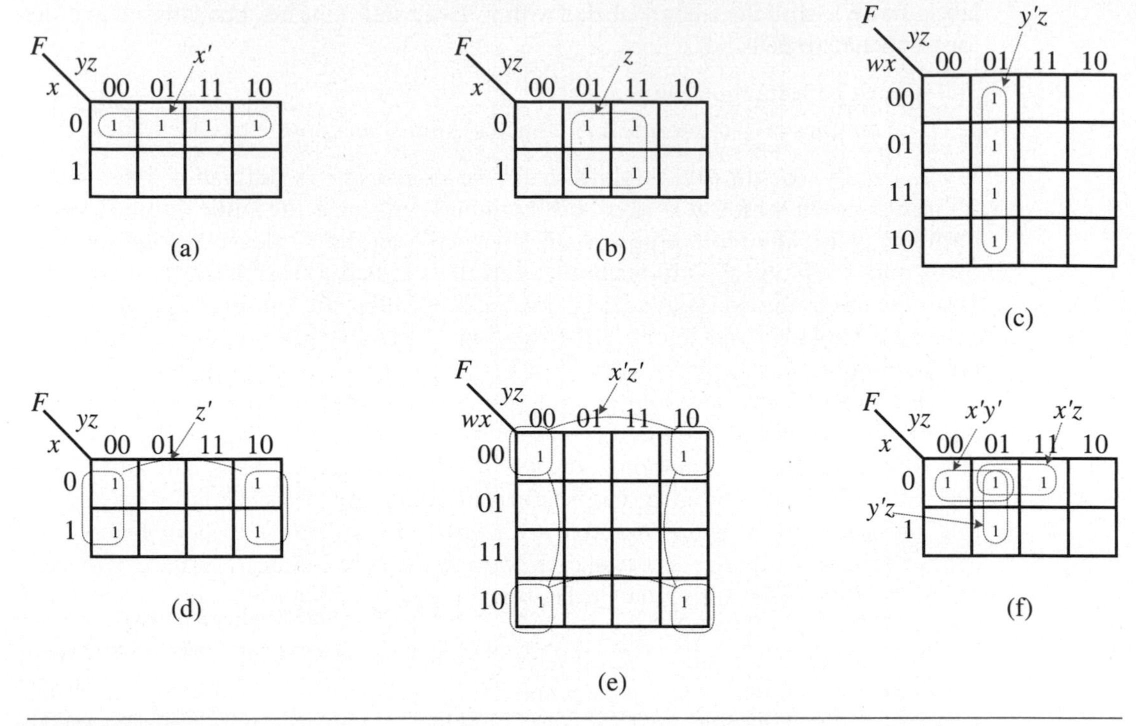

FIGURE 3.6 Examples of K-maps with 2-subcubes for: (a) 3 variables; (b) 3 variables; (c) 4 variables; (d) 3 variables with wraparound subcube; (e) 4 variables with wraparound subcube; (f) four adjacent minterms that cannot form a 2-subcube.

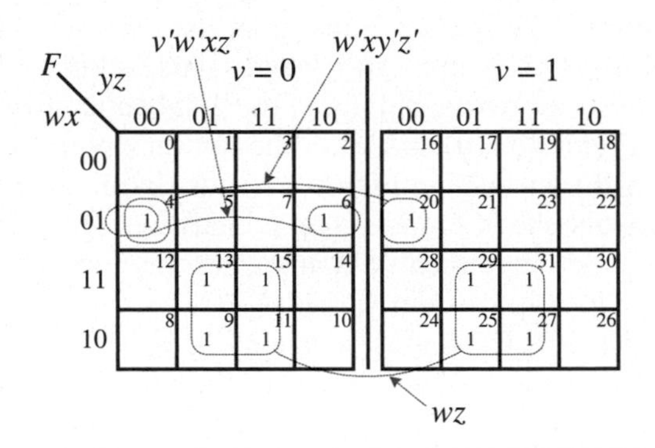

FIGURE 3.7 A 5-variable K-map with three wraparound subcubes.

in three dimensions, we also need to change the condition of the subcube shape to be a three-dimensional rectangle.

You can see that this visualization technique becomes almost impossible to work with as we increase the number of variables further. Thus, K-maps are not suitable for use in more realistic designs with many more variables; instead, tabular methods are used to reduce the size of equations.

The K-map method reduces a Boolean function from its canonical form to its standard form. The goal for the K-map method is to find as few subcubes as possible to cover all of the 1-minterms in the given function. This naturally implies that the size of the subcube should be as big as possible. The reasoning for this is that each subcube corresponds to a product term, and all of the subcubes (or product terms) must be ORed together to get the function. Larger subcubes require fewer AND gates because of fewer variables in the product term, and fewer subcubes will require fewer inputs to the OR gate.

The procedure for using the K-map method is as follows:

1. Draw the appropriate K-map for the given function and place a 1 in the squares that correspond to the function's 1-minterms.
2. For each 1-minterm, find the largest subcube that covers this 1-minterm. This largest subcube is known as a **prime implicant** (PI). By definition, a prime implicant is a subcube that is not contained within any other subcube. If more than one subcube is the same size as the largest subcube, then they are all prime implicants.
3. Look for 1-minterms that are covered by only one prime implicant. Since this prime implicant is the only subcube that covers this particular 1-minterm, this prime implicant must be in the final solution. This prime implicant is referred to as an **essential prime implicant** (EPI). By definition, an essential prime implicant is a prime implicant that includes a 1-minterm that is not included in any other prime implicant.
4. Create a minimal cover list by selecting the smallest possible number of prime implicants such that every 1-minterm is contained in at least one prime implicant. This minimal cover list must include all of the essential prime implicants plus zero or more of the remaining prime implicants. It is acceptable that a particular 1-minterm is covered in more than one prime implicant, but all 1-minterms must be covered.
5. The final minimized function is obtained by ORing all of the prime implicants from the minimal cover list.

Note that the final minimized function obtained by the K-map method may not be in its most reduced form. It is only in its most reduced *standard* form. Sometimes, it is possible to reduce the standard form further into a non-standard form using Boolean algebra.

EXAMPLE 3.6

Using K-map to minimize a 4-variable function

Use the K-map method to minimize a 4-variable (w, x, y, and z) function F with the 1-minterms: m_0, m_2, m_5, m_7, m_{10}, m_{13}, m_{14}, and m_{15}.

We start with the following 4-variable K-map with a 1 placed in each of the eight minterm squares:

F: wx \ yz	00	01	11	10
00	1 (0)	(1)	(3)	1 (2)
01	(4)	1 (5)	1 (7)	(6)
11	(12)	1 (13)	1 (15)	1 (14)
10	(8)	(9)	(11)	1 (10)

The prime implicants for each of the 1-minterms are shown in the following K-map and table:

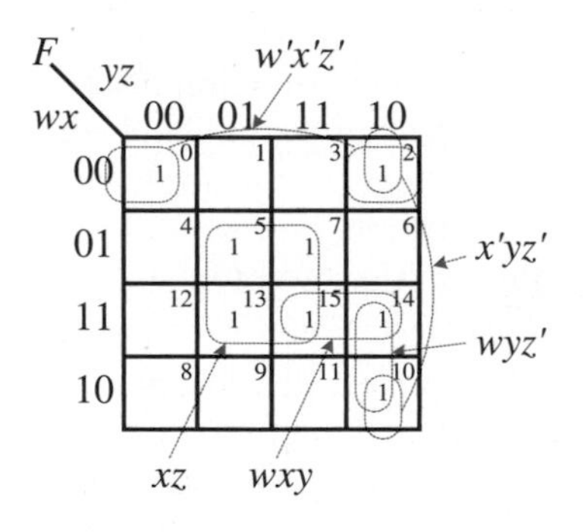

1-minterm	Prime Implicant	Essential PI
m_0	$w'x'z'$	$w'x'z'$
m_2	$w'x'z'$, $x'yz'$	
m_5	xz	xz
m_7	xz	xz
m_{10}	$x'yz'$, wyz'	
m_{13}	xz	xz
m_{14}	wyz', wxy	
m_{15}	xz	

For minterm m_0, there is only one prime implicant, $w'x'z'$. For minterm m_2, there are two 1-subcubes that cover it, and they are the largest. Therefore, m_2 has two prime implicants, $w'x'z'$ and $x'yz'$. When we consider m_{14}, again there are two 1-subcubes that cover it, and they are the largest. So m_{14} also has two prime implicants. Minterm m_{15}, however, has only one prime implicant, xz. Although the 1-subcube wxy also covers m_{15}, it is not a prime implicant for m_{15} because it is smaller than the 2-subcube xz.

From the K-map, we see that there are five prime implicants: $w'x'z'$, $x'yz'$, xz, wyz', and wxy. Of these five prime implicants, $w'x'z'$ and xz are essential prime implicants, since m_0 is covered only by $w'x'z'$, and m_5, m_7, and m_{13} are covered only by xz.

We start the minimal cover list by including the two essential prime implicants $w'x'z'$ and xz. These two subcubes will have covered the minterms m_0, m_2, m_5, m_7, m_{13}, and m_{15}. To cover the remaining two uncovered minterms, m_{10} and m_{14}, we want to use as few prime implicants as possible. Hence, we select the prime implicant wyz', which covers both of them.

Finally, the following reduced standard-form equation is obtained by ORing the two essential prime implicants and the one prime implicant in the minimal cover list.

$$F = w'x'z' + xz + wyz'$$

Notice that we can reduce this standard-form equation even further by factoring out the z' from the first and the last terms to get the nonstandard-form equation

$$F = z'(w'x' + wy) + xz$$

EXAMPLE 3.7

Using K-map to minimize a 5-variable function

Use the K-map method to minimize a 5-variable function F (v, w, x, y, and z) with the 1-minterms: $v'w'x'yz'$, $v'w'x'yz$, $v'w'xy'z$, $v'w'xyz$, $vw'x'yz'$, $vw'x'yz$, $vw'xyz'$, $vw'xyz$, $vwx'y'z$, $vwx'yz$, $vwxy'z$, and $vwxyz$.

First, we obtain the following K-map.

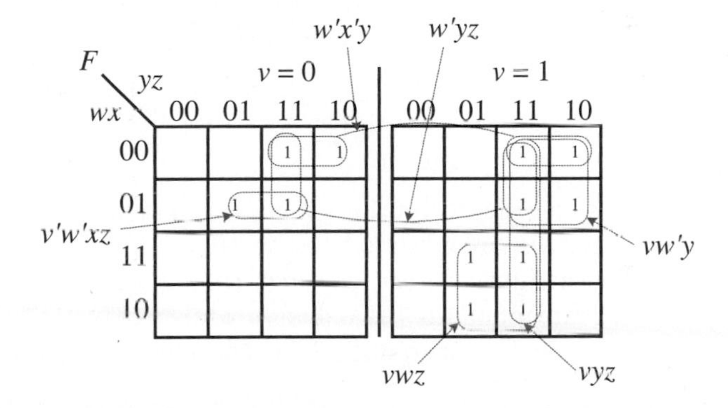

The list of prime implicants is: $v'w'xz$, $w'x'y$, $w'yz$, $vw'y$, vyz, and vwz. From this list of prime implicants, four of them, $v'w'xz$, $w'x'y$, $vw'y$, and vwz, are essential. These four essential prime implicants are able to cover all of the 1-minterms. Hence, the solution in standard form is

$$F = v'w'xz + w'x'y + vw'y + vwz$$

3.3.3 Don't-Cares

There are times when a function is not specified fully. In other words, there are some minterms for the function where we do not care whether their values are a 0 or a 1. When drawing the K-map for these "**don't-care**" minterms, we assign an "×" in that square instead of a 0 or a 1. Usually, a function can be reduced even further if we remember that these ×'s can be taken as either a 0 or a 1. As you recall when drawing K-maps, enlarging a subcube reduces the number of variables for that term. Thus, in drawing subcubes, some of them may be enlarged if we treat some of these ×'s as 1s. On the other hand, if some of these ×'s do not help to enlarge a subcube, then we would want to treat them as 0s so that we do not need to cover them. It is not necessary to treat all of the ×'s to be all 1s or all 0s; some can be taken as 1s and some as 0s.

For example, given a 4-variable function (w, x, y, z) having the following 1-minterms and don't-care minterms:

1-minterms: m_0, m_1, m_2, m_3, m_4, m_7, m_8, and m_9

×-minterms: m_{10}, m_{11}, m_{12}, m_{13}, m_{14}, and m_{15}

we obtain the following K-map with the prime implicants x', yz, and $y'z'$.

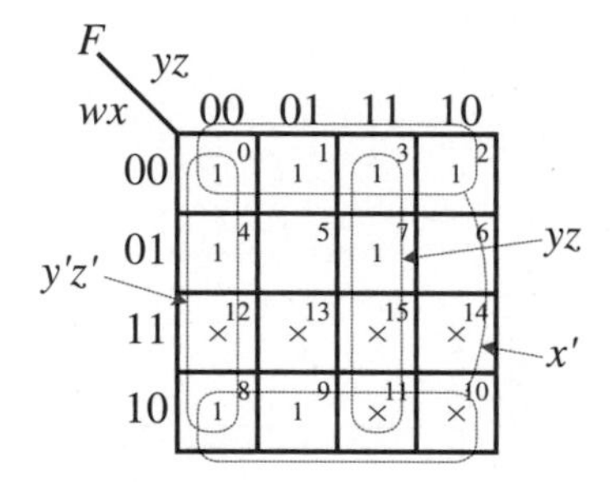

Notice that, in order to get the 4-subcube characterized by x', the two don't-care minterms m_{10} and m_{11} are taken to have the value 1. Similarly, the don't-care minterms m_{12} and m_{15} are assigned a 1 for the subcubes $y'z'$ and yz, respectively. On the other hand, the don't-care minterms m_{13} and m_{14} are taken to have the value 0, so that they do not need to be covered in the solution. The reduced standard form function as obtained from the K-map is, therefore,

$$F = x' + yz + y'z'$$

This equation can be reduced further by recognizing that $yz + y'z' = y \odot z$. Thus,

$$F = x' + (y \odot z)$$

3.3.4 **Tabulation Method**

K-maps are useful for manually obtaining the minimized standard-form Boolean function for maybe up to, at most, five variables. However, for functions with more than five variables, it becomes difficult to visualize how the minterms should be combined into subcubes. Moreover, the K-map algorithm is not as straightforward for converting to a computer program. There are **tabulation methods** that are better suited for programming the computer, and thus, can solve any function given in canonical form having any number of variables. One tabulation method is known as the **Quine-McCluskey** method. Example 3.8 illustrates the Quine-McCluskey algorithm.

EXAMPLE 3.8

Illustrating the Quine-McCluskey algorithm

We now illustrate the Quine-McCluskey algorithm using the same 4-variable function from Example 3.6 and repeated here.

$$F(w, x, y, z) = \Sigma(0, 2, 5, 7, 10, 13, 14, 15)$$

To construct thc initial table, the minterms are grouped according to the number of 1s in that minterm number's binary representation. For example, m_0 (0000) has no 1s; m_2 (0010) has one 1; m_5 (0101) has two 1s; and so on. Thus, the initial table of 0-subcubes (i.e., subcubes having only one minterm) as obtained from the function stated above is as follows.

Group	Subcube Minterms	Subcube Value				Subcube Covered
		w	x	y	z	
G_0	m_0	0	0	0	0	✔
G_1	m_2	0	0	1	0	✔
G_2	m_5	0	1	0	1	✔
	m_{10}	1	0	1	0	✔
G_3	m_7	0	1	1	1	✔
	m_{13}	1	1	0	1	✔
	m_{14}	1	1	1	0	✔
G_4	m_{15}	1	1	1	1	✔

Group G_0 contains all the minterms with no 1s (m_0); group G_1 contains all the minterms with one 1 (m_2); group G_2 contains all the minterms with two 1s (m_5 and m_{10}); group G_3 contains all the minterms with three 1s (m_7, m_{13}, and m_{14}); and finally, group G_4 contains all the minterms with four 1s (m_{15}). The "Subcube Value" column lists the binary values of the variables for each subcube minterm. The "Subcube Covered" column is filled in from the next step.

In Step 2, we construct a second table by combining those subcubes (minterms) in adjacent groups from the first table that differ in only one bit position, as shown next. For example, m_0 and m_2 differ in only the y bit. Therefore, in the second table, we have an entry for the 1-subcube containing the two minterms m_0 and m_2. A dash (–) is used in the bit position that is different in the two minterms. Since this 1-subcube covers the two individual minterms m_0 and m_2, we make a note of it by checking these two minterms in the “Subcube Covered” column in the previous table. This process is equivalent to saying that the two minterms m_0 $(w'x'y'z')$ and m_2 $(w'x'yz')$ can be combined together and are reduced to the one term, $w'x'z'$. The dash under the y column simply means that y can be either a 0 or a 1, and therefore, y can be discarded. Thus, this second table simply lists all of the 1-subcubes. This matching process to find subcubes to be combined must be done for all combinations of subcubes in all adjacent groups. Again, the “Subcube Covered” column in this second table will be filled in from the third step.

Group	Subcube Minterms	Subcube Value				Subcube Covered
		w	x	y	z	
G_0	m_0m_2	0	0	–	0	
G_1	m_2m_{10}	–	0	1	0	
G_2	m_5m_7	0	1	–	1	✔
	m_5m_{13}	–	1	0	1	✔
	$m_{10}m_{14}$	1	–	1	0	
G_3	m_7m_{15}	–	1	1	1	✔
	$m_{13}m_{15}$	1	1	–	1	✔
	$m_{14}m_{15}$	1	1	1	–	

In Step 3, we perform the same matching process as before, but from the second table. We look for subcubes in adjacent groups that differ in only one bit position. In the matching, the dash also must match. These subcubes are combined to create the next subcube table. The resulting table, however, is a table containing 2-subcubes. From the above 1-subcube table, we get the following 2-subcube table.

Group	Subcube Minterms	Subcube Value				Subcube Covered
		w	x	y	z	
G_2	$m_5m_7m_{13}m_{15}$	–	1	–	1	

From the 1-subcube table, subcubes m_5m_7 and $m_{13}m_{15}$ can be combined together to form the subcube $m_5m_7m_{13}m_{15}$ in the 2-subcube table, since they differ in only the w bit. Similarly, subcubes m_5m_{13} and m_7m_{15} from the 1-subcube table can also be combined together to form the subcube, $m_5m_7m_{13}m_{15}$, because they differ in only the y bit. From both of these combinations, the resulting subcube is the same. Therefore, we have the four checks in the 1-subcube table, but only one resulting subcube in the 2-subcube table. Notice that in the subcube $m_5m_7m_{13}m_{15}$, there are two dashes; one that is carried over from Step 2, and one for where the bit is different from the current step.

We continue to repeat the matching step as long as there are adjacent subcubes that differ in only one bit position. We stop when there are no more subcubes that can be combined. The prime implicants are those subcubes that are not covered, (i.e., those without a check mark in the "Subcube Covered" column in all of the tables). In the 2-subcube table there is a subcube that does not have a check mark, and it has the value $x = 1$ and $z = 1$; thus, this subcube is characterized by the prime implicant xz. The 1-subcube table has four subcubes that do not have a check mark; they are the four prime implicants: $w'x'z'$, $x'yz'$, wyz', and wxy. Together, these five prime implicants (xz, $w'x'z'$, $x'yz'$, wyz', and wxy) are exactly the same as those obtained in Example 3.6.

Note that not all of the prime implicants obtained may be necessary in the final reduced standard-form equation as we saw in Example 3.6. We still have to find the essential prime implicants and the minimal cover list that will cover all of the 1-minterms in the original function to arrive at the final minimized standard-form equation. Of the five prime implicants, only three of them ($w'x'z'$, xz, and wyz') are needed in the final minimized standard-form equation.

3.4 Timing Hazards and Glitches

As you probably know, things in practice don't always work according to what you learn in school. Hazards and glitches in circuits are such examples of things that might go awry. In our analysis of combinational circuits, we have performed only functional analysis. A functional analysis assumes that there is no delay for signals to pass from the input to the output of a gate. In other words, we look at a circuit only with respect to its logical operation as defined by the Boolean theorems. We have not considered the timing of the circuit. When a circuit is actually implemented, the timing of the circuit (i.e., the time for the signals to pass from the input of a logic gate to the output) is critical and must be treated with care. Otherwise, an actual implementation of the circuit may not work according to its functional analysis. **Timing hazards** are problems in a circuit as a result of timing issues. These problems can be observed only from a timing analysis of the circuit or from an actual implementation of the circuit. A functional analysis of the circuit will not reveal timing hazard problems.

A **glitch** is when a signal is expected to be stable (from a functional analysis), but it changes value for a brief moment and then goes back to what it is expected to be. For example, if a signal is expected to be at a stable 0, but instead, it goes up to a 1 and then quickly drops back to a 0. This sudden, unexpected transition of the signal is a glitch, and the circuit having this behavior contains a hazard.

Take, for example, the simple 2-to-1 multiplexer circuit shown in Figure 3.8(a). Let us assume that both d_0 and d_1 are at a constant 1 and that s goes from a 1 to a 0. For a functional analysis of the circuit, the output y should remain at a constant 1. However, if we perform a timing analysis of the circuit, we see something different in the timing diagram. Let us assume that all of the logic gates in the circuit have a delay of one time unit. The resulting timing trace is shown in Figure 3.8(b). At time t_0, s drops to a 0. Since it takes one time unit for s to be inverted through the NOT gate, s' changes to a 1 after one time unit at time t_1. At the same time, it takes the bottom AND gate one time unit

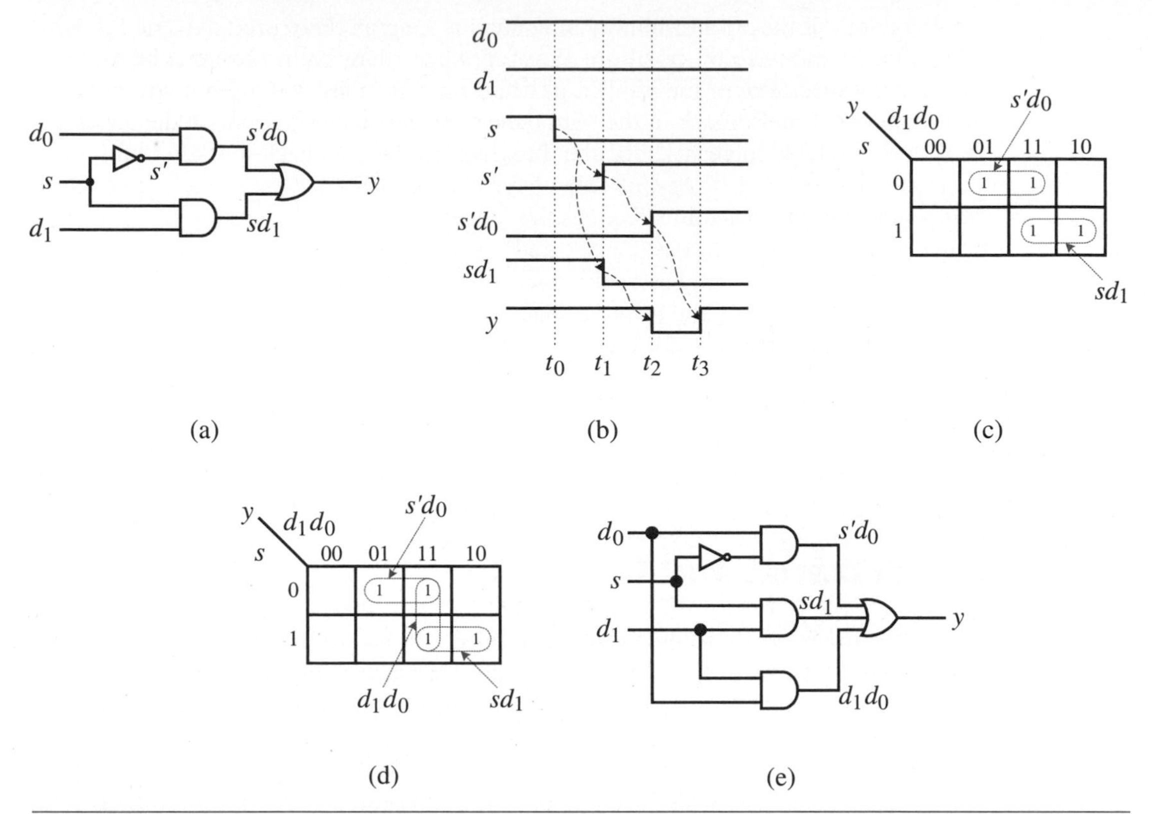

FIGURE 3.8 Example of a glitch: (a) 2-to-1 multiplexer circuit with glitches; (b) timing trace; (c) K-map with glitches; (d) K-map without glitches; (e) 2-to-1 multiplexer circuit without glitches.

for the output sd_1 to change to a 0 also at time t_1. However, the top AND gate will not see any input change until time t_1, and when it does, it takes another one time unit for its output $s'd_0$ to rise to a 1 at time t_2. Starting at time t_1, both inputs of the OR gate are 0, so after one time unit, the OR gate outputs a 0 at time t_2. At time t_2, when the top AND gate outputs a 1, the OR gate will take this 1 input and outputs a 1 after one time unit at t_3. So between times t_2 and t_3, output y unexpectedly drops to a 0 for one time unit and then rises back to a 1. Hence, the output signal y has a glitch, and the circuit has a hazard.

As you may have noticed, glitches in a signal are caused by multiple sources having paths of different delays driving that signal. These types of simple glitches can be solved easily using K-maps. A glitch generally occurs if, by simply changing one input, we have to go out of one prime implicant in a K-map and into an adjacent one (i.e., moving from one subcube to another). The glitch can be eliminated by adding an extra prime implicant, so that when going from one prime implicant to the adjacent one, we remain inside the third prime implicant.

Figure 3.8(c) shows the K-map with the two original prime implicants $s'd_0$ and sd_1 that correspond to the circuit in Figure 3.8(a). When we change s from a 1 to a 0, we have to go out of the prime implicant sd_1 and into the prime implicant $s'd_0$. Figure 3.8(d) shows the addition of the extra prime implicant d_1d_0. This time, when moving from the prime implicant sd_1 to the prime implicant $s'd_0$, we remain inside the new prime implicant d_1d_0. The 2-to-1 multiplexer circuit with the extra prime implicant d_1d_0 added, as shown in Figure 3.8(e), will prevent the glitch from happening.

3.4.1 Using Glitches

Sometimes, we can use glitches to our advantage, as shown in the following example.

EXAMPLE 3.9

A one-shot circuit using glitches

A circuit that outputs a single, short pulse when given an input of arbitrary time length is known as a **one-shot**. A one-shot circuit is used, for example, to generate a single, short 1 pulse when a key is pressed. Sometimes, when a key is pressed, we do not want to generate a continuous 1 signal for as long as the key is pressed; instead, we want the output signal to be just a single, short pulse, even if the key is still being pressed.

Since logic gates have an inherent signal delay, we can use this delay to determine the desired duration of the short pulse. This short pulse, of course, is really just a glitch in the circuit. Figure 3.9(a) shows a sample one-shot circuit using signal delays through three NOT gates; Figure 3.9(b) shows a sample timing trace for it.

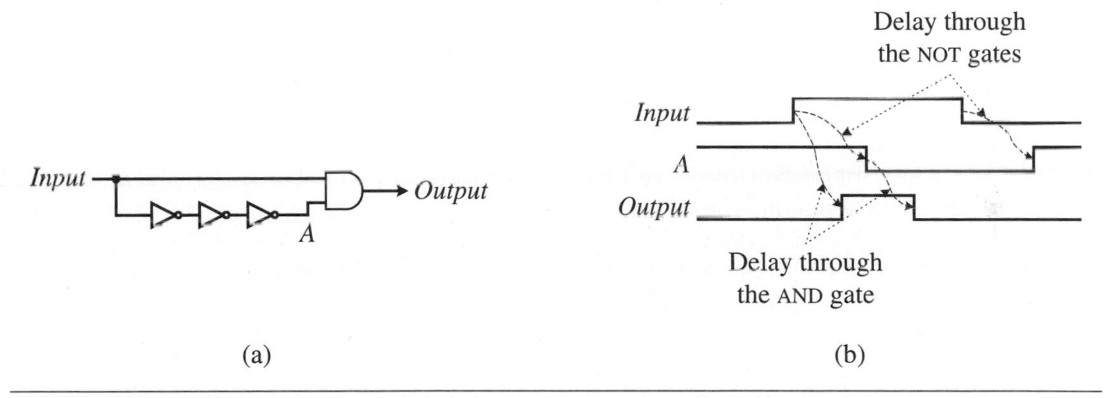

FIGURE 3.9 A one-shot circuit: (a) using signal delay through three NOT gates; (b) timing trace.

Initially, assume that the value for *Input* is a 0, and point *A* is a 1; therefore, the output of the AND gate is 0. When we set *Input* to a 1 momentarily, both inputs to the AND gate will be 1, and so after a delay through the AND gate, *Output* will be a 1. After a delay through the three NOT gates, with *Input* still at 1, point *A* will go to a 0, and *Output* will change back to a 0. When we set *Input* back to a 0, *Output* will continue to be a 0. After the delay through the NOT gates when point *A* goes back to a 1, *Output* remains at 0.

As a result, a glitch is created by the signal delay through the three NOT gates. This glitch, however, is the short 1 pulse that we wanted, and the length of this pulse is determined by the delay through the NOT gates. With this one-shot circuit, it does not matter how long the input key is pressed, the output signal will always be the same 1 pulse each time when the key is pressed.

3.5 BCD to 7-Segment Decoder

We will now design the circuit for a BCD to 7-segment decoder for driving a 7-segment LED display. The decoder converts a 4-bit **binary coded decimal** (BCD) input to seven output signals for turning on the seven LEDs in a 7-segment LED display. The 4-bit input encodes the binary representation of a decimal digit. Given the decimal digit input, the seven output lines are turned on in such a way so that the 7-segment LED displays the corresponding decimal digits from 0 to 9. The 7-segment LED display schematic with the names of each segment is shown next.

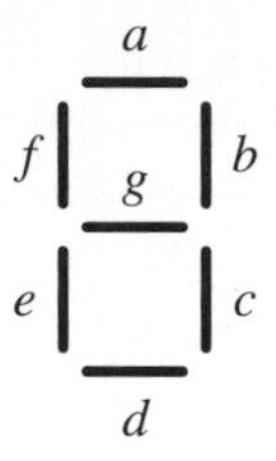

The operation of the BCD to 7-segment decoder is specified in the truth table shown in Figure 3.10. The four inputs to the decoder are i_3, i_2, i_1, and i_0, and the seven outputs for each of the seven LED segments are labeled a, b, c, d, e, f, and g. For each input combination, the corresponding digit to display on the 7-segment LED is shown in the "Display" column. The segments that need to be turned on for that digit will have a 1, while the segments that need to be turned off for that digit will have a 0. For example, for the 4-bit input 0000, which corresponds to the decimal digit 0, segments a, b, c, d, e, and f need to be turned on, while segment g needs to be turned off.

Notice that the input combinations from 1010 to 1111 are not used, and so don't-care values are assigned to all of the segments for these six combinations. Alternatively, you can assign 0s to these unused combinations to turn all of the LEDs off, or to decode them for the six hexadecimal digits from A to F.

From the truth table in Figure 3.10, we are able to specify seven equations that are dependent on the four inputs for each of the seven segments. For example, the canonical form equation for segment a is

$$a = i_3'i_2'i_1'i_0' + i_3'i_2'i_1i_0' + i_3'i_2'i_1i_0 + i_3'i_2i_1'i_0 + i_3'i_2i_1i_0' \\ + i_3'i_2i_1i_0 + i_3i_2'i_1'i_0' + i_3i_2'i_1'i_0$$

Inputs				Decimal Digit	Display	a	b	c	d	e	f	g
i_3	i_2	i_1	i_0									
0	0	0	0	0		1	1	1	1	1	1	0
0	0	0	1	1		0	1	1	0	0	0	0
0	0	1	0	2		1	1	0	1	1	0	1
0	0	1	1	3		1	1	1	1	0	0	1
0	1	0	0	4		0	1	1	0	0	1	1
0	1	0	1	5		1	0	1	1	0	1	1
0	1	1	0	6		1	0	1	1	1	1	1
0	1	1	1	7		1	1	1	0	0	0	0
1	0	0	0	8		1	1	1	1	1	1	1
1	0	0	1	9		1	1	1	0	0	1	1
Rest of the Combinations						×	×	×	×	×	×	×

FIGURE 3.10 Truth table for the BCD to 7-segment decoder.

Before implementing this equation directly in a circuit, we may want to simplify it first using the K-map method. The K-map for the equation for segment a is

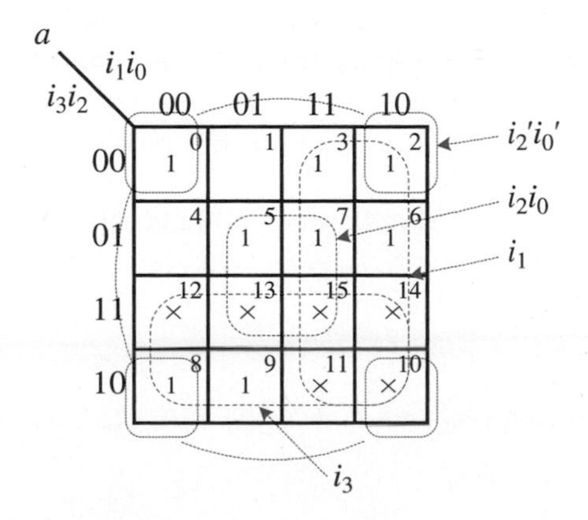

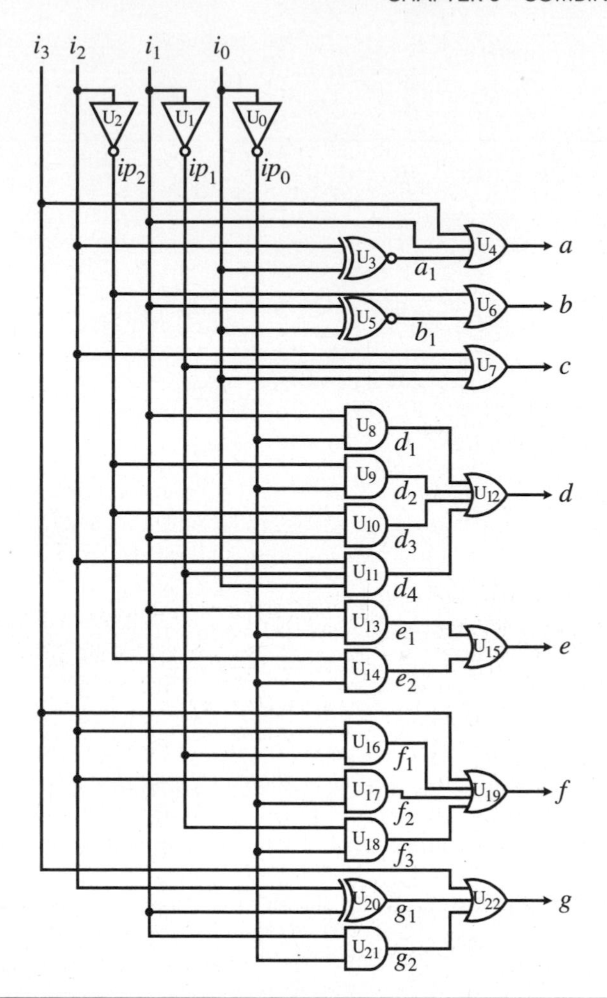

FIGURE 3.11 Circuit for the BCD to 7-segment decoder.

From evaluating the K-map, we derive the simpler equation for segment *a* as

$$a = i_3 + i_1 + i_2'i_0' + i_2i_0 = i_3 + i_1 + (i_2 \odot i_0)$$

Proceeding in a similar manner, we get the following remaining six equations.

$$b = i_2' + (i_1 \odot i_0)$$

$$c = i_2 + i_1' + i_0$$

$$d = i_1i_0' + i_2'i_0' + i_2'i_1 + i_2i_1'i_0$$

$$e = i_1i_0' + i_2'i_0'$$

$$f = i_3 + i_2 i_1' + i_2 i_0' + i_1' i_0'$$

$$g = i_3 + (i_2 \oplus i_1) + i_1 i_0'$$

From these seven simplified equations, we can now implement the circuit, as shown in Figure 3.11. The labeling of the nodes and gates in the drawing will be explained and used in Section 3.6.

3.6 Verilog and VHDL Code for Combinational Circuits

Writing Verilog or VHDL code to describe a digital circuit can be done using any one of three coding styles or levels: **structural**, **dataflow**, or **behavioral**. The choice of which coding style to use usually depends on what is known about the circuit and how much control you would like to have in terms of the final netlist generated by the synthesizer. Whereas writing code at the structural and dataflow level requires a good understanding of digital logic circuit design, writing code at the behavioral level requires only an understanding of the functionality of the circuit.

At the structural level, which is the lowest level, you first have to design the circuit manually. Having drawn the circuit, you use Verilog or VHDL to specify the components and gates that are needed by the circuit and how they are connected together by following your circuit exactly. Synthesizing a structural HDL description of a circuit will produce a netlist exactly like your original circuit. The advantage of working at the structural level is that you have full control as to what components are used and how they are connected together. The disadvantage, of course, is that you need to come up with the circuit manually, and so the full capabilities of the synthesizer are not used.

At the dataflow level, the circuit is defined using the built-in logic operators AND, OR, and NOT that are applied to the input signals. To work at this level, you typically need to have the Boolean equations for the circuit. Hence, the dataflow level is best suited for describing a circuit that already is expressed as a Boolean function. The equations are converted easily to the required HDL syntax using signal-assignment statements. All of the statements used in the structural and dataflow levels are executed concurrently, as opposed to statements in a computer program, which usually are executed in a sequential manner. In other words, the ordering of the concurrent HDL statements written in the structural or dataflow level does not matter—the results would be exactly the same.

Describing a circuit at the behavioral level is similar to writing a computer program. You have all of the standard high-level programming constructs, such as the FOR LOOP, WHILE LOOP, IF THEN ELSE, CASE, and variable assignments. The statements are enclosed in a sequential block and are executed sequentially.

3.6.1 Structural Verilog Code

Figure 3.12 shows the structural Verilog code for the BCD to 7-segment decoder based on the circuit shown in Figure 3.11. All of the names and labels used in the statements correspond to the labels in the circuit diagram shown in Figure 3.11. The code starts with declaring the input and output signals to the decoder. The `wire`

```
module bcd (
  input i0, i1, i2, i3,
  output a, b, c, d, e, f, g
);
  wire ip0,ip1,ip2,a1,b1,d1,d2,d3,d4,e1,e2,f1,f2,f3,g1,g2;

  // first parameter is the output
  // remaining parameters are the inputs
  not U0(ip0,i0);
  not U1(ip1,i1);
  not U2(ip2,i2);
  xnor U3(a1,i2,i0);
  or U4(a,i3,i1,a1);
  xnor U5(b1,i1,i0);
  or U6(b,ip2,b1);
  or U7(c,i2,ip1,i0);
  and U8(d1,i1,ip0);
  and U9(d2,ip2,ip0);
  and U10(d3,ip2,i1);
  and U11(d4,i2,ip1,i0);
  or U12(d,d1,d2,d3,d4);
  and U13(e1,i1,ip0);
  and U14(e2,ip2,ip0);
  or U15(e,e1,e2);
  and U16(f1,i2,ip1);
  and U17(f2,i2,ip0);
  and U18(f3,ip1,ip0);
  or U19(f,i3,f1,f2,f3);
  xor U20(g1,i2,i1);
  and U21(g2,i1,ip0);
  or U22(g,i3,g1,g2);
endmodule
```

FIGURE 3.12 Structural Verilog code for the BCD to 7-segment decoder.

command declares the internal signal names that will be used for making the connections between the various gates. Only the standard built-in logic gates are used. For each gate statement, the first parameter is the output from the gate, and the remaining parameters are the inputs to the gate. As you can see, this list of gate statements is a direct translation from the schematic circuit. For example, the first `not` gate statement in the code has i0 as the input and ip0 as the output. This matches the NOT gate labeled U_0 in the circuit diagram. The remaining gate statements are connected similarly, based on the circuit diagram.

All of the gate statements are executed concurrently, and therefore, the ordering of these statements is irrelevant. In other words, changing the ordering of these statements will still produce the same result. Any time when an input signal in a gate statement changes value (i.e., from a 0 to a 1 or vice versa) that statement is executed to produce an output value.

3.6.2 Structural VHDL Code

Figure 3.13 shows the structural VHDL code for the BCD to 7-segment decoder based on the circuit shown in Figure 3.11. The code starts with declaring and defining all of the components needed in the circuit. For this decoder circuit, only basic gates (such as the NOT gate, 2-input AND, 3-input AND, and so on) are used. The ENTITY statement is used to declare all of these components, and the ARCHITECTURE statement is used to define the operation of these components. Since we are using only simple gates, defining these components using the dataflow model is the simplest. For more complex components (as we will see in later

```
----------------- NOT gate -----------------------
LIBRARY IEEE;
USE IEEE.STD_LOGIC_1164.ALL;
ENTITY notgate IS PORT(
  i: IN STD_LOGIC;
  o: OUT STD_LOGIC);
END notgate;
ARCHITECTURE Dataflow OF notgate IS
BEGIN
  o <= NOT i;
END Dataflow;
----------------- 2-input AND gate ---------------
LIBRARY IEEE;
USE IEEE.STD_LOGIC_1164.ALL;
ENTITY and2gate IS PORT(
  i1, i2: IN STD_LOGIC;
  o: OUT STD_LOGIC);
END and2gate;
ARCHITECTURE Dataflow OF and2gate IS
BEGIN
  o <= i1 AND i2;
END Dataflow;
----------------- 3-input AND gate ---------------
LIBRARY IEEE;
USE IEEE.STD_LOGIC_1164.ALL;
ENTITY and3gate IS PORT(
  i1, i2, i3: IN STD_LOGIC;
  o: OUT STD_LOGIC);
END and3gate;
ARCHITECTURE Dataflow OF and3gate IS
BEGIN
  o <= (i1 AND i2 AND i3);
END Dataflow;
----------------- 2-input OR gate ----------------
LIBRARY IEEE;
USE IEEE.STD_LOGIC_1164.ALL;
```

FIGURE 3.13 Structural VHDL code for the BCD to 7-segment decoder.
(continued on next page)

```
ENTITY or2gate IS PORT(
  i1, i2: IN STD_LOGIC;
  o: OUT STD_LOGIC);
END or2gate;
ARCHITECTURE Dataflow OF or2gate IS
BEGIN
  o <= i1 OR i2;
END Dataflow;
----------------- 3-input OR gate ----------------
LIBRARY IEEE;
USE IEEE.STD_LOGIC_1164.ALL;
ENTITY or3gate IS PORT(
  i1, i2, i3: IN STD_LOGIC;
  o: OUT STD_LOGIC);
END or3gate;
ARCHITECTURE Dataflow OF or3gate IS
BEGIN
  o <= i1 OR i2 OR i3;
END Dataflow;
----------------- 4-input OR gate ----------------
LIBRARY IEEE;
USE IEEE.STD_LOGIC_1164.ALL;
ENTITY or4gate IS PORT(
  i1, i2, i3, i4: IN STD_LOGIC;
  o: OUT STD_LOGIC);
END or4gate;
ARCHITECTURE Dataflow OF or4gate IS
BEGIN
  o <= i1 OR i2 OR i3 OR i4;
END Dataflow;
----------------- 2-input XOR gate ---------------
LIBRARY IEEE;
USE IEEE.STD_LOGIC_1164.ALL;
ENTITY xor2gate IS PORT(
  i1, i2: IN STD_LOGIC;
  o: OUT STD_LOGIC);
END xor2gate;
ARCHITECTURE Dataflow OF xor2gate IS
BEGIN
  o <= i1 XOR i2;
END Dataflow;
----------------- 2-input XNOR gate --------------
LIBRARY IEEE;
USE IEEE.STD_LOGIC_1164.ALL;
ENTITY xnor2gate IS PORT(
  i1, i2: IN STD_LOGIC;
  o: OUT STD_LOGIC);
```

FIGURE 3.13 Structural VHDL code for the BCD to 7-segment decoder.
(continued on next page)

```
END xnor2gate;
ARCHITECTURE Dataflow OF xnor2gate IS
BEGIN
  o <= NOT(i1 XOR i2);
END Dataflow;
---------------- bcd entity --------------------
LIBRARY IEEE;
USE IEEE.STD_LOGIC_1164.ALL;
ENTITY bcd IS PORT(
  i0, i1, i2, i3: IN STD_LOGIC;
  a, b, c, d, e, f, g: OUT STD_LOGIC);
END bcd;

ARCHITECTURE Structural OF bcd IS
  COMPONENT notgate PORT(
    i: IN STD_LOGIC;
    o: OUT STD_LOGIC);
  END COMPONENT;
  COMPONENT and2gate PORT(
    i1, i2: IN STD_LOGIC;
    o: OUT STD_LOGIC);
  END COMPONENT;
  COMPONENT and3gate PORT(
    i1, i2, i3: IN STD_LOGIC;
    o: OUT STD_LOGIC);
  END COMPONENT;
  COMPONENT or2gate PORT(
    i1, i2: IN STD_LOGIC;
    o: OUT STD_LOGIC);
  END COMPONENT;
  COMPONENT or3gate PORT(
    i1, i2, i3: IN STD_LOGIC;
    o: OUT STD_LOGIC);
  END COMPONENT;
  COMPONENT or4gate PORT(
    i1, i2, i3, i4: IN STD_LOGIC;
    o: OUT STD_LOGIC);
  END COMPONENT;
  COMPONENT xor2gate PORT(
    i1, i2: IN STD_LOGIC;
    o: OUT STD_LOGIC);
  END COMPONENT;
  COMPONENT xnor2gate PORT(
    i1, i2: IN STD_LOGIC;
    o: OUT STD_LOGIC);
  END COMPONENT;

  SIGNAL ip0,ip1,ip2,a1,b1,d1,d2,d3,d4: STD_LOGIC;
  SIGNAL e1,e2,f1,f2,f3,g1,g2: STD_LOGIC;
```

FIGURE 3.13 Structural VHDL code for the BCD to 7-segment decoder.
(continued on next page)

```
BEGIN
  -- last parameter is the output
  -- remaining parameters are the inputs
  U0: notgate PORT MAP(i0,ip0);
  U1: notgate PORT MAP(i1,ip1);
  U2: notgate PORT MAP(i2,ip2);
  U3: xnor2gate PORT MAP(i2, i0, a1);
  U4: or3gate PORT MAP(i3, i1, a1, a);
  U5: xnor2gate PORT MAP(i1, i0, b1);
  U6: or2gate PORT MAP(ip2, b1, b);
  U7: or3gate PORT MAP(i2, ip1, i0, c);
  U8: and2gate PORT MAP(i1, ip0, d1);
  U9: and2gate PORT MAP(ip2, ip0, d2);
  U10: and2gate PORT MAP(ip2, i1, d3);
  U11: and3gate PORT MAP(i2, ip1, i0, d4);
  U12: or4gate PORT MAP(d1, d2, d3, d4, d);
  U13: and2gate PORT MAP(i1, ip0, e1);
  U14: and2gate PORT MAP(ip2, ip0, e2);
  U15: or2gate PORT MAP(e1, e2, e);
  U16: and2gate PORT MAP(i2, ip1, f1);
  U17: and2gate PORT MAP(i2, ip0, f2);
  U18: and2gate PORT MAP(ip1, ip0, f3);
  U19: or4gate PORT MAP(i3, f1, f2, f3, f);
  U20: xor2gate PORT MAP(i2, i1, g1);
  U21: and2gate PORT MAP(i1, ip0, g2);
  U22: or3gate PORT MAP(i3, g1, g2, g);
END Structural;
```

FIGURE 3.13 Structural VHDL code for the BCD to 7-segment decoder.

chapters), we want to choose the model that is best suited for the information that we have available for the circuit. The reason why the code shown in Figure 3.13 is structural is not because of how these components are defined, but rather on how these components are connected together to form the enclosing entity; in this case, the *bcd* entity. Notice that the LIBRARY and USE statements need to be repeated for every ENTITY declaration.

The actual structural code begins with the *bcd* ENTITY declaration. The *bcd* circuit shown in Figure 3.11 has four input signals: i_3, i_2, i_1, and i_0, and seven output signals: a, b, c, d, e, f, and g. These signals are declared in the PORT list using the keyword IN for the input signals, and OUT for the output signals; both of which are of type STD_LOGIC.

The ARCHITECTURE section begins by specifying the components needed in the circuit using the COMPONENT statement. The port list in the COMPONENT statements must match exactly the port list in the entity declarations of the components. They must match not only in the number, direction, and type of the signals but also in the names given to the signals. Note also that names in the component port list can be the same as the names in the *bcd* entity port list, but they are not the same signals. For example, the *and2gate* component port list and the *bcd* entity port list both have two signals called i_1 and i_2. References to these two signals in the body of the *bcd* architecture are for the signals declared in the *bcd* entity.

After the COMPONENT statements, the internal node signals are declared using the SIGNAL statement. The names listed are the same as the internal node names used in the circuit in Figure 3.11 for easy reference.

Following all of the declarations, the body of the architecture starts with the keyword BEGIN. For each gate used in the circuit, there is a corresponding PORT MAP statement. Each PORT MAP statement begins with an optional label (i.e., U_1, U_2, and so on) followed by the name of the component (as previously declared with the COMPONENT statements) to use. Again, the labels used in the PORT MAP statements correspond to the labels on the gates in the circuit in Figure 3.11. The parameter list in the PORT MAP statement matches the port list in the component declaration. For example, U_0 is instantiated with the component *notgate*. The first parameter in the PORT MAP statement is the input signal i_0, and the second parameter is the output signal ip_0. U_4 is instantiated with the 3-input OR gate. The three inputs are i_3, i_1, and a_1, and the output is a. Here, a_1 is the output from the 2-input XNOR gate of U_3. The rest of the PORT MAP statements in the program are obtained in a similar manner.

All of the PORT MAP statements are executed concurrently, and therefore, the ordering of these statements is irrelevant. In other words, changing the ordering of these statements will still produce the same result. Any time when a signal in a PORT MAP statement changes value (i.e., from a 0 to a 1 or vice versa) that PORT MAP statement is executed.

3.6.3 Dataflow Verilog Code

Figure 3.14 shows the dataflow Verilog code for the BCD to 7-segment decoder. This code uses the concurrent `assign` statements to assign values to the output signals based on the Boolean equations derived for the circuit in Section 3.5. The built-in logical operator symbols &, |, ~, and ^ are used for the corresponding AND, OR, NOT, and XOR operators. Seven concurrent `assign` statements are used: one for each of the seven Boolean equations, which corresponds to the seven LED segments a to g. For example, the equation for segment a is

$$a = i_3 + i_1 + (i_2 \odot i_0)$$

```
module bcd (
  input i0, i1, i2, i3,
  output a, b, c, d, e, f, g
);
  // bitwise ~=NOT; &=AND; |=OR; ^=XOR
  assign a = i3 | i1 | (i2 ~^ i0);
  assign b = ~i2 | ~(i1 ^ i0);
  assign c = i2 | ~i1 | i0;
  assign d = (i1 & ~i0) | (~i2 & ~i0) | (~i2 & i1) |
             (i2 & ~i1 & i0);
  assign e = (i1 & ~i0) | (~i2 & ~i0);
  assign f = i3 | (i2 & ~i1) | (i2 & ~i0) | (~i1 & ~i0) |
             (i2 & ~i0) | (~i1 & ~i0);
  assign g = i3 | (i2 ^ i1) | (i1 & ~i0);
endmodule
```

FIGURE 3.14 Dataflow Verilog code for the BCD to 7-segment decoder.

This is converted to the statement

assign $a = i3 \mid i1 \mid (i2 \sim\wedge i0);$

Proceeding in a similar manner, we obtain the signal assignment statements in the dataflow code for the remaining six equations.

Just like with the `gate` statement used in the structural level, these concurrent `assign` statements are executed concurrently, and therefore, the ordering of these statements is irrelevant. These statements are executed whenever one of its input value changes, and the result is assigned to the output signal that is on the left of the equal sign.

3.6.4 Dataflow VHDL Code

Figure 3.15 shows the dataflow VHDL code for the BCD to 7-segment decoder based on the Boolean equations derived in Section 3.5. The ENTITY declaration for this dataflow code is exactly the same as that for the structural code, since the interface for the decoder remains the same.

In the ARCHITECTURE section, seven concurrent signal assignment statements are used: one for each of the seven Boolean equations, which corresponds to the seven LED segments. For example, the equation for segment a is

$$a = i_3 + i_1 + (i_2 \odot i_0)$$

This is converted to the signal assignment statement:

a <= i3 OR i1 OR (i2 XNOR i0);

```
LIBRARY IEEE;
USE IEEE.STD_LOGIC_1164.ALL;

ENTITY bcd IS PORT(
  i0, i1, i2, i3: IN STD_LOGIC;
  a, b, c, d, e, f, g: OUT STD_LOGIC);
END bcd;

ARCHITECTURE Dataflow OF bcd IS
BEGIN
  a <= i3 OR i1 OR (i2 XNOR i0);                           -- seg a
  b <= (NOT i2) OR NOT (i1 XOR i0);                        -- seg b
  c <= i2 OR (NOT i1) OR i0;                               -- seg c
  d <= (i1 AND NOT i0) OR (NOT i2 AND NOT i0)              -- seg d
       OR (NOT i2 AND i1) OR (i2 AND NOT i1 AND i0);
  e <= (i1 AND NOT i0) OR (NOT i2 AND NOT i0);             -- seg e
  f <= i3 OR (i2 AND NOT i1)                               -- seg f
       OR (i2 AND NOT i0) OR (NOT i1 AND NOT i0);
  g <= i3 OR (i2 XOR i1) OR (i1 AND NOT i0);               -- seg g
END Dataflow;
```

FIGURE 3.15 Dataflow VHDL code for the BCD to 7-segment decoder.

Proceeding in a similar manner, we obtain the signal assignment statements in the dataflow code for the remaining six equations.

All of the signal assignment statements are executed concurrently, and therefore, the ordering of these statements is irrelevant. In other words, changing the ordering of these statements will still produce the same result. Any time when a signal on the right side of an assignment statement changes value (i.e., from a 0 to a 1 or vice versa) that assignment statement is executed.

3.6.5 **Behavioral Verilog Code**

The behavioral Verilog code for the BCD to 7-segment decoder is shown in Figure 3.16. For some variations, the input and output signals are declared slightly different from those in the previous sections. Instead of having individual input and output signals, they are declared as a vector. The input vector, *I*, is declared as an array of size 4 with a range from 3 to 0, whereas the output vector, *Segs*, is declared as an array of size 7 with a range from 0 to 6. For the *Segs* output vector, the leftmost bit with index 0 corresponds to segment *a*, the next bit with index 1 corresponds to segment *b*, and so on up to the rightmost bit with index 6 for segment *g*.

The `always` statement declares a process block in which the statements inside the block are executed sequentially, and therefore, changing the ordering of these statements will produce different results. Although the high-level statements inside this block are executed sequentially, the resulting circuit produced from the synthesis of this code inherently will operate in parallel. This is because no matter how you connect the gates up for a circuit, they all will operate together when power is applied to them.

```
module bcd (
  input [3:0] I,
  output reg [0:6] Segs
);

  always @ (I) begin
   case (I)
   0: Segs = 7'b1111110;
   1: Segs = 7'b0110000;
   2: Segs = 7'b1101101;
   3: Segs = 7'b1111001;
   4: Segs = 7'b0110011;
   5: Segs = 7'b1011011;
   6: Segs = 7'b1011111;
   7: Segs = 7'b1110000;
   8: Segs = 7'b1111111;
   9: Segs = 7'b1110011;
   default: Segs = 7'b0000000;
   endcase
 end
endmodule
```

FIGURE 3.16 Behavioral Verilog code for the BCD to 7-segment decoder.

The sensitivity list in the `always` statement consists of the variables listed inside the parentheses in the `always` statement. The `always` block is executed whenever any one of the variables in the sensitivity list changes value. In the code, only the `case` statement is inside the `always` block. The ten cases for the ten decimal digits are listed, and for each case a 7-bit value is assigned to the output signal *Segs*. A 1 bit will turn a LED on and a 0 bit will turn it off. For case 0, we want to display the decimal digit 0, so all of the segments are turned on except for segment *g*, hence the bit string 1111110 is assigned to the output signal *Segs*. To specify that it is a 7-bit binary value, we use the notation 7'b followed by the actual 7-bit bit string. The `default` case is taken if *I* is not a number between 0 and 9, and all of the segments will be turned off.

In Verilog, there are two different assignment statements that can be used inside the `always` block: a blocking assignment statement (using the = sign) and a non-blocking assignment statement (using the <= sign). The blocking assignment (=) statement should be used when modeling combinational circuits inside an `always` block. These two assignment statements are different from the `assign` statement used in the dataflow model.

3.6.6 **Behavioral VHDL Code**

The behavioral VHDL code for the BCD to 7-segment decoder is shown in Figure 3.17. For some variations, the port list for this entity is slightly different from the two entities in the previous sections. Instead of having the four separate input signals, i_0, i_1, i_2, and i_3, we have declared a vector, *I*, of length four. This vector, *I*, is declared with the type keyword STD_LOGIC_VECTOR, that is, a vector of type STD_LOGIC. The length of the vector is specified by the range (3 DOWNTO 0). The first number 3 in the range denotes the index of the most significant bit of the vector, and the second number 0 in the range denotes the index of the least significant bit of the vector. Likewise, the seven output signals, *a* to *g*, are replaced with the STD_LOGIC_VECTOR *Segs* of length 7. This time, however, the keyword TO is used in the range to mean that the most significant bit in the vector is index 0 and the least significant bit in the vector is index 6.

In the architecture section, a PROCESS statement is used. All of the statements inside the process block are executed sequentially, therefore changing the ordering of these statements will produce different results. Although the high-level statements inside this block are executed sequentially, the resulting circuit produced from the synthesis of this code inherently will operate in parallel. This is because no matter how you connect the gates up for a circuit, they all will operate together when power is applied to them. The process block itself, however, is treated as a single concurrent statement. Thus, the architecture section can have two or more process blocks together with other concurrent statements, and these all will execute concurrently.

The parenthesized list of signals after the PROCESS keyword is referred to as the sensitivity list. The purpose of the sensitivity list is that, when a value for any of the listed signals changes, the entire process block is executed from the beginning to the end.

```
LIBRARY IEEE;
USE IEEE.STD_LOGIC_1164.ALL;

ENTITY bcd IS PORT (
  I: IN STD_LOGIC_VECTOR (3 DOWNTO 0);
  Segs: OUT STD_LOGIC_VECTOR (0 TO 6));
END bcd;

ARCHITECTURE Behavioral OF bcd IS
BEGIN
  PROCESS(I)
  BEGIN
    CASE I IS
    WHEN "0000" => Segs <= "1111110";
    WHEN "0001" => Segs <= "0110000";
    WHEN "0010" => Segs <= "1101101";
    WHEN "0011" => Segs <= "1111001";
    WHEN "0100" => Segs <= "0110011";
    WHEN "0101" => Segs <= "1011011";
    WHEN "0110" => Segs <= "1011111";
    WHEN "0111" => Segs <= "1110000";
    WHEN "1000" => Segs <= "1111111";
    WHEN "1001" => Segs <= "1110011";
    WHEN OTHERS => Segs <= "0000000";
    END CASE;
  END PROCESS;
END Behavioral;
```

FIGURE 3.17 Behavioral VHDL code for the BCD to 7-segment decoder.

In the code, there is a CASE statement inside the process block. Depending on the value of *I*, one of the WHEN parts will be executed. A WHEN part consists of the keyword WHEN followed by a constant value for the variable *I* to match, followed by the symbol =>. The statement (or statements) after the symbol => is executed when *I* matches that corresponding constant. In the code, all of the WHEN parts contain one signal assignment statement. All of the signal assignment statements assign a string of seven bits to the output signal *Segs*. This string of seven bits corresponds to the on-off values of the seven segments *a* to *g*, as shown in the 7-segment decoder truth table of Figure 3.10. For example, looking at the truth table, we see that when *I* = "0000" (i.e., for the decimal digit 0) we want all of the segments to be on except for segment *g*. Recall that in the declaration of the *Segs* vector, the most significant bit, which is the leftmost bit in the bit string, is index 0, and the least significant bit, which is the rightmost bit, is index 6. In VHDL, the notation *Segs*(*n*) is used to denote the index *n* of the *Segs* vector. In the code, we have designated *Segs*(0) for segment *a*, *Segs*(1) for segment *b*, and so on to *Segs*(6) for segment *g*. So, in order to display the decimal digit 0, we need to assign the bit string "1111110" to *Segs*.

If the value of *I* does not match any of the WHEN parts, then the WHEN OTHERS part will be chosen. In this case, all of the segments will be turned off. Notice that for both the structural and the dataflow code, the segments are not all turned off when I is one of these values. Instead, a certain combination of LEDs is turned on because the K-maps assigned some of the don't-cares to 1s. If we assign all the don't-cares to 0, then all the LEDs will be turned off. An alternative to turning all of the segments off for the remaining six cases is to display the six letters A, b, C, d, E, and F for the six hexadecimal digits. The two letters b and d have to be displayed in lower case, because otherwise, it will be the same as the numbers 8 and 0, respectively.

3.7 PROBLEMS

3.1. Derive the truth table for the following circuits:

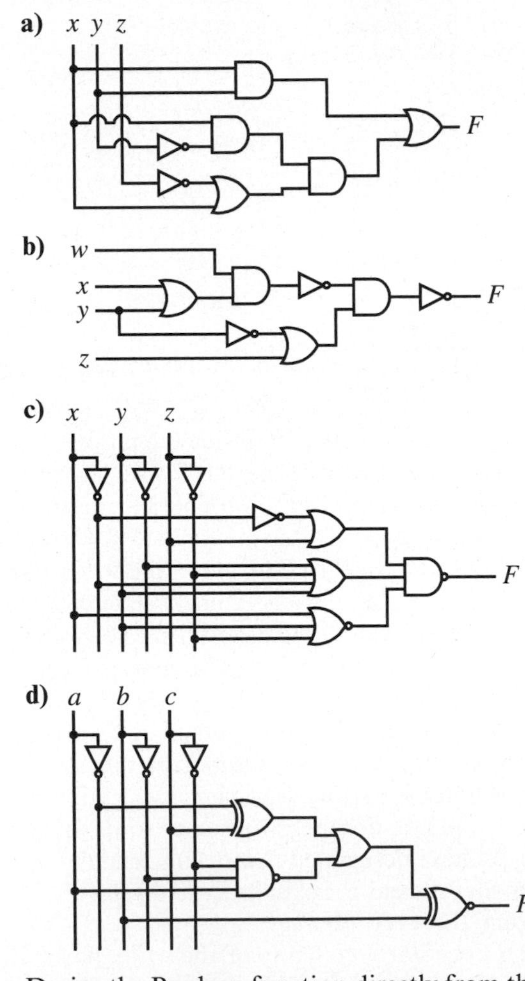

3.2. Derive the Boolean function directly from the circuits in Problem 3.1.

3.3. Draw the circuit diagram that implements the following truth tables.

a)

a	b	c	F
0	0	0	0
0	0	1	0
0	1	0	1
0	1	1	1
1	0	0	0
1	0	1	0
1	1	0	1
1	1	1	0

b)

w	x	y	z	F
0	0	0	0	0
0	0	0	1	0
0	0	1	0	1
0	0	1	1	0
0	1	0	0	1
0	1	0	1	1
0	1	1	0	0
0	1	1	1	1
1	0	0	0	0
1	0	0	1	1
1	0	1	0	1
1	0	1	1	0
1	1	0	0	1
1	1	0	1	1
1	1	1	0	0
1	1	1	1	1

c)

w	x	y	z	F_1	F_2
0	0	0	0	1	1
0	0	0	1	0	1
0	0	1	0	0	1
0	0	1	1	1	1
0	1	0	0	0	0
0	1	0	1	1	1
0	1	1	0	1	0
0	1	1	1	0	0
1	0	0	0	0	1
1	0	0	1	1	1
1	0	1	0	1	0
1	0	1	1	0	0
1	1	0	0	1	1
1	1	0	1	0	1
1	1	1	0	0	1
1	1	1	1	1	1

d)

N_3	N_2	N_1	N_0	F
0	0	0	0	0
0	0	0	1	0
0	0	1	0	1
0	0	1	1	1
0	1	0	0	0
0	1	0	1	0
0	1	1	0	1
0	1	1	1	0
1	0	0	0	0
1	0	0	1	0
1	0	1	0	1
1	0	1	1	1
1	1	0	0	1
1	1	0	1	0
1	1	1	0	0
1	1	1	1	1

3.4. Draw the circuit diagram that implements the following expressions:

a) $F(x, y, z) = \Sigma(0, 1, 6)$
b) $F(w, x, y, z) = \Sigma(0, 1, 6)$
c) $F(w, x, y, z) = \Sigma(2, 6, 10, 11, 14, 15)$
d) $F(x, y, z) = \Pi(0, 1, 6)$
e) $F(w, x, y, z) = \Pi(0, 1, 6)$
f) $F(w, x, y, z) = \Pi(2, 6, 10, 11, 14, 15)$

3.5. Draw the circuit diagram that implements the following Boolean functions using as few basic gates as possible, but without modifying the equation.

a) $F = xy' + x'y'z + xyz'$
b) $F = w'z' + w'xy + wx'z + wxyz$
c) $F = w'xy'z + w'xyz + wxy'z + wxyz$
d) $F = N_3'N_2'N_1N_0' + N_3'N_2'N_1N_0 + N_3N_2'N_1N_0' + N_3N_2'N_1N_0$
$+ N_3N_2N_1'N_0' + N_3N_2N_1N_0$
e) $F = [(x \odot y)' + (xyz)'](w' + x + z)$
f) $F = x \oplus y \oplus z$
g) $F = [w'xy'z + w'z(y \oplus x)]'$

3.6. Draw the circuit diagram that implements the Boolean functions in Problem 3.5 using only 2-input AND, 2-input OR, and NOT gates.

3.7. Design a circuit that inputs a 4-bit number. The circuit outputs a 1 if the input number is any one of the following numbers: 2, 3, 10, 11, 12, and 15. Otherwise, it outputs a 0.

3.8. Design a circuit that inputs a 4-bit number. The circuit outputs a 1 if the input number is greater than or equal to 5. Otherwise, it outputs a 0.

3.9. Design a circuit that inputs a 4-bit number. The circuit outputs a 1 if the input number has an even number of zeros. Otherwise, it outputs a 0.

3.10. Construct the following circuit. The circuit has five input signals and one output signal. The five input lines are labeled W, X, Y, Z, and E, and the output line is labeled F. E is used to enable (turn on) or disable (turn off) the circuit; thus, when $E = 0$, the circuit is disabled, and F is always 0. When $E = 1$, the circuit is enabled, and F is determined by the value of the four input signals, W, X, Y, and Z, where W is the most significant bit. If the value is odd, then $F = 1$, otherwise $F = 0$.

3.11. Draw the smallest circuit that inputs two 2-bit numbers. The circuit outputs a 2-bit number that represents the count of the number of even numbers in the inputs. The number 0 is taken as an even number. For example, if the two input numbers are 0 and 3, then the circuit outputs the number 1 in binary. If the two input numbers are 0 and 2, then the circuit outputs the number 2 in binary. Show your work by deriving the truth table, the equation, and finally the circuit. Minimize the equations as much as possible.

3.12. Derive and draw the circuit that inputs two 2-bit unsigned numbers. The circuit outputs a 3-bit signed number that represents the difference between the two input numbers (i.e., it is the result of the first number minus the second number). Derive the truth table and equations in canonical form.

3.13. Use Boolean algebra to show that the following circuit is equivalent to the NOT gate.

3.14. Construct a 4-input NAND gate circuit using only 2-input NAND gates.

3.15. Implement the following circuit using as few NAND gates (with any number of inputs) as possible.

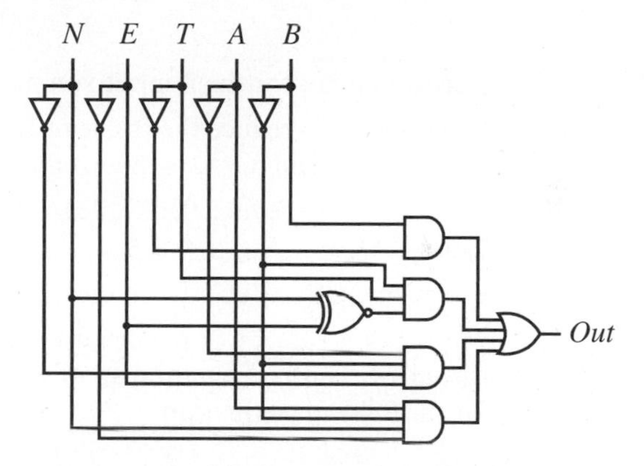

3.16. Draw the circuit diagram that implements the Boolean functions in Problem 3.5 using only 2-input NAND gates.

3.17. Draw the circuit diagram that implements the Boolean functions in Problem 3.5 using only 3-input NAND gates.

3.18. Draw the circuit diagram that implements the Boolean functions in Problem 3.5 using only 3-input NOR gates.

3.19. Convert the following circuit as is (i.e., do not reduce it first) to use only 2-input NOR gates.

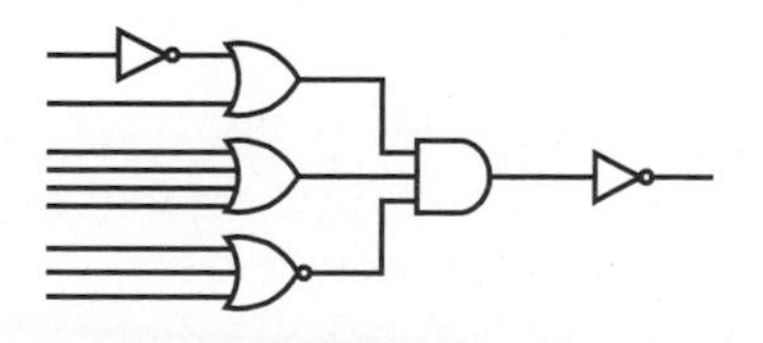

3.20. Convert the following full adder circuit to use only eleven 2-input NAND gates.

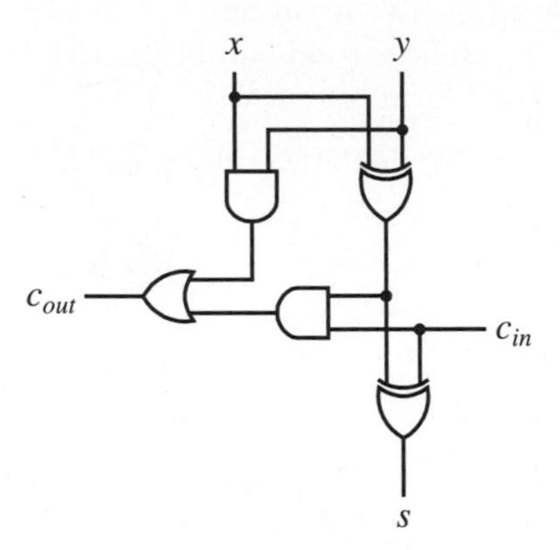

3.21. Derive a circuit for the 2-input XOR gate that uses only 2-input NAND gates.

3.22. Use K-maps to reduce the Boolean functions represented by the truth tables in Problem 3.3 to standard form.

3.23. Use K-maps to reduce the Boolean functions in Problem 3.4 to standard form.

3.24. Use K-maps to reduce the Boolean functions in Problem 3.5 to standard form.

3.25. List all of the PIs, EPIs, and all of the minimized standard form solutions for the following equation.

$$F(v, w, x, y, z) = \Pi(2, 3, 4, 5, 6, 7, 8, 9, 11, 13, 15, 18, 19, 20, 21, 22, 29, 30, 31)$$

3.26. Use K-maps to reduce the following 4-variable Boolean functions $F(w, x, y, z)$ to standard form:

a) 1-minterms: m_2, m_3, m_4, m_5
Don't-care minterms: m_{10}, m_{11}, m_{12}, m_{13}, m_{14}, m_{15}

b) 1-minterms: 1, 3, 4, 7, 9
Don't-care minterms: 0, 2, 13, 14, 15

c) 1-minterms: 2, 3, 8, 9
Don't-care minterms: 1, 5, 6, 7, 13, 15

3.27. Use K-maps to reduce the following 5-variable Boolean functions $F(v, w, x, y, z)$ to standard form:

a) 1-minterms: 1, 3, 4, 7, 9
Don't-care minterms: 0, 2, 13, 14, 15

b) 1-minterms: 2, 4, 10, 15, 16, 21, 26, 29
Don't-care minterms: 5, 7, 13, 18, 23, 24, 31

3.28. Use the Quine-McCluskey method to simplify the function $f(w, x, y, z) = \Sigma(0, 2, 5, 7, 13, 15)$. List all the PIs, EPIs, cover lists, and solutions.

3.29. Use the Quine-McCluskey method to reduce the Boolean functions in Problem 3.4 to standard form.

3.30. Write the function that eliminates the static hazard(s) in the function $F = w'z + xyz' + wx'y$.

3.31. Write the function that eliminates the static hazard(s) in the function $F = y'z' + wz + w'x'y$.

3.32. Write the complete structural Verilog code for the Boolean functions in Problem 3.4.

3.33. Write the complete dataflow Verilog code for the Boolean functions in Problem 3.4.

3.34. Write the complete structural VHDL code for the Boolean functions in Problem 3.4.

3.35. Write the complete dataflow VHDL code for the Boolean functions in Problem 3.4.

3.36. Write the behavioral Verilog code for converting an 8-bit binary number to a 3-digit decimal number to be displayed on three 7-segment LEDs. Another input signal, *signednumber*, is used to determine whether to interpret the binary number as a signed or unsigned number. If *signednumber* is a 1, the binary number is interpreted as a signed number, otherwise it is interpreted as an unsigned number. An optional negative sign is displayed on a fourth 7-segment LED. This circuit is used as the output circuit for many designs in later chapters.

3.37. Write the behavioral VHDL code for Problem 3.36.

CHAPTER 4

Standard Combinational Components

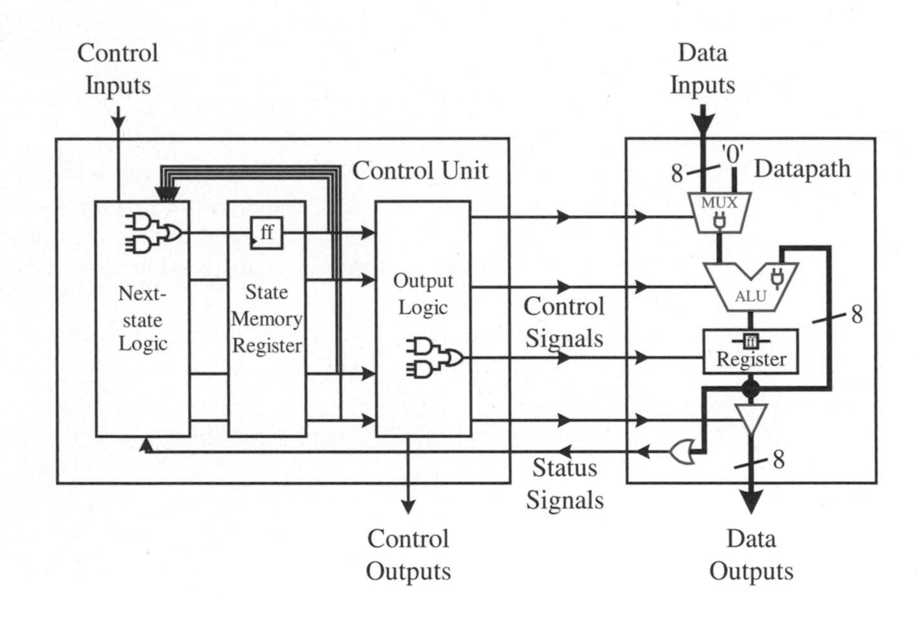

As with many construction projects it is often easier to build digital circuits in a hierarchical fashion. Initially, we use the most basic building blocks to build slightly larger building blocks, and then from these larger building blocks, we build yet larger building blocks, and so on. Similarly, in constructing large digital circuits, instead of starting with the basic logic gates as building blocks each time, we often start with larger building blocks. Many of these larger building blocks often are used over and over again in different digital circuits, and therefore, are considered as standard components for large digital circuits. In order to reduce the design time, these standard components are often made available in standard libraries so that they do not have to be redesigned each time that they are needed. For example, many digital circuits require the addition of two numbers, therefore, an adder circuit is considered a standard component and is available in most standard libraries.

Standard **combinational components** are combinational circuits that are available in standard libraries. These combinational components are used mainly in the construction of datapaths. Some standard combinational components that are typically used inside a microprocessor include the multiplexers, adders, subtractors, ALUs, comparators, tri-state buffers, and multipliers. Although the next-state logic and output logic circuits in the control unit are combinational circuits, they are not considered as standard combinational components because they are designed uniquely for a particular control unit to solve a specific problem and usually are not reused in another design.

In this chapter, we will design some standard combinational components. These components will be used in later chapters to build the datapath in the microprocessor.

4.1 Signal Naming Conventions

So far in our discussion, we have often used the words "high" and "low" to mean 1 and 0, or "on" and "off," respectively. However, this is somewhat arbitrary, and there is no reason why we can't say a 0 is a high or a 1 is off. In fact, many standard off-the-shelf components use what we call **negative logic** where 0 is for on and 1 is for off. Using negative logic usually is more difficult to understand because we are used to **positive logic** where 1 is for on and 0 is for off. In all of our discussions, we will use the more natural, positive logic.

Nevertheless, in order to prevent any confusion as to whether we are using positive logic or negative logic, we often use the words "assert," "de-assert," "active-high," and "active-low." Regardless of whether we are using positive or negative logic, **active-high** always means that a 1 (i.e., a high) will cause the signal to be active or enabled and that a 0 will cause the signal to be inactive or disabled. For example, if there is an active-high signal called *add* and we want to enable it (i.e., to make it do what it is intended for, which in this case is to add something), then we need to set this signal line to a 1. Setting this signal to a 0 will cause this signal to be disabled or inactive. An **active-low** signal, on the other hand, means that a 0 will cause the signal to be active or enabled, and that a 1 will cause the signal to be inactive or disabled. So if the signal *add* is an active-low signal, then we need to set it to a 0 to make it add something. When we name an active-low signal we typically append the prime (′) symbol after the name such as *add*′. This way, just by looking at the name and without needing any further explanation, we will know whether the signal is active high or low. For example, a signal named *count* is active-high so a 1 will cause the circuit to count and a 0 will cause the circuit not to count. On the other hand, a signal named *stop'* is active-low because of the prime symbol, and so a 0 will cause the circuit to stop and a 1 will cause the circuit to go.

We also use the word **assert** to mean to make a signal active or to enable the signal. To **de-assert** a signal is to disable the signal or to make it inactive. For example, to assert the active-high *add* signal line means to set the *add* signal to a 1. To de-assert an active-low line also means to set the line to a 1—since a 0 will enable the line (active-low)—and we want to disable (de-assert) it.

4.2 Multiplexer

The **multiplexer**, or **mux** for short, allows the selection of one input signal among n signals, where $n > 1$ and is a power of two. Select lines connected to the multiplexer determine which input signal is selected and passed to the output of the multiplexer. In general, an n-to-1 multiplexer has n data input lines, m select lines where $2^m = n$, and one output line. For a 2-to-1 multiplexer, there are two data input lines, d_0 and d_1, and one select line (since $2^1 = 2$), s, to select between the two data input lines. When $s = 0$, the input line d_0 is selected, and the data present on d_0 is passed to the output y. When $s = 1$, the input line d_1 is selected and the data on d_1 is passed to y. From this description of the 2-to-1 mux and from what we have already learned from the previous chapter, we can easily construct the truth table, followed by the equation, and then the circuit for it as shown in Figures 4.1(a), (b), and (c), respectively. The logic symbol for the 2-to-1 mux is shown in Figure 4.1(d). When we use this or any other components in a higher-level schematic circuit diagram, we simply use the logic symbol instead of drawing the detail circuit for the component. This way, the details of the component are hidden at this higher level.

Often, we might need to use a larger multiplexer. Constructing a larger-sized multiplexer, such as the 8-to-1 multiplexer, can be done similarly. In addition to having the eight data input lines, d_0 to d_7, the 8-to-1 multiplexer has three $(2^3 = 8)$ select lines, s_0, s_1, and s_2. Depending on the value of the three select lines, one of the eight data input lines will be selected, and the data on that input line will be passed to the output. For example, if the value of the select lines is 101 (which is decimal 5), then the data input line d_5 is selected, and the data that is present on d_5 will be passed to the output.

The truth table, circuit, and logic symbol for the 8-to-1 multiplexer are shown in Figure 4.2. The truth table is written in a slightly different format. Instead of including the *d*'s in the input columns and enumerating all $2^{11} = 2048$ rows (the eleven variables come from the eight *d*'s and the three *s*'s), the *d*'s are written in the entry under the output column. For example, when the select line value is 101, the entry under the output column is d_5, which means that y takes on the value of the input line d_5.

To understand the circuit in Figure 4.2(b), notice that each AND gate acts as a switch and is turned on by one unique combination of the three select lines. When a particular AND gate is turned on, the data at the corresponding d input are passed through that AND gate. The outputs of the remaining AND gates are all 0s. All of the outputs of the AND

s	d_1	d_0	y
0	0	0	0
0	0	1	1
0	1	0	0
0	1	1	1
1	0	0	0
1	0	1	0
1	1	0	1
1	1	1	1

(a)

$$\begin{aligned} y &= s'd_1'd_0 + s'd_1d_0 + sd_1d_0' + sd_1d_0 \\ &= s'd_0(d_1' + d_1) + sd_1(d_0' + d_0) \\ &= s'd_0 + sd_1 \end{aligned}$$

(b)

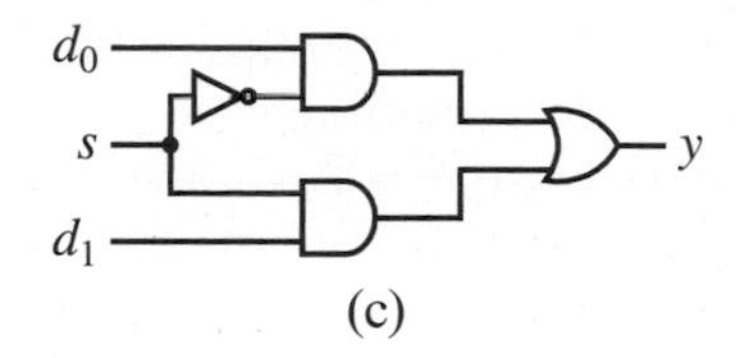

(c)

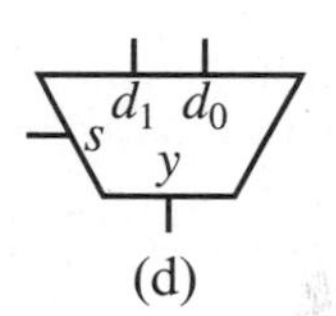

(d)

FIGURE 4.1 A 2-to-1 multiplexer: (a) truth table; (b) equation; (c) circuit; (d) logic symbol.

s_2	s_1	s_0	y
0	0	0	d_0
0	0	1	d_1
0	1	0	d_2
0	1	1	d_3
1	0	0	d_4
1	0	1	d_5
1	1	0	d_6
1	1	1	d_7

(a)

d_7 d_6 d_5 d_4 d_3 d_2 d_1 d_0
s_2
s_1
s_0
y

(b)

d_7 d_6 d_5 d_4 d_3 d_2 d_1 d_0
s_2
s_1
s_0
y

(c)

FIGURE 4.2 An 8-to-1 multiplexer: (a) truth table; (b) circuit; (c) logic symbol.

gates are ORed together to produce the final output. Since only one of the AND gates is turned on, therefore, only the value from that AND gate will be passed to the final y output.

Larger multiplexers also can be constructed from smaller multiplexers. For example, an 8-to-1 multiplexer can be constructed using seven 2-to-1 multiplexers, as shown in Figure 4.3. The four top-level 2-to-1 multiplexers provide the eight data inputs and

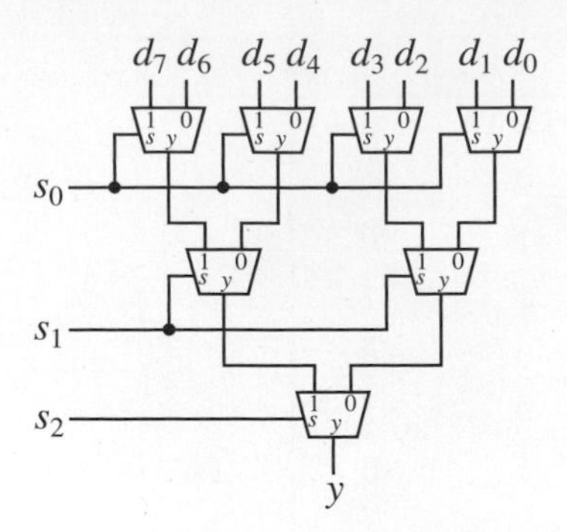

FIGURE 4.3 An 8-to-1 multiplexer implemented using seven 2-to-1 multiplexers.

all are switched by the same least significant select line s_0. This top level selects one from each group of two data inputs. The middle level then groups the four outputs from the top level again into groups of two, and selects one from each group using the middle select line s_1. Finally, the multiplexer at the bottom level uses the most significant select line s_2 to select one of the two outputs from the middle-level multiplexers.

Verilog Code for a Multiplexer

Figure 4.4 shows the behavioral Verilog code for an 8-bit wide 4-to-1 multiplexer. There are two select lines *S*, and four data input lines each being 8 bits wide. The data output line *Y* is also 8 bits wide. Notice that the data output line *Y* is declared with the `reg` keyword, whereas the input lines are not. The `always` statement is executed whenever the value for any one of the signals inside the sensitivity list changes. Statements inside

```
// A 4-to-1 8-bit wide multiplexer
module Multiplexer (
 input [1:0] S,        // 2 select lines
 input [7:0] D0,       // 4 data inputs, each is 8 bits wide
 input [7:0] D1,
 input [7:0] D2,
 input [7:0] D3,
 output reg [7:0] Y    // 8-bit wide output
);
 always @ (S or D0 or D1 or D2 or D3) begin
  case (S)
  0: Y = D0;
  1: Y = D1;
  2: Y = D2;
  default:
    Y = D3;
  endcase
 end
endmodule
```

FIGURE 4.4 Behavioral Verilog code for an 8-bit wide 4-to-1 multiplexer.

```
-- A 4-to-1 8-bit wide multiplexer
LIBRARY IEEE;
USE IEEE.STD_LOGIC_1164.ALL;

ENTITY Multiplexer IS PORT (
 S: IN STD_LOGIC_VECTOR(1 DOWNTO 0);          -- select lines
 D0: IN STD_LOGIC_VECTOR(7 DOWNTO 0);         -- data bus D0 input
 D1: IN STD_LOGIC_VECTOR(7 DOWNTO 0);         -- data bus D1 input
 D2: IN STD_LOGIC_VECTOR(7 DOWNTO 0);         -- data bus D2 input
 D3: IN STD_LOGIC_VECTOR(7 DOWNTO 0);         -- data bus D3 input
 Y: OUT STD_LOGIC_VECTOR(7 DOWNTO 0));        -- data bus output
END Multiplexer;

-- Behavioral level code
ARCHITECTURE Behavioral OF Multiplexer IS
BEGIN
 PROCESS (S,D0,D1,D2,D3)
 BEGIN
  CASE S IS
   WHEN "00" => Y <= D0;
   WHEN "01" => Y <= D1;
   WHEN "10" => Y <= D2;
   WHEN OTHERS => Y <= D3
  END CASE;
 END PROCESS;
END Behavioral;
```

FIGURE 4.5 Behavioral level VHDL code for an 8-bit wide 4-to-1 multiplexer.

the `always` block are executed sequentially. The `case` statement selects the value of the S signal and executes one of the four cases. If none of the cases matches, then the `default` case is selected.

VHDL Code for a Multiplexer

The behavioral level VHDL code for an 8-bit wide 4-to-1 multiplexer is shown in Figure 4.5. The PROCESS block is executed whenever one of the signals inside the sensitivity list changes value. Statements inside the PROCESS block are executed sequentially. The CASE statement is used to select between the four choices for S. If S is equal to 00, then the value $D0$ is assigned to Y. If S does not match any one of the cases, 00, 01, or 10, then the WHEN OTHERS clause will be selected.

4.3 Adder

An adder is for adding two n-bit binary numbers together to produce a sum.

4.3.1 Full Adder

To construct an adder for adding two n-bit binary numbers, $X = x_{n-1} \ldots x_0$ and $Y = y_{n-1} \ldots y_0$, we need to first consider the addition of a single bit slice, x_i with y_i, together with the carry-in bit, c_i, from the previous bit position on the right. The result from this addition is a sum bit, s_i, and a carry-out bit, c_{i+1}, for the next bit position. In other words, $s_i = x_i + y_i + c_i$, and $c_{i+1} = 1$ if there is a carry from the addition to the next bit on the left, otherwise, $c_{i+1} = 0$. Note that the + operator in this equation is for addition and not the logical OR operation.

For example, consider the following addition of the two 4-bit binary numbers, $X = 1001$ and $Y = 0011$.

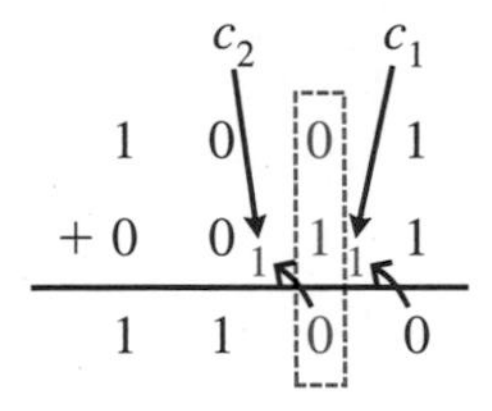

The result of the addition is 1100. The addition is performed just like that for decimal numbers, except that there is a carry whenever the sum is either a 2 or a 3 in decimal, since 2 is 10 in binary and 3 is 11. The most significant bit in the 10 or the 11 is the carry-out bit. Looking at the bit slice that is highlighted in blue, where $x_1 = 0$, $y_1 = 1$, and $c_1 = 1$, the addition for this bit slice is $x_1 + y_1 + c_1 = 0 + 1 + 1 = 10$. Therefore, the sum bit is $s_1 = 0$, and the carry-out bit is $c_2 = 1$.

The circuit for the addition of a single bit slice is known as a **full adder** (FA), and its truth table is shown in Figure 4.6(a). For each row in the truth table, the combined two output bits $c_{i+1}s_i$ is obtained by adding the three input bits, $c_{i+1}s_i = x_i + y_i + c_i$. The derivation and simplification of the equations for s_i and c_{i+1} are shown in Figure 4.6(b). From these two equations, we get the circuit for the full adder shown in Figure 4.6(c), and the logic symbol shown in Figure 4.6(d). The dataflow Verilog and VHDL code for the full adder is shown in Figures 4.7 and 4.8, respectively.

4.3.2 Ripple-Carry Adder

The full adder is for adding two operands that are only one bit wide. To add two operands that are, say, four bits wide, we connect four full adders together in series through their carry-in and carry-out signals. The resulting circuit, shown in Figure 4.9, is called a **ripple-carry adder** for adding two 4-bit operands.

The input for one carry-in line is connected to the carry-out line from the previous FA. Since an FA adds the three bits, x_i, y_i, and c_i, together, we need to set the first carry-in bit, c_0, to 0 in order to perform the addition correctly, otherwise, the sum will always be one more than the correct result. Moreover, the output signal c_{out} serves as an overflow signal, and is a 1 whenever there is an overflow in the addition.

Verilog Code for a 4-bit Adder

The behavioral Verilog code for the 4-bit adder is shown in Figure 4.10. The syntax [3:0] in the signal declaration denotes that the signals are 4 bits wide. Notice on the left side of the `assign` statement, *Cout* and *Sum* are concatenated together using the syntax `{Cout, Sum}`, resulting in a 5-bit output.

x_i	y_i	c_i	c_{i+1}	s_i
0	0	0	0	0
0	0	1	0	1
0	1	0	0	1
0	1	1	1	0
1	0	0	0	1
1	0	1	1	0
1	1	0	1	0
1	1	1	1	1

(a)

$$\begin{aligned} s_i &= x_i'y_i'c_i + x_i'y_ic_i' + x_iy_i'c_i' + x_iy_ic_i \\ &= (x_i'y_i + x_iy_i')c_i' + (x_i'y_i' + x_iy_i)c_i \\ &= (x_i \oplus y_i)c_i' + (x_i \oplus y_i)'c_i \\ &= x_i \oplus y_i \oplus c_i \\ c_{i+1} &= x_i'y_ic_i + x_iy_i'c_i + x_iy_ic_i' + x_iy_ic_i \\ &= x_iy_i(c_i' + c_i) + c_i(x_i'y_i + x_iy_i') \\ &= x_iy_i + c_i(x_i \oplus y_i) \end{aligned}$$

(b)

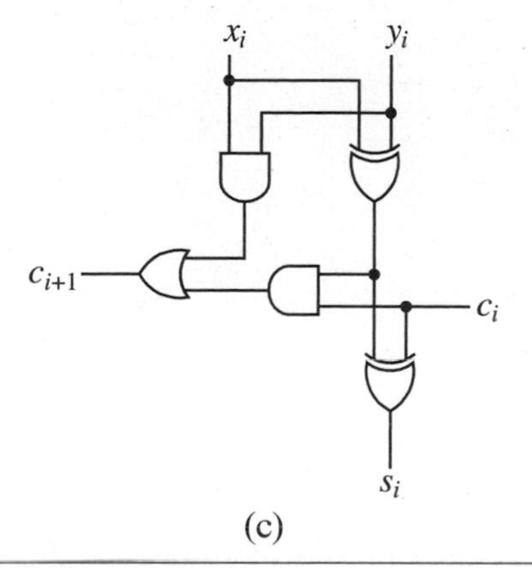

(c)

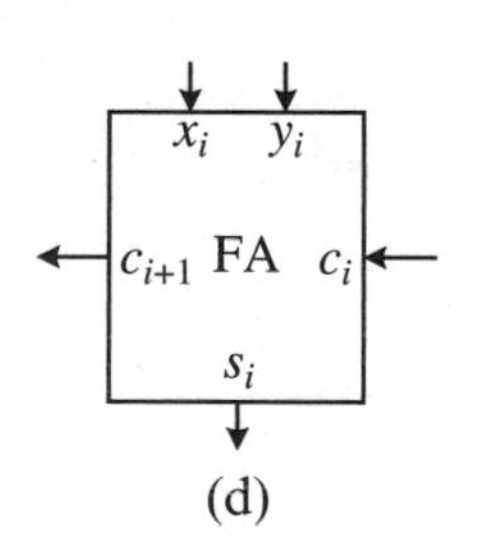

(d)

FIGURE 4.6 Full adder: (a) truth table; (b) equations for si and c_{i+1}; (c) circuit; (d) logic symbol.

```
module fa (
 input ci, xi, yi,
 output ci1, si
);
 // using the equations derived for the FA
 // bitwise &=AND; |=OR; ^=XOR
 assign ci1 = (xi & yi) | (ci & (xi ^ yi));
 assign si = xi ^ yi ^ ci;
endmodule
```

FIGURE 4.7 Dataflow Verilog code for a 1-bit full adder.

VHDL Code for a 4-bit Adder

The behavioral VHDL code for the 4-bit adder is shown in Figure 4.11. Unlike Verilog, VHDL does not provide a concatenation of the result, hence, we need first to zero extend the two input operands and then perform a 5-bit addition to store the result in a 5-bit vector *Temp*. The & symbol is used to concatenate a 0 bit to the 4-bit vector A or B,

```
LIBRARY IEEE;
USE IEEE.STD_LOGIC_1164.ALL;
ENTITY fa IS PORT (
 ci, xi, yi: IN STD_LOGIC;
 ci1, si: OUT STD_LOGIC);
END fa;

ARCHITECTURE Dataflow OF fa IS
BEGIN
 ci1 <= (xi AND yi) OR (ci AND (xi XOR yi));
 si <= xi XOR yi XOR ci;
END Dataflow;
```

FIGURE 4.8 Dataflow VHDL code for a 1-bit full adder.

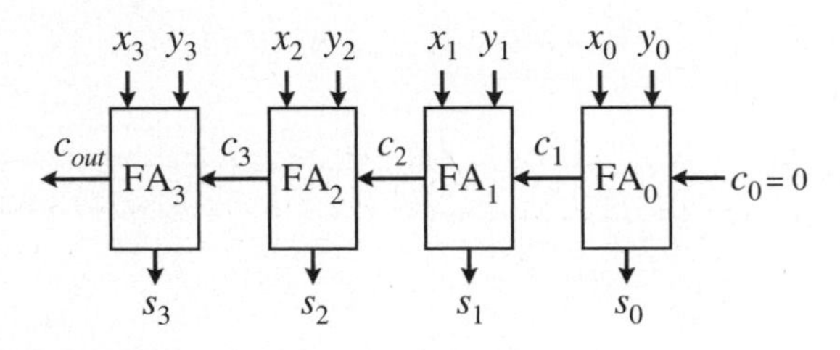

FIGURE 4.9 Ripple-carry adder.

```
module Adder4 (
 input [3:0] A, B,
 output [3:0] Sum,
 output Cout
 );
 assign {Cout,Sum} = A + B;
endmodule
```

FIGURE 4.10 Behavioral Verilog code for a 4-bit adder.

thus resulting in a 5-bit vector for the addition. We then extract the most significant bit from *Temp* for *Cout*, and the last four bits from *Temp* for *Sum*.

4.3.3 Carry-Lookahead Adder

The ripple-carry adder is slow because the carry-in signal for each FA is dependent on the carry-out signal from the previous FA. So before FA_i can output valid data, it must wait for FA_{i-1} to have valid data. Hence, the time needed for the adder to output valid data is dependent on the number of bits in the adder. In the **carry-lookahead adder**, each bit slice eliminates this dependency on the previous carry-out signal and instead uses the values of the two input operands, *X* and *Y*, directly to deduce the needed signals. This is

```
LIBRARY IEEE;
USE IEEE.STD_LOGIC_1164.ALL;
ENTITY Adder4 IS PORT (
 A, B: IN STD_LOGIC_VECTOR(3 DOWNTO 0);
 Sum: OUT STD_LOGIC_VECTOR(3 DOWNTO 0);
 Cout: OUT STD_LOGIC);
END Adder4;
ARCHITECTURE Behavioral OF Adder4 IS
 SIGNAL Temp : STD_LOGIC_VECTOR (4 DOWNTO 0);
BEGIN
 Temp <= ('0' & A) + ('0' & B);
 Cout <= Temp(4);
 Sum <= Temp(3 DOWNTO 0);
END Behavioral;
```

FIGURE 4.11 Behavioral VHDL code for a 4-bit adder.

possible from the following observations regarding the carry-out signal. For each FA_i, the carry-out signal, c_{i+1}, is set to a 1 if either one of the following two conditions is true:

$$x_i = 1 \text{ and } y_i = 1$$

or

$$(x_i = 1 \text{ or } y_i = 1) \text{ and } c_i = 1$$

In other words,

$$c_{i+1} = x_i y_i + (x_i + y_i)c_i \tag{4.1}$$

At first glance, this carry-out equation looks different from the carry-out equation

$$c_{i+1} = x_i y_i + c_i(x_i \oplus y_i)$$

deduced in Figure 4.6(b) for the full adder. However, they are functionally equivalent. (See Problem 4.13.)

If we let

$$g_i = x_i y_i$$

and

$$p_i = x_i + y_i$$

then Equation 4.1 can be rewritten as

$$c_{i+1} = g_i + p_i c_i \tag{4.2}$$

Using Equation 4.2 for c_{i+1}, we can recursively expand it to get the carry-out equations for any bit slice, c_i, that is dependent only on the two input operands, X and Y, and the initial carry-in bit, c_0. Using this technique, we get the following carry-out equations for the first four bit slices:

$$c_1 = g_0 + p_0c_0 \tag{4.3}$$

$$\begin{aligned} c_2 &= g_1 + p_1c_1 \\ &= g_1 + p_1(g_0 + p_0c_0) \\ &= g_1 + p_1g_0 + p_1p_0c_0 \end{aligned} \tag{4.4}$$

$$\begin{aligned} c_3 &= g_2 + p_2c_2 \\ &= g_2 + p_2(g_1 + p_1g_0 + p_1p_0c_0) \\ &= g_2 + p_2g_1 + p_2p_1g_0 + p_2p_1p_0c_0 \end{aligned} \tag{4.5}$$

$$\begin{aligned} c_4 &= g_3 + p_3c_3 \\ &= g_3 + p_3(g_2 + p_2g_1 + p_2p_1g_0 + p_2p_1p_0c_0) \\ &= g_3 + p_3g_2 + p_3p_2g_1 + p_3p_2p_1g_0 + p_3p_2p_1p_0c_0 \end{aligned} \tag{4.6}$$

Using Equations 4.3 to 4.6, we obtain the circuit for generating the carry-lookahead signals for c_1 to c_4, as shown in Figure 4.12(a). Note that each equation is translated to a three-level combinational logic—one level for generating the g_i and p_i, and two levels

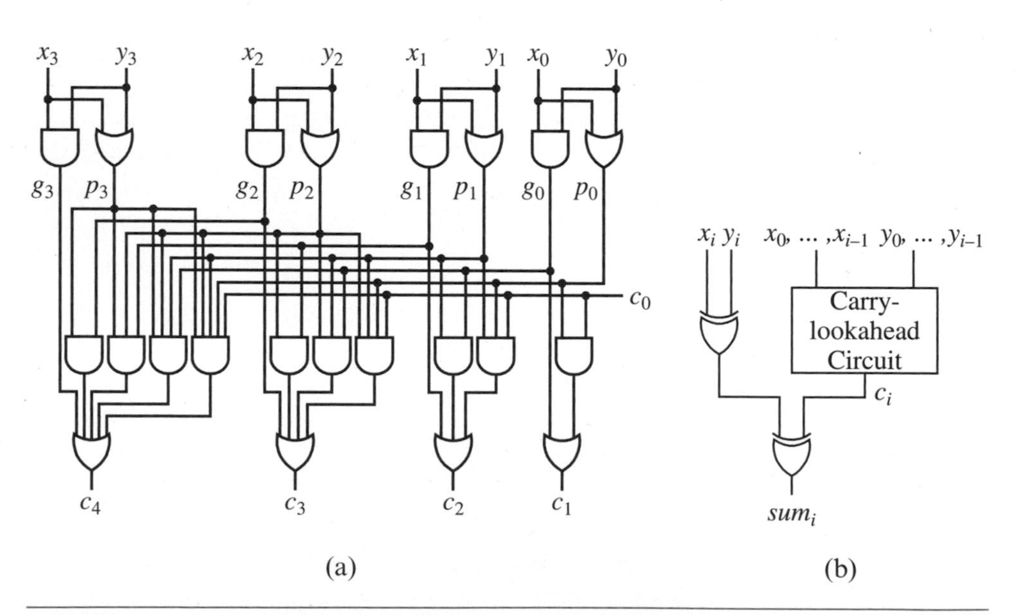

FIGURE 4.12 **(a) Circuit for generating the carry-lookahead signals, c_1 to c_4; (b) one bit slice of the carry-lookahead adder.**

(for the sum-of-products format) for generating the c_i expression. This carry-lookahead circuit can be reduced even further because we want c_0 to be a 0 when performing additions, and this 0 will cancel the rightmost product term in each equation. (See Problem 4.14.)

The FA for the carry-lookahead adder also can be made simpler, since it is no longer required to generate the carry-out signal for the next bit slice. In other words, the carry-in signal for the FA now comes from the new carry-lookahead circuit rather than from the carry-out signal of the previous bit slice. Thus, this FA only needs to generate the sum_i signal. Figure 4.12(b) shows one bit slice of the carry-lookahead adder. For an n-bit carry-lookahead adder, we use n bit slices. These n bit slices are not connected in series as with the ripple-carry adder; otherwise, it defeats the purpose of having the more complicated carry-out circuit.

4.4 Subtractor

We can construct a 1-bit subtractor circuit similar to the method used for constructing the full adder. However, instead of the sum bit, s_i, for the addition, we have a difference bit, d_i, for the subtraction, and instead of having carry-in and carry-out signals, we have borrow-in (b_i) and borrow-out (b_{i+1}) signals. So, when we subtract the ith bit of the two operands x_i and y_i, we get the difference of $d_i = x_i - y_i$. If, however, the previous bit on the right has to borrow from this i^{th} bit, then input b_i will be set to a 1, and the equation for the difference will be $d_i = x_i - b_i - y_i$. On the other hand, if the ith bit has to borrow from the next bit on the left for the subtraction, then the output b_{i+1} will be set to a 1. The value borrowed is a 2, and so the resulting equation for the difference will be $d_i = x_i - b_i + 2b_{i+1} - y_i$. Note that the symbols + and − used in this equation are for addition and subtraction, and not for logical operations. The term $2b_{i+1}$ is "2 multiply by b_{i+1}." Since b_{i+1} is a 1 when we have to borrow and we borrow a 2 each time, the equation just adds a 2 when there is a borrow. When there is no borrow, b_{i+1} is 0, and so the term $2b_{i+1}$ cancels out to 0.

For example, consider the following subtraction of the two 4-bit binary numbers, $X = 0100$ and $Y = 0011$:

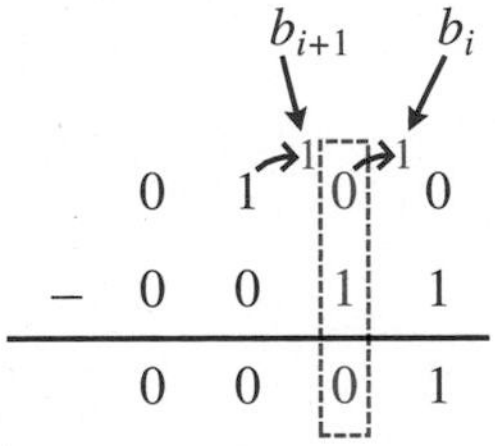

```
            b_{i+1}   b_i
               ↘       ↙
              →1  →1
      0   1   [0]   0
  −   0   0   [1]   1
  ----------------------
      0   0   [0]   1
```

Consider the bit position that is highlighted in blue. Since the subtraction for the previous bit on the right has to borrow, therefore b_i is a 1. Moreover, b_{i+1} is also a 1 because the current bit also has to borrow from the next bit on the left. When it borrows, it gets a 2. Therefore, $d_i = x_i - b_i + 2b_{i+1} - y_i = 0 - 1 + 2(1) - 1 = 0$.

The truth table for the 1-bit subtractor is shown in Figure 4.13(a), from which the equations for d_i and b_{i+1}, as shown in Figure 4.13(b), are derived. From these two equations, we get the circuit for the subtractor, as shown in Figure 4.13(c). Figure 4.13(d) shows the logic symbol for the subtractor.

Building a subtractor circuit for subtracting an n-bit operand can be done by daisy-chaining n 1-bit subtractor circuits together, similar to the ripple-carry adder

x_i	y_i	b_i	b_{i+1}	d_i
0	0	0	0	0
0	0	1	1	1
0	1	0	1	1
0	1	1	1	0
1	0	0	0	1
1	0	1	0	0
1	1	0	0	0
1	1	1	1	1

(a)

$$\begin{aligned} d_i &= x_i'y_i'b_i + x_i'y_ib_i' + x_iy_i'b_i' + x_iy_ib_i \\ &= (x_i'y_i + x_iy_i')b_i' + (x_i'y_i' + x_iy_i)b_i \\ &= (x_i \oplus y_i)b_i' + (x_i \oplus y_i)'b_i \\ &= x_i \oplus y_i \oplus b_i \\ b_{i+1} &= x_i'y_i'b_i + x_i'y_ib_i' + x_i'y_ib_i + x_iy_ib_i \\ &= x_i'b_i(y_i' + y_i) + x_i'y_i(b_i' + b_i) + y_ib_i(x_i' + x_i) \\ &= x_i'b_i + x_i'y_i + y_ib_i \end{aligned}$$

(b)

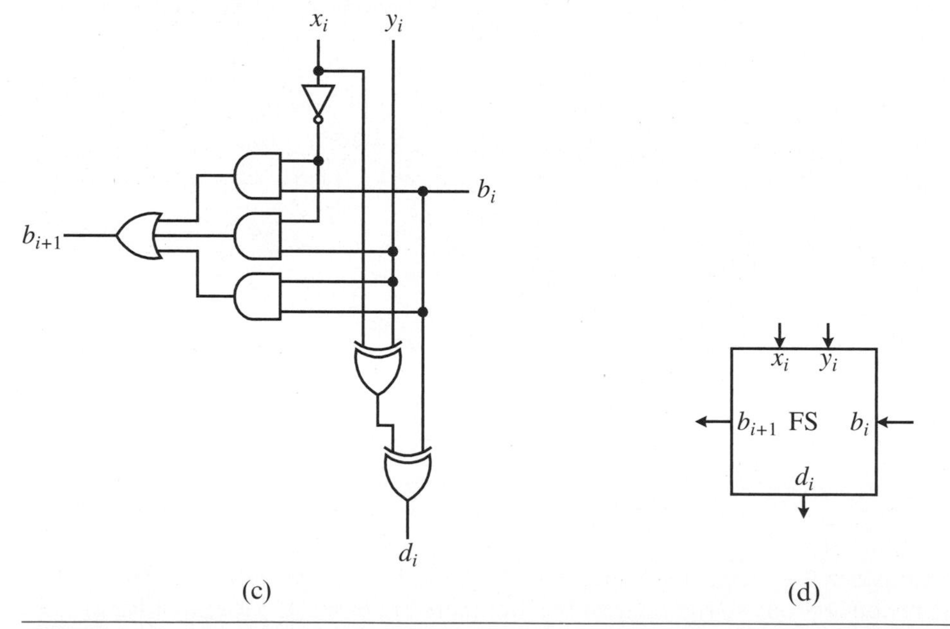

FIGURE 4.13 1-bit subtractor: (a) truth table; (b) equations for d_i and b_{i+1}; (c) circuit; (d) logic symbol.

circuit shown in Figure 4.9. Like the adder, the initial borrow bit, b_0, should be set to 0. The input operands are interpreted as unsigned numbers, and the last overflow bit, b_n, will be asserted when the result of the subtraction is negative. However, there is a much better subtractor circuit, as shown in the next section.

4.5 Adder-Subtractor Combination

It turns out that, instead of having to build a separate adder and subtractor units, we can modify the ripple-carry adder slightly to perform both operations. The modified circuit performs subtraction by adding the negated value of the second operand. In other words, instead of performing the subtraction $A - B$, the addition operation $A + (-B)$ is performed.

Recall that in two's complement representation, to negate a value involves inverting all the 0s to 1 and all the 1s to 0, and then adding a 1. Hence, we need to modify the adder circuit so that we selectively can do either one of two things: (1) not flip the bits and not add an extra 1 for the addition operation, or (2) flip the bits of the B operand and then add an extra 1 for the subtraction operation.

For this adder-subtractor combination circuit (in addition to the two input operands A and B), a select signal, s, is needed to select which operation to perform. The assignment of the two operations to the select signal s is shown in Figure 4.14(a). When $s = 0$, we want to perform an addition, and when $s = 1$, we want to perform a subtraction. When $s = 0$, B does not need to be modified, and like the adder circuit from Section 4.3.2, the initial carry-in signal c_0 needs to be set to a 0. On the other hand, when $s = 1$, we need to invert the bits in B and add a 1. The addition of a 1 is accomplished by setting the initial carry-in signal c_0 to a 1. Two circuits are needed for handling the above situations: one for inverting the bits in B and one for setting c_0. Both of these circuits are dependent on s.

The truth table for these two circuits is shown in Figure 4.14(b). In this truth table, the input variable b_i is the ith bit of the B operand. The output variable y_i is the output from the circuit that either inverts or does not invert the bits in B. So when s is a 0, y_i is the same as b_i, but when s is a 1, y_i is the inverse of b_i. Furthermore, $c_0 = s$. From this truth table, we can conclude that the circuit for y_i is just a 2-input XOR gate, while the circuit for c_0 is just a direct connection from s. Putting everything together, we obtain the adder-subtractor combination circuit (for four bits), as shown in Figure 4.14(c). The logic symbol for the circuit is shown in Figure 4.14(d).

Notice that the adder-subtractor circuit in Figure 4.14(c) has two different overflow signals, *Unsigned_Overflow* and *Signed_Overflow*. This is because the circuit can deal with both signed and unsigned numbers. To determine whether there is an overflow for unsigned numbers, the *Unsigned_Overflow* signal is obtained by XORing C_{out} with s because it is equal to C_{out} for additions but it is the inverse of C_{out} for subtractions. However, to determine whether there is an overflow for signed numbers, the *Signed_Overflow* signal is obtained by XORing C_{out} with c_3. The reason for this is explained below.

For example, the valid range for a 4-bit *signed* number goes from -2^3 to $2^3 - 1$ (i.e., from -8 to 7). Adding the two signed numbers, $4 + 5 = 9$ should result in a

s	Function	Operation
0	Add	$F = A + B$
1	Subtract	$F = A + B' + 1$

(a)

s	b_i	y_i	c_0
0	0	0	0
0	1	1	0
1	0	1	1
1	1	0	1

(b)

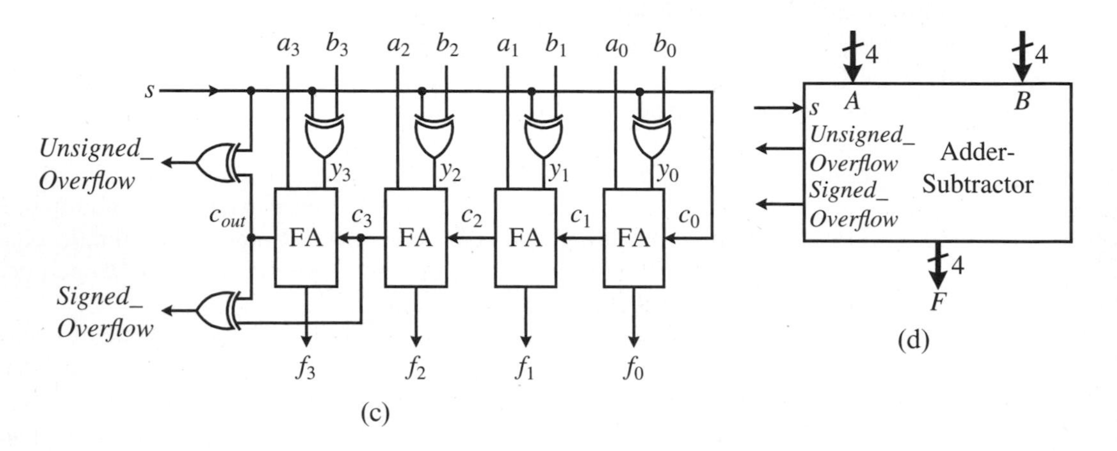

FIGURE 4.14 Adder-subtractor combination: (a) operation table; (b) truth table for y_i and c_0; (c) circuit; (d) logic symbol.

signed number overflow, since 9 is outside the range. However, the valid range for a 4-bit *unsigned* number goes from 0 to $2^4 - 1$ (i.e., 0 to 15). If we treat the two numbers 4 and 5 as unsigned numbers, then the result of adding these two unsigned numbers, 9, is inside the range. So when adding the two numbers 4 and 5, the *Unsigned_Overflow* signal should be de-asserted, while the *Signed_Overflow* signal should be asserted. Performing the addition of 4 + 5 in binary as shown next:

$$
\begin{array}{rcccc}
 & & c_3 & & \\
 & 0 & 1 & 0 & 0 \\
+ & 0 & 1 & 0 & 1 \\
\hline
0 & 1 & 0 & 0 & 1
\end{array}
$$

Unsigned Overflow → 0

0 XOR 1 = 1 ← *Signed Overflow*

we get 0100 + 0101 = 1001, which produces a 0 for the *Unsigned_Overflow* signal. However, the addition produces a 1 for c_3, and XORing these two values, 0 for *Unsigned_Overflow* and 1 for c_3, results in a 1 for the *Signed_Overflow* signal.

For another example, adding the two 4-bit signed numbers, $-4 + (-3) = -7$ should not result in a signed overflow. Performing the arithmetic in binary, $-4 = 1100$ and $-3 = 1101$, as shown next:

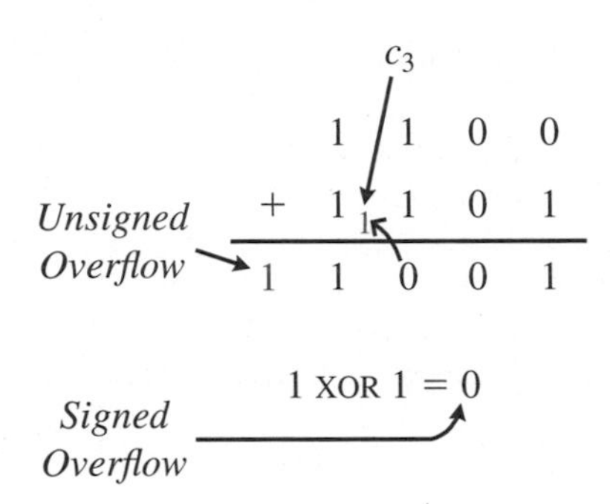

we get 1100 + 1101 = 11001, which produces a 1 for both *Unsigned_Overflow* and c_3. XORing these two values together gives a 0 for the *Signed_Overflow* signal. On the other hand, if we treat the two binary numbers, 1100 and 1101, as unsigned numbers, then we are adding 12 + 13 = 25. Since 25 is outside the unsigned number range, and so the *Unsigned_Overflow* signal is correct with a 1.

HDL Code for an Adder-Subtractor Combination

The behavioral Verilog and behavioral VHDL code for the 4-bit adder-subtractor combination circuit are shown in Figures 4.15 and 4.16, respectively. The basic structure of the code is similar to that of the adder shown in Figure 4.11, except that there is the `if-else` statement to check whether to perform the addition or the subtraction.

The *Unsigned_Overflow* bit is obtained by performing the addition or subtraction operation using $n + 1$ bits. The two input operands A and B are first zero extended before the operation is performed. The result of the operation is stored in the $n + 1$ bit vector, *TempF*. The most significant bit of this vector, *TempF*(n), is already the correct *Unsigned_Overflow* signal for both additions and subtractions. Notice here that we do not need to XOR this overflow bit with the select signal as in the case for the circuit shown in Figure 4.14 because the code is actually performing a subtraction and not an addition with a negative number. (See Problem 4.8.)

In the circuit shown in Figure 4.14, the *Signed_Overflow* bit is obtained by XORing c_{out} and c_3. However, both Verilog and VHDL do not provide a construct that can easily extract the carry bit c_3 that is in the middle of a bit vector, so we use a different

```
module AddSub (
 input S, // select 0=add, 1=subtract
 input [3:0] A, B,
 output reg [3:0] F,
 output reg Unsigned_Overflow,
 output reg Signed_Overflow
);
 reg [4:0] TempF;

 always @ (S or A or B) begin
  if (S == 0) begin // addition
   // zero extend A and B before add
   TempF = {1'b0,A} + {1'b0,B};
   F = TempF[3:0];
   Unsigned_Overflow = TempF[4];
    // Signed overflow = MSB of A'B'F + ABF'
   Signed_Overflow = ((!A[3]) & (!B[3]) & (TempF[3])) +
      ((A[3]) & (B[3]) & (!TempF[3]));
  end else begin  // subtract
   // zero extend A and B before subtract
   TempF = {1'b0,A} - {1'b0,B};
   F = TempF[3:0];
   Unsigned_Overflow = TempF[4];
    // Signed overflow = MSB of AB'F' + A'BF
   Signed_Overflow = ((A[3]) & (!B[3]) & (!TempF[3])) +
      ((!A[3]) & (B[3]) & (TempF[3]));
  end
 end
endmodule
```

FIGURE 4.15 Behavioral Verilog code for a 4-bit adder-subtractor combination component.

method. It turns out that for addition, the signed overflow can occur only if $A_n'B_n'F_n$ or $A_nB_nF_n'$ is a 1 where A_n, B_n, and F_n are the most significant bit of the two input operands A and B, and F is the result of the addition. Thus, the Boolean equation for the *Signed_Overflow* bit for addition is

$$Signed\_Overflow = (A_n' \cdot B_n' \cdot F_n) + (A_n \cdot B_n \cdot F_n')$$

Similarly, for subtraction, the signed overflow can only occur if $A_nB_n'F_n'$ or $A_n'B_nF_n$ is a 1. Thus, the Boolean equation for the *Signed_Overflow* bit for subtraction is

$$Signed\_Overflow = (A_n \cdot B_n' \cdot F_n') + (A_n' \cdot B_n \cdot F_n)$$

```
LIBRARY IEEE;
USE IEEE.STD_LOGIC_1164.ALL;
USE IEEE.STD_LOGIC_UNSIGNED.ALL; -- need this to do + and -
ENTITY AddSub IS
PORT(S: IN STD_LOGIC; -- select subtract signal
 A, B: IN STD_LOGIC_VECTOR(3 DOWNTO 0);
 F: OUT STD_LOGIC_VECTOR(3 DOWNTO 0);
 Unsigned_Overflow: OUT STD_LOGIC;
 Signed_Overflow: OUT STD_LOGIC);
END AddSub;
ARCHITECTURE Behavioral OF AddSub IS
 SIGNAL TempF : STD_LOGIC_VECTOR(4 DOWNTO 0);
BEGIN
 PROCESS(S, A, B)
 BEGIN
  IF (S = '0') THEN   -- addition
   TempF <= ('0' & A) + ('0' & B);
   F <= TempF(3 DOWNTO 0);
   Unsigned_Overflow <= TempF(4);
    -- Signed overflow = most significant bit of A'B'F + ABF'
   Signed_Overflow <= ((NOT A(3)) AND (NOT B(3)) AND
        (TempF(3))) OR ((A(3)) AND (B(3)) AND (NOT TempF(3)));
  ELSE      -- subtraction
   TempF <= ('0' & A) - ('0' & B);
   F <= TempF(3 DOWNTO 0);
   Unsigned_Overflow <= TempF(4);
    -- Signed overflow = most significant bit of AB'F' + A'BF
   Signed_Overflow <= ((A(3)) AND (NOT B(3)) AND
        (NOT TempF(3))) OR ((NOT A(3)) AND (B(3)) AND
(TempF(3)));
  END IF;
 END PROCESS;
END Behavioral;
```

FIGURE 4.16 Behavioral VHDL code for a 4-bit adder-subtractor combination component.

4.6 Arithmetic Logic Unit

The **arithmetic logic unit** (ALU) is one of the main components inside a microprocessor. It is responsible for performing arithmetic and logic operations, such as addition, subtraction, logical AND, and logical OR. The ALU, however, is not used to perform multiplications or divisions because these operations are much more complex. It turns out that, in constructing the circuit for the ALU, we can use the same idea as for constructing the adder-subtractor combination circuit, as discussed in the previous section. Again, we will use the ripple-carry adder as the building block and then insert some combinational logic circuitry in front of the two input operands to each full adder. This way, the primary inputs will be modified accordingly,

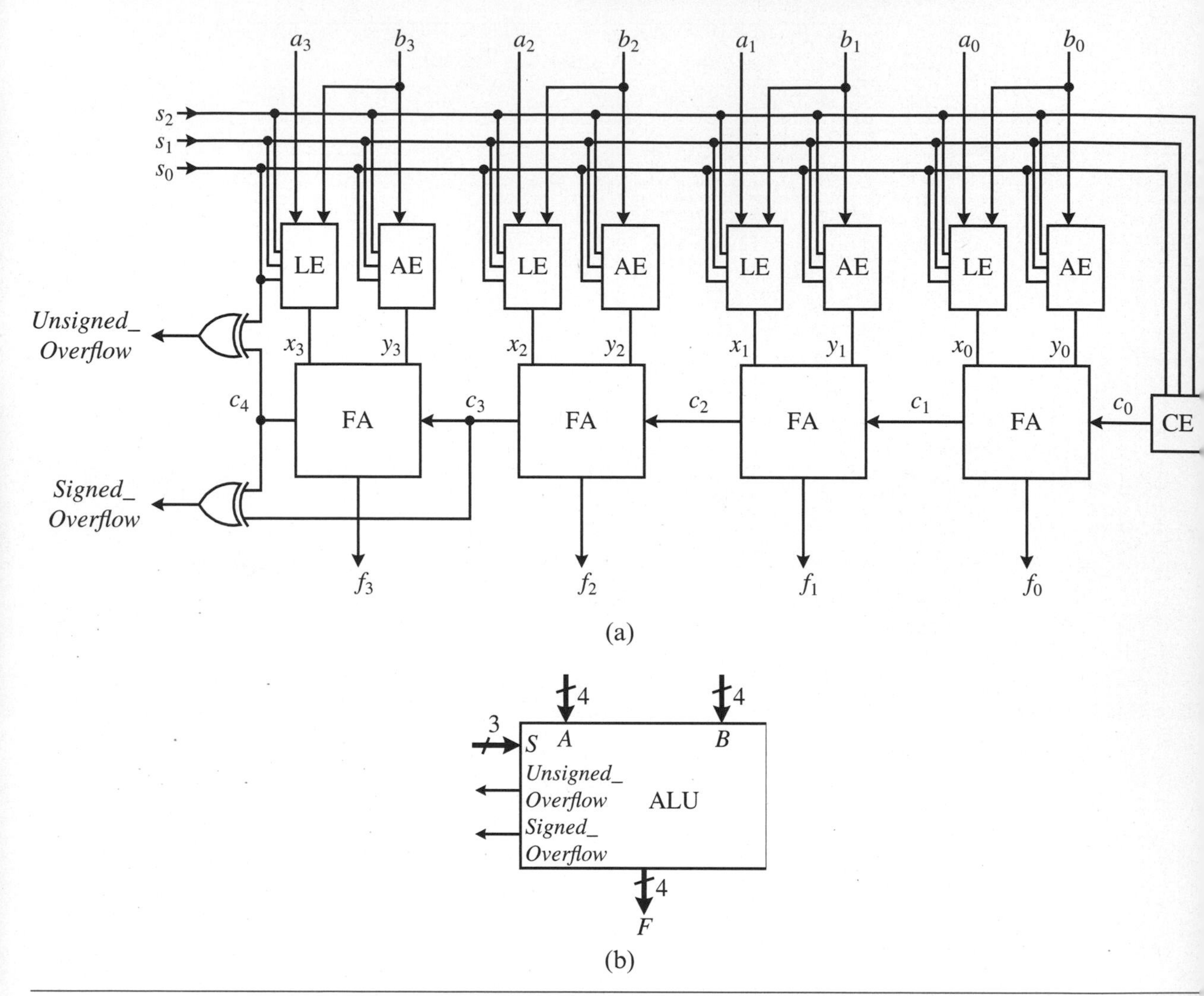

FIGURE 4.17 4-bit ALU: (a) circuit; (b) logic symbol.

depending on the operations being performed before being passed to the full adder. The general, overall circuit for a 4-bit ALU is shown in Figure 4.17(a) and its logic symbol in Figure 4.17(b).

As we can see in Figure 4.17(a), the two combinational circuits above each FA are labeled LE and AE. The logic extender (LE) is for manipulating all logical operations; whereas, the arithmetic extender (AE) is for manipulating all arithmetic operations. The LE performs the actual logical operations on the two primary operands, a_i and b_i, before passing the result to the first operand, x_i, of the FA. On the other hand, the AE modifies only the second operand, b_i, and passes it to the second operand, y_i, of the FA where the actual arithmetic operation is performed.

We saw from the adder-subtractor circuit that, to perform additions and subtractions, we need to modify only y_i (the second operand to the FA) so that all operations can be done with additions. Thus, the AE takes only the second operand of the primary input, b_i, as its input and modifies the value depending on the operation being performed. Its output is y_i, and it is connected to the second operand input of the FA. As in the adder-subtractor circuit, the addition is performed in the FA. When arithmetic operations are being performed, the LE must pass the first operand unchanged from the primary input a_i to the output x_i for the FA.

Unlike the AE (where it modifies only the B operand), the LE performs the actual logical operations. Thus, for example, if we want to perform the operation A OR B, the LE for each bit slice will take the corresponding bits, a_i and b_i, and OR them together. Hence, one bit from both operands, a_i and b_i, are inputs to the LE. The output of the LE is passed to the first operand, x_i, of the FA. Since this value is already the result of the logical operation, we do not want the FA to modify it but to simply pass it on to the primary output, f_i. This is accomplished by setting both the second operand, y_i, of the FA and c_0 to 0, since adding a 0 will not change the resulting value.

The combinational circuit labeled CE (for carry extender) is for modifying the primary carry-in signal, c_0, so that arithmetic operations are performed correctly. Logical operations do not use the carry signal, so c_0 is set to 0 for all logical operations.

In the circuit shown in Figure 4.17, three select lines, s_2, s_1, and s_0, are used to select the operations of the ALU. With these three select lines, the ALU circuit can implement up to eight different operations. Suppose that the operations that we want to implement in our ALU are as defined in Figure 4.18(a). The x_i column shows the values that the LE must generate for the different operations. The y_i column shows the values that the AE must generate. The c_0 column shows the carry signals that the CE must generate.

For example, for the pass-through operation, the value of a_i is passed through without any modifications to x_i. For the AND operation, x_i gets the result of a_i AND b_i. As mentioned before, both y_i and c_0 are set to 0 for all of the logical operations, because we do not want the FA to change the results. The FA is used only to pass the results from the LE straight through to the output, F. For the subtraction operation, instead of subtracting B, we want to add $-B$. Changing B to $-B$ in two's complement format requires flipping the bits of B and then adding a 1. Thus, y_i gets the inverse of b_i, and the 1 is added through the carry-in, c_0. To increment A, we set y_i to all 0s, and add the 1 through the carry-in, c_0. To decrement A, we add a -1 instead. Negative one in two's complement format is a bit string with all 1s. Hence, we set y_i to all 1s and the carry-in c_0 to 0. For all the arithmetic operations, we need the first operand, A, unchanged for the FA. Thus, x_i gets the value of a_i for all arithmetic operations.

Figures 4.18(b), (c), and (d) show the truth tables for the LE, AE, and CE, respectively. The LE circuit is derived from the x_i column of Figure 4.18(b); the AE circuit is derived from the y_i column of Figure 4.18(c); and the CE circuit is derived from the c_0 column of Figure 4.18(d). Notice that x_i is dependent on five variables, s_2, s_1, s_0, a_i, and b_i; whereas, y_i is dependent on only four variables, s_2, s_1, s_0, and b_i; and c_0 is dependent

s_2	s_1	s_0	Operation Name	Operation	x_i (LE)	y_i (AE)	c_0 (CE)
0	0	0	Pass	Pass A to output	a_i	0	0
0	0	1	AND	A AND B	a_i AND b_i	0	0
0	1	0	OR	A OR B	a_i OR b_i	0	0
0	1	1	NOT	A'	a_i'	0	0
1	0	0	Addition	$A + B$	a_i	b_i	0
1	0	1	Subtraction	$A - B$	a_i	b_i'	1
1	1	0	Increment	$A + 1$	a_i	0	1
1	1	1	Decrement	$A - 1$	a_i	1	0

(a)

s_2	s_1	s_0	x_i
0	0	0	a_i
0	0	1	$a_i b_i$
0	1	0	$a_i + b_i$
0	1	1	a_i'
1	×	×	a_i

(b)

s_2	s_1	s_0	b_i	y_i
0	×	×	×	0
1	0	0	0	0
1	0	0	1	1
1	0	1	0	1
1	0	1	1	0
1	1	0	0	0
1	1	0	1	0
1	1	1	0	1
1	1	1	1	1

(c)

s_2	s_1	s_0	c_0
0	×	×	0
1	0	0	0
1	0	1	1
1	1	0	1
1	1	1	0

(d)

FIGURE 4.18 ALU operations: (a) function table; (b) LE truth table; (c) AE truth table; (d) CE truth table.

on only the three select lines, s_2, s_1, and s_0. The K-maps, equations, and schematics for these three circuits are shown in Figure 4.19.

The *Unsigned_Overflow* signal must be XORed correctly with the select lines similar to the adder-subtractor combination circuit. In the operational table given in Figure 4.18(a), there are three select lines, s_2, s_1, and s_0. We see that the select lines for the addition and subtraction operations differ only in the s_0 bit. Thus the *Unsigned_Overflow* signal is c_4 XOR s_0. This, of course, will be different if you assign different operations to the select lines.

The *Signed_Overflow* signal is obtained exactly like in the adder-subtractor combination where the C_{out} bit is XORed with the c_3 bit.

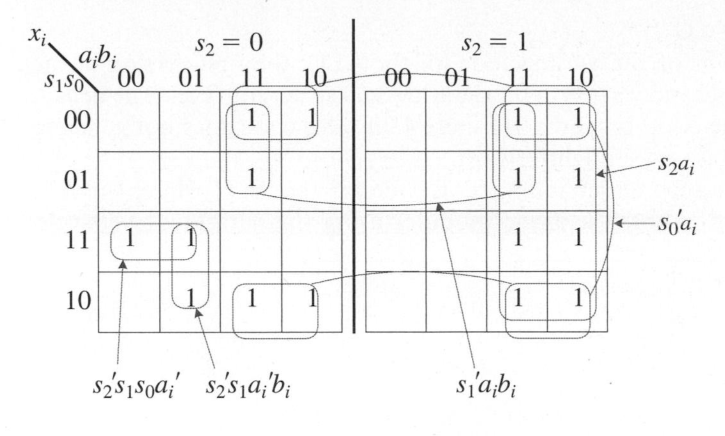

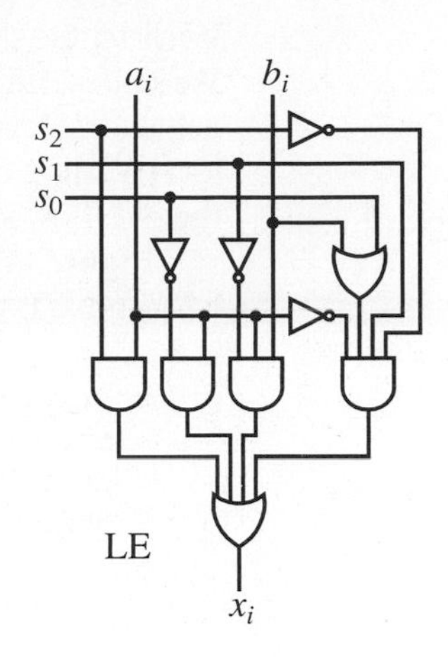

$$x_i = s_2a_i + s_0'a_i + s_1'a_ib_i + s_2's_1a_i'b_i + s_2's_1s_0a_1'$$
$$= s_2a_i + s_0'a_i + s_1'a_ib_i + s_2's_1a_i'(b_i + s_0)$$

(a)

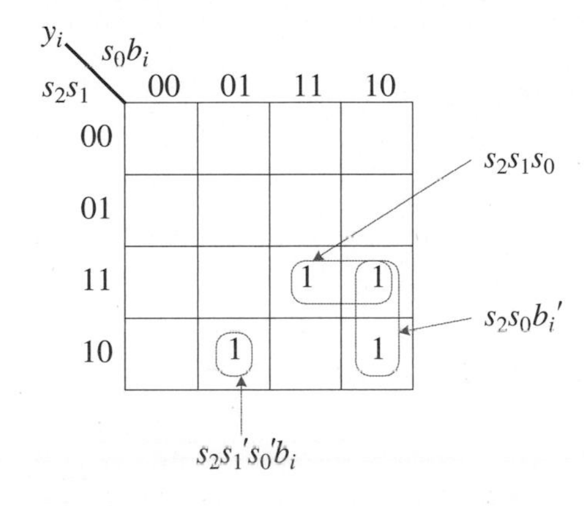

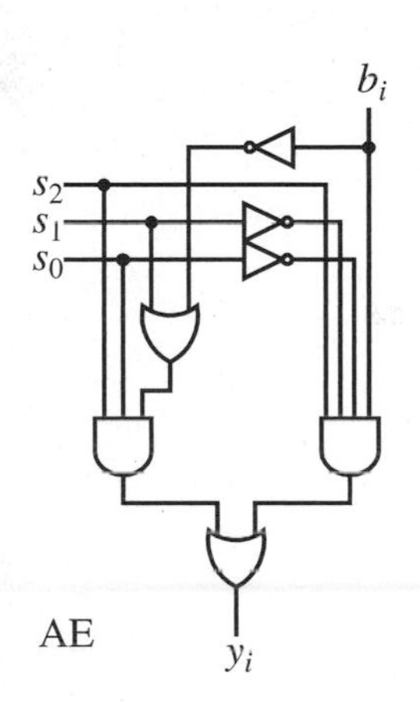

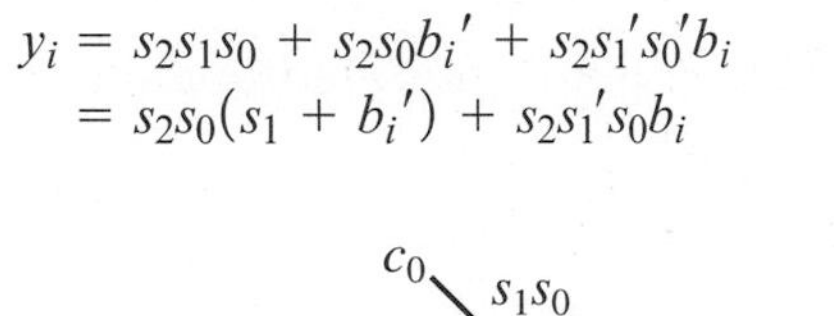

$$y_i = s_2s_1s_0 + s_2s_0b_i' + s_2s_1's_0'b_i$$
$$= s_2s_0(s_1 + b_i') + s_2s_1's_0b_i$$

(b)

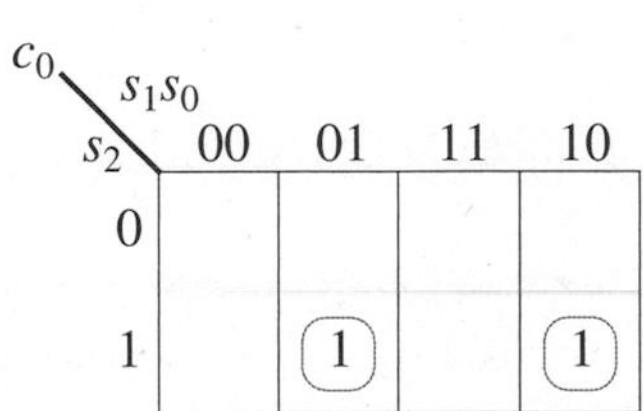

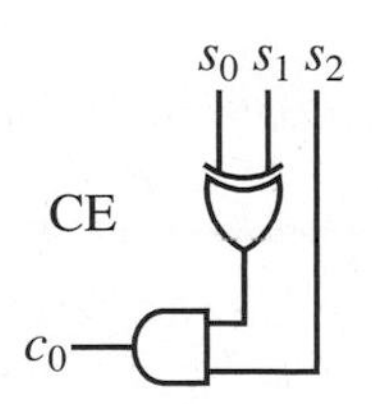

$$c_0 = s_2s_1's_0 + s_2s_1s_0'$$
$$= s_2(s_1 \oplus s_0)$$

(c)

FIGURE 4.19 K-maps, equations, and schematics for: (a) LE; (b) AE; and (c) CE.

Verilog Code for an ALU

We will write two versions of the Verilog code for the ALU: the first version at the behavioral level and the second version at the dataflow and structural level. The behavioral Verilog code for the ALU is shown in Figure 4.20. This code does not generate the overflow signal, so it is very straightforward.

Figure 4.21 lists the second version Verilog code for the ALU. Here, the AE, LE, and CE are defined in three separate modules using the dataflow level style.

```
module alu (
 input [2:0] S,
 input [n-1:0] A, B,
 output reg [n-1:0] F
);
 parameter n = 4;

 always @ (S or A or B) begin
  case (S)
  0: F = A;
  1: F = A & B;
  2: F = A | B;
  3: F = ~A;     // bitwise NOT
  4: F = A + B;
  5: F = A - B;
  6: F = A + 1;
  7: F = A - 1;
  endcase
 end
endmodule
```

FIGURE 4.20 Behavioral Verilog code for a 4-bit ALU.

```
module LE (
 input [2:0] s,
 input ai, bi,
 output xi
);
 // using the equation derived for the LE
 assign xi = (s[2] & ai) | (~s[0] & ai) | (~s[1] & ai & bi) |
    (~s[2] & s[1] & ~ai & (bi | s[0]));
endmodule

module AE (
 input [2:0] s,
 input bi,
 output yi
);
 // using the equation derived for the AE
 assign yi = (s[2] & s[0] & (s[1] | ~bi)) | (s[2] & ~s[1]
    & ~s[0] & bi);
endmodule
```

FIGURE 4.21 Dataflow and structural Verilog code for a 4-bit ALU.
(continued on next page)

```
module CE (
 input [2:0] s,
 output c0
);
 // using the equation derived for the CE
 assign c0 = (s[2] & (s[1] ^ s[0]));
endmodule
module FA (
 input ci, xi, yi,
 output ci1, fi
);
 // using the equations derived for the FA
 // bitwise &=AND; |=OR; ^=XOR
 assign ci1 = (xi & yi) | (ci & (xi ^ yi));
 assign fi = xi ^ yi ^ ci;
endmodule
module bitslice (
 input [2:0] s,
 input ai, bi,
 input ci,
 output ci1, fi
);
 wire xi, yi;
 // each bit slice consists of the LE, AE and FA
 LE U2(s, ai, bi, xi);
 AE U1(s, bi, yi);
 FA U0(ci, xi, yi, ci1, fi);
endmodule
module alu (
 input [2:0] S,
 input [n-1:0] A, B,
 output [n-1:0] F,
 output Unsigned_Overflow, Signed_Overflow
);
 parameter n = 4;

 wire [n:0] C;
 // only correct for this one
 assign Unsigned_Overflow = C[4] ^ S[0];
 assign Signed_Overflow = C[4] ^ C[3];

 // top level: connect the four bit slices and the CE together
 bitslice U3(S, A[3], B[3], C[3], C[4], F[3]);
 bitslice U2(S, A[2], B[2], C[2], C[3], F[2]);
 bitslice U1(S, A[1], B[1], C[1], C[2], F[1]);
 bitslice U0(S, A[0], B[0], C[0], C[1], F[0]);
 CE U4(S, C[0]);
endmodule
```

FIGURE 4.21 Dataflow and structural Verilog code for a 4-bit ALU.

The equations derived earlier for these three modules are used in the respective `assign` statements. The structural level is used for each bit slice to connect the AE, LE, and FA together. Finally, four bit slices and the CE are connected together using the structural level. The *Unsigned_Overflow* signal is assigned the value of C[4] XOR S[0]. This is only correct for this given set of operations for the ALU as shown in Figure 4.18(a) because the add and subtract operations differ only by the S[0] bit.

VHDL Code for an ALU

The behavioral VHDL code for the ALU is shown in Figure 4.22, and a sample simulation trace for all the operations using the two inputs 5 and 3 is shown in Figure 4.23.

```
LIBRARY IEEE;
USE IEEE.STD_LOGIC_1164.ALL;
USE IEEE.STD_LOGIC_UNSIGNED.ALL; -- needed for doing + and -
ENTITY alu IS PORT (
 S: IN STD_LOGIC_VECTOR(2 DOWNTO 0); -- select for operations
 A, B: IN STD_LOGIC_VECTOR(3 DOWNTO 0); -- input operands
 F: OUT STD_LOGIC_VECTOR(3 DOWNTO 0)); -- output
END alu;
ARCHITECTURE Behavioral OF alu IS
BEGIN
 PROCESS(S, A, B)
 BEGIN
   CASE S IS
   WHEN "000" =>  -- pass A through
    F <= A;
   WHEN "001" =>  -- AND
    F <= A AND B;
   WHEN "010" =>  -- OR
    F <= A OR B;
   WHEN "011" =>  -- NOT A
    F <= NOT A;
   WHEN "100" =>  -- add
    F <= A + B;
   WHEN "101" =>  -- subtract
    F <= A - B;
   WHEN "110" =>  -- increment
    F <= A + 1;
   WHEN OTHERS => -- decrement
    F <= A - 1;
   END CASE;
 END PROCESS;
END Behavioral;
```

FIGURE 4.22 Behavioral VHDL code for a 4-bit ALU.

Pass *A* AND OR NOT *A* Add Subtract Increment Decrement

Name: 200.0ns 400.0ns 600.0ns 800.0

S 0 1 2 3 4 5 6 7

A 5

B 3

F 5 1 7 A 8 2 6 4

FIGURE 4.23 Sample simulation trace with the two input operands, 5 and 3, for all of the eight operations.

4.7 Decoder

A **decoder**, also known as a **demultiplexer**, asserts one out of n output lines, depending on the value of an m-bit binary input data. In general, an m-to-n decoder has m input lines, $A_{m-1}, \ldots, A_0$, and n output lines, $Y_{n-1}, \ldots, Y_0$, where $n = 2^m$. In addition, it has an enable line, E, for enabling the decoder. When the decoder is disabled with E set to 0, all of the output lines are de-asserted. When the decoder is enabled, then the output line whose index is equal to the value of the input binary data is asserted. For example, for a 3-to-8 decoder, if the input address is 101, then the output line Y_5 is asserted (set to 1 for active-high), while the rest of the output lines are de-asserted (set to 0 for active-high).

A decoder is used in a system having multiple components, and we want only one component to be selected or enabled at any one time. For example, in a large memory system with multiple memory chips, only one memory chip is enabled at a time. One output line from the decoder is connected to the enable input on each memory chip. Thus, an address presented to the decoder will enable that corresponding memory chip. The truth table, circuit, and logic symbol for a 3-to-8 decoder are shown in Figure 4.24.

E	A_2	A_1	A_0	Y_7	Y_6	Y_5	Y_4	Y_3	Y_2	Y_1	Y_0
0	x	x	x	0	0	0	0	0	0	0	0
1	0	0	0	0	0	0	0	0	0	0	1
1	0	0	1	0	0	0	0	0	0	1	0
1	0	1	0	0	0	0	0	0	1	0	0
1	0	1	1	0	0	0	0	1	0	0	0
1	1	0	0	0	0	0	1	0	0	0	0
1	1	0	1	0	0	1	0	0	0	0	0
1	1	1	0	0	1	0	0	0	0	0	0
1	1	1	1	1	0	0	0	0	0	0	0

(a)

FIGURE 4.24 A 3-to-8 decoder: (a) truth table; (b) circuit; (c) logic symbol.
(continued on next page)

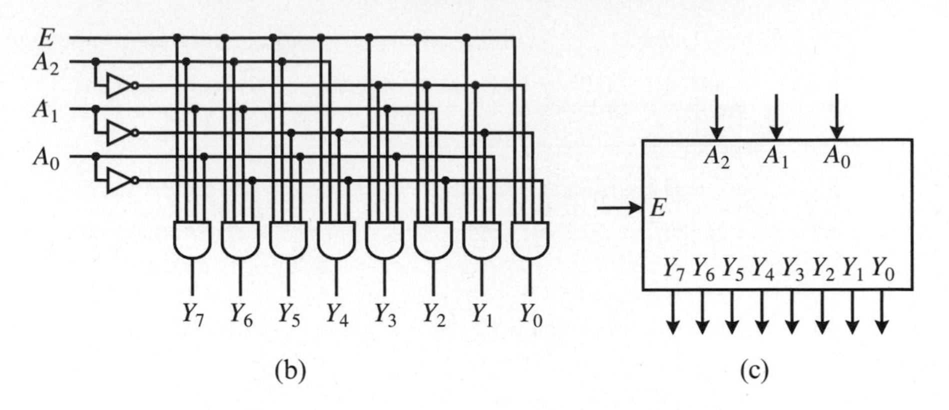

FIGURE 4.24 A 3-to-8 decoder: (a) truth table; (b) circuit; (c) logic symbol.

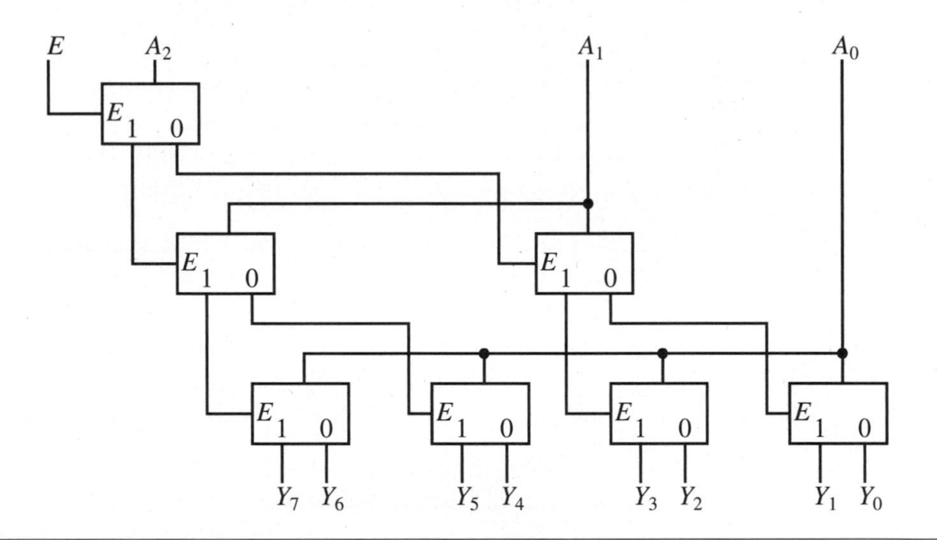

FIGURE 4.25 A 3-to-8 decoder implemented with seven 1-to-2 decoders.

A larger size decoder can be implemented using several smaller decoders. For example, Figure 4.25 uses seven 1-to-2 decoders to implement a 3-to-8 decoder. The correct operation of this circuit is left as an exercise for the reader.

HDL Code for a Decoder

The behavioral Verilog and VHDL codes for the 3-to-8 decoder are shown in Figures 4.26 and 4.27, respectively.

```
module Decoder (
 input E,
 input [2:0] A,
 output reg [7:0] Y
);
 always @ (E or A) begin
  if (E) begin
   case (A)
   0: Y = 8'b00000001;
   1: Y = 8'b00000010;
   2: Y = 8'b00000100;
   3: Y = 8'b00001000;
   4: Y = 8'b00010000;
   5: Y = 8'b00100000;
   6: Y = 8'b01000000;
   7: Y = 8'b10000000;
   endcase
  end else begin
   Y = 0;
  end
 end
endmodule
```

FIGURE 4.26 Behavioral Verilog code for a 3-to-8 decoder.

```
LIBRARY IEEE;
USE IEEE.STD_LOGIC_1164.ALL;
ENTITY Decoder IS PORT(
 E: IN STD_LOGIC;          -- enable
 A: IN STD_LOGIC_VECTOR(2 DOWNTO 0);
 Y: OUT STD_LOGIC_VECTOR(7 DOWNTO 0));
END Decoder;
ARCHITECTURE Behavioral OF Decoder IS
BEGIN
 PROCESS (E, A)
 BEGIN
  IF (E = '0') THEN             -- disabled
   Y <= (OTHERS => '0');        -- 8-bit vector of 0
  ELSE
   CASE A IS                    -- enabled
    WHEN "000" => Y <= "00000001";
    WHEN "001" => Y <= "00000010";
    WHEN "010" => Y <= "00000100";
    WHEN "011" => Y <= "00001000";
    WHEN "100" => Y <= "00010000";
```

FIGURE 4.27 Behavioral VHDL code for a 3-to-8 decoder. *(continued on next page)*

```
        WHEN "101" => Y <= "00100000";
        WHEN "110" => Y <= "01000000";
        WHEN "111" => Y <= "10000000";
        WHEN OTHERS => NULL;
      END CASE;
    END IF;
  END PROCESS;
END Behavioral;
```

FIGURE 4.27 Behavioral VHDL code for a 3-to-8 decoder.

4.8 Tri-State Buffer

A **tri-state** buffer, as the name suggests, has three states: 0, 1, and a third state denoted by Z. The value Z represents a high-impedance state, which, for all practical purposes, acts like a switch that is opened or a wire that is cut. Tri-state buffers are used to connect several devices to the same bus. A **bus** is one or more wires for transferring signals. If two or more devices are connected directly to a bus without using tri-state buffers, signals will get corrupted on the bus because the devices are always outputting either a 0 or a 1. However, with a tri-state buffer in between, devices that are not using the bus can disable the tri-state buffer so that it acts as if those devices are physically disconnected from the bus. At any one time, only one active device will have its tri-state buffers enabled, and thus, use the bus.

The truth table and symbol for the tri-state buffer are shown in Figures 4.28(a) and (b). The active-high enable line E turns the buffer on or off. When E is de-asserted with a 0, the tri-state buffer is disabled, and the output y is in its high-impedance Z state. When E is asserted with a 1, the buffer is enabled, and the output y follows the input d.

A circuit consisting of only logic gates cannot produce the high-impedance state required by the tri-state buffer, since logic gates can output only a 0 or a 1. To provide the high-impedance state, the tri-state buffer circuit uses two transistors in conjunction with logic gates, as shown in Figure 4.28(c). The top PMOS transistor is enabled with a

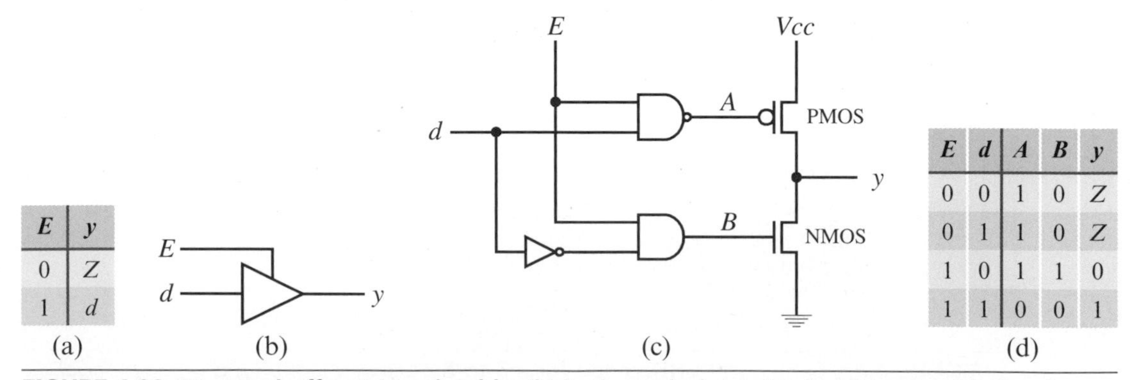

E	y
0	Z
1	d

E	d	A	B	y
0	0	1	0	Z
0	1	1	0	Z
1	0	1	1	0
1	1	0	0	1

FIGURE 4.28 Tri-state buffer: (a) truth table; (b) logic symbol; (c) circuit; (d) truth table for the control portion of the tri-state buffer circuit.

0 at the node labeled A, and when it is enabled, a 1 signal from Vcc passes down through the transistor to y. The bottom NMOS transistor is enabled with a 1 at the node labeled B, and when it is enabled, a 0 signal from ground passes up through the transistor to y. When the two transistors are disabled (with $A = 1$ and $B = 0$), they both will output a high impedance Z value; so y will have a Z value. Refer to the online chapter on Implementation Technologies for a detailed discussion on how transistors work.

Having the two transistors, we need a circuit that will control these two transistors so that together they realize the tri-state buffer function. The truth table for this control circuit is shown in Figure 4.28(d). The truth table is derived as follows: When $E = 0$ (it does not matter what the input d is), we want both transistors to be disabled so that the output y has the Z value. The PMOS transistor is disabled when the input $A = 1$; whereas, the NMOS transistor is disabled when the input $B = 0$. When $E = 1$ and $d = 0$, we want the output y to be a 0. To get a 0 on y, we need to enable the bottom NMOS transistor (setting B to 1) and disable the top PMOS transistor (setting A to 1) so that a 0 will pass through the NMOS transistor to y. To get a 1 on y for when $E = 1$ and $d = 1$, we need to do the reverse by enabling the top PMOS transistor (setting A to 0) and disabling the bottom NMOS transistor (setting B to 0).

From the truth table, we obtain the resulting circuit shown in Figure 4.28(c). When $E = 0$, the output of the NAND gate is a 1 regardless of what the other input is, and so the top PMOS transistor is turned off. Similarly, the output of the AND gate is a 0, and so the bottom NMOS transistor also is turned off. Thus, when $E = 0$, both transistors are off, so the output y is in the Z state.

When $E = 1$, the outputs of both the NAND and AND gates are equal to d'. So if $d = 0$, the output of the two gates are both 1, so the bottom transistor is turned on while the top transistor is turned off. Thus, y will have the value 0, which is equal to d. On the other hand, if $d = 1$, the top transistor is turned on while the bottom transistor is turned off, and y will have the value 1, which again is equal to d.

HDL Code for a Tri-state Buffer

The Verilog code for a 4-bit wide tri-state buffer is shown in Figure 4.29. A conditional `assign` statement is used to generate the output signal. If the condition `(E)` is true, that is, $E = 1$, then the first value D is assigned to Y, otherwise, the second value `{n{1'bz}}` is assigned to Y. The term `{n{1'bz}}` means n bits of the z value. The `parameter` statement declares n and initializes it to 4.

```
module TriState_Buffer
#(parameter n = 4)    // allow n to be changed externally
(
 input E,
 input [n-1:0] D,
 output [n-1:0] Y
);
 assign Y = (E) ? D : {n{1'bz}};
endmodule
```

FIGURE 4.29 Verilog code for a 4-bit wide tri-state buffer.

```
LIBRARY IEEE;
USE IEEE.STD_LOGIC_1164.ALL;
ENTITY TriState_Buffer IS
GENERIC (n: INTEGER := 4);
PORT (
 E: IN STD_LOGIC;
 D: IN STD_LOGIC_VECTOR(n-1 DOWNTO 0);
 Y: OUT STD_LOGIC_VECTOR(n-1 DOWNTO 0));
END TriState_Buffer;
ARCHITECTURE Behavioral OF TriState_Buffer IS
BEGIN
 Y <= D WHEN (E = '1') ELSE (OTHERS => 'Z');
END Behavioral;
```

FIGURE 4.30 VHDL code for a 4-bit wide tri-state buffer.

The behavioral VHDL code for a 4-bit wide tri-state buffer is shown in Figure 4.30. A conditional assignment statement is used to: generate the output signal. If the condition (E = '1') is true then the value D is assigned to Y, otherwise, the value `(OTHERS => 'Z')` in the `ELSE` part is assigned to Y. The clause `(OTHERS => 'Z')` means a bit vector with all Z values. The `GENERIC` statement declares n and initializes it to 4.

4.9 Comparator

Quite often, we need to compare two values for their arithmetic relationship (equal, greater, less than, etc.). A **comparator** is a circuit that compares two binary values and indicates whether the relationship is true or false. To compare whether a value is equal or not equal to a constant value, a simple AND gate can be used. For example, to compare a 4-bit variable x with the constant 3, the circuit in Figure 4.31(a) can be used. The AND gate outputs a 1 when the input is equal to the value 3. Since 3 is 0011 in binary, therefore, x_3 and x_2 must be inverted. A simple NOR gate also can be used to test for whether a variable x is equal to 0. The NOR gate outputs a 1 when all of its inputs are 0s.

The XOR and XNOR gates can be used for comparing inequality and equality, respectively, between two values. The XOR gate outputs a 1 when its two input values are different. Hence, we can use one XOR gate for comparing each bit pair of the two operands. A 4-bit inequality comparator is shown in Figure 4.31(b). Four XOR gates are used, with each one comparing the same bit from the two operands. The outputs of the XOR gates are ORed together so that if any bit pair is different, then the two operands are different, and the resulting output is a 1. Similarly, an equality comparator can be constructed using XNOR gates instead, since the XNOR gate outputs a 1 when its two input values are the same.

To compare the greater-than or less-than relationships, we can construct a truth table and build the circuit from it. For example, to compare whether a 4-bit value X is less than five, we get the truth table, equation, and circuit shown in Figure 4.31(c).

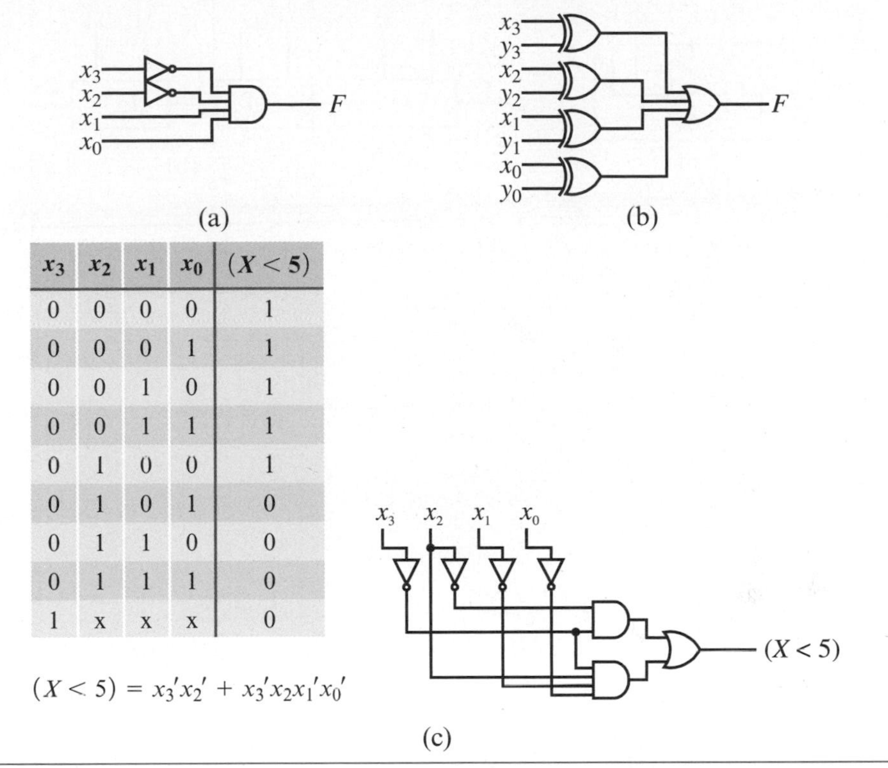

x_3	x_2	x_1	x_0	$(X < 5)$
0	0	0	0	1
0	0	0	1	1
0	0	1	0	1
0	0	1	1	1
0	1	0	0	1
0	1	0	1	0
0	1	1	0	0
0	1	1	1	0
1	x	x	x	0

$$(X < 5) = x_3'x_2' + x_3'x_2x_1'x_0'$$

FIGURE 4.31 Simple 4-bit comparators for: (a) X=3; (b) X≠Y; (c) X<5.

Instead of constructing a comparator for a fixed number of bits for the input values, we often prefer to construct an **iterative circuit** by constructing a 1-bit slice comparator and then daisy-chaining n of them together to make an n-bit comparator. The 1-bit slice comparator will have (in addition to the two input operand bits, x_i and y_i) a p_i bit that keeps track of whether all the previous bit pairs compared so far are true or false for that particular relationship. The circuit outputs a 1 if $p_i = 1$, and the relationship is true for the current bit pair, x_i and y_i. Figure 4.32(a) shows a 1-bit slice comparator for the equal relationship. If the current bit pair, x_i and y_i, is equal, the XNOR gate will output a 1. Hence, $p_{i+1} = 1$ if the current bit pair is equal and the previous bit pair, p_i, is a 1. To obtain a 4-bit iterative equality comparator, we connect four 1-bit equality comparators in series, as shown in Figure 4.32(b). The initial p_0 bit must be set to a 1. Thus, if all four bit pairs are equal, then the last bit, p_4, will be a 1; otherwise, p_4 will be a 0.

Building an iterative comparator circuit for the greater-than relationship $X > Y$ is slightly more difficult. The 1-bit slice comparator circuit for the condition $x_i > y_i$ is constructed as follows. In addition to the two operand input bits, x_i and y_i, there are also two status input bits, g_{in} and e_{in}. Here, g_{in} is a 1 if the condition $x_i > y_i$ is true for the previous bit slice; otherwise, g_{in} is a 0. Furthermore, e_{in} is a 1 if the condition $x_i = y_i$ is true; otherwise, e_{in} is a 0. The circuit also has two status output bits, g_{out} and e_{out}, having the same meaning as the g_{in} and e_{in} signals. These two input and two output status bits allow the bit slices to be daisy-chained together. Following the above description of the 1-bit slice circuit, we obtain the truth table shown in Figure 4.33(a). The equations for e_{out} and g_{out} are shown in Figure 4.33(b), and the 1-bit slice circuit in Figure 4.33(c).

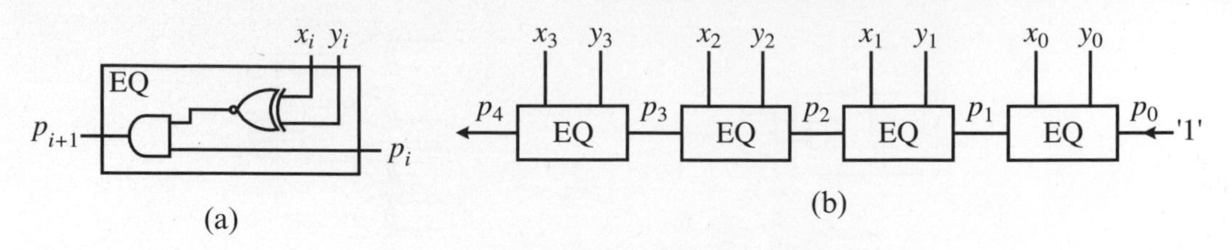

FIGURE 4.32 Iterative comparators: (a) 1-bit slice for $x_i = y_i$; (b) 4-bit $X = Y$.

g_{in}	e_{in}	x_i	y_i	Meaning	g_{out}	e_{out}
0	0	×	×	<	0	0
0	1	0	0	=	0	1
0	1	0	1	<	0	0
0	1	1	0	>	1	0
0	1	1	1	=	0	1
1	0	×	×	>	1	0
1	1	×	×	Invalid	1	1

(a)

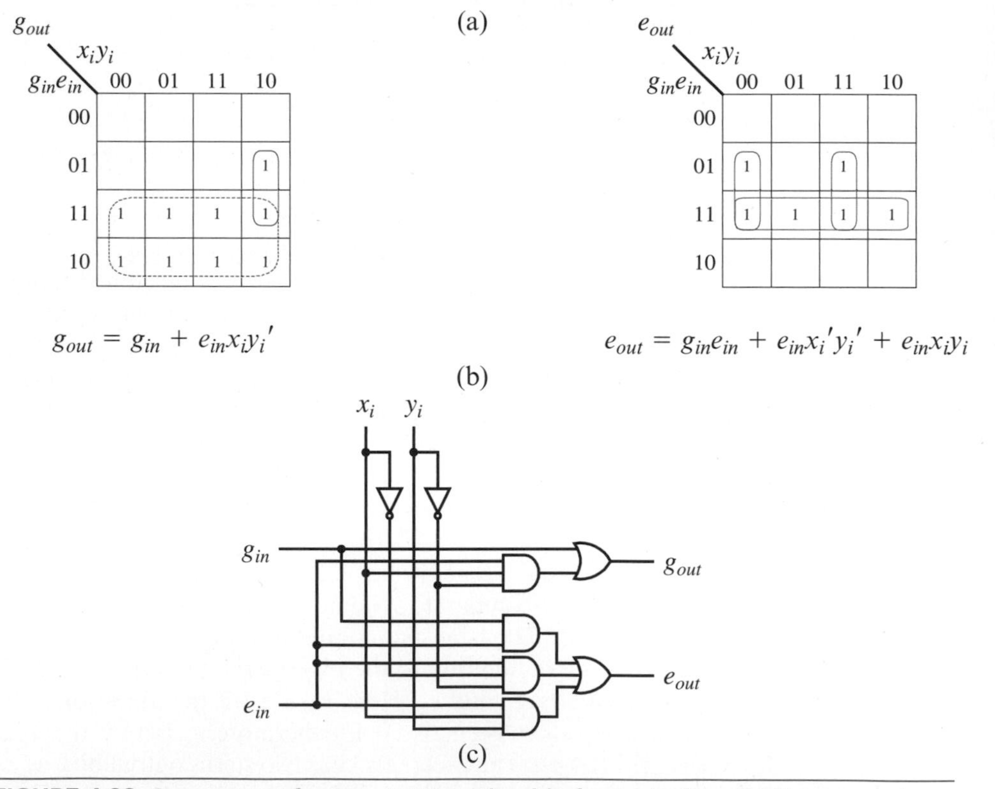

FIGURE 4.33 Comparator for $X > Y$: (a) truth table for 1-bit slice; (b) K-maps and equations for g_{out} and e_{out}; (c) circuit for 1-bit slice; (d) 4-bit $X > Y$ comparator circuit; (e) operational table. *(continued on next page)*

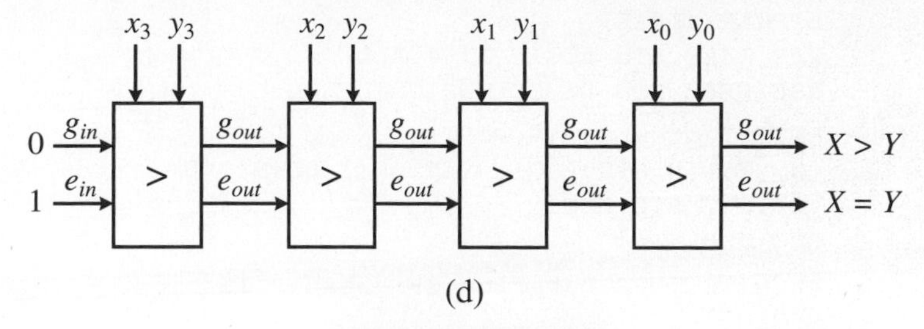

(d)

Condition	e_{out}	g_{out}
Invalid	1	1
$X = Y$	1	0
$X > Y$	0	1
$X < Y$	0	0

(e)

FIGURE 4.33 Comparator for $X > Y$: (a) truth table for 1-bit slice; (b) K-maps and equations for g_{out} and e_{out}; (c) circuit for 1-bit slice; (d) 4-bit $X > Y$ comparator circuit; (e) operational table.

```
module greater (
 input [3:0] X,Y,
 output G
);
 assign G = (X > Y) ? 1'b1 : 1'b0;
endmodule
```

FIGURE 4.34 Verilog code for a 4-bit greater-than comparator.

In order for the bit slices to operate correctly, we need to perform the comparisons from the most significant bit to the least significant bit. The complete 4-bit iterative comparator circuit for the condition $X > Y$ is shown in Figure 4.33(d). The initial values for g_{in} and e_{in} must be set to $g_{in} = 0$ and $e_{in} = 1$.

If $X = Y$, then the last e_{out} is a 1; otherwise, e_{out} is a 0. If the last e_{out} is a 0, then the last g_{out} can be either a 1 or a 0. If $X > Y$ then g_{out} is a 1; otherwise, g_{out} is a 0. Notice that both e_{out} and g_{out} cannot both be 1s. The operation of this comparator circuit is summarized in Figure 4.33(e).

HDL Code for a 4-bit Greater-than Comparator

Figure 4.34 shows the Verilog code for a 4-bit greater-than comparator. The conditional `assign` statement is used. The syntax `1'b1` means one 1 bit, and 1'b0 means one 0 bit.

Figure 4.35 shows the VHDL code for a 4-bit greater-than comparator. The conditional signal assignment statement is used. The value '1' is assigned to G when the condition $(X > Y)$ is true, otherwise the value '0' is assigned to G.

```
LIBRARY IEEE;
USE IEEE.STD_LOGIC_1164.ALL;
USE IEEE.STD_LOGIC_UNSIGNED.ALL;

ENTITY greater IS PORT (
 X, Y: IN STD_LOGIC_VECTOR(3 DOWNTO 0);
 G: OUT STD_LOGIC);
END VHDL;

ARCHITECTURE Dataflow OF greater IS
BEGIN
  G <= '1' WHEN (X > Y) ELSE '0';
END Dataflow;
```

FIGURE 4.35 VHDL code for a 4-bit greater-than comparator.

4.10 **Shifter**

The **shifter** is used for shifting bits in a binary string one position either to the left or to the right. The operations for the shifter are referred to either as **shifting** or **rotating**, depending on how the end bits are shifted in or out. For a shift operation, the bit at the end is discarded and does not wrap around. For a rotate operation, the bit at the end wraps back around to the other end. Figure 4.36 shows six different shift and rotate operations.

For example, for the "Shift left with 0" operation, all of the bits are shifted one position to the left. The original leftmost bit is shifted out (i.e., discarded) and the rightmost bit is filled with a 0. For the "Rotate left" operation, all of the bits are shifted one position to the left. However, instead of discarding the leftmost bit, it is shifted in as the rightmost bit (i.e., it rotates around).

For each bit position, a multiplexer is used to move a bit from either the left or the right to the current bit position. The size of the multiplexer will determine the number of operations that can be implemented. For example, we can use a 4-to-1 multiplexer to implement the four operations, as specified by the operation table shown in Figure 4.37(a). Two select lines, s_1 and s_0, are needed to select between the four different operations. For a 4-bit operand, we need to use four 4-to-1 multiplexers, as shown in Figure 4.37(b). How the inputs to the multiplexers are connected will depend on the given operations.

In this example, when $s_1 = s_0 = 0$, we want to pass the bit straight through without shifting (i.e., we want the value from in_i to pass to out_i). Given $s_1 = s_0 = 0$, d_0 of the multiplexer is selected, hence, in_i is connected to d_0 of MUX$_i$, which outputs to out_i. For $s_1 = 0$ and $s_0 = 1$, we want to shift left (i.e., we want the value from in_i to pass to out_{i+1}). With $s_1 = 0$ and $s_0 = 1$, d_1 of the multiplexer is selected, hence, in_i is connected to d_1 of MUX$_{i+1}$, which outputs to out_{i+1}. For this selection, we also want to shift in a 0 bit, so d_1 of MUX$_0$ is connected directly to a 0. The two remaining operations are connected in a similar manner.

HDL Code for a Shifter

The behavioral Verilog code for an 8-bit shifter having the functions as defined in Figure 4.37(a) is shown in Figure 4.38.

Operation	Comment	Example
Shift left with 0	Shift bits one position to the left. The leftmost bit is discarded and the rightmost bit is filled with a 0.	10110100 X01101000←
Shift left with 1	Same as above, except that the rightmost bit is filled with a 1.	10110100 X01101001←
Shift right with 0	Shift bits one position to the right. The rightmost bit is discarded and the leftmost bit is filled with a 0.	10110100 →01011010X
Shift right with 1	Same as above, except that the leftmost bit is filled with a 1.	10110100 →11011010X
Rotate left	Shift bits one position to the left. The leftmost bit is moved to the rightmost bit position.	10110100 01101001
Rotate right	Shift bits one position to the right. The rightmost bit is moved to the leftmost bit position.	10110100 01011010

FIGURE 4.36 Shifter and rotator operations.

s_1	s_0	**Operation**
0	0	Pass through
0	1	Shift left and fill with 0
1	0	Shift right and fill with 0
1	1	Rotate right

(a)

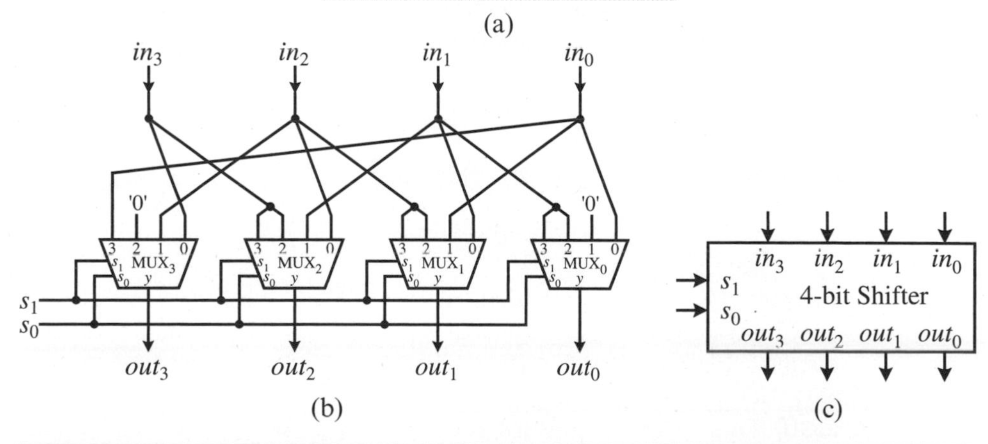

FIGURE 4.37 A 4-bit shifter: (a) operation table; (b) circuit; (c) logic symbol.

```
module shifter (
 input [1:0] S,
 input [7:0] data_in,
 output reg [7:0] data_out
);
 always @ (S or data_in) begin
  case (S)
  0: data_out = data_in;      // pass through
  1: data_out = data_in << 1;    // shift left
  2: data_out = data_in >> 1;    // shift right
  3: data_out = {data_in[0], data_in[7:1]}; // rotate right
  endcase
 end

endmodule
```

FIGURE 4.38 Behavioral Verilog code for an 8-bit shifter having the operations as defined in Figure 4.37(a).

```
LIBRARY IEEE;
USE IEEE.STD_LOGIC_1164.ALL;
USE IEEE.STD_LOGIC_UNSIGNED.ALL;
ENTITY shifter IS PORT (
 S: IN STD_LOGIC_VECTOR(1 DOWNTO 0); -- select for operations
 data_in: IN STD_LOGIC_VECTOR(7 DOWNTO 0); -- input
 data_out: OUT STD_LOGIC_VECTOR(7 DOWNTO 0)); -- output
END shifter;
ARCHITECTURE Behavioral OF shifter IS
BEGIN
 PROCESS(S, data_in)
 BEGIN
   CASE S IS
   WHEN "00" =>                           -- pass through
    data_out <= data_in;
   WHEN "01" =>                           -- shift left with 0
    data_out <= data_in(6 DOWNTO 0) & '0';
   WHEN "10" =>                           -- shift right with 0
    data_out <= '0' & data_in(7 DOWNTO 1);
   WHEN OTHERS =>                         -- rotate right
    data_out <= data_in(0) & data_in(7 DOWNTO 1);
   END CASE;
 END PROCESS;
END Behavioral;
```

FIGURE 4.39 Behavioral VHDL code for an 8-bit shifter having the operations as defined in Figure 4.37(a).

The behavioral VHDL code for an 8-bit shifter having the functions as defined in Figure 4.37(a) is shown in Figure 4.39.

4.11 Multiplier

In grade school, we were taught to multiply two numbers using a shift-and-add procedure. Regardless of whether the two numbers are in decimal or binary, we use the same shift-and-add procedure for multiplying them. In fact, multiplying with binary numbers is even easier, because you are always multiplying with either a 0 or a 1. Figure 4.40(a) shows the multiplication of two 4-bit unsigned binary numbers—the multiplicand M ($m_3m_2m_1m_0$) with the multiplier Q ($q_3q_2q_1q_0$)—to produce the resulting product P ($p_7p_6p_5p_4p_3p_2p_1p_0$). The bit width of the product P is always twice the bit width of the operands. Since we are working with binary numbers, we always will be multiplying either with a 1 or a 0, therefore, the intermediate products always will either be the same as the multiplicand (if the multiplier bit is a 1) or zero (if the multiplier bit is a 0).

We can derive a combinational multiplication circuit based on this shift-and-add procedure, as shown in Figure 4.40(b). Each intermediate product is obtained by ANDing the multiplicand M with one bit of the multiplier q_i. Since q_i is always a 1 or a 0, the output of the AND gates is always either m_i or 0. For example, bit zero of the first intermediate product is obtained by ANDing m_0 with q_0; bit one is obtained by ANDing m_1 with q_0; and so on. Hence, the four bits for the first intermediate product are m_3q_0, m_2q_0, m_1q_0, and m_0q_0; the four bits for the second intermediate product are m_3q_1, m_2q_1, m_1q_1, and m_0q_1; and so on.

Multiple adders are used to sum all of the intermediate products together to give the final product. Each intermediate product is shifted over to the correct bit position for the addition. For example, p_0 is just m_0q_0; p_1 is the sum of m_1q_0 and m_0q_1; p_2 is the sum of m_2q_0, m_1q_1, and m_0q_2; and so on. The four FAs (1-bit adders) in each row are connected, as in the ripple-carry adder with each carry-out signal connected to the carry-in of the next FA. The carry-out of the last FA is connected to the input of the last FA in the row below. The last carry-out from the last row of FAs is the value for p_7 of the final product. As in the ripple-carry adder, all of the initial carry-in, c_0, are set to a 0.

HDL Code for a Multiplier

The behavioral Verilog code for a 4-bit multiplier $M \times Q$ giving the 8-bit product P is shown in Figure 4.41.

Multiplicand (M)		1101
Multiplier (Q)		× 1011
		1101
Intermediate products	{	1101
		0000
		+ 1101
Product (P)		10001111

				m_3	m_2	m_1	m_0
			×	q_3	q_2	q_1	q_0
				m_3q_0	m_2q_0	m_1q_0	m_0q_0
			m_3q_1	m_2q_1	m_1q_1	m_0q_1	
		m_3q_2	m_2q_2	m_1q_2	m_0q_2		
+	m_3q_3	m_2q_3	m_1q_3	m_0q_3			
p_7	p_6	p_5	p_4	p_3	p_2	p_1	p_0

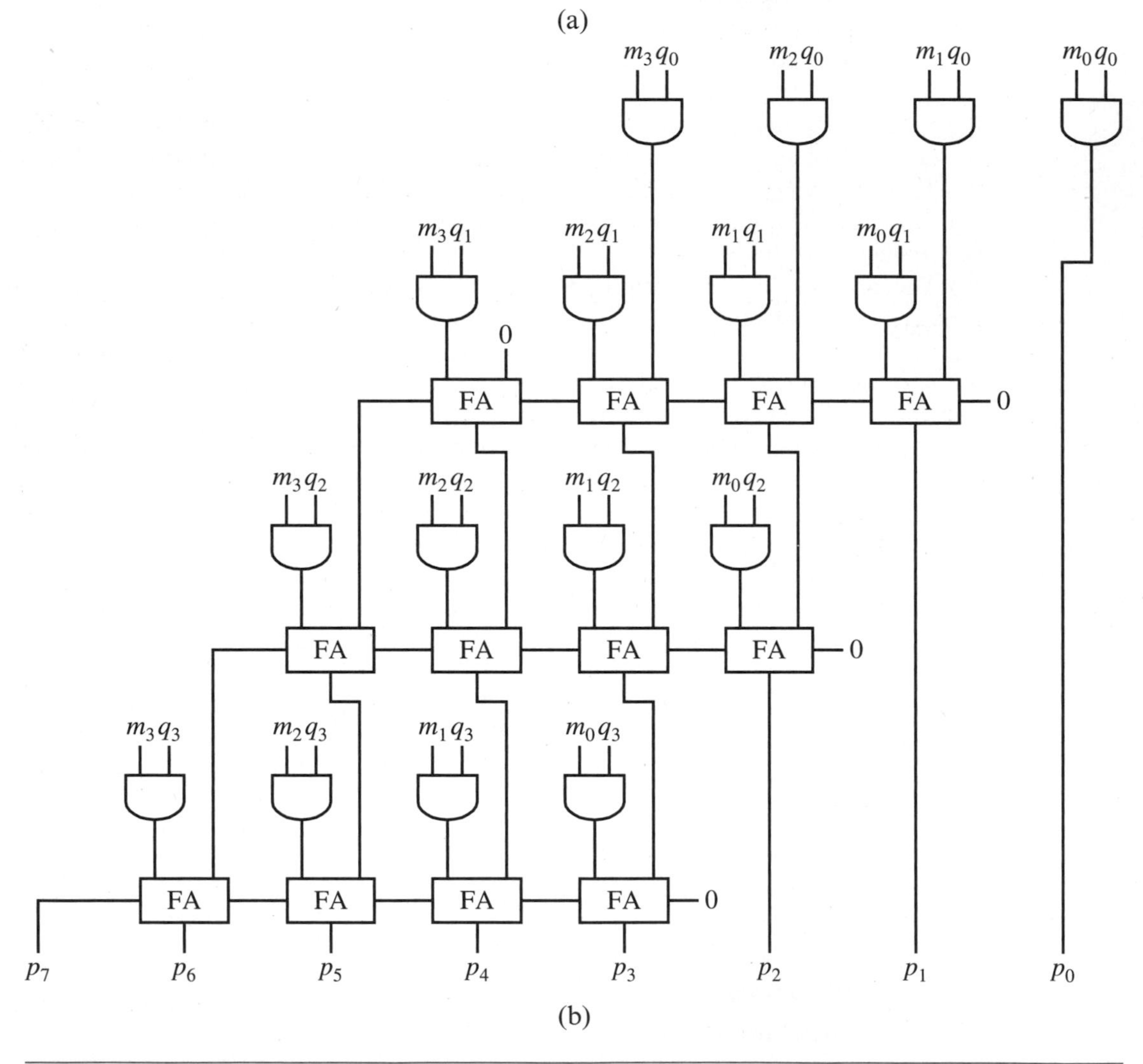

FIGURE 4.40 Multiplication: (a) method; (b) circuit.

```
module multiplier (
 input [3:0] M, Q,
 output [7:0] P
);
 assign P = M * Q;

endmodule
```

FIGURE 4.41 Behavioral Verilog code for a 4-bit multiplier.

```
LIBRARY IEEE;
USE IEEE.STD_LOGIC_1164.ALL;
USE IEEE.NUMERIC_STD.ALL; -- needed for UNSIGNED
ENTITY multiplier IS PORT (
 M, Q: IN STD_LOGIC_VECTOR(3 DOWNTO 0);
 P: OUT STD_LOGIC_VECTOR(7 DOWNTO 0));
END multiplier;
ARCHITECTURE Behavioral OF multiplier IS
BEGIN
 P <= STD_LOGIC_VECTOR(UNSIGNED(M) * UNSIGNED(Q));
END Behavioral;
```

FIGURE 4.42 Behavioral VHDL code for a 4-bit multiplier.

The behavioral VHDL code for a 4-bit multiplier $M \times Q$ giving the 8-bit product P is shown in Figure 4.42.

4.12 PROBLEMS

4.1. Draw the complete circuit for a 4-bit wide 4-to-1 multiplexer circuit using only AND, OR, and NOT gates. 4-bit wide means that each data input signal is four bits and the output Y is four bits.

4.2. Draw the circuit for a 16-to-1 multiplexer using only 4-to-1 multiplexers.

4.3. Draw the circuit for a 16-to-1 multiplexer using only 2-to-1 multiplexers.

4.4. Write the complete dataflow Verilog code for an 8-bit wide 4-to-1 multiplexer circuit.

4.5. Write the complete dataflow VHDL code for an 8-bit wide 4-to-1 multiplexer circuit.

4.6. Write the complete structural Verilog code for the FA circuit shown in Figure 4.6(c).

4.7. Write the complete structural VHDL code for the FA circuit shown in Figure 4.6(c).

4.8. For the subtractor circuit discussed in Section 4.4, the last overflow bit, b_n, is a 1 when the result of the subtraction is a negative number. This overflow bit is correct when the input operands are interpreted as unsigned numbers. Verify that this is true. Using four bits, show examples where this overflow bit is not correct when the input operands are interpreted as signed numbers.

4.9. The following shows eight different 4-bit binary calculations, and whether there is an overflow error if the binary numbers are interpreted as either signed or unsigned numbers.

	4-bit Calculations	Signed Interpretation	Range is −8 to 7 Overflow	Unsigned Interpretation	Range is 0 to 15 Overflow
a.	0011 + 0100 = 0111	3 + 4 = 7 ✔	0	3 + 4 = 7 ✔	0
b.	1111 + 1110 = 1101	−1 + (−2) = −3 ✔	0	15 + 14 = 13 ✘	1
c.	0111 + 0110 = 1101	7 + 6 = −3 ✘	1	7 + 6 = 13 ✔	0
d.	1000 + 1001 = 0001	−8 + (−7) = 1 ✘	1	8 + 9 = 1 ✘	1
e.	0111 − 0110 = 0001	7 − 6 = 1 ✔	0	7 − 6 = 1 ✔	0
f.	0000 − 0111 = 1001	0 − 7 = −7 ✔	0	0 − 7 = 9 ✘	1
g.	1000 − 0001 = 0111	(−8) − (1) = 7 ✘	1	8 − 1 = 7 ✔	0
h.	0001 − 1000 = 1001	1 − (−8) = −7 ✘	1	1 − 8 = 9 ✘	1

1) Zero extend the operands to five bits and perform the same operations. Verify that the most significant bit of the result is already the correct **unsigned** overflow bit.
2) Sign extend the operands to five bits and perform the same operations. Verify that XORing the two most significant bits of the result is the correct **signed** overflow bit.
3) In the 4-bit calculations, verify that the **signed** overflow bit can be obtained by the Boolean equations:

$$Signed\_Overflow = (A_n' \cdot B_n' \cdot F_n) + (A_n \cdot B_n \cdot F_n') \text{ for additions}$$

$$Signed\_Overflow = (A_n \cdot B_n' \cdot F_n') + (A_n' \cdot B_n \cdot F_n) \text{ for subtractions}$$

where A_n, B_n, and F_n are the most significant bit of A, B, and F.

4) If we perform the same calculations using the adder-subtractor combination circuit shown in Figure 4.14, the **unsigned** overflow bit needs to be the XOR of the most significant bit of the result with the select signal. Verify that this is correct.

4.10. Show that when adding two n-bit signed numbers, $A_{n-1} \ldots A_0$ and $B_{n-1} \ldots B_0$, producing the result, $S_{n-1} \ldots S_0$, the *Signed_Overflow* flag can be deduced by the equation:

$$Signed\_Overflow = A_{n-1} \text{ XOR } B_{n-1} \text{ XOR } S_{n-1} \text{ XOR } S_n$$

4.11. To get the *Signed_Overflow* bit for the 4-bit adder-subtractor combination circuit, we need to XOR the *Unsigned_Overflow* bit with the carry bit, c_3, from the second-to-last bit slice. Write the Verilog code segment to first get this c_3 bit and then output the *Signed_Overflow* signal.

4.12. Repeat Problem 4.11, but use VHDL.

4.13. Use a truth table to show that the following equation is true:

$$x_iy_i + c_i(x_i + y_i) = x_iy_i + c_i(x_i \oplus y_i)$$

4.14. The carry-lookahead circuit shown in Figure 4.12(a) can be reduced because c_0 is a 0. Derive the carry-lookahead equations for c_1 to c_4, and draw this simpler circuit for when c_0 is a 0.

4.15. Draw the smallest possible complete circuit for a 2-bit carry-lookahead adder.

4.16. Draw the complete circuit for a 4-bit carry-lookahead adder.

4.17. Derive the carry-lookahead equation and circuit for c5.

4.18. Draw the complete 4-bit ALU circuit having the following operations. Use K-maps to reduce all of the equations to standard form.

s_2	s_1	s_0	Operations
0	0	0	$B - 1$
0	0	1	A NOR B
0	1	0	$A - B$
0	1	1	A XNOR B
1	0	0	1
1	0	1	A NAND B
1	1	0	$A + B$
1	1	1	A'

4.19. Draw the complete 4-bit ALU circuit having the following operations. Don't-care values are assigned to unused select combinations. Use K-maps to reduce all of the equations to standard form.

s_2	s_1	s_0	Operations
0	0	0	Pass A through the LE
0	0	1	Pass B through the LE
0	1	0	NOT A
0	1	1	NOT B
1	0	0	$A - B$
1	0	1	$B - A$
1	1	0	$B + 1$

4.20. Draw the complete 4-bit ALU circuit having the following operations. Use K-maps to reduce all of the equations to standard form.

s_2	s_1	s_0	**Operations**
0	0	0	*A* plus *B*
0	0	1	Increment *A*
0	1	0	Increment *B*
0	1	1	Pass *A*
1	0	0	$A - B$
1	0	1	*A* XOR *B*
1	1	0	*A* AND *B*

4.21. Draw the complete 4-bit ALU circuit having the following operations. Use K-maps to reduce all of the equations to standard form.

s_2	s_1	s_0	**Operations**
0	0	0	Pass *A*
0	0	1	Pass *B* through the AE
0	1	0	*A* plus *B*
0	1	1	A'
1	0	0	*A* XOR *B*
1	0	1	*A* NAND *B*
1	1	0	$A - 1$
1	1	1	$A - B$

4.22. Given the following K-maps for the LE, AE, and CE of an ALU, determine the ALU operations assigned to each of the select line combinations.

LE

	$s_2 = 0$				$s_2 = 1$			
s_1s_0 \ a_ib_i	00	01	11	10	00	01	11	10
00		1	1					
01	1				1	1		1
11	1		1		1	1		
10			1	1			1	1

AE

s_1s_0 \ s_2b_i	00	01	11	10
00	1	1		
01				
11				
10	1		1	

CE

s_2 \ s_1s_0	00	01	11	10
0				1
1	1			

4.23. A four-function ALU has the following equations for its LE, AE, and CE:

$$x_i = a_i + s_1{}'s_0b_i$$

$$y_i = s_1{}'s_0{}' + s_1s_0b_i{}'$$

$$c_0 = s_1s_0$$

Determine the four functions in the correct order that are implemented in this ALU. Show all of your work.

4.24. Draw the circuit for the 2-to-4 decoder with enable.

4.25. Derive the truth table for a 3-to-8 decoder with enable where the enable and output signals use negative logic, and the address inputs use positive logic.

4.26. Draw the circuit for the 4-to-16 decoder using only 2-to-4 decoders.

4.27. Derive the truth table, equation, and circuit for comparing two unsigned 2-bit operands for the less-than relationship.

4.28. Draw a 4-bit iterative comparator circuit that tests for the greater-than-or-equal-to relationship.

4.29. Draw a 4-bit shifter circuit for the following operational table.

s_2	s_1	s_0	Operation
0	0	0	Pass through
0	0	1	Rotate left
0	1	0	Shift right and fill with 1
0	1	1	Not used
1	0	0	Shift left and fill with 0
1	0	1	Pass through
1	1	0	Rotate right
1	1	1	Shift right and fill with 0

4.30. Draw a 4-bit shifter circuit for the following operational table. Use only the basic gates AND, OR, and NOT (i.e., do not use multiplexers).

s_1	s_0	Operation
0	0	Shift left and fill with 0
0	1	Shift right and fill with 0
1	0	Rotate left
1	1	Rotate right

4.31. Draw a 4-bit shifter circuit for the following operation table using only six 2-to-1 multiplexers.

s_1	s_0	Operation
0	0	Shift left and fill with 0
0	1	Shift right and fill with 0
1	0	Rotate left
1	1	Rotate right

4.32. Derive the truth table for the following combinational circuit. Write also the operation name for each row in the table.

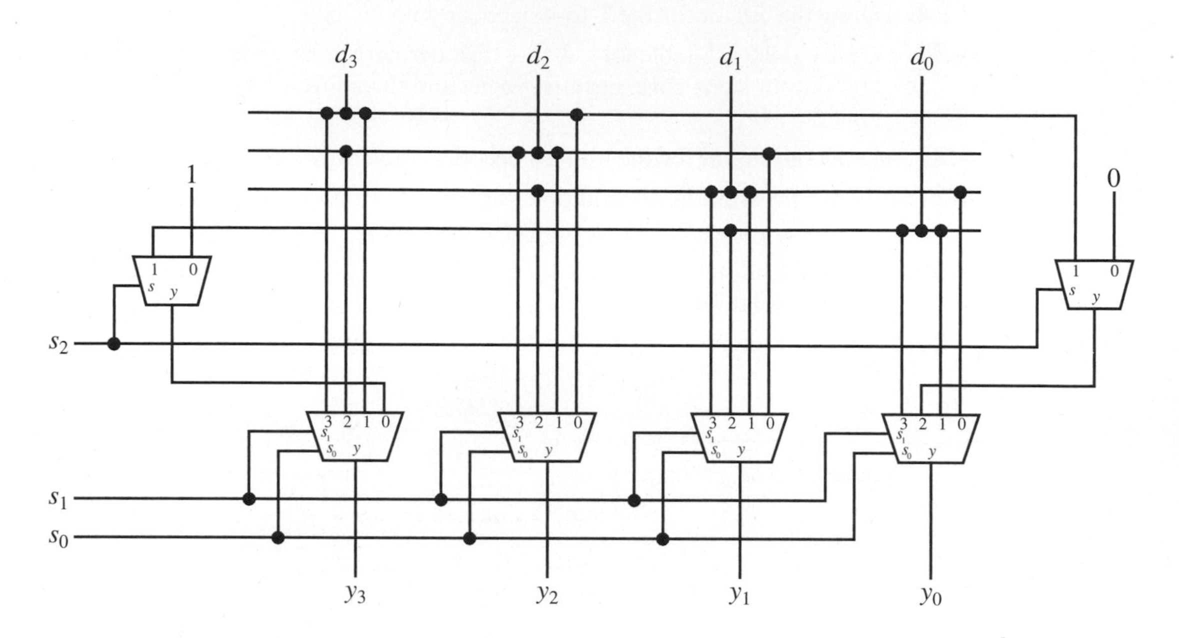

4.33. A **barrel shifter** is a shifter that can shift or rotate the data by any number of bits in a single operation. The select lines for a barrel shifter are used, not to determine what kind of operations (shift or rotate) to perform as for the general shifter, but rather, to determine how many bits to move. Draw a 4-bit barrel shifter circuit for the rotate left operation according to the following operation table:

Select $s_1\ s_0$	Operation	Output $out_3\ out_2\ out_1\ out_0$
00	No rotation	$in_3\ in_2\ in_1\ in_0$
01	Rotate left by 1 bit position	$in_2\ in_1\ in_0\ in_3$
10	Rotate left by 2 bit positions	$in_1\ in_0\ in_3\ in_2$
11	Rotate left by 3 bit positions	$in_0\ in_3\ in_2\ in_1$

When $s_1s_0 = 00$, no rotation is performed (i.e., a pass through). When $s_1s_0 = 01$, the data bits are rotated one position to the left. When $s_1s_0 = 10$, the data bits are rotated two positions to the left.

4.34. Draw a 4-bit barrel shifter circuit as in Problem 4.33, but for the rotate right operation.

4.35. Implement the 4-bit multiplier circuit shown in Figure 4.40(b) and verify that it works correctly.

CHAPTER 5

Sequential Circuits

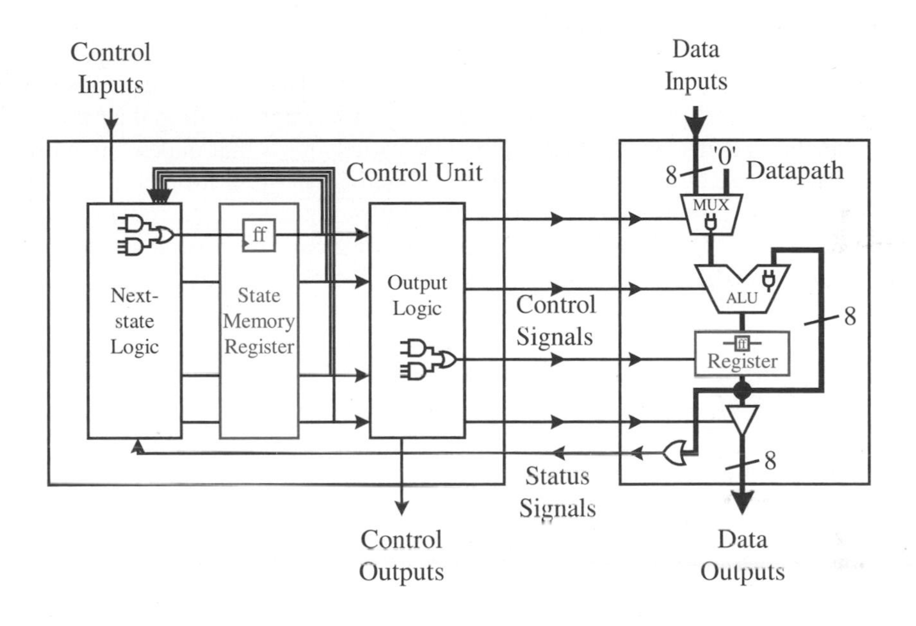

So far, we have been looking at the design of combinational circuits. We will now turn our attention to the design of **sequential circuits**. Recall that the outputs of sequential circuits are dependent not only on their current inputs (as in combinational circuits), but also on all their past inputs. Because of their need to remember the history of inputs, sequential circuits must contain memory elements.

The car security system from Section 2.11 is an example of a combinational circuit. In that example, the siren is turned on when the master switch is on and someone opens the door. If you close the door afterward, then the siren will turn off immediately. For a more realistic car security system, we would like the siren to remain on even if you close the door after it was first triggered. In order for this modified system to work correctly, the siren must be dependent not only on the master, door, and vibration switches, but also on whether the siren is currently on or off. In other words, this modified system is a sequential circuit that is dependent on both the current and the past inputs to the system.

In order to remember this history of inputs, sequential circuits must have memory elements. Memory elements, however, are just like combinational circuits in the sense that they are made up of the same basic logic gates. What makes them different is in the way these logic gates are connected together. In order for a circuit to "remember" its current value, we have to connect the output of a logic gate either directly or indirectly back to the input of that same gate. We call this a **feedback loop** circuit, and it forms the basis for all memory elements. Combinational circuits do not have any feedback loops.

Latches and **flip-flops** are the basic memory elements for storing information. Hence, they are the fundamental building blocks for all sequential circuits. A single latch or flip-flop can store only one bit of information. This bit of information that is stored in a latch or flip-flop is referred to as the **state** of the latch or flip-flop. Hence, a single latch or flip-flop can be in either one of two states: 0 or 1. We say that a latch or a flip-flop changes state when its content changes from a 0 to a 1 or vice versa. This state value is always available at the output. Consequently, the content of a latch or a flip-flop is the state value, and is always equal to its output value.

The main difference between a latch and a flip-flop is that for a latch, its state or output is constantly affected by its input as long as its enable signal is asserted. In other words, when a latch is enabled, its state changes immediately when its input changes. When a latch is disabled, its state remains constant, thereby, remembering its previous value. On the other hand, a flip-flop changes state only at the active edge of its enable signal, that is, at precisely the moment when its enable signal changes either from a 0 to a 1 (referred to as the rising edge of the signal), or from a 1 to a 0 (the falling edge). However, after the rising or falling edge of the enable signal, and during the time when the enable signal is at a constant 1 or 0, the flip-flop's state remains constant, even if the input changes.

In a microprocessor system, we usually want changes to occur at precisely the same moment. Hence, flip-flops are used more often than latches, because they all can be synchronized to change only at the active edge of the enable signal. This enable signal for the flip-flops is usually the global controlling clock signal. As a result, all flip-flops in the system will change state synchronously at the active edge of the clock.

Historically, there are four main types of flip-flops: SR, D, JK, and T. The main differences between them are the number of inputs they have and how their contents change based on their inputs. Any given sequential circuit can be built using any of these types of flip-flops (or combinations of them). However, selecting one type of flip-flop over another type to use in a particular sequential circuit can affect the overall size of the circuit. Today, sequential circuits are designed mainly with D flip-flops because of their ease of use. This is simply a tradeoff issue between ease of circuit design versus circuit size. With the much larger capacity FPGAs (field-programmable gate arrays) available today, the small circuit size reduction is insignificant and does not justify the unnecessary increase in complexity in the circuit design process. Thus, we will focus only on the D flip-flop and how it is used in building larger sequential circuits.

In this chapter, we will look at how latches and flip-flops are designed and how they work. Because flip-flops are at the heart of all sequential circuits, a good understanding of their design and operation is very important in the design of microprocessors. We then will use the flip-flops to build larger components, such as the registers and counters.

5.1 Bistable Element

Let us look at the inverter. If you provide the inverter input with a 1, the inverter will output a 0. If you do not provide the inverter with an input (i.e., neither a 0 nor a 1), the inverter will not have a value to output. If you want to construct a memory circuit using the inverter, you would want the inverter to continue to output the 0 (i.e., to remember the 0) even after you remove the 1 input. In order for the inverter to continue to output a 0, you need the inverter to self-provide its own input. In other words, you want the output to feed back the 0 to the input. However, you cannot connect the output of the inverter directly back to its input, because you will have a 0 connected to a 1, and therefore would create a short circuit.

The solution is to connect two inverters in series in a loop, as shown in Figure 5.1. This circuit is called a **bistable element**, and it is the simplest memory circuit. The bistable element has two symmetrical nodes labeled Q and Q', both of which can be viewed as either an input or an output signal. Because Q and Q' are symmetrical, we can arbitrarily use Q as the state variable, so that the state of the circuit is the value at Q. Let us assume that Q originally has the value 0 when power is first applied to the circuit. Because Q is the input to the bottom inverter, therefore, Q' is a 1. A 1 going to the input of the top inverter will produce a 0 at the output Q, which is what we started off with. Hence, the value at Q will remain at a 0 indefinitely. Similarly, if we power up the circuit with $Q = 1$, we will

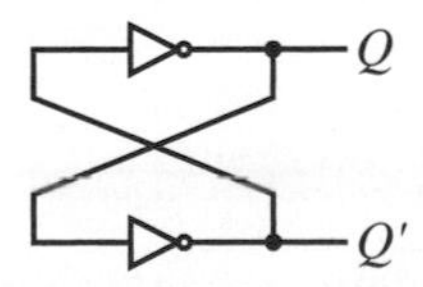

FIGURE 5.1 Bistable element circuit.

get $Q' = 0$, and again, we get a stable situation with Q remaining at a 1 indefinitely. Thus, the circuit has two stable states: $Q = 0$ and $Q = 1$; hence, the name "bistable."

We say that the bistable element has memory because it can remember its state (i.e., keep the value at Q constant) indefinitely. Unfortunately, we cannot change its state (i.e., cannot change the value at Q). We cannot just input a different value to Q, because it will create a short circuit by connecting a 0 to a 1. For example, let us assume that Q is currently 0. If we want to change the state, we need to set Q to a 1, but in so doing we will be connecting a 1 to a 0, thus creating a short. Another way of looking at this problem is that we can think of both Q and Q' as being the primary outputs, which means that the circuit does not have any external inputs. Therefore, there is no way for us to input a different value.

5.2 **SR Latch**

In order to change the state for the bistable element, we need to add external inputs to the circuit. The simplest way to add extra inputs is to replace the two inverters with two NAND gates, as shown in Figure 5.2(a). This circuit is called an **SR latch**. In addition to the two outputs Q and Q', there are two inputs S' and R' for *set* and *reset*, respectively. Just like the bistable element, the SR latch can be in one of two states: a set state when $Q = 1$, or a reset state when $Q = 0$. Following the signal naming convention, the

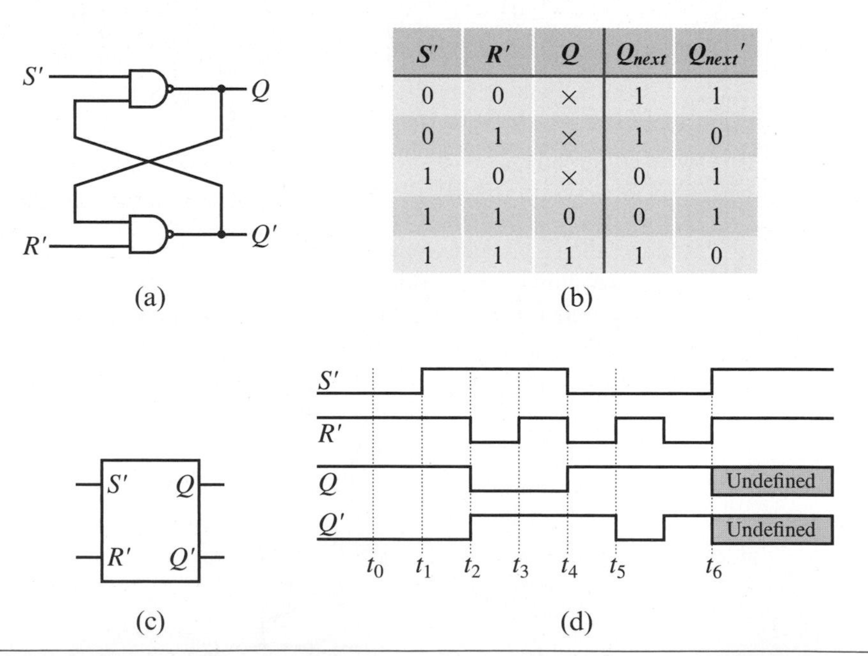

S'	R'	Q	Q_{next}	Q_{next}'
0	0	×	1	1
0	1	×	1	0
1	0	×	0	1
1	1	0	0	1
1	1	1	1	0

FIGURE 5.2 SR latch: (a) circuit using NAND gates; (b) truth table; (c) logic symbol; (d) sample trace.

primes in S and R denote that these inputs are active-low (i.e., a 0 asserts them, and a 1 de-asserts them).

To make the SR latch go to the set state (i.e., set Q to a 1), we simply assert the S' input by setting it to 0 (and de-asserting R'). It doesn't matter what the other input to the top NAND gate is, because 0 NAND anything gives a 1, hence $Q = 1$, and the latch is set. If S' remains at 0 so that Q (which is connected to one input of the bottom NAND gate) remains at 1, and if we now de-assert R' (i.e., set R' to a 1), then the output of the bottom NAND gate will be 0, and so, $Q' = 0$. This situation is shown in Figure 5.2(d) at time t_0. From this current situation, if we now de-assert S' so that $S' = R' = 1$, the latch will remain in the set state because Q' (the second input to the top NAND gate) is 0, which will keep $Q = 1$, as shown at time t_1. At time t_2, we reset the latch by making $R' = 0$ (and S' remains at a 1). With R' being a 0, Q' will go to a 1. At the top NAND gate, 1 NAND 1 is 0, thus forcing Q to go to 0. If we de-assert R' next so that, again, we have $S' = R' = 1$, this time the latch will remain in the reset state, as shown at time t_3.

Notice the two times (at t_1 and t_3) when both S' and R' are de-asserted (i.e., $S' = R' = 1$). At t_1, Q is at a 1; whereas, at t_3, Q is at a 0. Why is this so? What is different between these two times? The difference is in the value of Q immediately before those times. The value of Q right before t_1 is 1; whereas, the value of Q right before t_3 is 0. When both inputs are de-asserted, the SR latch remembers its previous state. Previous to t_1, Q has the value 1, so at t_1, Q remains at a 1. Similarly, previous to t_3, Q has the value 0, so at t_3, Q remains at a 0.

If both S' and R' are asserted (i.e., $S' = R' = 0$), then both Q and Q' are equal to a 1, as shown at time t_4, because 0 NAND anything gives a 1. Note that there is nothing wrong with having Q equal to Q'. Because we named these two points Q and Q', we like them to be inverses of each other, however, we could have used another name say, P instead of Q'.

After time t_4, if one of the input signals is de-asserted earlier than the other, the latch will end up in the state forced by the signal that is de-asserted later, as shown at time t_5. At t_5, R' is de-asserted first, so the latch goes into the set state with $Q = 1$, and $Q' = 0$.

A problem exists if both S' and R' are de-asserted at exactly the same time, as shown at time t_6. Let us assume for a moment that both gates have exactly the same delay and that the two wire connections between the output of one gate to the input of the other gate also have exactly the same delay. Immediately before time t_6, both Q and Q' are at a 1. If at time t_6 we set S' and R' to a 1 at exactly the same time, then both NAND gates will perform a 1 NAND 1 and will both output a 0 at exactly the same time. The two 0s will be fed back to the two gate inputs at exactly the same time, because the two wire connections have the same delay. This time around, the two NAND gates will perform a 1 NAND 0 and will both produce a 1, again at exactly the same time. This time, two 1s will be fed back to the inputs, which again will produce a 0 at the outputs, and so on. This oscillating behavior, called the **critical race**, will continue indefinitely until one outpaces the other. If the two gates or the two connecting wires do not have exactly the same delay, then the situation is similar to de-asserting one input before the other, and so, the latch will go into one state or the other. However, because we do not know which gate or wire has a shorter delay, therefore we do not know which state the latch will end up in. Thus, the latch's next state is undefined.

Of course, in practice, it is next to impossible to manufacture two gates and make the two connections with precisely the same delay. Furthermore, both S' and R' need to be de-asserted at exactly the same time. Nevertheless, if this circuit is used to control some mission-critical device, we don't want even this slim chance to happen.

In order to avoid this non-deterministic behavior, we must make sure that the two inputs are never de-asserted at the same time. However, we do want the situation when both of them are de-asserted, as in times t_1 and t_3, so that the circuit can remember its current content. We want to de-assert one input after de-asserting the other, but just not de-asserting both of them at exactly the same time. In practice, it is very difficult to guarantee that these two signals are never de-asserted at the same time, so we relax the condition slightly by not having both of them to be asserted together. In other words, if one is asserted, then the other one cannot be asserted. Therefore, if both of them are never asserted at the same time, then they cannot be de-asserted at the same time. A minor side benefit for not having both of them asserted together is that Q and Q' are never equal to each other. Recall that, from the names that we have given these two nodes, we do want them to be inverses of each other.

From the above analysis, we obtain the truth table in Figure 5.2(b) for the NAND implementation of the SR latch. In the truth table, Q and Q_{next} actually represent the same point in the circuit. The difference is that Q is the current value at that point, while Q_{next} is the new value to be updated in the next time period. Another way of looking at it is that Q is the input to a gate, and Q_{next} is the output from a gate. In other words, the signal Q goes into a gate, propagates through the two gates, and arrives back at Q as the new signal Q_{next}. Figure 5.2(c) shows the logic symbol for the SR latch.

The SR latch also can be implemented using NOR gates, as shown in Figure 5.3(a). The truth table for this implementation is shown in Figure 5.3(b). From the truth table, we see that the main difference between this implementation and the NAND implementation is that, for the NOR implementation, the S and R inputs are active-high, so that setting S to 1 will set the latch, and setting R to 1 will reset the latch. However, just like the NAND implementation, the latch is set when $Q = 1$ and reset when $Q = 0$. The latch remembers its previous state when $S = R = 0$. When $S = R = 1$, both Q and Q' are 0.

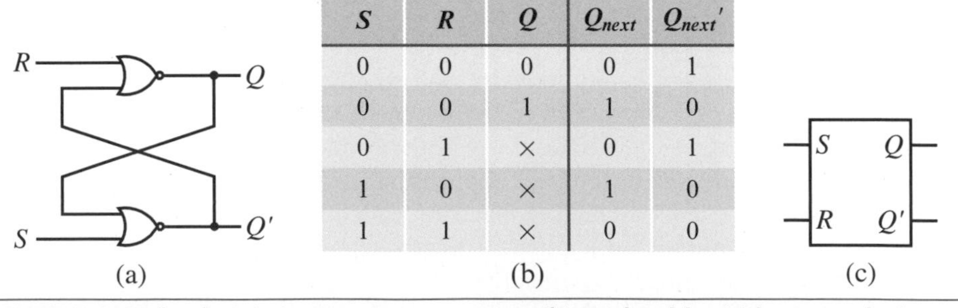

S	R	Q	Q_{next}	Q_{next}'
0	0	0	0	1
0	0	1	1	0
0	1	×	0	1
1	0	×	1	0
1	1	×	0	0

FIGURE 5.3 SR latch: (a) circuit using NOR gates; (b) truth table; (c) logic symbol.

The logic symbol for the SR latch using NOR implementation is shown in Figure 5.3(c). The only difference in this symbol is that neither *S* nor *R* has a prime in its name.

5.3 Car Security System—Version 2

In Section 2.11, we designed a combinational circuit for a car security system in which the siren will turn on when the master switch is on and either the door switch or the vibration switch is also on. However, as soon as both the door and the vibration switches are off, the siren will turn off immediately, even though the master switch is still on. In reality, what we really want is for the siren to remain on after it has been turned on, even after both the door and vibration switches are off. In order to do so, we need to remember the state of the siren. In other words, for the siren to remain on, it should be dependent not only on whether the door or the vibration switch is on, but also on the fact that the siren is currently on.

We can use the state of a SR latch to remember the state of the siren (i.e., the output of the latch will drive the siren). The state of the latch is driven by the conditions of the input switches. The modified circuit, as shown in Figure 5.4, has an SR latch, in addition to its original combinational circuit, in order to remember the current state of the siren. The latch is set from the output of the combinational circuit at *S*. Because *S*, the output from the combinational circuit, is active-high and the latch's set signal S' is active-low, therefore an inverter is needed to connect between these two points. The latch's reset is connected to the master switch so that the siren can be turned off immediately by the master switch. The siren is now connected to the output of the latch at *Q* instead of from the output of the combinational circuit at *S*.

A sample timing trace for the operation of this circuit is shown in Figure 5.5. At time 0, the siren is off, even though the door switch is on, because the master switch is off. At time 200 ns, the master switch is turned on, but the siren remains off. At time 300 ns, the siren is turned on by the door switch because the master switch is also on. At time 500 ns, both the door and the vibration switches are off, but the siren remains on because it remembers that it was turned on previously and the master switch is still on. Finally, the siren is turned off by the master switch at time 600 ns.

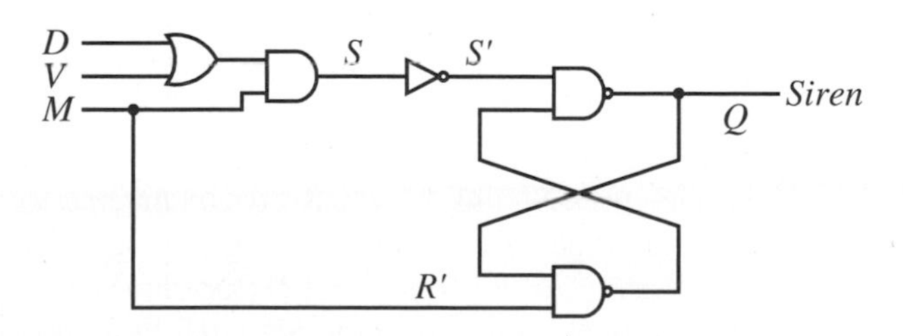

FIGURE 5.4 Modified car security system circuit with memory.

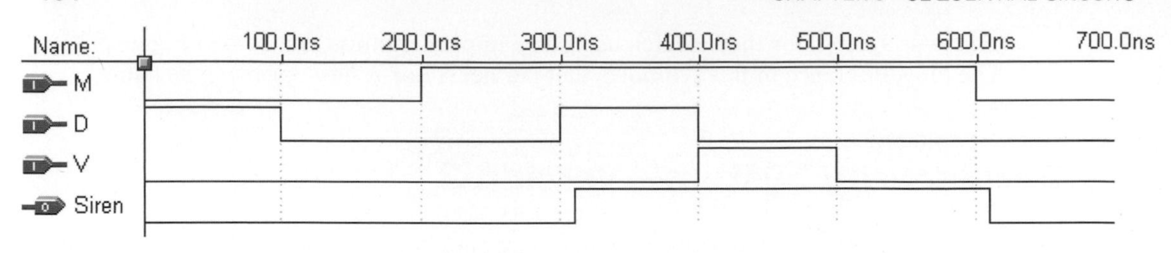

FIGURE 5.5 Sample timing trace for the modified car security system circuit with memory.

5.4 SR Latch with Enable

The SR latch is sensitive to its inputs all the time. In other words, Q will always change when either S or R is asserted. It is sometimes useful to be able to disable the inputs so that asserting them will not cause the latch to change state, but to keep its current state. Because this is achieved by de-asserting both S and R, so what we want is just one enable signal that will de-assert them both. The **SR latch with enable** (also known as a **gated SR latch**) shown in Figure 5.6(a) accomplishes this goal by adding two extra NAND gates to the original NAND-gate implementation of the latch. These two new NAND gates are controlled by the enable input, E, which determines whether the latch is enabled or disabled. When $E = 1$, the latch is enabled and the circuit behaves like the normal NAND-gate implementation of the SR latch, except that the new S and R inputs are active-high rather than active-low. When $E = 0$, the latch is disabled because $S' = R' = 1$, and the latch will remain in its previous state, regardless of the S and R inputs. The truth table and the logic symbol for the SR latch with enable are shown in Figures 5.6(b) and (c), respectively.

A typical operation of the latch is shown in the sample trace in Figure 5.6(d). Between t_0 and t_1, $E = 0$, so changing the S and R inputs does not affect the output. Between t_1 and t_2, $E = 1$, and the trace is similar to the trace of Figure 5.2(d), except that the input signals are inverted.

5.5 D Latch

Recall from Section 5.2 that the disadvantage with the SR latch is that we need to ensure that the two inputs S' and R', are never de-asserted at exactly the same time, and we said that we can guarantee this by not having both of them asserted at the same time. This situation is prevented in the **D latch** by adding an inverter between the original S' and R' inputs. This way, S' and R' will always be inverses of each other, and so, they will never be asserted together. The circuit using NAND gates and the inverter is shown in Figure 5.7(a). There is now only one input D (for data). When $D = 0$, then $S' = 1$ and $R' = 0$, so this is similar to resetting the SR latch by making $Q = 0$. Similarly, when $D = 1$, then $S' = 0$ and $R' = 1$, and Q will be set to 1. From this observation, we see

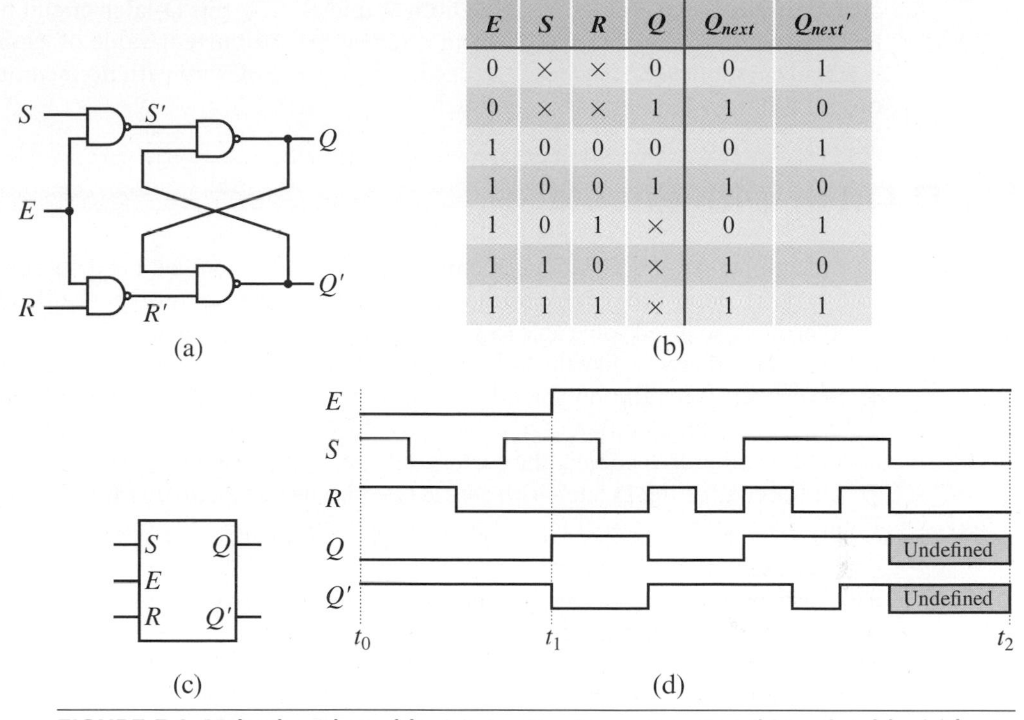

E	S	R	Q	Q_{next}	Q_{next}'
0	×	×	0	0	1
0	×	×	1	1	0
1	0	0	0	0	1
1	0	0	1	1	0
1	0	1	×	0	1
1	1	0	×	1	0
1	1	1	×	1	1

FIGURE 5.6 SR latch with enable: (a) circuit using NAND gates; (b) truth table; (c) logic symbol; (d) sample trace.

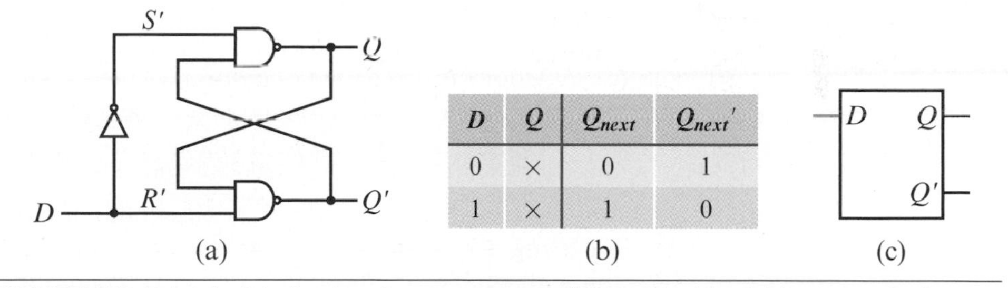

D	Q	Q_{next}	Q_{next}'
0	×	0	1
1	×	1	0

FIGURE 5.7 D latch: (a) circuit using NAND gates; (b) truth table; (c) logic symbol.

that Q_{next} always gets the same value as the input D and is independent of the current value of Q. Hence, we obtain the truth table for the D latch, as shown in Figure 5.7(b).

Comparing the truth table for the D latch shown in Figure 5.7(b) with the truth table for the SR latch shown in Figure 5.2(b), it is obvious that we have eliminated not just one, but three rows, where $S' = R'$. The reason for adding the inverter to the SR-latch circuit was to eliminate the row where $S' = R' = 0$. However, we still need to have the other two rows where $S' = R' = 1$ in order for the circuit to remember its current value. By not being able to set both S' and R' to 1, this D-latch circuit has now lost its ability to remember. Q_{next} cannot remember the current value of Q, instead, it will always follow D. The result is like having a piece of wire with no memory—the output is always the same as the input.

5.6 D Latch with Enable

In order to make the D latch remember the current value, we need to connect Q (the current state value) back to the input D, thus creating another feedback loop. Furthermore, we need to be able to select whether to loop Q back to D or input a new value for D. Otherwise, like the bistable element, we will not be able to change the state of the circuit. We could do something similar to what we did with the SR latch, but we will do something different. One way to achieve this is to use a 2-input multiplexer to select whether to feedback the current value of Q or pass an external input back to D. The circuit for the **D latch with enable** (also known as a **gated D latch**) is shown in Figure 5.8(a). The external input becomes the new D input, the output of the multiplexer is connected to the original D input, and the select line of the multiplexer is the enable signal E.

When the enable signal E is asserted ($E = 1$), the external D input passes through the multiplexer, and so Q_{next} (i.e., the output Q) follows the D input. On the other hand, when E is de-asserted ($E = 0$), the current value of Q loops back as the input to the circuit, and so Q_{next} retains its last value independent of the D input.

When the latch is enabled, the latch is said to be opened, and the path from the input D to the output Q is transparent. In other words, Q follows D. Because of this characteristic, the D latch with enable circuit is often referred to as a **transparent latch**. When the latch is disabled, it is closed, and the latch remembers its current state. The truth table and the logic symbol for the D latch with enable are shown in Figures 5.8(b) and (c). A sample trace for the operation of the D latch with enable is shown in Figure 5.8(d). Between t_0 and t_1, the latch is enabled with $E = 1$, so the output Q follows the input D. Between t_1 and t_2, the latch is disabled, so Q remains stable even when D changes.

An alternative way to construct the D latch with enable circuit is shown in Figure 5.9. Instead of using a 2-input multiplexer, as shown in Figure 5.8(a), we start with the SR latch with enable circuit of Figure 5.6(a), and connect the S and R inputs together with an inverter. The functional operations of these two circuits (Figures 5.8(a) and 5.9) are identical.

5.7 Verilog and VHDL Code for Memory Elements

Neither Verilog nor VHDL has any explicit object for defining a memory element. Instead, the semantics of the language provide for signals to be interpreted as a memory element. In other words, the memory element is declared depending on how these signals are assigned.

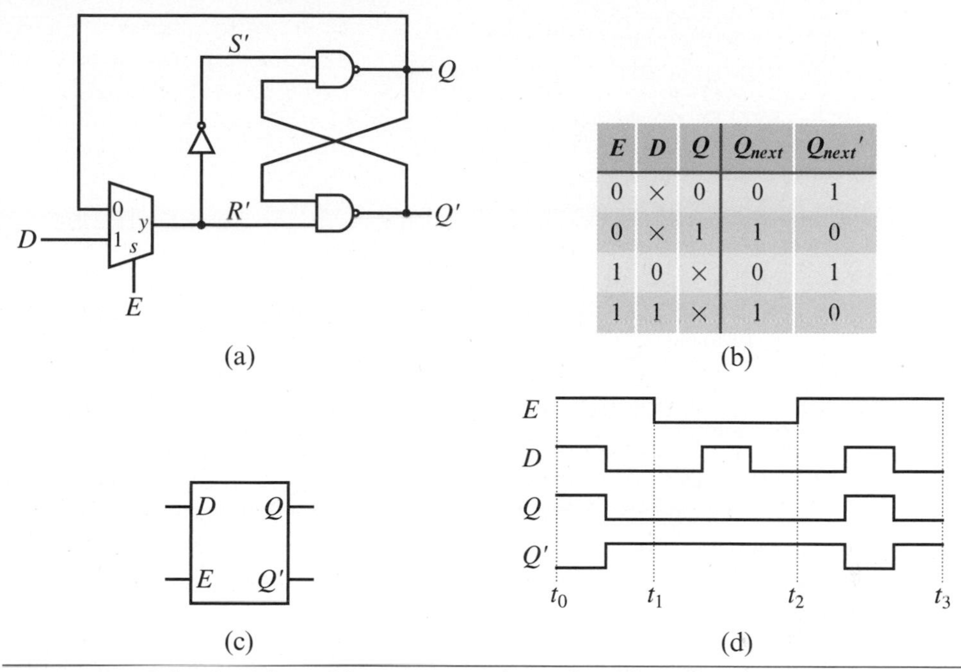

E	D	Q	Q_{next}	Q_{next}'
0	×	0	0	1
0	×	1	1	0
1	0	×	0	1
1	1	×	1	0

FIGURE 5.8 D latch with enable: (a) circuit; (b) truth table; (c) logic symbol; (d) sample trace.

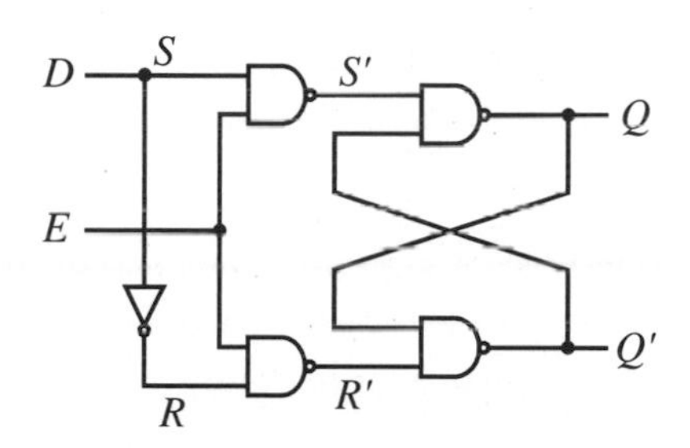

FIGURE 5.9 D latch with enable circuit using four NAND gates.

Consider the VHDL code in Figure 5.10. If *Enable* is 1, then *Q* gets the value of *D*; otherwise, *Q* gets a 0. In this code, *Q* is assigned a value for all possible outcomes of the test in the IF statement. With this construct, a combinational circuit (a 2-to-1 mux) is produced.

If we remove the ELSE and the statement in the ELSE part, as shown in Figure 5.11, then we have a situation where no value is assigned to *Q* if *Enable* is not 1. The key point here is that VHDL semantics stipulate that, in cases where the code does not specify a value for a signal, the signal should retain its current value. In other words, the signal must remember its current value, and in order to do so, a memory element is implied.

```
LIBRARY IEEE;
USE IEEE.STD_LOGIC_1164.ALL;

ENTITY no_memory_element IS PORT (
  D, Enable: IN STD_LOGIC;
  Q: OUT STD_LOGIC);
END no_memory_element;

ARCHITECTURE Behavior OF no_memory_element IS
BEGIN
  PROCESS(D, Enable)
  BEGIN
    IF (Enable = '1') THEN
      Q <= D;
    ELSE
      Q <= '0';
    END IF;
  END PROCESS;
END Behavior;
```

FIGURE 5.10 Sample VHDL description of a combinational circuit.

5.7.1 VHDL Code for a D Latch with Enable

Figure 5.11 shows the VHDL code for a D latch with enable. If *Enable* is 1, then *Q* gets the value of *D*. However, if *Enable* is not 1, the code does not specify what *Q* should be, therefore, *Q* retains its current value by using a memory element. The process sensitivity list includes both *D* and *Enable*, because either one of these signals can cause a change in the value of the *Q* output.

```
LIBRARY IEEE;
USE IEEE.STD_LOGIC_1164.ALL;

ENTITY D_latch_with_enable IS PORT (
  D, Enable: IN STD_LOGIC;
  Q: OUT STD_LOGIC);
END D_latch_with_enable;

ARCHITECTURE Behavior OF D_latch_with_enable IS
BEGIN
  PROCESS(D, Enable)
  BEGIN
    IF (Enable = '1') THEN
      Q <= D;
    END IF;
  END PROCESS;
END Behavior;
```

FIGURE 5.11 VHDL code for a D latch with enable.

```
module D_latch_with_enable (
  input D,
  input Enable,
  output reg Q
  );

  always @(D or Enable)
    if (Enable == 1'b1) begin
     Q <= D;// assign value from D to Q
    end
endmodule
```

FIGURE 5.12 Verilog code for a D latch with enable.

5.7.2 **Verilog Code for a D Latch with Enable**

Figure 5.12 shows the Verilog code for a D latch with enable. The sensitivity list in the `always` statement includes both *D* and *Enable*, because either one of these signals can cause a change in the value of the *Q* output. Just like VHDL, Verilog does not have any explicit object for defining a memory element. The statement *Q* <= *D* assigns the value of *D* to *Q*. Because of the `if` statement, this assignment is done only if the *Enable* signal is a 1. If *Enable* is 0 then no value is assigned to *Q*, and this implies that a memory element (a latch in this case) is needed to remember the current value of *Q*.

On the other hand, if there is an `else` part to the `if` statement, and *Q* is also assigned a value in the `else` part, then a combination circuit (a 2-to-1 mux, in this case) will be used for *Q*. Signals that are driven from inside an `always` block must be of type `reg`. So the use of the `reg` keyword in the declaration of the *Q* signal does not always mean that *Q* will use a memory element. Whether *Q* will use a memory element depends only on the structure of the `if` statement.

Note also the use of the non-blocking assignment (<=) statement, which is used when modeling sequential circuits inside an `always` block.

5.8 **Clock**

Latches are known as **level-sensitive** because their outputs are affected by their inputs as long as they are enabled. Their memory state can change during this entire time when the enable signal is asserted. In a computer circuit, however, we do not want the memory state to change at different times when the enable signal is asserted. Instead, we like to synchronize all of the state changes to happen at precisely the same moment and at regular intervals. In order to achieve this, two things are needed: (1) a synchronizing signal, and (2) a memory circuit that is not level-sensitive. The synchronizing signal, of course, is the **clock**, and the non-level-sensitive memory circuit is the flip-flop.

The clock is simply a very regular square wave signal, as shown in Figure 5.13. We call the edge of the clock signal when it changes from 0 to 1 the **rising edge**. Conversely,

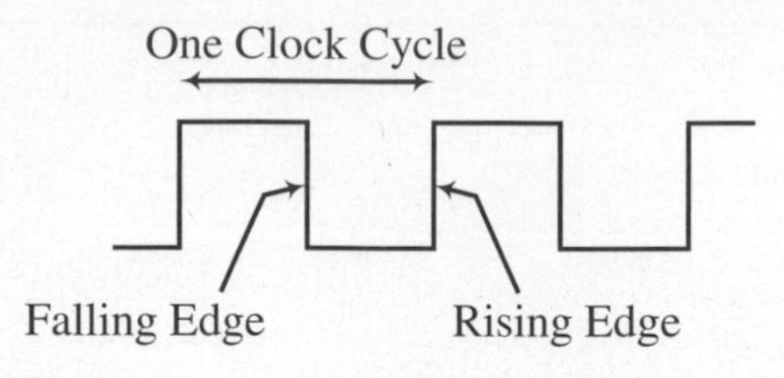

FIGURE 5.13 Clock signal.

the **falling edge** of the clock is the edge when the signal changes from 1 to 0. We will use the symbol ↑ to denote the rising edge and ↓ for the falling edge. In a computer circuit, either the rising edge or the falling edge of the clock can be used as the synchronizing signal for writing data into a memory element, and this edge signal is referred to as the **active edge** of the clock. In all of our examples, we will use the rising clock edge as the active edge. Therefore, at every rising edge, data will be clocked or stored into the memory element.

A **clock cycle** is the time from one rising edge to the next rising edge or from one falling edge to the next falling edge. The **speed** of the clock, measured in hertz (Hz), is the number of clock cycles per second. Typically, the clock speed for a microprocessor in an embedded system runs around 50 MHz, while the microprocessor in a personal computer runs upward of 2 GHz and higher. A clock **period** is the time for one clock cycle (seconds per cycle), so it is just the inverse of the clock speed.

The maximum speed of the clock is determined by how fast a circuit can produce valid results. For example, a 2-to-1 mux will have valid results at its output much sooner than, say, an ALU can. Of course, we want the clock speed to be as fast as possible, but it can only be as fast as the slowest circuit in the entire system. We want the clock period to be the time it takes for the slowest data manipulation circuit (such as the ALU) to get its input from a memory element, operate on the data, and then write the data back into a memory element as depicted in Figure 5.14.

Because data is written into memory elements at the rising clock edge, therefore, all data manipulations must be completed before the rising clock edge so that the correct results are stored. Furthermore, shortly after the data is written at the rising clock edge, it will be available for reading soon after the rising clock edge. This

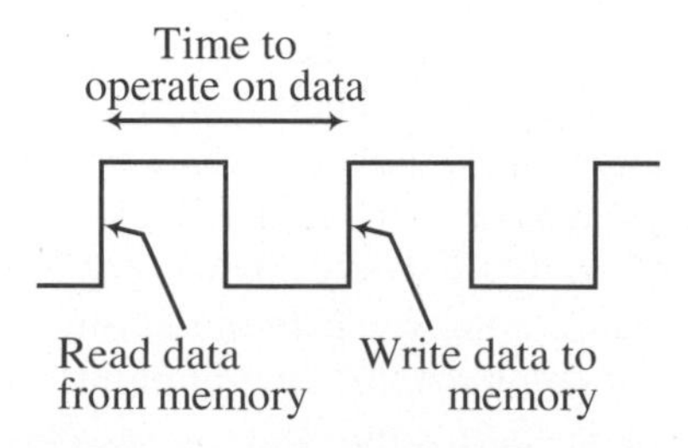

FIGURE 5.14 Relationship between the clock period and the time to operate on the data.

```
module clock_divider
// 50M/8 = 4 Hz output clock
#(parameter [24:0] half = 25'd6250000)
(
  input clock_in,           // 50MHz input clock
  output reg clock
  );

  reg [24:0] count;

  always @(posedge clock_in) begin
    if (count == half) begin
      count <= 25'd0;
      clock <= ~clock;
      end
    else begin
      count <= count+1;
      end
  end
endmodule
```

FIGURE 5.15 Behavioral Verilog description of a clock divider circuit.

sequence of events is referred to as a **register transfer** because data read from a register (or memory element) is operated on and then written back to a register. This will be covered in more detail in later sections.

Verilog Code for a Clock Divider

Figure 5.15 shows a behavioral Verilog description of a clock divider circuit that slows down a 50 MHz input clock signal to 4 Hz.

VHDL Code for a Clock Divider

Figure 5.16 shows the same clock divider circuit but written in behavioral VHDL.

5.9 D Flip-Flop

We mentioned in the last section that in a computer system, we need to synchronize all memory state changes to happen at precisely the same moment. A **flip-flop** can do just that. Unlike the latch, a flip-flop is not level-sensitive, but rather **edge-triggered**. In other words, data gets stored into a flip-flop only at the active edge of the clock. An **edge-triggered D flip-flop** achieves this by combining in series a pair of D latches. Figure 5.17(a) shows a **positive edge-triggered D flip-flop** where two D latches are connected in series. A clock signal *Clk* is connected to the *E* input of the two latches: one directly, and one through an inverter.

The first latch is called the master latch. The master latch is enabled when $Clk = 0$ because of the inverter, and so QM follows the primary input D. However, the signal

```
LIBRARY IEEE;
USE  IEEE.STD_LOGIC_1164.ALL;

ENTITY clock_divider IS
-- 50M/8 = 4 Hz output clock
GENERIC (half: INTEGER := 50000000/8);
PORT (
  clock_in: IN STD_LOGIC;     -- 50MHz input clock
  clock: BUFFER STD_LOGIC);
END clock_divider;

ARCHITECTURE Behavior OF clock_divider IS
  SIGNAL count: INTEGER RANGE 0 TO half;
BEGIN
  PROCESS
  BEGIN
    WAIT UNTIL clock_in'EVENT and clock_in = '1';
    IF (count = half) THEN
      count <= 0;
      clock <= NOT clock;  -- toggle the clock
    ELSE
      count <= count + 1;
    END IF;
  END PROCESS;
END Behavior;
```

FIGURE 5.16 Behavioral VHDL description of a clock divider circuit.

at QM cannot pass over to the primary output Q, because the second latch (called the slave latch) is disabled when $Clk = 0$. When $Clk = 1$, the master latch is disabled, but the slave latch is enabled so that the output from the master latch, QM, is transferred to the primary output Q. The slave latch is enabled all the while that $Clk = 1$, but its content changes only at the rising edge of the clock, because once Clk is 1, the master latch is disabled, and the input to the slave latch, QM, will be constant. Therefore, when $Clk = 1$ and the slave latch is enabled, the primary output Q will not change because the input QM is not changing.

The circuit shown in Figure 5.17(a) is called a positive edge-triggered D flip-flop because the primary output Q on the slave latch changes only at the rising edge of the clock. If the slave latch is enabled when the clock is low (i.e., with the inverter output connected to the E of the slave latch), then it is referred to as a **negative edge-triggered** flip-flop. The circuit is also referred to as a **master-slave** D flip-flop because of the two D latches used in the circuit.

Figure 5.17(b) shows the operation table for the D flip-flop. The ↑ symbol signifies the rising edge of the clock. When Clk is at either 0 or 1, the flip-flop retains its current value (i.e., $Q_{next} = Q$). Q_{next} changes and follows the primary input D only at the rising edge of the clock. The logic symbol for the positive edge-triggered D flip-flop is shown in Figure 5.17(c). The small triangle at the clock input indicates that the circuit is triggered by the edge of the signal, and so it is a flip-flop. Without the small triangle, the symbol would be that for a latch. If there is a circle in front of the clock

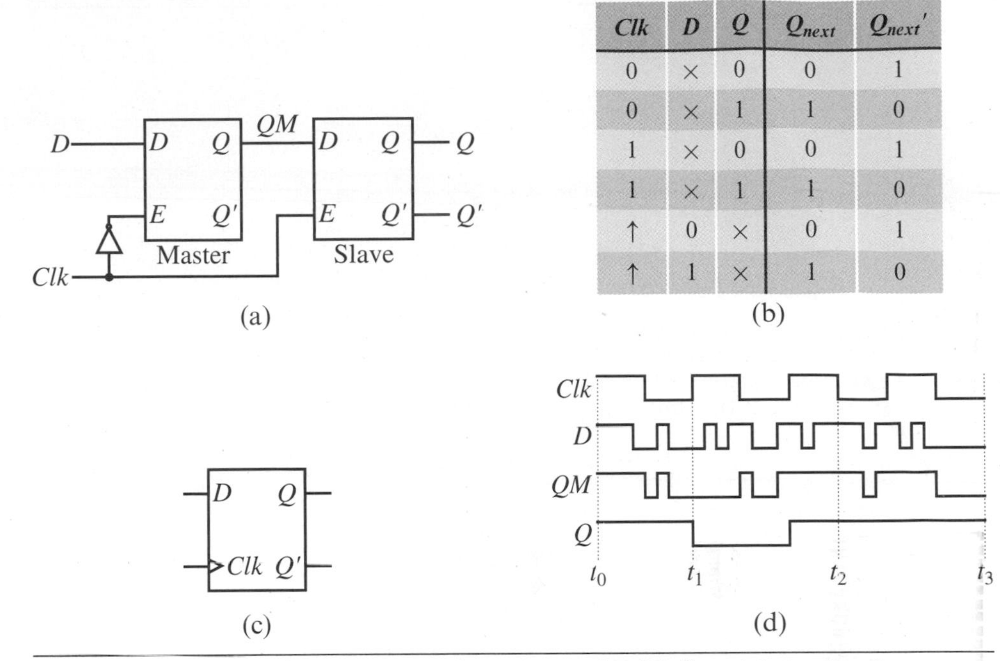

Clk	*D*	*Q*	Q_{next}	Q_{next}'
0	×	0	0	1
0	×	1	1	0
1	×	0	0	1
1	×	1	1	0
↑	0	×	0	1
↑	1	×	1	0

FIGURE 5.17 **Master-slave positive edge-triggered D flip-flop: (a) circuit using D latches; (b) operation table; (c) logic symbol; (d) sample trace.**

line, then the flip-flop is triggered by the falling edge of the clock, making it a negative edge-triggered flip-flop. Figure 5.17(d) shows a sample trace for the D flip-flop. Notice that when $Clk = 0$, *QM* follows *D*, and the output of the slave latch, *Q*, remains constant. On the other hand, when $Clk = 1$, *Q* follows *QM*, and the output of the master latch, *QM*, remains constant.

Verilog Code for a D Flip-Flop

Figure 5.18 shows the behavioral Verilog code for a positive edge-triggered D flip-flop. To be a positive edge-triggered D flip-flop, *Q* must follow *D* only at the rising edge of the clock as specified here by the `always @ (posedge Clock)` statement. The `posedge` keyword means at the rising edge of the *Clock* signal. All the statements inside the body of this `always` statement are executed at every rising edge of the clock. There is only one statement inside this `always` block, and that is to assign the value from the *D* input to the *Q* output.

VHDL Code for a D Flip-Flop

Figure 5.19 shows the behavioral VHDL code for a positive edge-triggered D flip-flop. To be a positive edge-triggered D flip-flop, *Q* must follow *D* only at the rising edge of the clock, as specified here by the condition "*Clock*' EVENT AND *Clock* = '1'." The 'EVENT attribute refers to any changes in the qualifying *Clock* signal. Therefore, when this happens and the resulting *Clock* value is a 1, we have, in effect, a condition for a positive

```
module D_flipflop (
  input Clock,
  input D,
  output reg Q
);

  // execute on rising clock edge
  always @(posedge Clock) begin
      // assign value from D to Q at every rising clock edge
      Q <= D;
  end

endmodule
```

FIGURE 5.18 Behavioral Verilog code for a D flip-flop.

```
LIBRARY IEEE;
USE IEEE.STD_LOGIC_1164.ALL;

ENTITY D_flipflop IS PORT (
  Clock: IN STD_LOGIC;
  D: IN STD_LOGIC;
  Q: OUT STD_LOGIC);
END D_flipflop;

ARCHITECTURE Behavior OF D_flipflop IS
BEGIN
  PROCESS(Clock)                          -- sensitivity list is used
  BEGIN
    IF (Clock'EVENT AND Clock = '1') THEN
      Q <= D;
    END IF;
  END PROCESS;
END Behavior;
```

FIGURE 5.19 Behavioral VHDL code for a positive edge-triggered D flip-flop.

or rising clock edge. Again, the code does not specify what is assigned to Q when the condition in the IF statement is false, so it implies the use of a memory element. Note also that the process sensitivity list contains only the clock signal, because it is the only signal that can cause a change in the Q output.

Figure 5.20 compares the different operations between a latch and a flip-flop. In Figure 5.20(a), we have a D latch with enable, a positive edge-triggered D flip-flop, and a negative edge-triggered D flip-flop, all having the same D input and controlled by the same clock signal. Figure 5.20(b) shows a sample trace of the circuit's operations. Notice that the gated D latch, Q_a, follows the D input as long as the clock is high (between times t_0 and t_1, and times t_2 and t_3). The positive edge-triggered flip-flop, Q_b, follows the D input only at the rising edge of the clock at time t_2, while the negative edge-triggered flip-flop, Q_c, follows the D input only at the falling edge of the clock at times t_1 and t_3.

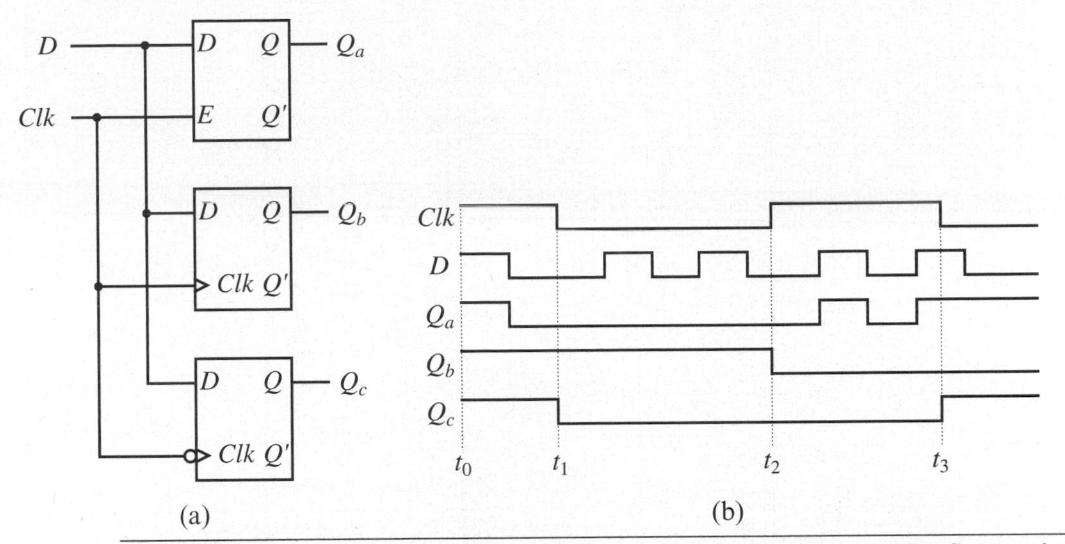

FIGURE 5.20 Comparison of a gated latch, a positive edge-triggered flip-flop, and a negative edge-triggered flip-flop: (a) circuit; (b) sample trace.

5.9.1 Alternative Smaller Circuit

Not all master-slave flip-flops are edge-triggered. For instance, using two SR latches to construct a master-slave flip-flop results in a flip-flop that is level-sensitive. Conversely, an edged-triggered D flip-flop can be constructed using SR latches instead of the master-slave D latches.

The circuit shown in Figure 5.21 shows how a positive edge-triggered D flip-flop can be constructed using three interconnected SR latches. The advantage of this circuit is that it uses only six NAND gates as opposed to eleven gates for the master-slave D flip-flop shown in Figure 5.17(a). The operation of the circuit is as follows. When $Clk = 0$, the outputs of Gates 2 and 3 will be 1 (because 0 NAND $x = 1$). With $n_2 = n_3 = 1$, this will keep the output latch (composed of Gates 5 and 6) in its current state. At the same time, $n_4 = D'$ because one input to Gate 4 is n_3, which is a 1 (1 NAND $x = x'$). Similarly, $n_1 = D$ because $n_2 = 1$, and the other input to Gate 1 is n_4, which is D' (again 1 NAND $x = x'$).

When *Clk* changes to 1, n_2 will be equal to D' because 1 NAND $n_1 = n_1'$, and $n_1 = D$. Similarly, n_3 will be equal to D when *Clk* changes to 1 because the other two inputs to Gate 3 are both D'. Therefore, if $Clk = 1$ and $D = 0$, then n_2 (which is equal to D') will be 1, and n_3 (which is equal to D) will be 0. With $n_2 = 1$ and $n_3 = 0$, this will de-assert S' and assert R', thus resetting the output latch Q to 0. On the other hand, if $Clk = 1$ and $D = 1$, then n_2 (which is equal to D') will be 0 and n_3 (which is equal to D) will be 1. This will assert S' and de-assert R', thus setting the output latch Q to 1. So at the rising edge of the *Clk* signal, Q will follow D.

The setting and resetting of the output latch occurs only at the rising edge of the *Clk* signal, because when *Clk* is at a 1 and remains at a 1, changing D will not change

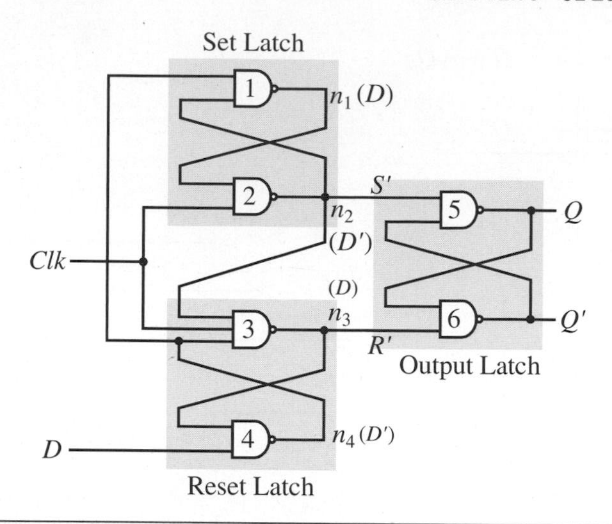

FIGURE 5.21 Positive edge-triggered D flip-flop using SR latches.

n_2 or n_3. The reason, as noted in the previous paragraph, is that n_2 and n_3 are always inverses of each other. Furthermore, the following argument shows that both n_2 and n_3 will remain constant even if D changes. Let us first assume that n_2 is a 0. If $n_2 = 0$, then n_3 (the output of Gate 3) will always be a 1 (because 0 NAND $x = 1$), regardless of what n_4 (the third input to Gate 3) may be. Hence, if n_4 (the output of Gate 4) cannot affect n_3, then D (the input to Gate 4) also cannot affect either n_2 or n_3. On the other hand, if $n_2 = 1$, then $n_3 = 0$ ($n_3 = n_2'$). With a 0 from n_3 going to the input of Gate 4, the output of Gate 4 at n_4 will always be a 1 (0 NAND $x = 1$), regardless of what D is. With the three inputs to Gate 3 being all 1s, n_3 will continue to be 0. Therefore, as long as $Clk = 1$, changing D will not change n_2 or n_3. And if n_2 and n_3 remain stable, then Q also will remain stable for the entire time that Clk is 1.

5.10 D Flip-Flop with Enable

So far, with the construction of the different memory elements, it seems like every time we add a new feature we also lose a feature that we need. The careful reader will have noticed that, in building the D flip-flop, we have again lost the most important property of a memory element—it can no longer remember its current content. At every active edge of the clock, the D flip-flop will load in a new value. So how do we get it to remember its current value and not load in a new value?

The answer, of course, is exactly the same as what we did with the D latch, and that is by adding an enable input, E, through a 2-input multiplexer, as shown in Figure 5.22(a). When $E = 1$, the primary input D signal will pass to the D input of the flip-flop, thus updating the content of the flip-flop at the active edge. When $E = 0$, the current content of the flip-flop at Q is passed back to the D input of the

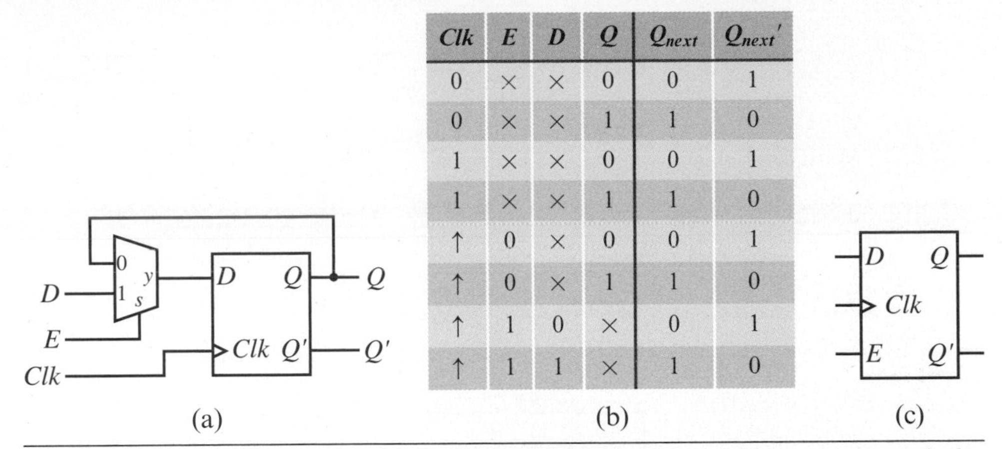

Clk	*E*	*D*	*Q*	Q_{next}	Q_{next}'
0	×	×	0	0	1
0	×	×	1	1	0
1	×	×	0	0	1
1	×	×	1	1	0
↑	0	×	0	0	1
↑	0	×	1	1	0
↑	1	0	×	0	1
↑	1	1	×	1	0

FIGURE 5.22 D flip-flop with enable: (a) circuit; (b) operation table; (c) logic symbol.

flip-flop, thus keeping its current value. Notice that changes to the flip-flop value occur only at the active edge of the clock. Here, we use the rising clock edge as the active edge. The operation table and the logic symbol for the D flip-flop with enable are shown in Figures 5.22(b) and (c), respectively.

5.10.1 Asynchronous Inputs

Flip-flops (as we have seen so far) change states only at the rising or falling edge of a synchronizing clock signal. Many circuits require the initialization of flip-flops to a known state that is independent of the clock signal. Sequential circuits that change states whenever a change in input values occurs that is independent of the clock are referred to as **asynchronous** sequential circuits. **Synchronous** sequential circuits, on the other hand, change states only at the active edge of the clock signal. Asynchronous inputs usually are available for both flip-flops and latches, and they are used to either set or clear the storage element's content that is independent of the clock.

Figure 5.23(a) shows a gated D latch with asynchronous active-low *Set'* and *Clear'* inputs, and (b) is its logic symbol. Figure 5.23(c) is the circuit for the edge-triggered D flip-flop with asynchronous *Set'* and *Clear'* inputs, and (d) is its logic symbol. When *Set'* is asserted (set to 0) the content of the storage element is set to 1 immediately (i.e., without having to wait for the next rising clock edge), and when *Clear'* is asserted (set to 0) the content of the storage element is set to 0 immediately.

Verilog Code for a D Flip-Flop with Enable and Clear

Figure 5.24 shows the behavioral Verilog code for a positive edge-triggered D flip-flop with synchronous *Enable* and asynchronous active-high *Clear*. The reason why the *Clear* signal is asynchronous is because it is included in the sensitivity list, whereas the *Enable* signal is synchronous because it is not included in the sensitivity list. When

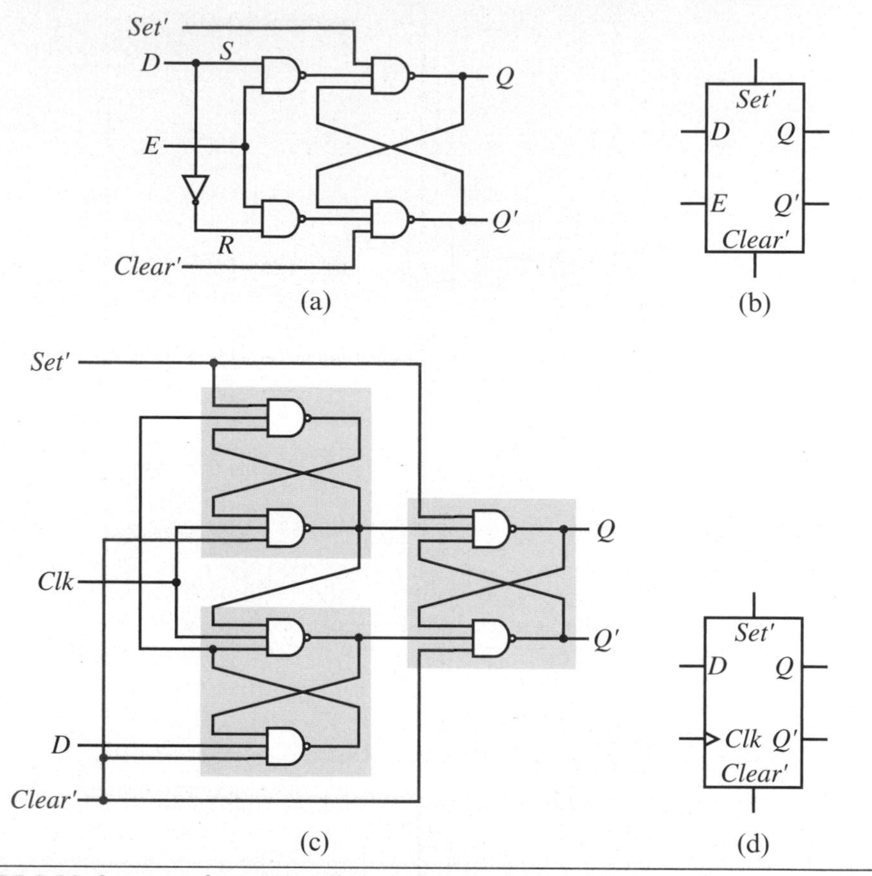

FIGURE 5.23 Storage elements with asynchronous inputs: (a) D latch with active-low *Set'* and *Clear'* (b) logic symbol for (a); (c) edge-triggered D flip-flop with active-low *Set'* and *Clear'* (d) logic symbol for (c).

the *Clear* signal changes value, the `always` block will execute immediately with the assignment statement to clear *Q*. On the other hand, when the *Enable* signal changes value, the `always` block will not execute immediately, rather, it will execute at the rising clock edge. Hence, the assignment of *D* into *Q* happens only at the rising clock edge and *Enable* is asserted.

VHDL Code for a D Flip-Flop with Enable and Clear

Figure 5.25 shows the behavioral VHDL code for a positive edge-triggered D flip-flop with synchronous *Enable* and asynchronous active-high *Clear*. The asynchronous *Clear* input is checked independently of the clock event. When the *Clear* input is asserted with a 1 (active-high) *Q* is reset to 0 immediately. If *Enable* is asserted with a 1, then *Q* follows *D* at the rising edge of the clock; otherwise, *Q* keeps its previous content.

```
module D_flipflop (
  input Clock,
  input Clear,
  input Enable,
  input D,
  output reg Q
);

  // execute on rising clock edge or Clear
  always @(posedge Clock or posedge Clear) begin
    if (Clear) begin
      Q <= 0;     // assign 0 to Q on clear
    end else if (Enable) begin
      // assign value from D to Q only if Enable is asserted
      Q <= D;
    end
  end

endmodule
```

FIGURE 5.24 **Behavioral Verilog code for a D flip-flop with synchronous active-high *Enable* and asynchronous *Clear* inputs.**

```
LIBRARY IEEE;
USE IEEE.STD_LOGIC_1164.ALL;

ENTITY D_flipflop IS PORT (
  Clock: IN STD_LOGIC;
  Clear: IN STD_LOGIC;
  Enable: IN STD_LOGIC;
  D: IN STD_LOGIC;
  Q: OUT STD_LOGIC);
END D_flipflop;

ARCHITECTURE Behavioral OF D_flipflop IS
BEGIN
  PROCESS(Clock,Clear)
  BEGIN
    IF (Clear = '1') THEN
      Q <= '0';
    ELSIF (Clock'EVENT AND Clock = '1') THEN
      IF (Enable = '1') THEN
        Q <= D;
      END IF;
    END IF;
  END PROCESS;
END Behavioral;
```

FIGURE 5.25 **Behavioral VHDL code for a positive edge-triggered D flip-flop with synchronous active-high *Enable* and asynchronous *Clear* inputs.**

5.11 Description of a Flip-Flop

Combinational circuits can be described with either a truth table or a Boolean equation. To describe the operation of a flip-flop or any sequential circuit in general, we use a characteristic table, a characteristic equation, or a state diagram, as discussed in the following subsections.

5.11.1 Characteristic Table

The **characteristic table** specifies the functional behavior of the flip-flop. It is a simplified version of the flip-flop's operational table by listing only how the state changes at the active clock edge as shown in Figure 5.26(a). The table has the flip-flop's input signal D, and current state Q listed in the input columns, and the next state Q_{next} listed in the output column. Q_{next}' is always assumed to be the inverse of Q_{next}, so it is not necessary to include this output column. The clock signal is not included in the table either, because it is an automatic signal that we do not modify. Nevertheless, the clock signal is always assumed to exist. Furthermore, because all state changes for a flip-flop (i.e., changes to Q_{next}) occur at the active edge of the clock, it is therefore not necessary to list the situations from the operation table for when the clock is at a constant value. From the operation table for the D flip-flop shown in Figure 5.17(b), we see that there are only two rows where Q_{next} is affected during the rising clock edge. Hence, these are the only two rows inserted into the characteristic table.

The characteristic table is used in the analysis of sequential circuits to answer the question of what is the next state, Q_{next}, when given the current state Q, and input signal D (for the D flip-flop).

5.11.2 Characteristic Equation

The **characteristic equation** is simply the Boolean equation that is derived directly from the characteristic table. Like the characteristic table, the characteristic equation specifies the flip-flop's next state Q_{next}, as a function of its current state Q, and input signal D. The D flip-flop characteristic table has only one 1-minterm, which results in the simple characteristic equation for the D flip-flop shown in Figure 5.26(b). This equation does not include the variable Q because Q contains only "don't care" values.

5.11.3 State Diagram

A **state diagram** is a graph with nodes and directed edges connecting the nodes, as shown in Figure 5.26(c). The state diagram graphically portrays the operation of the flip-flop. The nodes are labeled with the states of the flip-flop, and the directed edges are labeled with the input signals that cause the transition to go from one state of the flip-flop to the next. The state diagram for the D flip-flop has two states, $Q = 0$ and $Q = 1$, which correspond to the two values that the flip-flop can contain. The operation of the D flip-flop is such that when it is in state 0, it will change to state 1 if the input D is a 1; otherwise, if the input D is a 0, then it will remain in state 0. Hence, there is an edge labeled $D = 1$ that goes from state $Q = 0$ to $Q = 1$, and a second edge labeled $D = 0$ that goes from state $Q = 0$ back to itself. Similarly, when the flip-flop

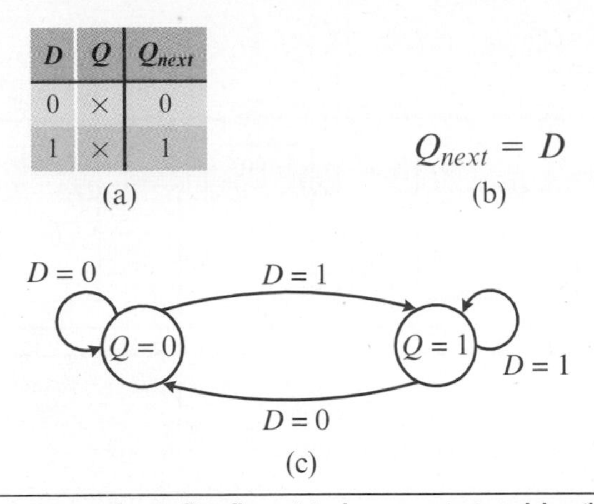

FIGURE 5.26 Description of a D flip-flop: (a) characteristic table; (b) characteristic equation; (c) state diagram.

is in state 1, it will change to state 0 if the input D is a 0; otherwise, it will remain in state 1. These two conditions correspond to the remaining two edges that go out from state $Q = 1$ in the state diagram; one edge going to state $Q = 0$ with the label $D = 0$, and the other edge going back to itself with the label $D = 1$.

5.12 **Register**

A flip-flop can store only one bit of data. When we want to store a byte of data, we need to combine eight flip-flops together and have them work together as a unit. A **register** is just a circuit with two or more D flip-flops connected together in such a way that they all work together as one unit and are synchronized by the same clock and enable signals. The only difference is that each flip-flop in the group is used to store a different bit of the data.

Figure 5.27(a) shows a 4-bit register with synchronous *Load* and asynchronous *Clear*. Four D flip-flops with active-high enable and asynchronous clear are used. Notice in the circuit that the control inputs *Clk*, *E*, and *Clear* for all of the flip-flops are connected in common to *Clock*, *Load*, and *Clear*, respectively, of the register signals; so that when a particular input is asserted, all of the flip-flops will behave in exactly the same way. The 4-bit input data is connected to D_0 through D_3, while Q_0 through Q_3 serve as the 4-bit output data for the register. When the active-high *Load* signal is asserted (i.e., $Load = 1$), the data presented on the D lines are stored into the register (the four flip-flops) at the next rising edge of the clock signal. When *Load* is de-asserted, the content of the register remains unchanged. The register can be cleared asynchronously (i.e., setting all of the Q_is to 0 immediately, without having to wait for the next active clock edge) by asserting the *Clear* line. The content of the register is always available on the Q output lines, so no control line is required for reading the data from the register. Figures 5.27(b) and (c) show the operation table and the logic symbol, respectively, for this 4-bit register.

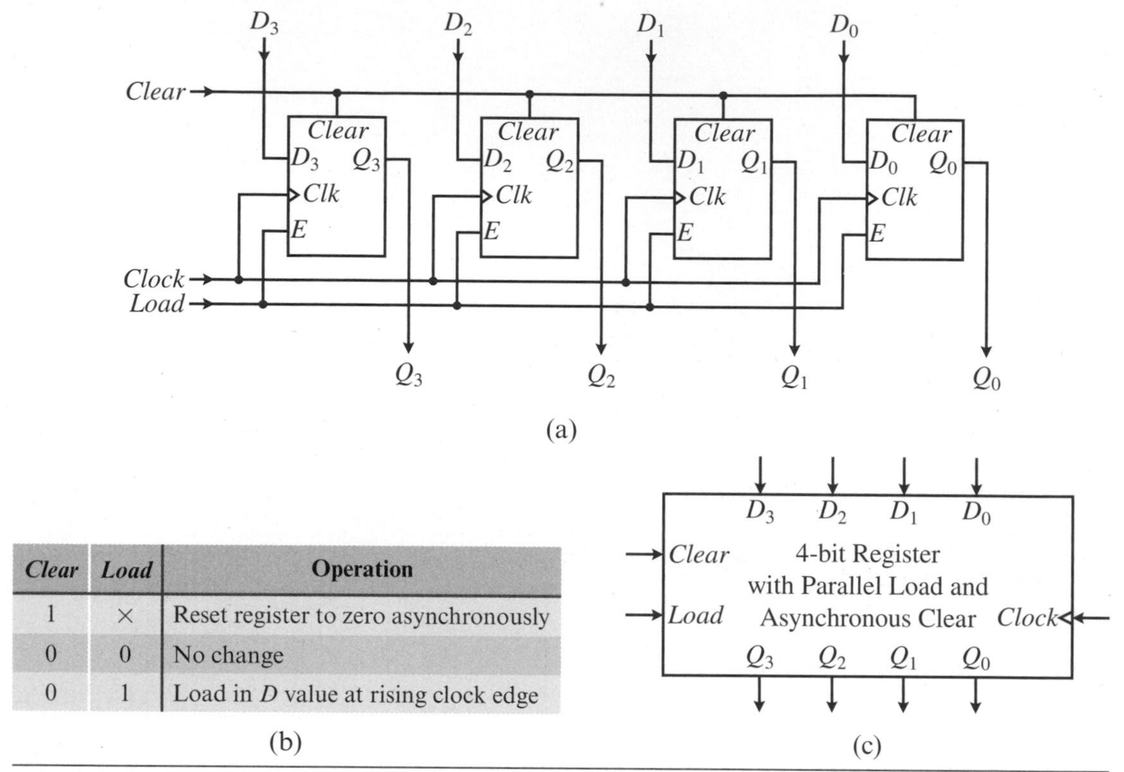

Clear	*Load*	Operation
1	×	Reset register to zero asynchronously
0	0	No change
0	1	Load in *D* value at rising clock edge

FIGURE 5.27 A 4-bit register with synchronous *Load* and asynchronous *Clear*: (a) circuit; (b) operation table; (c) logic symbol.

Verilog Code for a Register

Figure 5.28 shows the Verilog code for the 4-bit register with active-high synchronous *Load* and asynchronous *Clear* signals. Notice that the coding is similar to that for the single D flip-flop; the main difference is that the data inputs and outputs are 4 bits wide.

VHDL Code for a Register

Figure 5.29 shows the VHDL code for the 4-bit register with active-high synchronous *Load* and asynchronous *Clear* signals.

5.13 **Register File**

When we want to store several numbers concurrently in a digital circuit, we can use several individual registers in the circuit. However, there are times when we want to treat these registers as a unit, similar to addressing the individual locations of an array

```
module register
#(parameter n = 4)      // allow n to be changed
(
  input Clock, Clear, Load,
  input [n-1:0] D,
  output reg [n-1:0] Q
);
  always @ (posedge Clock or posedge Clear) begin
    if (Clear == 1) begin
      Q <= {n{1'b0}};   // n bits of 0
    end else if (Load) begin
      Q <= D;
    end
  end

endmodule
```

FIGURE 5.28 Behavioral Verilog code for a 4-bit register.

```
LIBRARY IEEE;
USE IEEE.STD_LOGIC_1164.ALL;

ENTITY Reg IS
GENERIC (n: INTEGER := 4);
PORT (
  Clock, Clear, Load: IN STD_LOGIC;
  D: IN STD_LOGIC_VECTOR(n-1 DOWNTO 0);
  Q: OUT STD_LOGIC_VECTOR(n-1 DOWNTO 0));
END Reg;

ARCHITECTURE Behavior OF Reg IS
BEGIN
  PROCESS(Clock, Clear)
  BEGIN
    IF (Clear = '1') THEN
      Q <= (OTHERS => '0');
    ELSIF (Clock'EVENT AND Clock = '1') THEN
      IF (Load = '1') THEN
        Q <= D;
      END IF;
    END IF;
  END PROCESS;
END Behavior;
```

FIGURE 5.29 Behavioral VHDL code for a 4-bit register.

or memory. So, instead of having several individual registers, we want to have an array of registers. This array of registers is known as a **register file**. In a register file, all the respective control signals for the individual registers are connected in common. Furthermore, all of the respective data input and output lines for all of the registers also are connected in common. For example, the *Load* lines for all of the registers are connected together, and all of the D_3 data lines for all of the registers are connected together. So the register file has only one set of input lines and one set of output lines for all of the registers. In addition, address lines are used to specify which register in the register file is to be accessed.

In a microprocessor circuit requiring an ALU, the register file usually is used for the source operands of the ALU. Because the ALU usually takes two input operands, we like the register file to be able to output two values from possibly two different locations of the register file at the same time. So, a typical register file will have one write port and two read ports. All three ports will have their own enable and address lines. When the read enable line is de-asserted, the read port will output a 0. On the other hand, when the read enable line is asserted, the content of the register specified by the read address lines is passed to the output port. The write enable line is used to load a value into the register specified by the write address lines.

The logic symbol for a 4 × 8 register file (four registers, each being 8 bits wide) is shown in Figure 5.30. The 8-bit write port is labeled *In*, and the two 8-bit read ports are labeled *Port A* and *Port B*. *WE* is the active-high write enable line. To write a value into the register file, this line must be asserted. The WA_1 and WA_0 are the two address lines for selecting the write location. Because there are four locations in this register file, two address lines are needed. The *RAE* line is the read enable line for *Port A*. The two read address lines for *Port A* are RAA_1 and RAA_0. For *Port B*, we have the *Port B* enable line, *RBE*, and the two address lines, RBA_1 and RBA_0.

The register circuit from Figure 5.27 does not have any control for the reading of the data to the output port. In order to control the output of data, we can use a 2-input AND gate to enable or disable each of the data output lines, Q_i. We want to control all the data output lines together, therefore, one input from all of the 2-input AND gates are connected in common. When this common input is set to a 0, all of the AND gates will output a 0. When this common input is set to a 1, the output for all the AND gates will be the value from the other input. An alternative to using AND gates to control the read ports is to use tri-state buffers. Instead of outputting a 0 when disabled, the tri-state buffers will have a high impedance and output the Z value.

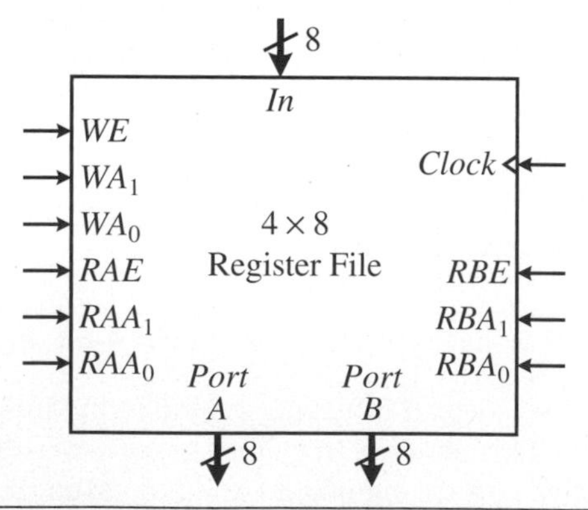

FIGURE 5.30 Logic symbol for a 4 × 8 register file.

Our register file has two read ports, that is, two output controls for each register. So, instead of having just one 2-input AND gate per output line, Q_i, we need to connect two AND gates to each output line: one for *Port A*, and one for *Port B*. An 8-bit wide register file cell circuit will have eight AND gates for *Port A*, and another eight AND gates for *Port B*, as shown in Figure 5.31. *AE* and *BE* are the read enable signals for *Port A* and *Port B*, respectively. For each read port, the read enable signal is connected in common to one input of all of the eight AND gates. The second input from each of the eight AND gates connects to the eight output lines, Q_0 to Q_7.

For a 4 × 8 register file, we need to use four 8-bit register file cells. In order to select which register file cell we want to access, three decoders are used to decode the addresses: WA_1, WA_0, RAA_1, RAA_0, RBA_1, and RBA_0. One decoder is used for the write addresses, WA_1 and WA_0; one for the *Port A* read addresses, RAA_1 and RAA_0; and one for the *Port B* read addresses, RBA_1 and RBA_0. The decoders' outputs are used to assert the individual register file cell's write line, *Load*, and read enable lines, *AE*, and *BE*. The complete circuit for the 4 × 8 register file is shown in Figure 5.32. The respective read ports from each register file cell are connected to the external read port through a 4-input × 8-bit OR gate.

For example, to read from Register 3 through *Port B*, the *RBE* line has to be asserted, and the *Port B* address lines RBA_1 and RBA_0 have to be set to 11 (for Register 3). The data from Register 3 will be available immediately on *Port B*. To write a value to Register 2, the write address lines WA_1 and WA_0 are set to 10, and then the write enable line *WE* is asserted. The data at input *D* is then written into Register 2 at the next active (rising) clock edge. Because all three decoders can be enabled at the same time, the two read operations and the write operation can occur simultaneously.

In terms of the timing issues, the data on the read ports is available immediately after the read enable line is asserted, whereas, the write occurs at the next active (rising) edge of the clock. Because of this, the same register can be accessed for both reading and writing at the same time; in other words, the read and write enable lines can be asserted at the same time using the same read and write address. When this happens, the value that is currently in the register is read through the read port, and a new value will be written into the register at the next rising clock edge. This timing is shown in Figure 5.33. The important point to remember is that when the read and write operations are asserted in the same clock cycle on the same register, the read operation

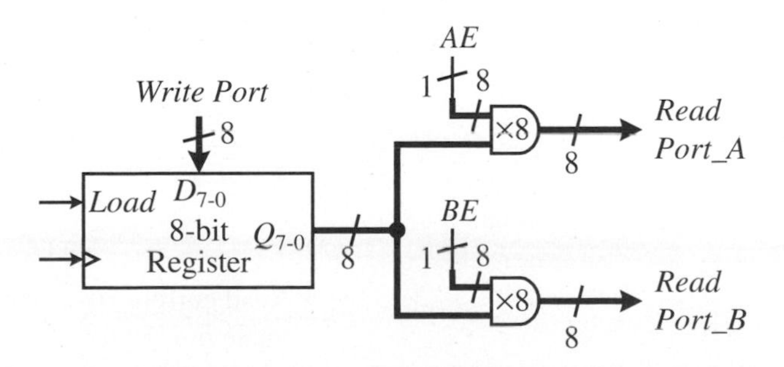

FIGURE 5.31 An 8-bit wide register file cell with one write port and two read ports.

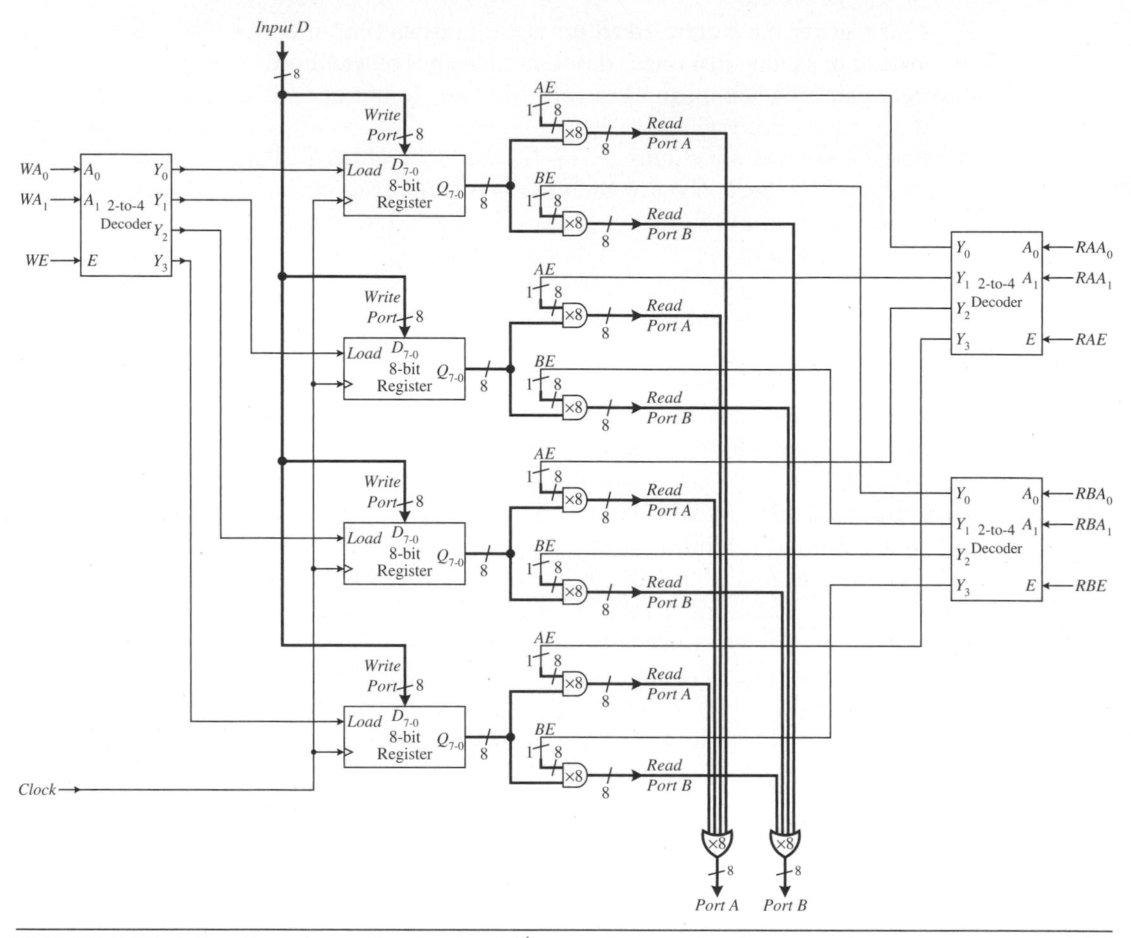

FIGURE 5.32 A 4 × 8 register file circuit with one write port and two read ports.

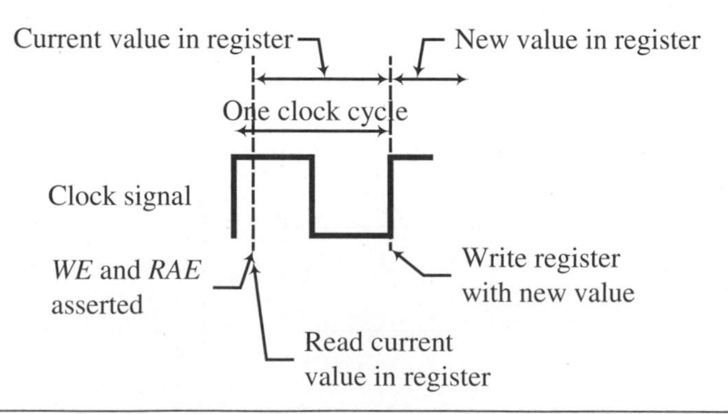

FIGURE 5.33 Read and write timings for a register file cell.

always reads the current value stored in the register and never the new value that is to be written in by the write operation. The new value written in is available only after the next rising clock edge.

Verilog Code for a Register File

The Verilog code for the 4 × 8 register file is shown in Figure 5.34. Four internal registers are used to save the contents of the four register file locations. The main code is composed of three `always` blocks: two for the two read ports and one for the write

```
// a 4-location x 8-bit register file
module regfile (
  input Clock,
  input WE,
  input [1:0] WA,
  input [7:0] D,
  input RAE, RBE,
  input [1:0] RAA, RBA,
  output reg [7:0] PortA,
  output reg [7:0] PortB
);
  reg [7:0]    reg0, reg1, reg2, reg3;

  // write
  always @ (posedge clock) begin
    if (WE)
      case (WA)
        0: reg0 <= D;
        1: reg1 <= D;
        2: reg2 <= D;
        3: reg3 <= D;
      endcase
  end

  // continuously output Port A
  always @ (RAA, RAE) begin
    if (RAE) begin
      case (RAA)
        0: PortA <= reg0;
        1: PortA <= reg1;
        2: PortA <= reg2;
        3: PortA <= reg3;
        default: PortA <= 8'h00;
      endcase
    end else begin
      PortA <= 8'h00;
    end
  end
```

FIGURE 5.34 Verilog code for a 4 × 8 register file with one write port and two read ports. *(continued on next page)*

```
  // output Port B
  always @ (RBA, RBE) begin
    if (RBE) begin
      case (RBA)
        0: PortB <= reg0;
        1: PortB <= reg1;
        2: PortB <= reg2;
        3: PortB <= reg3;
        default: PortB <= 8'h00;
      endcase
    end else begin
      PortB <= 8'h00;
    end
  end

endmodule
```

FIGURE 5.34 Verilog code for a 4 × 8 register file with one write port and two read ports.

port. These three `always` blocks are similar to three concurrent statements in that they are executed in parallel.

VHDL Code for a Register File

The VHDL code for the 4 × 8 register file is shown in Figure 5.35. The main code is composed of three processes: the write process and the two read port processes. These three processes are similar to three concurrent statements in that they are executed in parallel. The write process is sensitive to the clock, and because of the IF *clock* statement in the process, a write occurs only at the rising edge of the clock signal. The two read port processes are not sensitive to the clock but only to the read enable and read address signals. So the read data is available immediately when these lines are asserted. The function CONV_INTEGER(WA) converts the STD_LOGIC_VECTOR *WA* to an integer so that the address can be used as an index into the *RF* array.

5.14 Memories

Memories are large storage areas used in a computer system to store both data and instructions. They are usually separate components located external to the microprocessor. Registers and register files, on the other hand, are usually much smaller in capacity and are incorporated as part of the datapath inside the microprocessor. There are basically two types of memories, volatile and nonvolatile. Volatile memories are those that will lose their contents when power is removed, whereas nonvolatile memories will retain their contents even without power. Some common nonvolatile memories include read only memory (ROM), electrically erasable programmable read only memory (EEPROM), and flash memory. Common volatile memories include static and

```
LIBRARY IEEE;
USE IEEE.STD_LOGIC_1164.ALL;
USE IEEE.STD_LOGIC_UNSIGNED.ALL; -- needed for CONV_INTEGER()

ENTITY regfile IS PORT(
  Clock: IN STD_LOGIC;                            --clock
  WE: IN STD_LOGIC;                               --write enable
  WA: IN STD_LOGIC_VECTOR(1 DOWNTO 0);            --write address
  D: IN STD_LOGIC_VECTOR(7 DOWNTO 0);             --input
  RAE, RBE: IN STD_LOGIC;                --read enable ports A & B
                                         --read address ports A & B
  RAA, RBA: IN STD_LOGIC_VECTOR(1 DOWNTO 0);
  --output ports A & B
  PortA, PortB: OUT STD_LOGIC_VECTOR(7 DOWNTO 0));
END regfile;

ARCHITECTURE Behavioral OF regfile IS
  SUBTYPE reg  IS STD_LOGIC_VECTOR(7 DOWNTO 0);
  TYPE  regArray IS ARRAY(0 TO 3) OF reg;
  SIGNAL  RF: regArray;                  --register file contents
BEGIN
  WritePort: PROCESS (clock)
  BEGIN
    IF (clock'EVENT AND clock = '1') THEN
      IF (WE = '1') THEN
        -- fn to convert from vector to integer
        RF(CONV_INTEGER(WA)) <= D;
      END IF;
    END IF;
  END PROCESS;

  ReadPortA: PROCESS (RAA, RAE)
  BEGIN
   -- Read Port A
   IF (RAE = '1') THEN
      -- fn to convert from vector to integer
      PortA <= RF(CONV_INTEGER(RAA));
   ELSE
      PortA <= (OTHERS => '0');
    END IF;
  END PROCESS;

  ReadPortB: PROCESS (RBE, RBA)
  BEGIN
    -- Read Port B
```

FIGURE 5.35 **VHDL code for a 4 × 8 register file with one write port and two read ports.** *(continued on next page)*

```
      IF (RBE = '1') THEN
        -- fn to convert from vector to integer
        PortB <= RF(CONV_INTEGER(RBA));
      ELSE
        PortB <= (OTHERS => '0');
      END IF;
    END PROCESS;
  END Behavioral;
```

FIGURE 5.35 VHDL code for a 4 × 8 register file with one write port and two read ports.

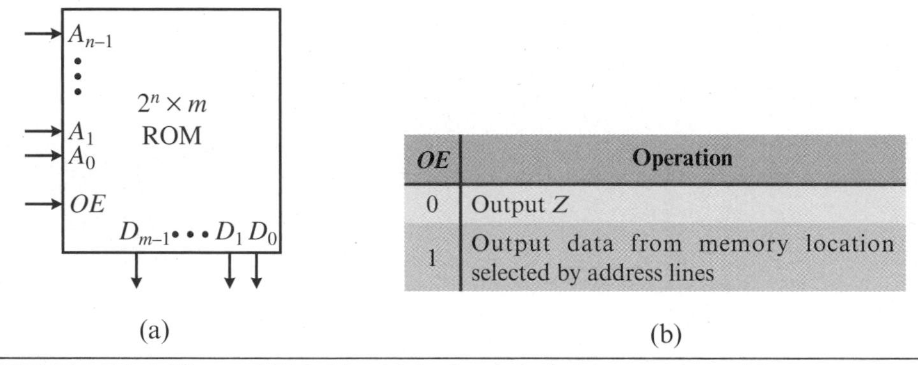

OE	Operation
0	Output *Z*
1	Output data from memory location selected by address lines

FIGURE 5.36 A $2^n \times m$ ROM chip: (a) logic symbol; (b) operation table.

dynamic random access memory (RAM.) Both types of memories are usually needed in a computer system.

We can make memory the same way we make the register file but with more storage locations. However, there are several reasons why we do not want to. One reason is that we usually want a lot of memory and we want it to be inexpensive, so we need to make each memory cell as small as possible. The tradeoff for this is that memory access is usually much slower than register access. Another reason is that we want to use a common bidirectional data bus both to read data from and to write data to the memory. This implies that the memory circuit should have just one data port (and not two or three like the register file) for both reading and writing of data.

5.14.1 ROM

ROM is nonvolatile. Because ROM retains its memory contents even when power is removed, it typically is used to store the boot loader for the operating system, or the entire operating system if it is small enough. When a computer system is first turned on, it needs to already have instructions to execute, and therefore, the computer system needs to have nonvolatile memory containing the initial instructions for it to execute.

The address of this memory needs to start at zero because this is the value in the program counter (PC) when it is reset or initialized. ROM can only be read from; no write operations can be performed on it.

The logic symbol for a typical ROM chip is shown in Figure 5.36(a). It has a set of data lines, D_i, and a set of address lines, A_i. The number of data lines is dependent on how many bits are used for storing data in each memory location. The number of address lines is dependent on how many locations are in the memory chip. For example, a 512-byte memory chip will have eight data lines (8 bits = 1 byte) and nine address lines ($2^9 = 512$).

The data lines output the data from the memory location that is specified by the address lines. In addition to the data and address lines, there is also an output enable, *OE*, control line. Data from a memory location is available on the data lines only when *OE* is asserted. There is no clock signal, so the output of the data is not synchronized to the clock.

HDL Code for ROM

The behavioral Verilog and VHDL descriptions of a ROM chip are shown in Figures 5.37 and 5.38, respectively. At the behavioral level, the HDL code simply uses a two-dimensional array to construct the ROM and to specify its size. In the two figures, a 16-location × 8-bit wide ROM is created. Assignment statements are used to initialize each of the memory locations with an 8-bit bit string. In this ROM description, we have initialized the ROM with the instructions that will be executed by our EC-1 microprocessor to be discussed in Chapter 8.

```
module rom
#(parameter size=4)
(
  input [size-1:0] Address,
  input OE,
  output [7:0] Data
);

  reg [7:0] mem[0:2**size-1];

  // initialize ROM with EC-1 countdown program
  initial begin
    mem[0] <= 8'b01100000;  // IN A
    mem[1] <= 8'b10000000;  // OUT A
    mem[2] <= 8'b10100000;  // DEC A
    mem[3] <= 8'b11000001;  // JNZ 0001
    mem[4] <= 8'b11111111;  // HALT
  end

  assign Data = (OE) ? mem[Address] : 8'bz;

endmodule
```

FIGURE 5.37 Verilog code for a 16 × 8 ROM.

```
LIBRARY IEEE;
USE IEEE.STD_LOGIC_1164.ALL;
USE IEEE.STD_LOGIC_UNSIGNED.ALL;    -- needed for CONV_INTEGER()

ENTITY rom IS
GENERIC (size: INTEGER := 4);
PORT(
  Address: IN STD_LOGIC_VECTOR(size-1 DOWNTO 0);
  OE: IN STD_LOGIC;
  Data: OUT STD_LOGIC_VECTOR(7 DOWNTO 0));
END rom;

ARCHITECTURE Behavioral OF rom IS
  TYPE mem_array IS ARRAY(0 TO (2**size)-1)OF
    STD_LOGIC_VECTOR(7 DOWNTO 0);
  -- initialize ROM with EC-1 countdown program
  CONSTANT mem: mem_array := (
        "01100000",  -- IN A
        "10000000",  -- OUT A
        "10100000",  -- DEC A
        "11000001",  -- JNZ 0001
        "11111111",  -- HALT
        "00000000","00000000","00000000","00000000","00000000",
        "00000000","00000000","00000000","00000000","00000000",
        "00000000"
        );
BEGIN
  Data <= mem(CONV_INTEGER(Address)) WHEN OE = '1' ELSE
    (OTHERS => 'Z');
END Behavioral;
```

FIGURE 5.38 VHDL code for a 16 × 8 ROM.

5.14.2 **RAM**

Unlike ROM, RAM is volatile, and capable of both reading and writing data. The logic symbol, showing all of the connections for a typical RAM chip is shown in Figure 5.39(a). There is a set of data lines, D_i, and a set of address lines, A_i. The bidirectional data lines serve for both input and output of the data to the location that is specified by the address lines. The number of data lines is dependent on how many bits are used for storing data in each memory location. The number of address lines is dependent on how many locations are in the memory chip. For example, a 512-byte memory chip will have eight data lines (8 bits = 1 byte) and nine address lines ($2^9 = 512$).

The operation of the RAM chip is shown in Figure 5.39(b). In addition to the data and address lines, there are usually two control lines: chip enable *CE*, and write enable *WR*. In order for a microprocessor to access memory, either with the read operation or with the write operation, the *CE* line must first be asserted. Asserting the *CE* line enables the entire memory chip. The active-high *WR* line selects which of the two

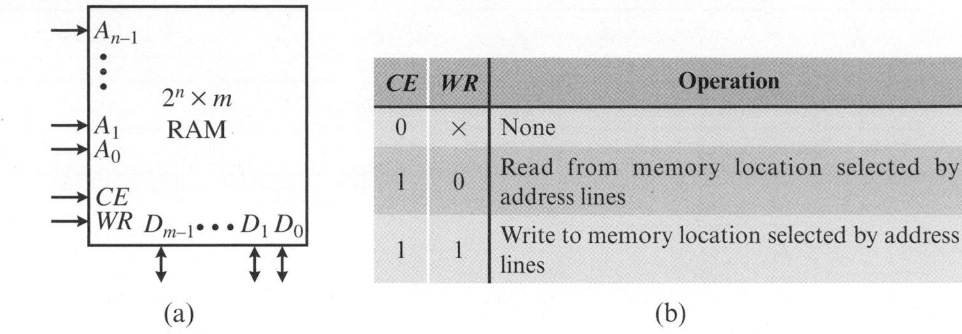

CE	*WR*	Operation
0	×	None
1	0	Read from memory location selected by address lines
1	1	Write to memory location selected by address lines

FIGURE 5.39 **A $2^n \times m$ RAM chip: (a) logic symbol; (b) operation table.**

memory operations is to be performed. Setting WR to a 0 selects the read operation, and data from the memory is retrieved. Setting WR to a 1 selects the write operation, and data from the microprocessor is written into the memory. The memory location in which the read and write operations are to take place, of course, is selected by the value of the address lines.

Notice in Figure 5.39(a) that the RAM chip does not require a clock signal, because neither the read operation nor the write operation is synchronized to the global system clock. Instead, the data operations are synchronized to the two control lines, CE and WR. Figure 5.40(a) shows the timing diagram for a memory write operation. The write operation begins with a valid address on the address lines, followed immediately by the CE line being asserted. Shortly after, valid data must be present on the data lines, and then the WR line is asserted. As soon as the WR line is asserted, the data that is on the data lines is written into the memory location that is addressed by the address lines.

A memory read operation also begins with setting a valid address on the address lines, followed by CE going high. The WR line then is pulled low, and shortly after, valid data from the addressed memory location are available on the data lines. The timing diagram for the read operation is shown in Figure 5.40(b).

Each bit in a static RAM chip is stored in a memory cell similar to the circuit shown in Figure 5.41(a). The main component in the cell is a D latch with enable. A tri-state buffer is connected to the output of the D latch so that it can be read from selectively. The *Cell enable* signal is used to enable the memory cell for both reading and writing. For reading, the *Cell enable* signal is used to enable the tri-state buffer. For writing, the *Cell enable* and *Write enable* signals are used together to enable the D latch so that the data on the *Input* line is latched into the cell. The logic symbol for the memory cell is shown in Figure 5.41(b).

To create a 4×4 static RAM chip, we need 16 memory cells forming a 4×4 grid, as shown in Figure 5.42. Each row forms a single storage location, and the number of memory cells in a row determines the bit width of each location. So all of the memory cells in a row are enabled with the same address. Again, a decoder is used to decode the address lines, A_0 and A_1. In this example, a 2-to-4 decoder is used to decode the four address locations. The CE signal is for enabling the chip, specifically to enable the read

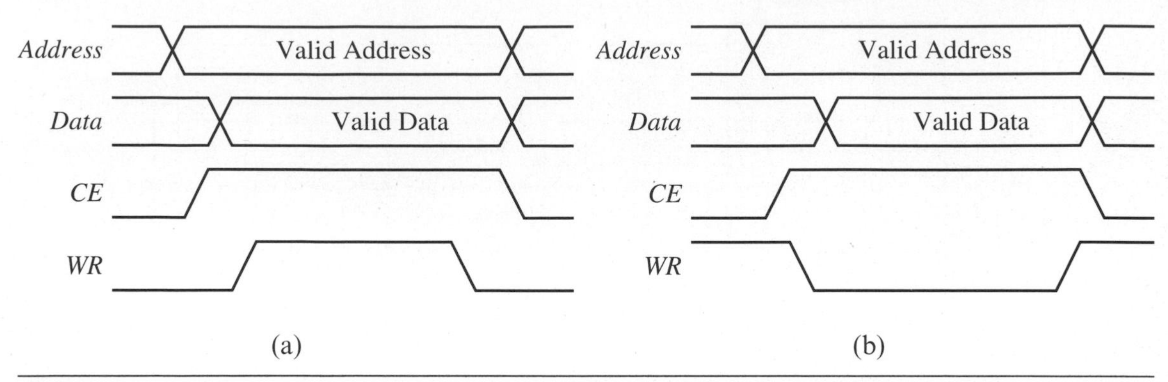

FIGURE 5.40 Memory timing diagram: (a) write operation; (b) read operation.

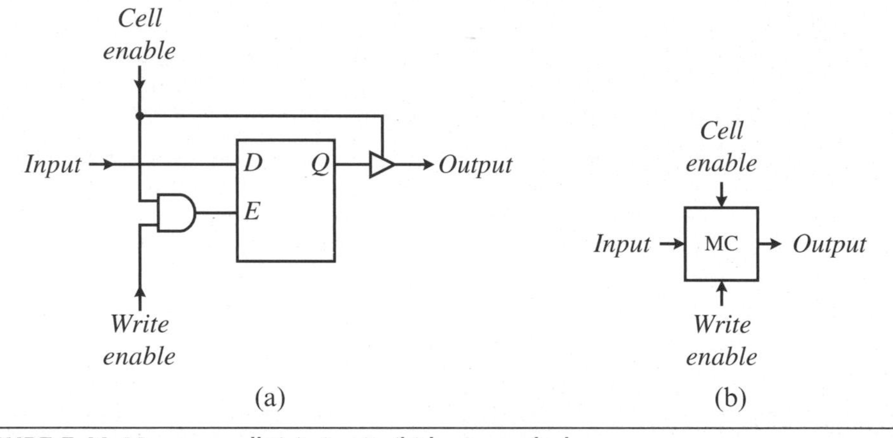

FIGURE 5.41 Memory cell: (a) circuit; (b) logic symbol.

and write functions through the two AND gates. The internal *WE* signal, asserted when both the *CE* and *WR* signals are asserted, is used to assert the *Write enable* signals for all of the memory cells. The data comes in from the external data bus, D_i, through the input buffer and to the *Input* line of each memory cell. An input buffer is used for each data line so that the external signal coming in needs to drive only one device (the buffer) rather than having to drive several devices (i.e., all of the memory cells in the same column). The row of memory cells into which data are actually written depend on the given address. The read operation requires *CE* to be asserted and *WR* to be de-asserted. Which will assert the internal *RE* signal, which in turn will enable the four output tri-state buffers at the bottom of the circuit diagram. Again, the location that is read from is selected by the address lines.

HDL Code for RAM

The behavioral Verilog and VHDL code for a typical RAM with a single bidirectional data bus are shown in Figures 5.43 and 5.44, respectively. The bidirectional data bus is declared as `inout`. The bidirectional data bus is tri-stated when the RAM is disabled with $CE = 0$.

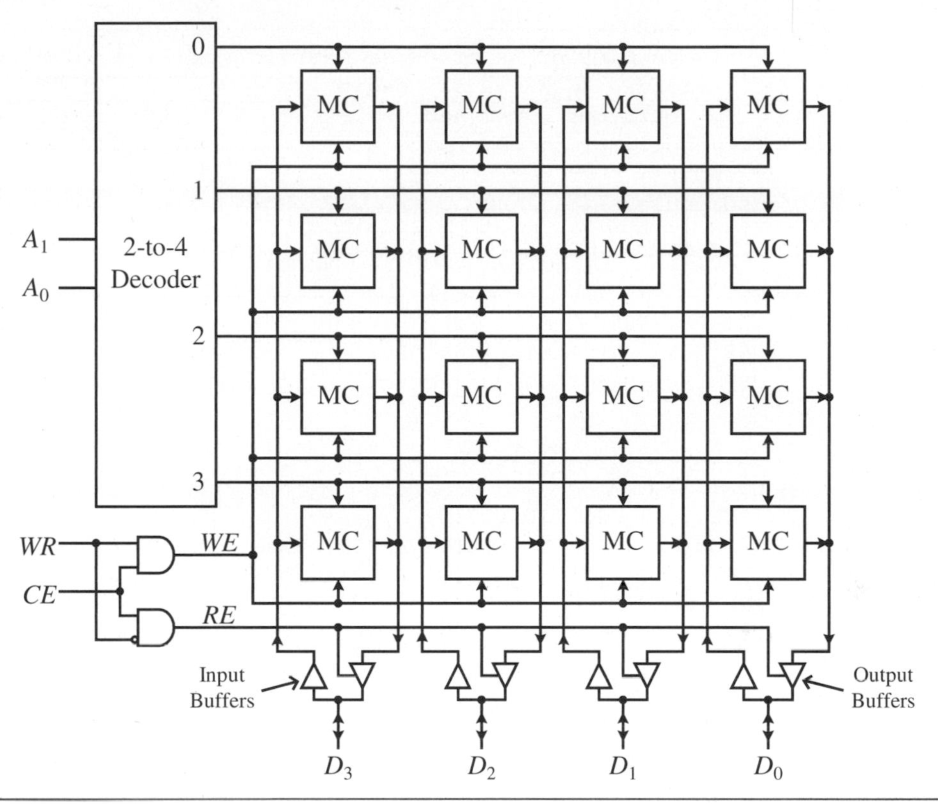

FIGURE 5.42 A 4 × 4 RAM chip circuit.

```
module ram
#(parameter size=5)
(
  input CE, WR, OE,  // chip enable; write; output enable
  input [size-1:0] Address,
  inout [7:0] Data  // bi-directional data bus
);

  reg [7:0] mem[0:2**size-1];
  reg [7:0] data_out;  // internal data

  // Tri-state buffer control
  assign Data = (CE && OE && ~WR) ? data_out : 8'bz;

  // write block
  always @(CE or WR or Data or Address) begin
    if (CE && WR) begin
      mem[Address] = Data;
    end
  end
```

FIGURE 5.43 Verilog code for a typical RAM. *(continued on next page)*

```
  // read
  always @(CE or WR or OE or Address) begin
    if (CE && ~WR && OE) begin
      data_out = mem[Address];
    end
  end
endmodule
```

FIGURE 5.43 Verilog code for a typical RAM.

```
LIBRARY IEEE;
USE IEEE.STD_LOGIC_1164.ALL;
USE IEEE.STD_LOGIC_unsigned.ALL;  -- needed for CONV_INTEGER()

ENTITY ram IS
GENERIC (size: INTEGER := 5);
PORT (
  CE, WR:  IN STD_LOGIC;
  Address: IN STD_LOGIC_VECTOR(size-1 DOWNTO 0);
  Data: INOUT STD_LOGIC_VECTOR(7 DOWNTO 0));
END ram;

ARCHITECTURE Behavioral OF ram IS
  TYPE memtype IS ARRAY(0 TO 2**size-1) OF
    STD_LOGIC_VECTOR(7 DOWNTO 0);
  -- initialize all bits to 0
  SIGNAL mem: memtype := (OTHERS => (OTHERS => '0'));
BEGIN
  PROCESS(CE,WR,Address)
  BEGIN
    IF (CE = '0') THEN
      Data <= (OTHERS => 'Z');
    ELSE
      IF (WR'EVENT AND WR = '0') THEN
        mem(CONV_INTEGER(Address)) <= Data;   -- write
      END IF;
      IF (WR = '0') THEN
        Data <= mem(CONV_INTEGER(Address));   -- read
      ELSE
        Data <= (OTHERS => 'Z');
      END IF;
    END IF;
  END PROCESS;
END Behavioral;
```

FIGURE 5.44 VHDL code for a typical RAM.

A computer system with an operating system allows the user to enter program instructions into the RAM, and then run the program. Our EC-2 computer system to be discussed in Chapter 8, however, does not have an operating system, so our RAM needs to be initialized with our program instructions on reset. Furthermore, to simplify the construction of our EC-2 microprocessor, we will use a RAM with separate read and write data ports. The behavioral Verilog and VHDL description of the RAM used in our EC-2 are shown in Figures 5.45 and 5.46, respectively.

5.15 Shift Registers

Similar to the combinational shifter and rotator circuits, there are the equivalent sequential shifter and rotator circuits. The circuits for the shift and rotate operations are constructed exactly the same. The only difference in the sequential version is that the operations are performed on the value that is stored in a register rather than directly

```
module ram(
#(parameter size=5)
  input Clock,
  input Reset,
  input WE,
  input [size-1:0] Address,
  input [7:0] D,
  output reg [7:0] Q
);

  reg [7:0] mem[2**size-1:0];

  always @(posedge Clock or posedge Reset) begin
    // this reset block and the Reset signal
    // is only needed to initialize the RAM locations
    if (Reset) begin
      // initialize RAM with EC-2 countdown program
      mem[0] <= 8'b10000000;  // IN A
      mem[1] <= 8'b01111111;  // SUB A,11111
      mem[2] <= 8'b10100100;  // JZ 00100
      mem[3] <= 8'b11000001;  // JPOS 00001
      mem[4] <= 8'b11111111;  // HALT
      mem[31]<= 8'b00000001;  // storage for the constant 1
    end else begin
      // write
      if (WE)
        mem[Address] <= D;
    end
  end  // always

  // read
  always @ (Address) begin
    Q <= mem[Address];
  end
endmodule
```

FIGURE 5.45 Verilog code for the 32 × 8 RAM used in the EC-2 microprocessor.

on the input value. The main usage for a shift register is to convert a serial-data input stream to a parallel-data output or vice versa. For a serial-to-parallel data conversion, the bits are shifted into the register at each clock cycle, and when all the bits (usually eight bits) are shifted in, the 8-bit register can be read to produce the 8-bit parallel

```
LIBRARY IEEE;
USE IEEE.STD_LOGIC_1164.ALL;
USE IEEE.STD_LOGIC_UNSIGNED.ALL;    -- needed for CONV_INTEGER()

ENTITY ram IS
GENERIC (size: INTEGER := 5);
PORT(
  Clock: IN STD_LOGIC;
  Reset: IN STD_LOGIC;
  WE: IN STD_LOGIC;
  Address: IN STD_LOGIC_VECTOR(size-1 DOWNTO 0);
  D: IN STD_LOGIC_VECTOR(7 DOWNTO 0);
  Q: OUT STD_LOGIC_VECTOR(7 DOWNTO 0));
END ram;

ARCHITECTURE Behavioral OF ram IS
  TYPE mem_type IS ARRAY(0 to (2**size)-1) OF
    STD_LOGIC_VECTOR(7 DOWNTO 0);
  SIGNAL mem: mem_type;
BEGIN
  PROCESS (Clock, Reset) IS
  BEGIN
    -- this reset block and the Reset signal
    -- is only needed to initialize the RAM locations
    IF (Reset = '1') THEN
      -- initialize RAM with EC-2 countdown program
      mem(0) <= "10000000";    -- IN A
      mem(1) <= "01111111";    -- SUB A, 11111
      mem(2) <= "10100100";    -- JZ 00100
      mem(3) <= "11000001";    -- JPOS 00001
      mem(4) <= "11111111";    -- HALT
      mem(31) <= "00000001";   -- storage for the constant 1
    ELSIF RISING_EDGE(Clock) THEN
      -- write
      IF (WE = '1') THEN
        mem(CONV_INTEGER(Address)) <= D;
      END IF;
    END IF;
  END PROCESS;

  -- read
  Q <= mem(CONV_INTEGER(Address));
END Behavioral;
```

FIGURE 5.46 **VHDL code for the 32 × 8 RAM used in the EC-2 microprocessor.**

output. For a parallel-to-serial conversion, the 8-bit register is first loaded with the input data. The bits are then shifted out individually, one bit per clock cycle, on the scrial output linc.

5.15.1 **Serial-to-Parallel Shift Register**

Figure 5.47(a) shows a 4-bit serial-to-parallel shift register. The input data bits come in on the *Serial_in* line at a rate of one bit per clock cycle. When *Shift* is asserted, the data bits are loaded in one bit at a time. In the first clock cycle, the first bit from the serial input stream, *Serial_in*, gets loaded into Q_3, while the original bit in Q_3 is loaded into Q_2, Q_2 is loaded into Q_1, and so on. In the second clock cycle, the bit that is in Q_3 (i.e., the first bit from the *Serial_in* line) gets loaded into Q_2, while Q_3 is loaded with the second bit from the *Serial_in* line. This continues for four clock cycles until four bits are shifted into the four flip-flops, with the first bit in Q_0, second bit in Q_1, and so on. These four bits then are available for parallel reading through the output Q. Figures 5.47(b) and (c) show the operation table and the logic symbol, respectively, for this shift register.

HDL Code for a Shift Register

The behavioral Verilog code for a 4-bit right-shift register is shown in Figure 5.48. The code for the actual shifting is performed in the concatenation operation { *Serial_in*, $Q[3:1]$ } where the most significant bit, *Serial_in*, is concatenated with the three upper bits of Q. The least significant bit of Q is discarded.

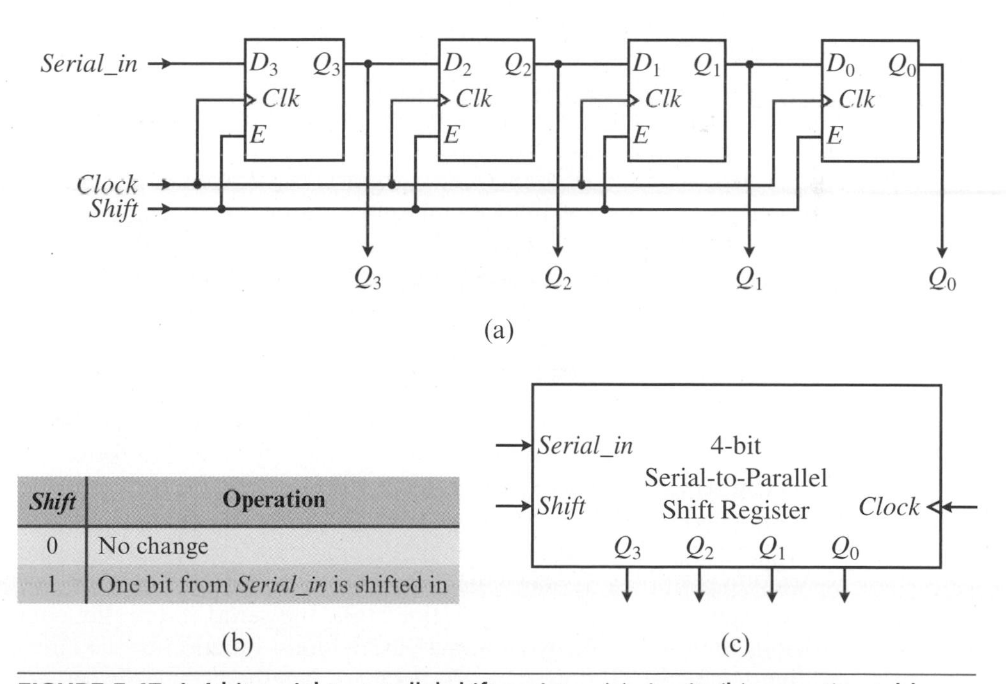

Shift	Operation
0	No change
1	Onc bit from *Serial_in* is shifted in

(b)

FIGURE 5.47 A 4-bit serial-to-parallel shift register: (a) circuit; (b) operation table; (c) logic symbol.

```
module ShiftReg (
  input Serial_in,
  input Clock,
  input Shift,
  output reg [3:0] Q
);

  always @(posedge Clock) begin
    if (Shift) begin
      Q <= { Serial_in, Q[3:1] };
    end
  end  // always

endmodule
```

FIGURE 5.48 Behavioral Verilog code for a 4-bit right-shift register.

The structural VHDL code for a 4-bit serial-to-parallel shift register is shown in Figure 5.49. The code is written at the structural level. The operation of a D flip-flop with enable is first defined. The ARCHITECTURE section for the *ShiftReg* entity uses four PORT MAP statements to instantiate four D flip-flops. These four flip-flops then are connected together using the internal signal N_0, N_1, N_2, and N_3 such that the output of one flip-flop is connected to the input of the next flip-flop. These four internal signals also connect to the four output signals Q_0 to Q_3 for the register output. We cannot use the output signals Q_0 to Q_3 to connect the four flip-flops together directly, because output signals cannot be read.

A sample simulation trace of the serial-to-parallel shift register is shown in Figure 5.50. At the first rising clock edge at time 100 ns, the *Serial_in* bit is a 0, so there is no change in the 4 bits of Q, because they are initialized to 0. At the next rising clock edge at time 300 ns, the *Serial_in* bit is a 1, and it is shifted into the leftmost bit of Q. Hence, Q has the value of 1000. At time 500 ns, another 1 bit is shifted in, giving Q the value of 1100. At time 700 ns, a 0 bit is shifted in, giving Q the value of 0110. Notice that as bits are shifted in, the rightmost bits are lost. At time 900 ns, *Shift* is de-asserted, so the 1 bit in the *Serial_in* line is not shifted in. Finally, at time 1.1 μs, another 1 bit is shifted in.

5.15.2 Serial-to-Parallel and Parallel-to-Serial Shift Register

For both the serial-to-parallel and parallel-to-serial operations, we perform the same left-to-right shifting of bits through the register. The only difference between the two operations is whether we want to perform a parallel read after the shifting or a parallel write before the shifting. For the serial-to-parallel operation, we want to perform a parallel read after the bits have been shifted in. On the other hand, for the parallel-to-serial operation, we want to perform a parallel write first and then shift the bits out as a serial stream.

We can implement both operations into the serial-to-parallel circuit from the previous section simply by adding a parallel load function to the circuit, as shown in Figure 5.51(a). The four multiplexers work together to select whether we want the flip-flops to retain the current value, load in a new value, or shift the bits to the right by one bit position. The operation of this circuit is dependent on the two select lines,

```
-- D flip-flop with enable
LIBRARY IEEE;
USE IEEE.STD_LOGIC_1164.ALL;

ENTITY D_flipflop IS
  PORT(D, Clock, E : IN STD_LOGIC;
      Q : OUT STD_LOGIC);
END D_flipflop;

ARCHITECTURE Behavior OF D_flipflop IS
BEGIN
  PROCESS(Clock)
  BEGIN
    IF (Clock'EVENT AND Clock = '1') THEN
      IF (E = '1') THEN
        Q <= D;
      END IF;
    END IF;
  END PROCESS;
END Behavior;

-- 4-bit shift register
LIBRARY IEEE;
USE IEEE.STD_LOGIC_1164.ALL;
ENTITY ShiftReg IS
  PORT(Serial_in, Clock, Shift : IN STD_LOGIC;
    Q : OUT STD_LOGIC_VECTOR(3 DOWNTO 0));
END ShiftReg;

ARCHITECTURE Structural OF ShiftReg IS
  SIGNAL N0, N1, N2, N3 : STD_LOGIC;
  COMPONENT D_flipflop PORT (D, Clock, E : IN STD_LOGIC;
      Q : OUT STD_LOGIC);
  END COMPONENT;

BEGIN
  U1: D_flipflop PORT MAP (Serial_in, Clock, Shift, N3);
  U2: D_flipflop PORT MAP (N3, Clock, Shift, N2);
  U3: D_flipflop PORT MAP (N2, Clock, Shift, N1);
  U4: D_flipflop PORT MAP (N1, Clock, Shift, N0);
  Q(3) <= N3;
  Q(2) <= N2;
  Q(1) <= N1;
  Q(0) <= N0;
END Structural;
```

FIGURE 5.49 Structural VHDL code for a 4-bit serial-to-parallel shift register.

$SHSel_1$ and $SHSel_0$, that control which input of the multiplexers is selected. The operation table and logic symbol are shown in Figures 5.51(b) and (c), respectively. The behavioral VHDL code and a sample simulation trace for this shift register are shown in Figures 5.52 and 5.53, respectively.

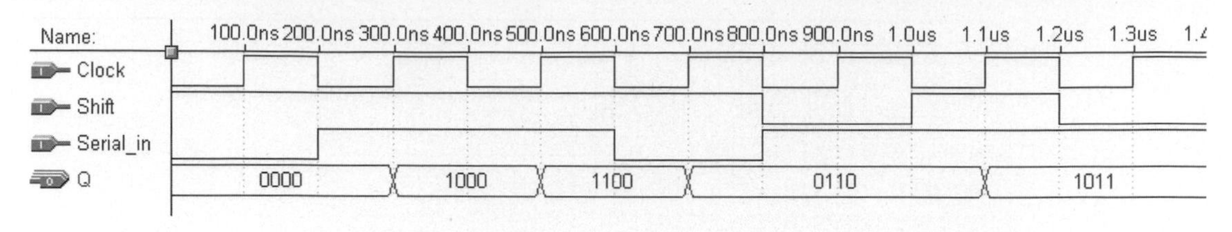

FIGURE 5.50 Sample simulation trace for the 4-bit serial-in-parallel-out shift register.

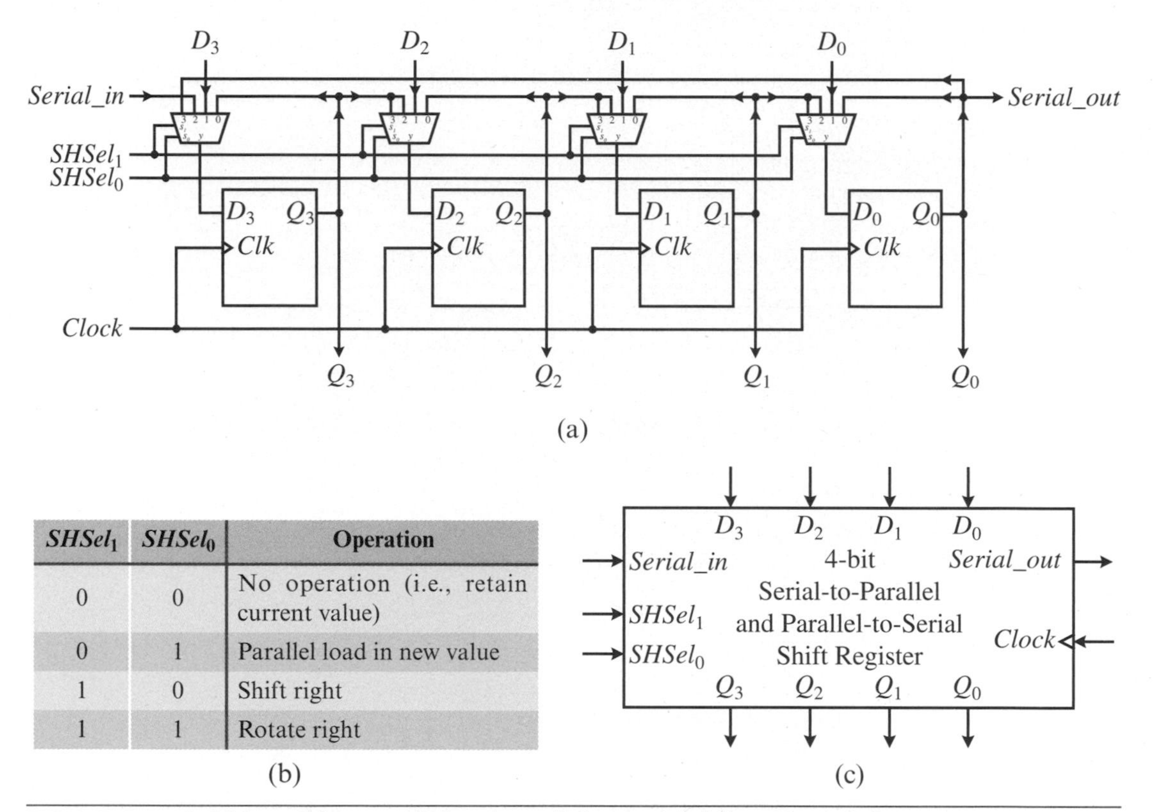

$SHSel_1$	$SHSel_0$	Operation
0	0	No operation (i.e., retain current value)
0	1	Parallel load in new value
1	0	Shift right
1	1	Rotate right

(b)

FIGURE 5.51 A 4-bit serial-to-parallel and parallel-to-serial shift register: (a) circuit; (b) operational table; (c) logic symbol.

5.15.3 Linear Feedback Shift Register

A linear feedback shift register (LFSR) is a special type of shift register where the input value to be shifted in is dependent on the register's current value. The LFSR is commonly used for generating pseudorandom numbers in digital circuits. The initial value in the register is called the seed, and it should be a non-zero number. The input

```
LIBRARY IEEE;
USE IEEE.STD_LOGIC_1164.ALL;

ENTITY shiftreg IS PORT (
  Clock: IN STD_LOGIC;
  SHSel: IN STD_LOGIC_VECTOR(1 DOWNTO 0);
  Serial_in: IN STD_LOGIC;
  D: IN STD_LOGIC_VECTOR(3 DOWNTO 0);
  Serial_out: OUT STD_LOGIC;
  Q: OUT STD_LOGIC_VECTOR(3 DOWNTO 0));
END shiftreg;

ARCHITECTURE Behavioral OF shiftreg IS
  SIGNAL content: STD_LOGIC_VECTOR(3 DOWNTO 0);
BEGIN
  PROCESS(Clock)
  BEGIN
    IF (Clock'EVENT AND Clock='1') THEN
      CASE SHSel IS
      WHEN "01" =>  -- load
        content <= D;
      WHEN "10" =>  -- shift right, pad with bit from Serial_in
        content <= Serial_in & content(3 DOWNTO 1);
      WHEN OTHERS =>
        NULL;
      END CASE;
    END IF;
  END PROCESS;

  Q <= content;
  Serial_out <= content(0);
END Behavioral;
```

FIGURE 5.52 Behavioral VHDL code for a 4-bit serial-to-parallel and parallel-to-serial shift register.

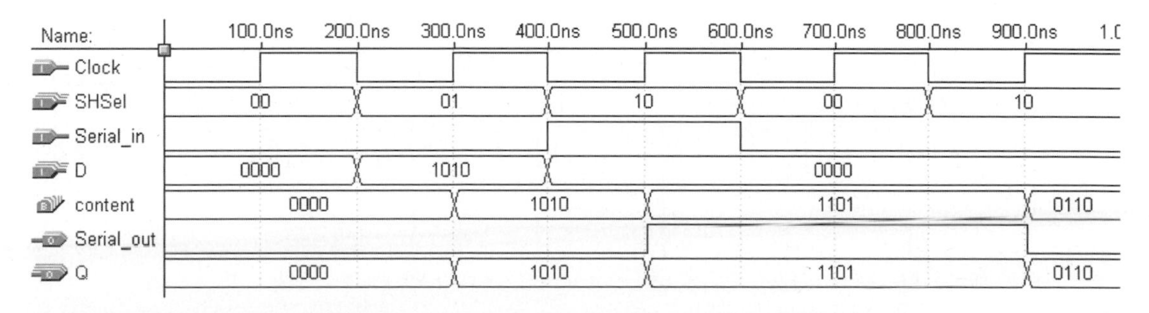

FIGURE 5.53 Sample trace for the 4-bit serial-to-parallel and parallel-to-serial shift register.

bit to shift in is driven by the XOR of some bits in the overall shift register value. The bits to be XORed must be well chosen in order to produce a sequence of numbers that appears random and has a very long cycle. An n-bit LFSR will have a maximum cycle length of $2^n - 1$ different combinations unless it contains all zeros, in which case it will never change because 0 XOR 0 is a 0.

The circuit for a 4-bit LFSR and an 8-bit LFSR are shown in Figures 5.54(a) and (b), respectively. To get the maximum cycle length, the bits to be XORed for the 4-bit LFSR must be bits 3 and 4, whereas for the 8-bit LFSR, the bits to be XORed must be bits 4, 5, 6, and 8. Notice that the bit numbering starts counting from 1 on the left and moving to the right, but reading the value from the register is still the same as before where the rightmost bit is the least significant.

Given an initial seed of 0101 for the 4-bit LFSR, the sequence of numbers produced by the 4-bit LFSR is:

0101, 1010, 1101, 1110, 1111, 0111, 0011, 0001, 1000, 0100, 0010, 1001, 1100, 0110, and 1011.

After 1011, the sequence repeats with 0101 as the next number. This sequence has the maximum cycle length of 15 numbers. If different bits are XORed, the circuit will produce a different sequence and it may not have the maximum cycle length.

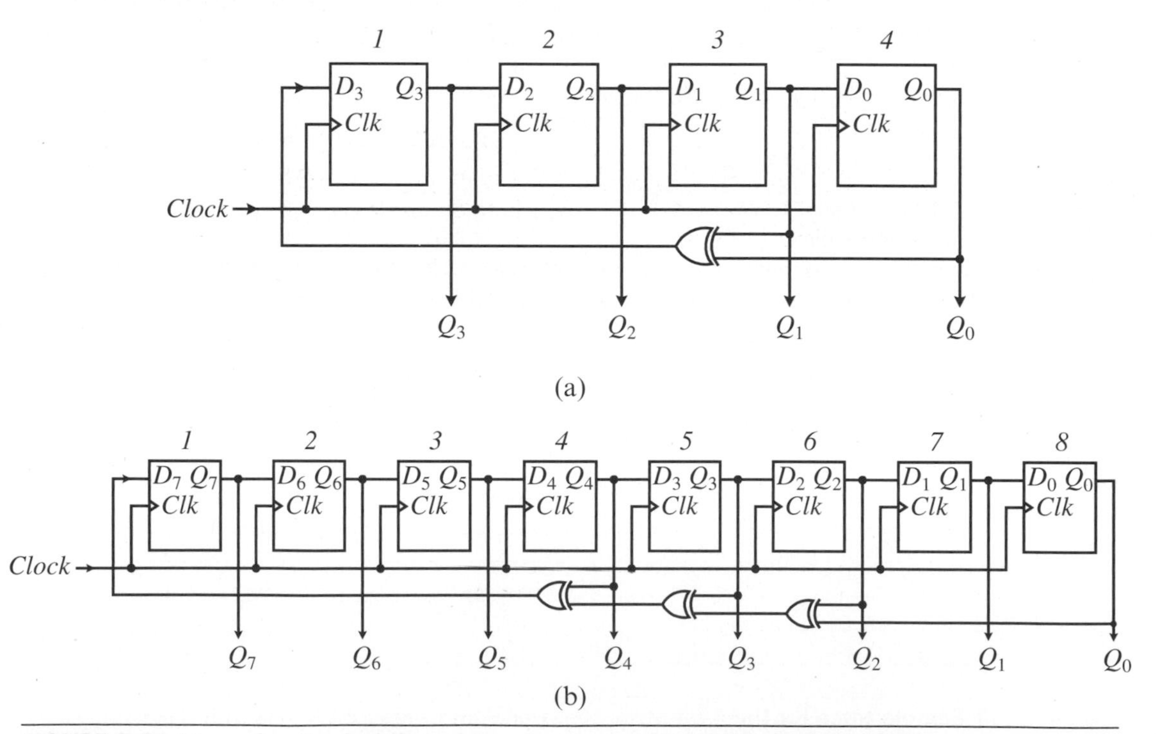

FIGURE 5.54 **Linear feedback shift register circuits: (a) 4-bit LFSR; (b) 8-bit LFSR.**

The following table shows the terms to be XORed to get the maximum cycle lengths for higher bit lengths,

Bit length	Terms to XOR
8	4, 5, 6, and 8
9	5 and 9
10	7 and 10
11	9 and 11
12	4, 10, 11, and 12
13	8, 11, 12, and 13

Bit length	Terms to XOR
14	2, 12, 13, and 14
15	14 and 15
16	11, 13, 14, and 16
17	14 and 17
18	11 and 18
19	14, 17, 18, and 19

5.16 Counters

Counters, as the name suggests, are for counting a sequence of values. However, there are many different types of counters, depending on the total number of count values, the sequence of values that it outputs, whether it counts up or down, and so on. The simplest is a modulo-n counter that counts the decimal sequence 0, 1, 2, . . . up to n-1 and back to 0.

5.16.1 Binary Up Counter

An n-bit binary up counter can be constructed using a modified n-bit register in which the data inputs for the register come from an adder. To get to the next up-count sequence from the value that is stored in a register, we simply have to add a 1 to it. We can use the full adder discussed in Section 4.2.1 as the input to the register, but we can do better by making it smaller. The full adder adds two operands plus the carry. But what we want is just to add a 1, so the second operand to the full adder is always a 1. Because the 1 also can be added in via the carry-in signal of the adder, we really do not need the second operand input. This modified adder that adds only one operand with the carry-in is called a **half adder** (HA); its truth table is shown in Figure 5.55(a). We have a as the only input operand, c_{in} and c_{out} are the carry-in and carry-out signals, respectively, and s is the sum of the addition. In the truth table, we are simply adding a plus c_{in} to give the sum s and possibly a carry-out, c_{out}. From the truth table, we obtain the two equations for c_{out} and s shown in Figure 5.55(b). The HA circuit is shown in Figure 5.55(c) and its logic symbol in (d).

Several HAs can be daisy-chained together, just like with the full adders to form an n-bit adder. The single operand input a comes from the register. The initial carry-in signal c_0 is used as the count enable signal, because a 1 on c_0 will result in incrementing a 1 to the register value, and a 0 will not. The resulting 4-bit binary up-counter circuit is shown in Figure 5.56(a), along with its operation table and logic symbol in (b) and (c), respectively. As long as *Count* is asserted, the counter will increment by 1 on each clock

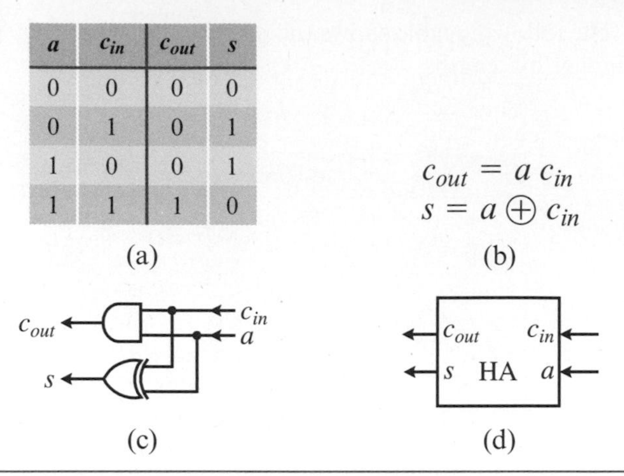

a	c_{in}	c_{out}	s
0	0	0	0
0	1	0	1
1	0	0	1
1	1	1	0

(a)

$$c_{out} = a\, c_{in}$$
$$s = a \oplus c_{in}$$

(b)

FIGURE 5.55 Half adder: (a) truth table; (b) equations; (c) circuit; (d) logic symbol.

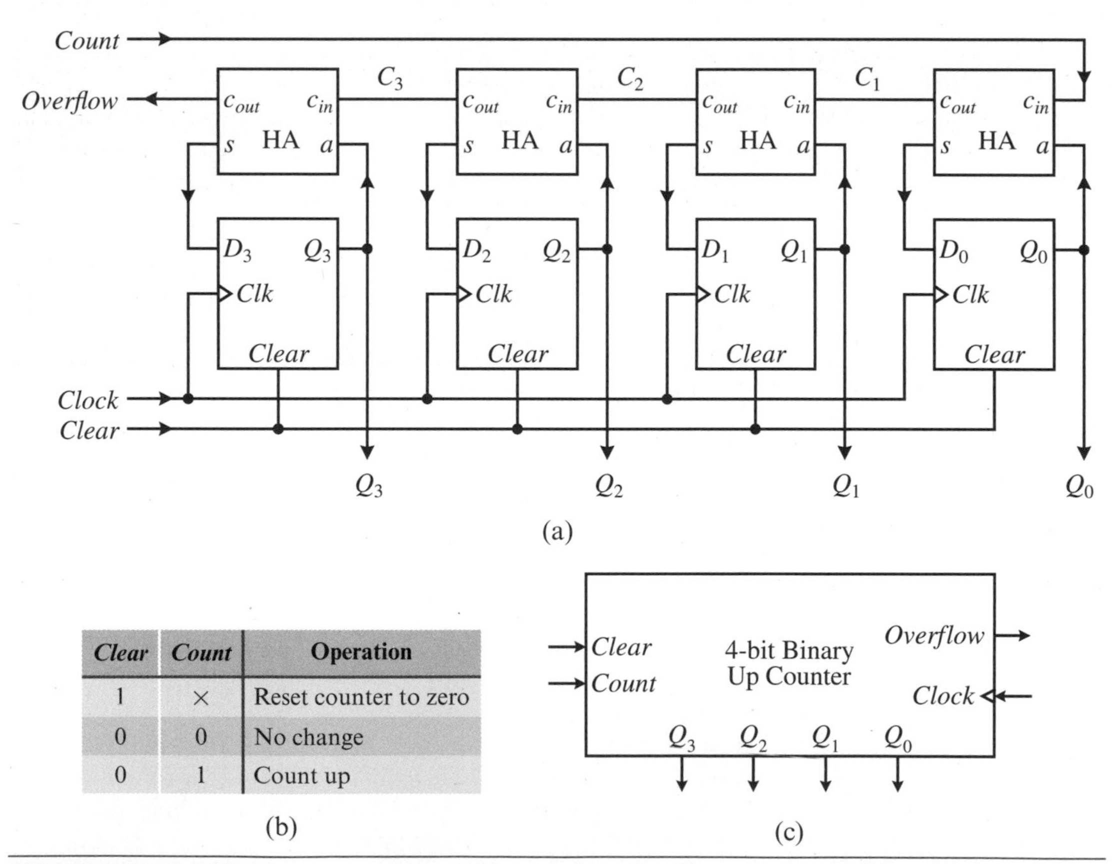

Clear	*Count*	Operation
1	×	Reset counter to zero
0	0	No change
0	1	Count up

(b)

FIGURE 5.56 A 4-bit binary up counter with asynchronous *Clear*: (a) circuit; (b) operation table; (c) logic symbol.

pulse until *Count* is de-asserted. When the count reaches $2^n - 1$ (which is equivalent to the binary number with all 1s), the next count will go back to 0, because adding a 1 to a binary number with all 1s will result in an overflow on the *Overflow* bit, and all the counter bits will reset to 0. The *Clear* signal allows an asynchronous reset of the counter to 0.

5.16.2 **Binary Up Counter with Parallel Load**

To make the binary counter more versatile, we need to be able to start the count sequence with any number other than zero. This is accomplished easily by modifying our counter circuit to allow it to load in an initial value. With the value loaded into the register, we can now count starting from this new value. The modified counter circuit is shown in Figure 5.57(a). The only difference between this circuit and the up counter circuit

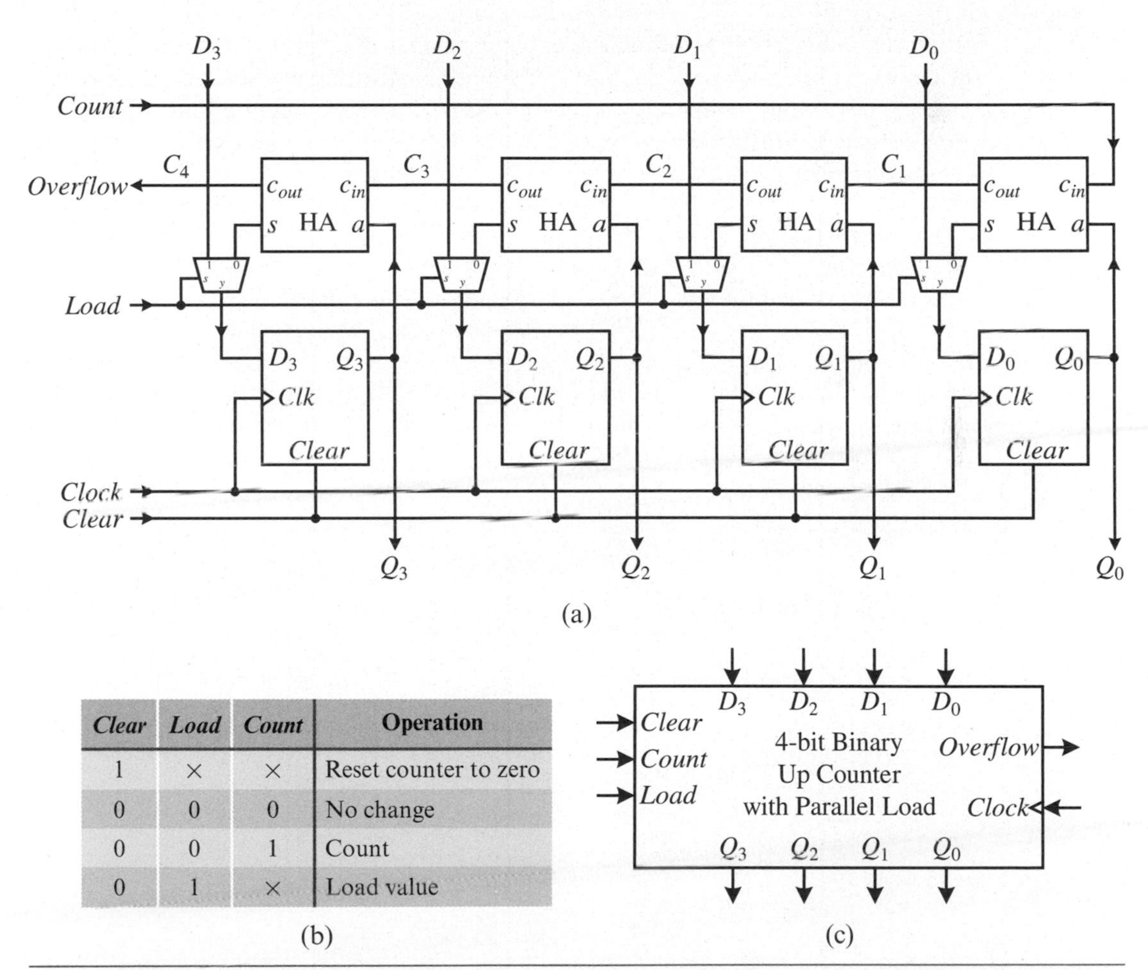

Clear	*Load*	*Count*	Operation
1	×	×	Reset counter to zero
0	0	0	No change
0	0	1	Count
0	1	×	Load value

(b)

FIGURE 5.57 A 4-bit binary up counter with parallel *Load* and asynchronous *Clear*: (a) circuit; (b) operation table; (c) logic symbol.

shown in Figure 5.56(a) is that a 2-input multiplexer is added between the *s* output of the HA and the D_i input of the flip-flop. By doing this, the D_i input of the flip-flop can be selected from either an external input value (if *Load* is asserted) or the next count value from the HA output (if *Load* is de-asserted). If the HA output is selected, then the circuit works exactly like before. If the external input is selected, then whatever value is presented on the input data lines will be loaded into the register. The operational table and logic symbol for this circuit are shown in Figures 5.57(b) and (c), respectively.

We have kept the *Clear* line, so that the counter still can be initialized to 0 at anytime. There is, however, a timing difference between asserting the *Clear* line to reset the counter to 0, as opposed to loading in a 0 by asserting the *Load* line and setting the data input to 0. In the first case, the counter is reset to 0 immediately after the *Clear* is asserted, while the latter case will reset the counter to 0 at the next rising edge of the clock.

This counter can start with whatever value is loaded into the register, but it will always count up to $2^n - 1$, where n is the number of bits for the register. This is when the register contains all 1s. When the counter reaches the end of the count sequence, it will always cycle back to 0, and not to the initial value that was loaded in. However, we can add a simple comparator to the Q_i output of this counter circuit so that the count sequence can start or end with any number in between, and cycle back to the new starting value.

HDL Code for a Counter

The behavioral Verilog code for a 4-bit binary up-down counter is shown in Figure 5.58. This counter will count continuously either up or down depending on the *Up* input signal. The actual count of the counter is stored in the internal register *value*. On every positive edge of the clock, the counter *value* either will be incremented or decremented by one. The internal counter value is then assigned to the output signal, *Q*. The counter *value* is also reset to zero when the asynchronous *Clear* signal is asserted.

```
module counter(
  input Clock,
  input Clear,
  input Up,
  output [3:0] Q
);

  reg [3:0] value = 0;

  always @(posedge Clock or posedge Clear) begin
    if (Clear)
      value <= 0;
    else if (Up)
      value <= value + 1;
    else
      value <= value - 1;
  end

  assign Q = value;

endmodule
```

FIGURE 5.58 Behavioral Verilog code for a 4-bit binary up-down counter.

The behavioral VHDL code for a 4-bit binary up counter is shown in Figure 5.59. The statement USE IEEE.STD_LOGIC_UNSIGNED.ALL is needed in order to perform additions on STD_LOGIC_VECTORS. The internal signal *value* is used to store the current count. When *Clear* is asserted, *value* is assigned the value "0000" using the expression OTHERS => '0'. Otherwise, if *Count* is asserted, then *value* will be incremented by 1 on the next rising clock edge. Furthermore, the count in *value* is assigned to the counter output *Q*, using the concurrent statement *Q* <= *value*, because it is outside the PROCESS block. A sample simulation trace is shown in Figure 5.60.

```
LIBRARY IEEE;
USE IEEE.STD_LOGIC_1164.ALL;
-- need this to add STD_LOGIC_VECTORs
USE IEEE.STD_LOGIC_UNSIGNED.ALL;

ENTITY counter IS PORT (
  Clock: IN STD_LOGIC;
  Clear: IN STD_LOGIC;
  Count: IN STD_LOGIC;
  Q   : OUT STD_LOGIC_VECTOR(3 DOWNTO 0));
END counter;

ARCHITECTURE Behavioral OF counter IS
  SIGNAL value: STD_LOGIC_VECTOR(3 DOWNTO 0);
BEGIN
  PROCESS (Clock, Clear)
  BEGIN
    IF (Clear = '1') THEN
      -- 4-bit vector of 0, same as "0000"
      value <= (OTHERS => '0');
    ELSIF (Clock'EVENT AND Clock='1') THEN
      IF (Count = '1') THEN
        value <= value + 1;
      END IF;
    END IF;
  END PROCESS;

  Q <= value;
END Behavioral;
```

FIGURE 5.59 Behavioral VHDL code for a 4-bit binary up counter.

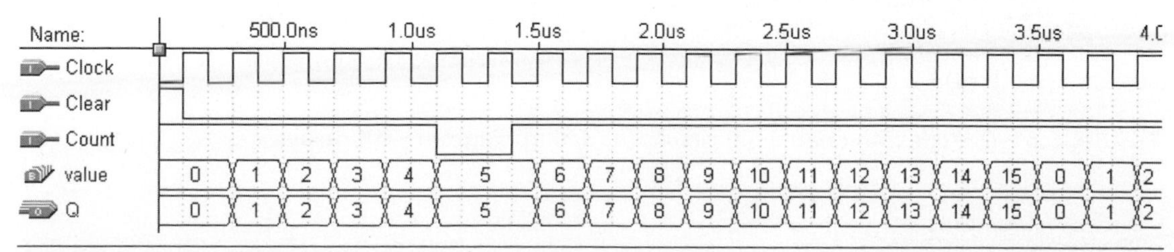

FIGURE 5.60 Simulation trace for the 4-bit binary up counter.

5.17 **Timing Issues**

So far in our discussion of latches and flip-flops, we have ignored timing issues and the effects of propagation delays. In practice, timing issues are very important in the correct design of sequential circuits. Considering the D latch with enable circuit from Section 5.6 the circuit is redrawn as shown in Figure 5.61(a). Signals from the inputs require some delay to propagate through the gates and finally to reach the outputs.

Assuming that the propagation delay for the inverter is 1 nanosecond (ns), and 2 ns for the NAND gates, the timing trace diagram would look like Figure 5.61(b) with the signal delays taken into consideration. The arrows denote which signal edge causes another signal edge. The number next to an arrow denotes the number of nanoseconds in delay for the resulting signal to change.

At time t_1, signal D drops to 0. This causes R to rise to 1 after a 1 ns delay through the inverter. The D edge also causes S' to rise to 1, but after a delay of 2 ns through the NAND gate. After that, R' drops to 0 at 2 ns after R rises to 1. This in turn causes Q' to rise to 1 after 2 ns, followed by Q dropping to 0.

At time t_2, signal E drops to 0, disabling the circuit. As a result, when D rises to 1 at time t_3, neither Q nor Q' is affected.

At time t_4, signal E rises to 1 and re-enables the circuit. This causes S' to drop to 0 after 2 ns. R' remains unchanged at 1 because the two inputs to the NAND gate, E and R, are 1 and 0, respectively. With S' asserted and R' de-asserted, the latch is set with Q rising to 1 at 2 ns after S' drops to a 0. This is followed by Q' dropping to 0 after another 2 ns.

Furthermore, for the D-latch circuit to latch in the data from input D correctly, there is a critical window of time right before and right after the falling edge of the

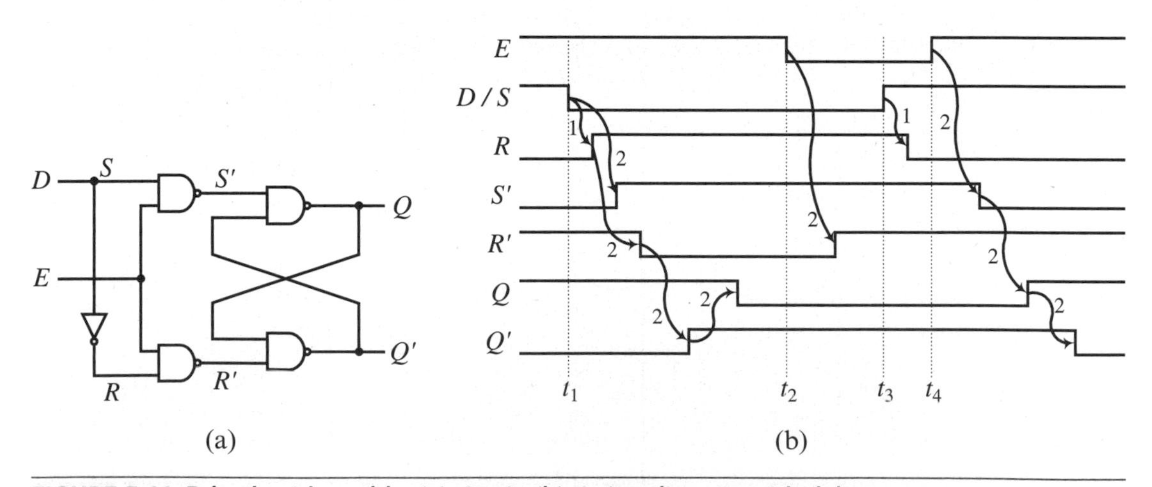

FIGURE 5.61 **D latch with enable: (a) circuit; (b) timing diagram with delays.**

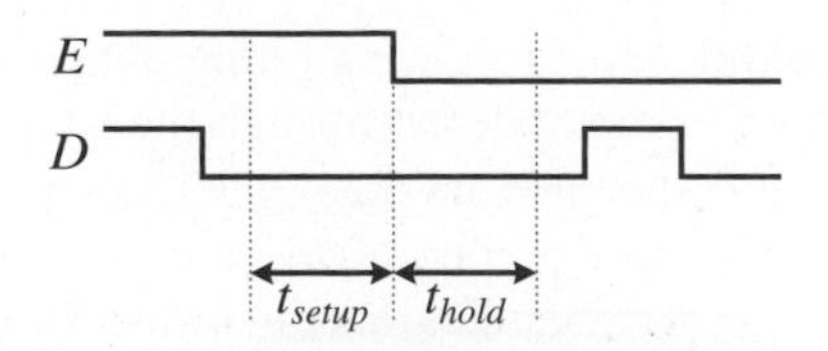

FIGURE 5.62 Setup and hold times for the gated *D* latch.

enable signal, *E*, that must be observed. Within this time frame, the input signal, *D*, must not change. As shown in Figure 5.62, the time before the falling edge of *E* is referred to as the **setup time**, t_{setup}, and the time after the falling edge of *E* is referred to as the **hold time**, t_{hold}. The length of these two times is dependent on the implementation and manufacturing process and can be obtained from the component data sheet.

5.18 PROBLEMS

5.1. Draw an SR latch with enable similar to that shown in Figure 5.6, but using NOR gates to implement the SR latch. Derive the truth table for this circuit.

5.2. Draw the D latch using NOR gates.

5.3. Draw the master-slave negative edge-triggered D flip-flop circuit.

5.4. Derive the truth table for a negative edge-triggered D flip-flop.

5.5. In the clock divider HDL code shown in Section 5.8, what should the value for *half* be in order to generate:

a) A 4 Hz output clock from the 50 MHz input clock source?

b) A 1 MHz output clock from the 50 MHz input clock source?

5.6. Complete the following truth table for the D latch with asynchronous *Set'* and *Clear'* circuit shown in Figure 5.23(a).

Clear'	*Set'*	*E*	*D*	*Q*	*Q'*
1	0	0	×		
0	1	0	×		
1	0	1	0		
0	1	1	1		
1	1	1	0		
1	1	1	1		
0	0	×	×		

5.7. Complete the following timing diagram for the following circuit. Assume that the signal delay through the NOR gates is 3 ns, and the delay through the NOT gate is 1 ns.

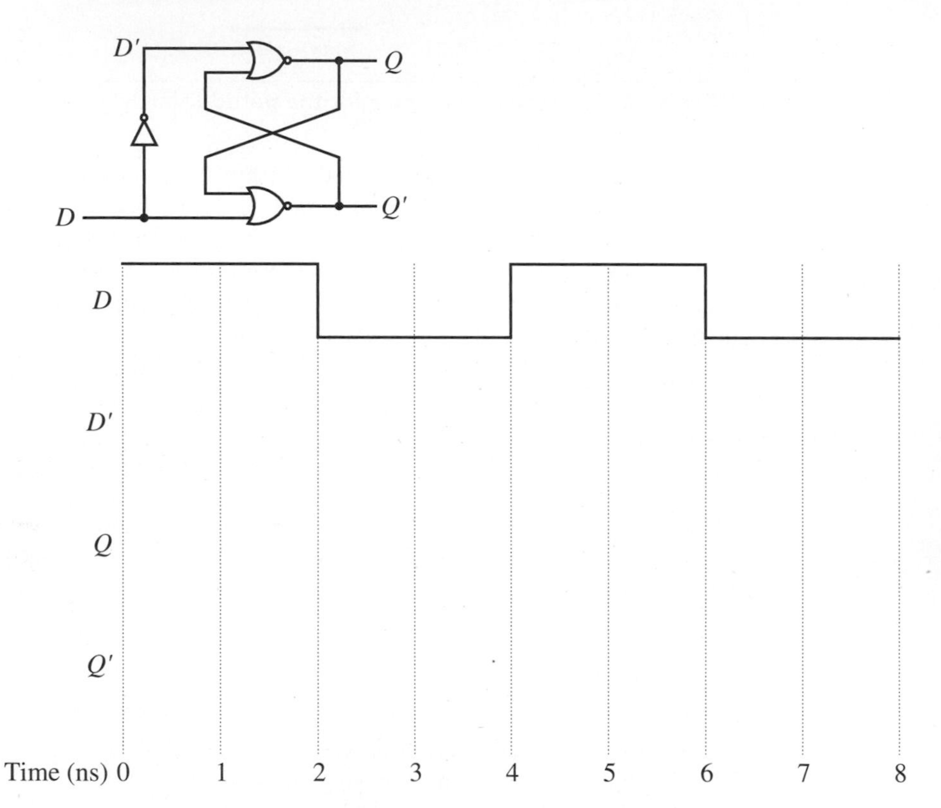

5.8. Traditionally, in addition to the D flip-flop, there are also three other types of flip-flops: SR, JK, and T. These names are given based on the input signals that they have. For example, the JK flip-flop has two inputs, *J* and *K*. The characteristic table, equation, and state diagram for these flip-flops are shown on the next page. They differ only in how *D* changes, that is, the input(s) to the *D* signal of the D flip-flop based on their respective characteristic equations. Draw the circuits for each of these three flip-flops. They all will use the D flip-flop with an extra combinational circuit (based on their respective characteristic equations) that generates the correct input signal for *D*. Hint: the characteristic equation for the D flip-flop is just $Q_{next} = D$, so the external *D* input connects directly to the *D* signal of the D flip-flop.

SR flip-flop

S	R	Q	Q_{next}	Q_{next}'
0	0	0	0	1
0	0	1	1	0
0	1	0	0	1
0	1	1	0	1
1	0	0	1	0
1	0	1	1	0
1	1	0	×	×
1	1	1	×	×

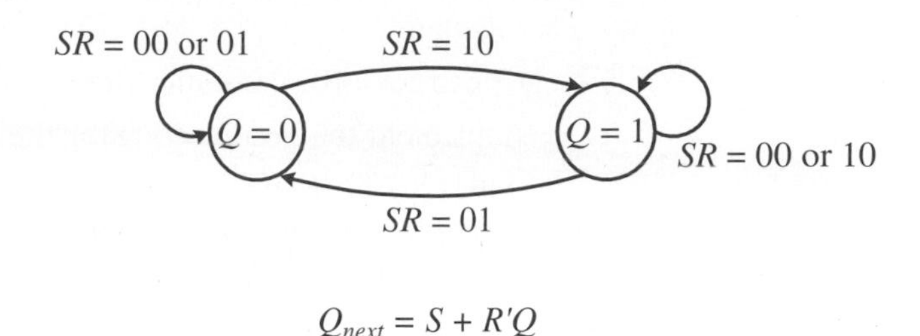

$Q_{next} = S + R'Q$

JK flip-flop

J	K	Q	Q_{next}	Q_{next}'
0	0	0	0	1
0	0	1	1	0
0	1	0	0	1
0	1	1	0	1
1	0	0	1	0
1	0	1	1	0
1	1	0	1	0
1	1	1	0	1

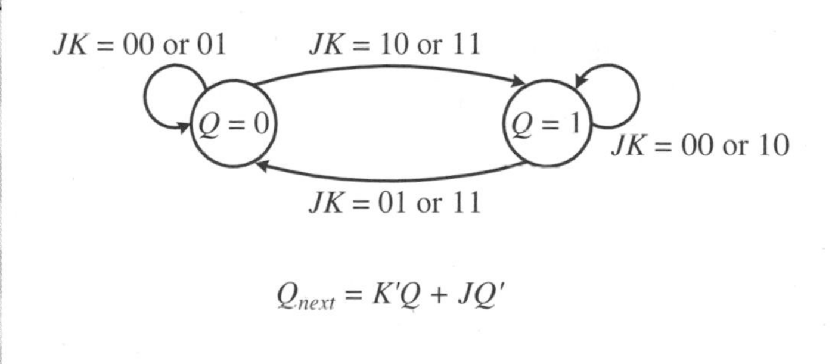

$Q_{next} = K'Q + JQ'$

T flip-flop

T	Q	Q_{next}	Q_{next}'
0	0	0	1
0	1	1	0
1	0	1	0
1	1	0	1

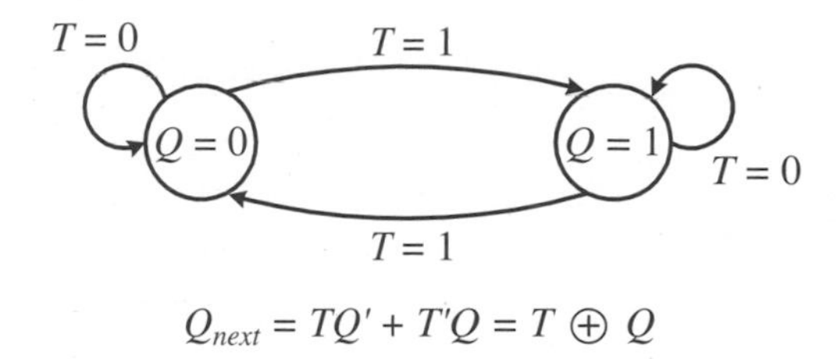

$Q_{next} = TQ' + T'Q = T \oplus Q$

5.9. Modify the 4-bit binary up counter circuit shown in Figure 5.56 so that it can count either up or down. Instead of a half adder, you will need to derive a half adder/subtractor (HAS) circuit.

5.10. A binary coded decimal (BCD) up counter uses four bits to count the decimal digits from 0 to 9 and then cycles back to 0.

a) Start with the 4-bit binary up counter with parallel load circuit shown in Figure 5.57, and then add extra circuitry to it so that it is a BCD counter.

b) What is the difference between using the *Clear* signal and the *Load* signal?

5.11. Construct a BCD counter that counts from 3 to 8, and back to 3.

5.12. Implement the Verilog code for the RAM shown in Figure 5.43 and verify that it works correctly.

5.13. Implement the VHDL code for the RAM shown in Figure 5.44 and verify that it works correctly.

CHAPTER 6

Finite-State Machines

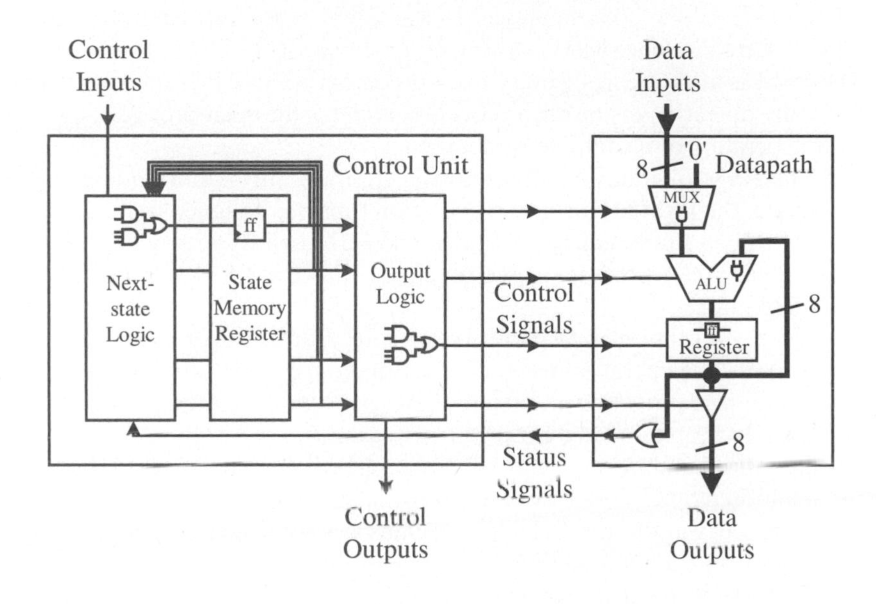

In Chapter 5, we looked at the design and operation of flip-flops—the most fundamental memory element used in microprocessor circuits. We saw that a single flip-flop is capable of remembering only one bit of information or one bit of history. In order for a microprocessor circuit to remember more inputs and a longer history, it must contain more flip-flops. This collection of (D) flip-flops used to remember the complete history of past inputs is referred to as the **state memory**. The entire content of the state memory at a particular instance of time forms a unique binary encoding that represents the complete history of inputs up to that time. We refer to this binary encoding at any instance of time as the **state** of the system at that time; different encodings, therefore, represent different states.

The state memory is one of the three main components in the controller circuit inside all microprocessors. Because the size of the state memory is finite, the total number of different states that it can represent is also finite; hence, this controller circuit is called a **finite-state machine** (FSM). A general overview of an FSM is shown in Figure 6.1. The FSM is at the heart of every microprocessor because it is this circuit that controls the entire operation of the microprocessor, and it is the microprocessor that controls the entire operation of a computer system.

The FSM operates by continuously stepping through a sequence of states. In each state, the FSM performs the operation that was assigned to that state. Although there is only a finite number of states, the FSM can go to any of these states more than once, and so the sequence of states that the FSM can go through can be infinitely long.

If we want the FSM to perform, say, four different operations, then we will need four states—one operation per state. We can reduce the number of states by assigning more than one operation to a state if the operations can be performed in parallel. However, to keep things simple for now, we simply will assign one operation per state. On the other hand, there might be an operation where we may want to repeat it for, say, a hundred times. Instead of assigning this same operation to one hundred different states, we will want to use just one state and have some form of looping capabilities to repeat that state a hundred times.

The operations that an FSM performs are realized by the output signals that the **output logic circuit** generates. Recall that the outputs of sequential circuits are dependent

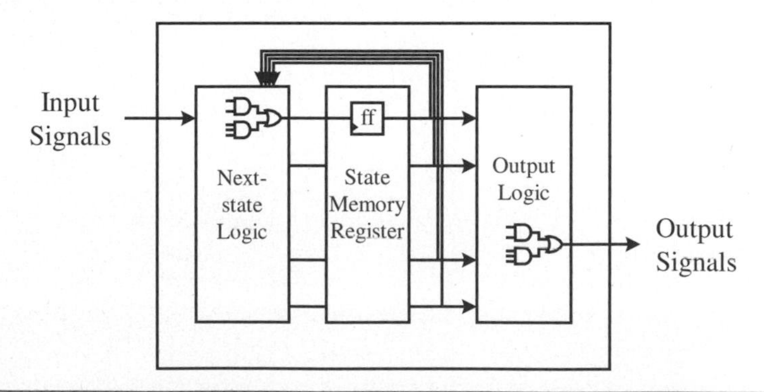

FIGURE 6.1 Finite-state machine overview.

on their past and current inputs, and because all of the inputs are remembered in the form of states in the state memory, we can say that the outputs are dependent on the content of the state memory. Therefore, the output logic circuit is dependent on the content of the state memory, and may or may not be dependent on the current inputs. The output logic circuit is a combinational circuit, and the output signals that it generates constitute the actions or operations that are performed by the FSM. Hence, an FSM can perform different operations in different states simply by generating different output signals.

Thus, an FSM operates by transitioning from one state to the next, generating different output signals in each state. The **next-state logic circuit** inside the FSM is responsible for determining what the next state to go to is. Based on the current state that the FSM is in (i.e., the past inputs) and the current inputs, the next-state logic will determine what the next state should be. This statement, in fact, is equivalent to saying that the outputs are dependent on the past and current inputs, since a state is used to remember the past inputs, and it also determines the outputs to be generated. The next-state logic circuit is a combinational circuit that takes the contents of the state memory flip-flops and the current inputs as its inputs. The outputs from the next-state logic circuit are used to change the contents of the state memory flip-flops. The FSM changes state when the contents of the state memory change, and this happens at the active (rising) edge of every clock cycle, since values are written into a flip-flop at the active clock edge.

The speed at which an FSM sequences through the states is determined by the speed of the clock signal. The state memory flip-flops are always enabled, so at every active edge of the clock, a new value is stored into the flip-flops. The limiting factor for the clock speed is in the time that it takes to perform all of the operations that are assigned to a particular state. All data operations assigned to a state must finish their operations within one clock period so that the results can be written into registers at the next active clock edge.

FSMs are the key to understanding how microprocessors are capable of controlling so many different things. They are responsible for determining when various data manipulations are to be performed, when control signals are to be generated, and the sequence in which the operations are to be performed. In order to be able to design and construct a microprocessor, it is important that we understand the operation and construction of FSMs. In this chapter, we will first look at how to describe precisely the operation of FSMs using state diagrams. Next, we will look at the analysis and synthesis of FSMs. Finally, we will give several complete examples of FSM constructions.

6.1 Finite-State Machine Models

In the introduction, we mentioned that the output logic circuit is dependent on the content of the state memory, and may or may not be dependent on the current inputs. The fact that the output logic may or may not be dependent on the current inputs gives rise to two different FSM models as shown in Figure 6.2.

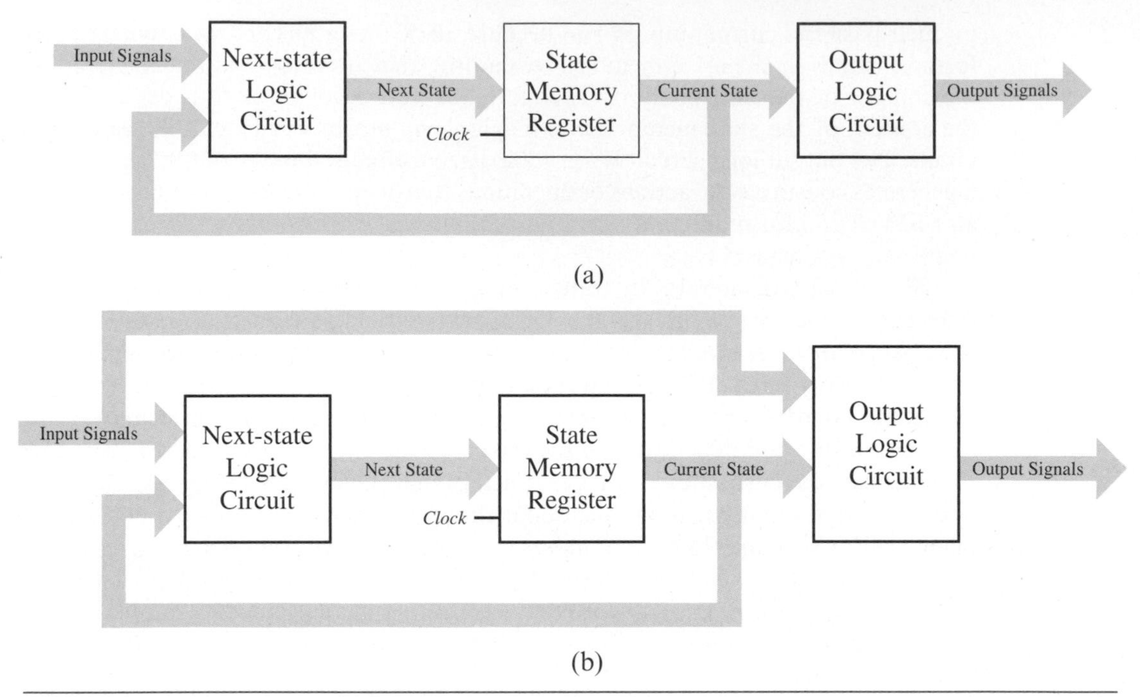

FIGURE 6.2 Finite-state machine models: (a) Moore FSM; (b) Mealy FSM.

Figure 6.2(a) shows the block diagram for the **Moore** FSM, where its outputs are dependent only on its current state (i.e., on the content of the state memory). Figure 6.2(b) shows the block diagram for the **Mealy** FSM, where its outputs are dependent on both the current state of the machine and the current inputs. The only difference between these two figures is that, for the Moore FSM, the output logic circuit only has the current state as its input; whereas, for the Mealy FSM, the output logic circuit has both the current state and the input signals as its inputs.

In both models, there are the three components: the next-state logic circuit, the state memory, and the output logic circuit. Both the next-state logic circuit and the output logic circuit are combinational circuits, whereas, the state memory is a sequential circuit composed of one or more D flip-flops.

The inputs to the next-state logic circuit are the primary input signals and the current state of the FSM. The next-state logic circuit generates values to change the contents of the state memory. Because the state memory is made up of one or more D flip-flops, and the content of the D flip-flop changes to whatever value is at its *D* input, in order to change a state, the next-state logic circuit simply has to generate values for all of the *D* inputs for all of the flip-flops. These *D* input values are referred to as the **excitation** values, since they "excite" or cause the D flip-flops to change states.

Recall that the D flip-flop stores a new value at every active edge of the clock signal, therefore, the contents stored in the state memory will change at every active clock edge. This means that the FSM will change to a new state at the beginning of every clock cycle. The state of the D flip-flop is just the value at its Q output, therefore, the binary encoding for the current state of the FSM is composed of all of the Q values from all of the D flip-flops.

Given just the current state information for the Moore model, or both the current state and input signals for the Mealy model, the output logic circuit will generate the appropriate control output signals to control the various operations intended by the FSM for that state.

Figures 6.3(a) and (b) show a sample circuit of a Moore FSM and a Mealy FSM, respectively. The two circuits are identical except for their outputs. For the Moore FSM, the output circuit is a 2-input AND gate that gets its input values from the outputs of the two D flip-flops. Remember that the state of the FSM is represented by the content of the state memory, which is the content of the D flip-flops as represented by the value at the Q (and Q') output. Hence, this output circuit is dependent only on the current state of the FSM.

For the Mealy FSM, the output circuit is a 3-input AND gate. In addition to getting its two inputs from the D flip-flops, the third input to this AND gate is connected to the primary input, C. With this one extra connection, this output circuit

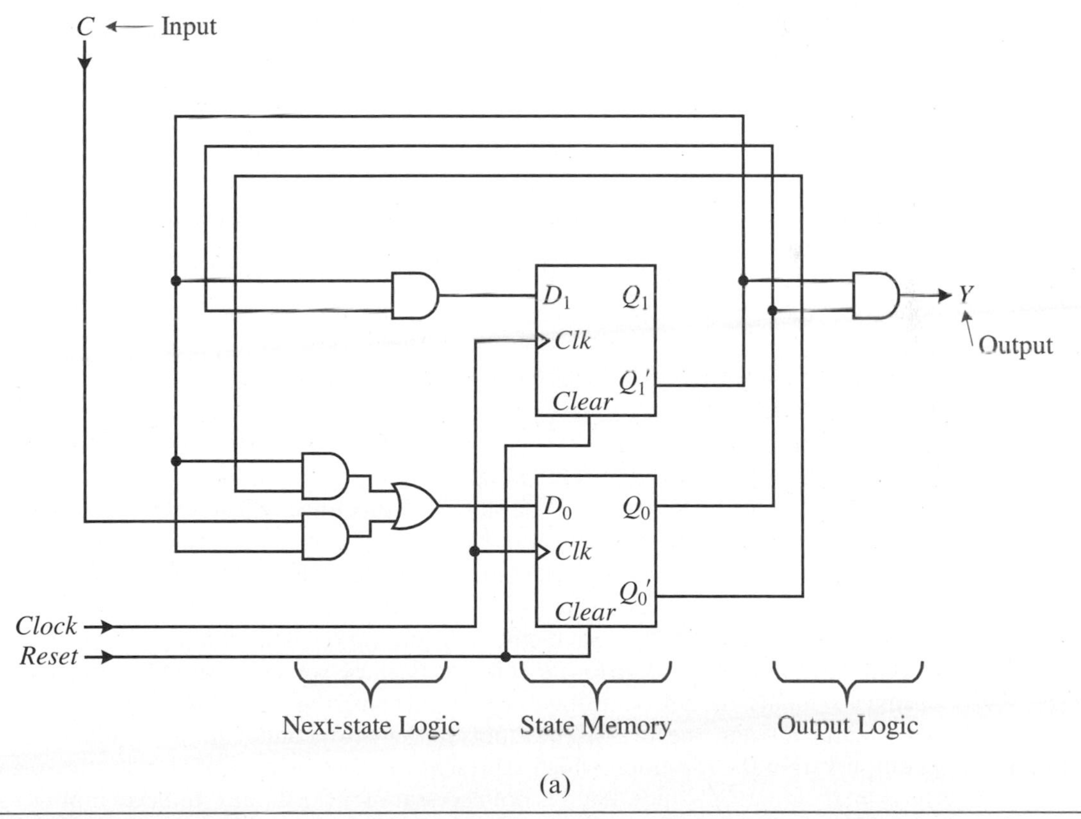

FIGURE 6.3 Sample finite-state machine circuits: (a) Moore FSM; (b) Mealy FSM. *(continued on next page)*

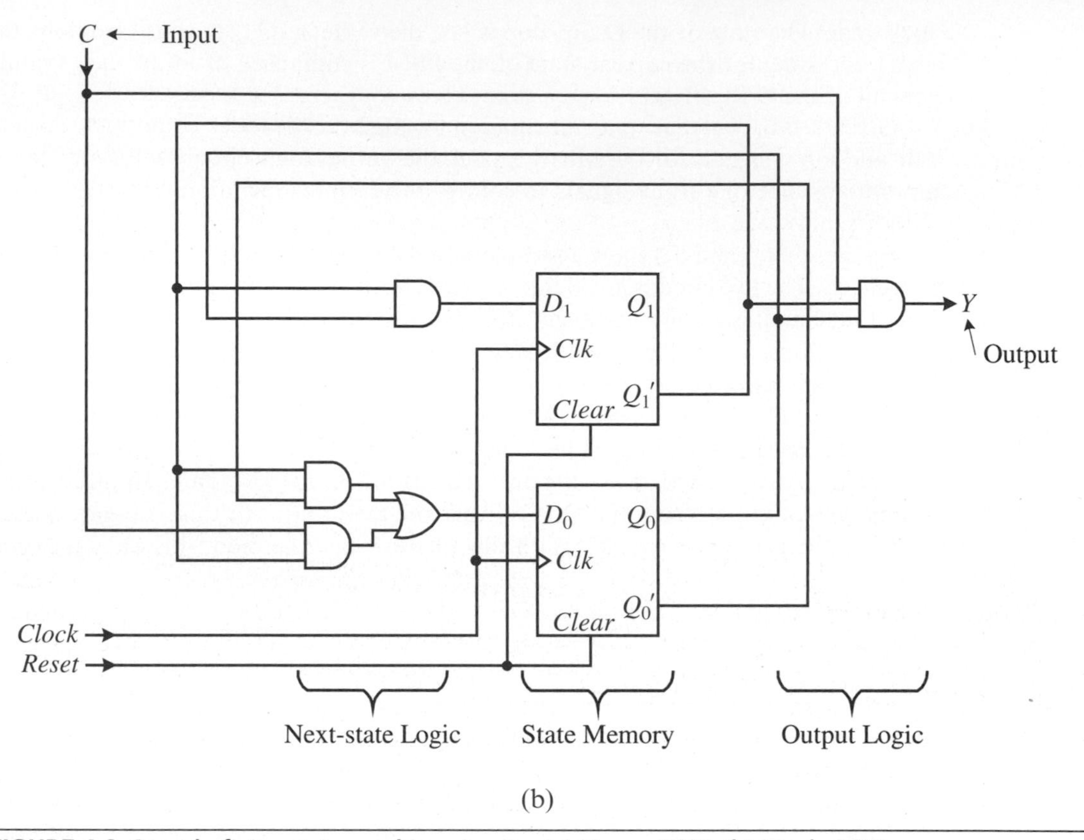

FIGURE 6.3 **Sample finite-state machine circuits: (a) Moore FSM; (b) Mealy FSM.**

is dependent on both the current state and the primary input, thus making it a Mealy FSM.

For both circuits, the state memory consists of the two D flip-flops. With two flip-flops, four different combinations of values can be represented. Hence, this FSM can be in any one of four different states. The state that this FSM will go to next depends on the value at the *D* inputs of the flip-flops because whatever values are at the *D* inputs will be stored into the flip-flops at the next active clock edge, and are made available at the *Q* outputs as the new state value.

Every D flip-flop in the state memory requires a combinational next-state circuit to generate a next-state value for its *D* input. Because we have two D flip-flops (each having a *D* input), the next-state logic circuit will consist of two combinational circuits: one for input D_0 and one for D_1. The inputs to these two combinational circuits are the *Q* outputs from the flip-flops, which represent the current state of the flip-flops, and the primary input *C*. Notice that it is not necessary for the input *C* to be an input to all of the combinational next-state circuits. In this sample circuit, only the bottom combinational circuit is dependent on the input *C*. Just like with any other combinational circuit as discussed in Chapter 3, the next-state and output logic circuits are derived from either a truth table or a Boolean equation.

At this point, it is not obvious why we have this distinction between Moore and Mealy FSMs. It turns out that these two models provide a tradeoff between the size of the state memory and next-state logic circuit with that of the output logic circuit. Comparing the two FSMs, a Moore FSM typically will have a larger state memory and therefore a larger next-state logic circuit, whereas, a Mealy FSM typically will have a larger output logic circuit. We will show examples of this tradeoff later in Sections 6.6.5 and 6.6.6.

6.2 State Diagrams

State diagrams are used to describe precisely the operation of FSMs. A state diagram is a deterministic graph with nodes and directed edges for connecting the nodes. There is one node for every state of the FSM, and each node is labeled with either its state name or its state encoding. Every state transition of the FSM has a directed edge connecting two nodes. The directed edge originates from the node for the current state that the FSM is transitioning from and goes to the node for the next state that the FSM is transitioning to. Edges may or may not have labels on them. Edges for unconditional transitions from one state to another will not have a label. In this case, only one edge can originate from that node. Conditional transitions from a state will have two outgoing edges for each input signal condition. The two edges from this node must be labeled with the corresponding input signal conditions: one edge with the label for when the condition is true, and the other edge with the label for when the condition is false. If there is more than one input signal, then all of the possible input conditions must be labeled on the outgoing edges from the node. The state diagram is deterministic because from any node, it should show exactly which node to go to next for any input combination.

Figure 6.4(a) shows a sample state diagram having four states, one input signal C, and one output signal Y. The four states are labeled with the four encoded binary values 00, 01, 10, and 11. There are three unconditional transitions (i.e., edges with no labels) from state 00 to 01, 10 to 00, and 11 to 00. There is one conditional transition from state 01 to either 10 or 11 depending on the input condition C. In this book, we use the notation of a conditional test enclosed in parentheses to denote a conditional label for an edge. The edge going from state 01 to 10 in Figure 6.4(a), has the conditional label $(C = 0)$ on it. This means that if the input condition $(C = 0)$ is true, then the transition from state 01 to 10 is made. Otherwise, if $(C = 0)$ is false, that is, $(C = 1)$ is true, then the transition from 01 to 11 is made. Note that the conditional labels $(C = 0)'$ and $(C = 1)$ mean the same thing and can be used interchangeably.

The output signal Y in Figure 6.4(a) is labeled inside or next to each node denoting that the output is dependent only on the current state. For example, when the FSM is in state 01, the output Y is set to 1; whereas, in state 11, Y is set to 0.

The operation of the FSM based on the state diagram in Figure 6.4(a) goes as follows. After reset, the FSM starts from state 00. In this book, we will always use state 0

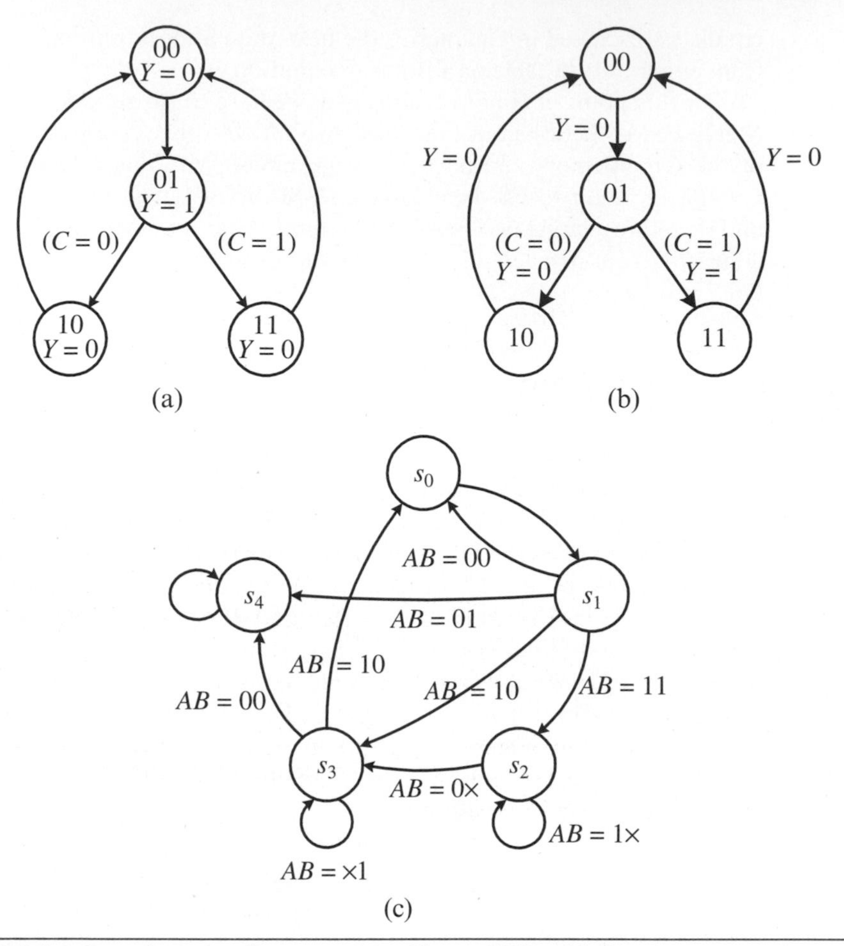

FIGURE 6.4 Sample state diagrams: (a) a Moore FSM with four states, one input signal *C*, and one output signal *Y*; (b) a Mealy FSM with four states, one input signal *C*, and one output signal *Y*; (c) an FSM with five states and two input signals *A* and *B*.

as the starting or reset state unless stated otherwise. When it is in state 00, it outputs a 0 for *Y*. At the next active clock edge, the FSM unconditionally transitions to state 01 and outputs a 1 for *Y*. Next, the FSM goes either to state 10 or 11 at the next active clock edge, depending on the condition $(C = 0)$. If the condition $(C = 0)$ is true, then the FSM will go to state 10 and outputs a 0 for *Y*; otherwise, it will go to state 11 and also outputs a 0 for *Y*. From either state 10 or 11, the FSM unconditionally transitions back to state 00 at the next clock cycle. The FSM always goes to a new state at the beginning of the next active clock edge.

Figure 6.4(b) shows a slightly different state diagram from the one in Figure 6.4(a). Instead of labeling the output signal *Y* inside or next to a node, it is labeled on the edges. What this means is that the output is dependent on both the current state (i.e., the state from which the edge originates) and the input signal C. For example, when the FSM is in state 01, if the FSM takes the left edge for the condition $(C = 0)$ to state 10, then it will output a 0 for Y. However, if the FSM takes the right edge for the condition $(C = 1)$ to state 11, then it will output a 1 for Y.

Figure 6.4(c) shows a state diagram having five states, two input signals, and no output signals. In practice, all FSMs should have output signals; otherwise, they don't do anything useful. The five states in this state diagram are given the symbolic state names of s_0, s_1, s_2, s_3, and s_4. The two input signals are A and B. Again, we use the state name with subscript 0, namely s_0, as the starting state. From state s_0, there is one unconditional edge going to state s_1. This unlabeled edge is equivalent to having the label $AB = \times\times$, where the $\times$ denotes "don't-care." This means that A can either be a 0 or a 1, and B also can be either a 0 or a 1; so this edge is taken for any combination of the two input signals. From state s_1, there are four outgoing edges labeled with the four different combinations of the two input signals. For example, when in state s_1, if the input signals AB are 11 (i.e., $A = 1$ and $B = 1$), then the FSM will go to state s_2 in the next clock cycle. State s_2 has only two outgoing edges. However, the two labels on them cover the four possible input conditions, because B is a don't-care (denoted by the $\times$) in both cases. State s_3 has three outgoing edges, but again, the labels on them cover all four input conditions.

As you can see, a state diagram is similar to a computer program flowchart in which the nodes are for the statements or data operations, and the edges are for the control of the program sequence. Because of this similarity, we should be able to convert any program to a state diagram. Example 6.1 shows how to convert a simple C-style pseudocode to a state diagram.

EXAMPLE 6.1

Converting pseudocode to a state diagram

Derive the state diagram based on the following pseudocode.

```
x = 5
WHILE (x ≠ 0){
   OUTPUT x
   x = x - 1
   }
```

The pseudocode has three data-operation statements and one conditional test. Each data-operation statement is assigned to a node (state), as shown in Figure 6.5(a). Each node is given a symbolic name for the state and is annotated with the statement to be executed in that state. At this point, instead of labeling the nodes with the actual binary encoding for the state, it is better to just give it a name. The actual encoding of the state can be done later during the synthesis process.

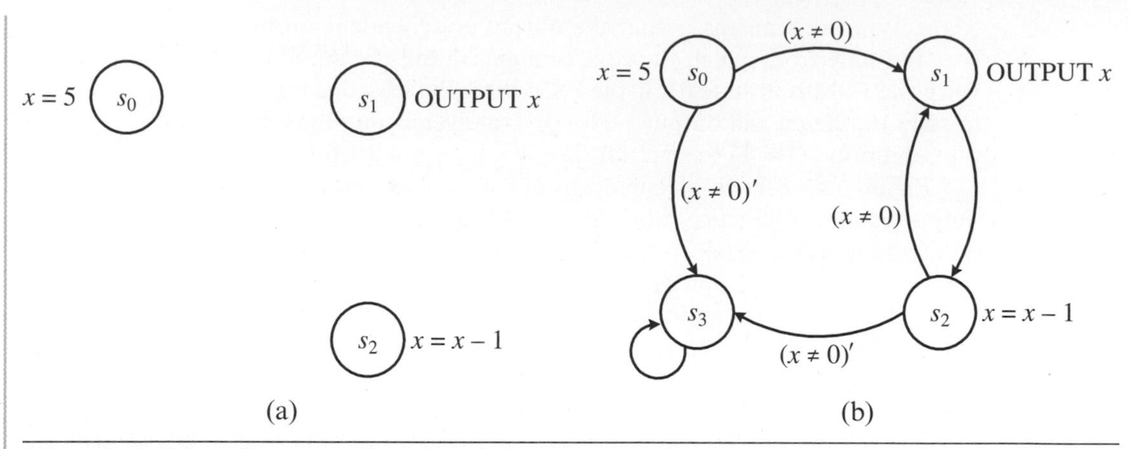

FIGURE 6.5 State diagram for Example 6.1: (a) data operations assigned to nodes; (b) complete state diagram with the transitional edges.

Next, we assign directional edges to the diagram based on the sequence of execution. Starting from state s_0 (where the statement $x = 5$ is executed), the program then tests for the condition $(x \neq 0)$. If the condition is true, then the output statement is executed; otherwise, the loop (and the program) is terminated. Referring to Figure 6.5(b), there are two outgoing edges from state s_0. The edge from s_0 to s_1 has the label $(x \neq 0)$, which means that if the condition $(x \neq 0)$ is true, then this edge is taken, and so it will go to state s_1 to execute the output statement. On the other hand, if the condition is false, the loop will be terminated. Because there is no statement after the loop, we have to add an extra no-operation state, s_3, to the state diagram for it to go to. The edge from s_0 to s_3 is labeled $(x \neq 0)'$, meaning that the edge is taken when the condition $(x \neq 0)$ is false.

After executing the output statement in state s_1, the decrement statement is executed. This sequence is reflected in the unconditional edge going from state s_1 to s_2. After executing the decrement statement in s_2, the condition $(x \neq 0)$ in the WHILE loop is tested again. If the condition is true, it will take the edge with the label $(x \neq 0)$ back to state s_1 to repeat the loop. If the condition is false, it will take the edge with the label $(x \neq 0)'$ to state s_3. From state s_3, it unconditionally loops back to itself; thus, going nowhere and doing nothing to represent that the program has halted.

6.3 Analysis of Finite-State Machines

We often are given an FSM circuit and need to know its operation. The **analysis of finite-state machines** is the process in which we are given an FSM circuit (such as the ones in Figure 6.3), and we want to obtain a precise description of the operation of the circuit by deriving the state diagram for it. The steps for the analysis of FSM circuits are as follows:

1. Derive the next-state equations from the combinational next-state logic circuit.
2. Derive the next-state table from the next-state equations.

3. Derive the output equations from the combinational output logic circuit.
4. Derive the output table from the output equations.
5. Draw the state diagram from the next-state table and the output table.

The following subsections explain these steps in detail.

6.3.1 Next-State Equations

The **next-state equations** are the equations derived from the next-state logic circuit in the FSM. Because the next-state logic circuit is a combinational circuit, deriving the next-state equations is simply an analysis of a combinational circuit, as discussed in Chapter 3. These equations provide the values to the D input signals of the D flip-flops, and are dependent on the current state and the primary inputs of the FSM. The current state is determined by the current contents of the D flip-flops (i.e., the flip-flops' output signals Q and Q'). Because the D flip-flop has only one data input D, there is one equation for each D flip-flop in the state memory.

Given the FSM circuit shown in Figure 6.6, we obtain the following two next-state equations as derived from the next-state logic circuit for the two D flip-flops used in the circuit. Equation 6.1 is from the next-state logic circuit for the D_1 input of flip-flop 1

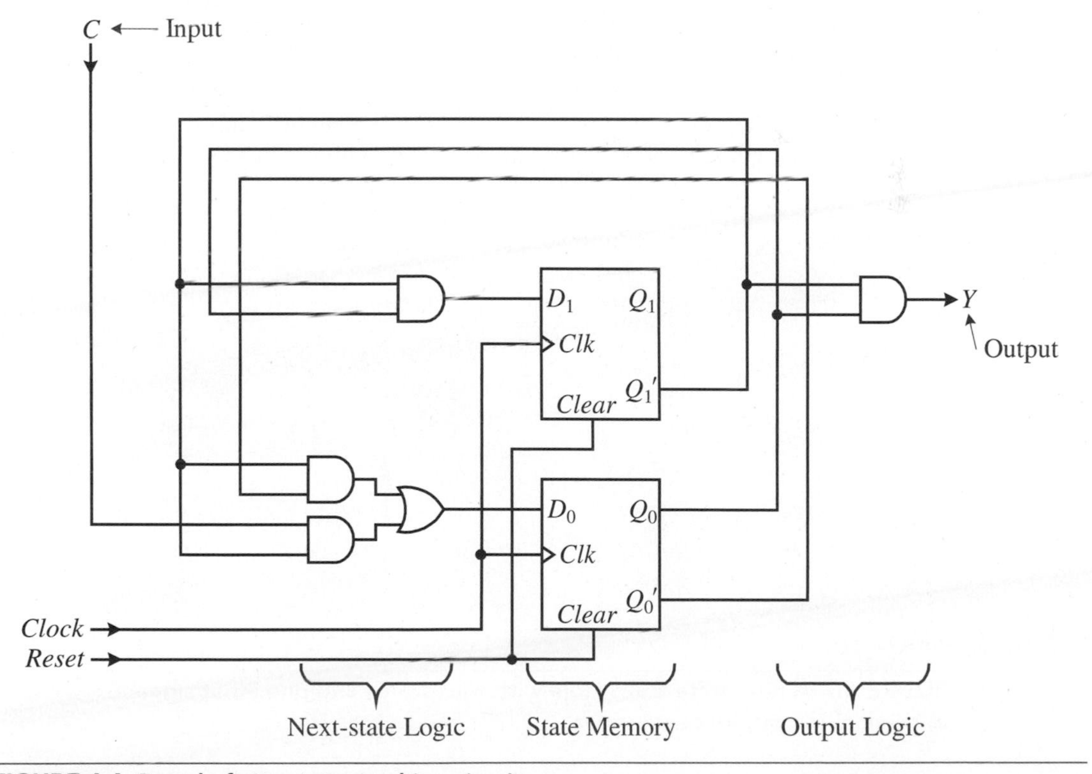

FIGURE 6.6 Sample finite-state machine circuit.

(the flip-flop on the top), and Equation 6.2 is from the next-state circuit for the D_0 input of flip-flop 0 (the flip-flop on the bottom).

$$D_1 = Q_1'Q_0 \tag{6.1}$$

$$D_0 = Q_1'Q_0' + CQ_1' \tag{6.2}$$

Recall from Chapter 5 that the characteristic equation for the D flip-flop is

$$Q_{next} = D$$

therefore, by substitution, we get the two final next-state equations as follows:

$$Q_{1next} = D_1 = Q_1'Q_0 \tag{6.3}$$

$$Q_{0next} = D_0 = Q_1'Q_0' + CQ_1' \tag{6.4}$$

6.3.2 **Next-State Table**

The **next-state table** is simply the truth table as derived from the next-state equations. For every combination of the current state values (Q) and input values, it lists what the next-state values (Q_{next}) should be. These next-state values are obtained by substituting the current state and input values into the appropriate next-state equations.

Figure 6.7(a) shows the truth table as obtained from the two next-state equations 6.3 and 6.4. There are three input variables C, Q_1, and Q_0, and two output variables Q_{1next} and Q_{0next}. For the next-state table, we want to use a slightly different format of the truth table as shown in Figure 6.7(b). This new format allows us to more easily see what

CQ_1Q_0	$Q_{1next}Q_{0next}$
000	01
001	10
010	00
011	00
100	01
101	11
110	00
111	00

(a)

Current State Q_1Q_0	Next State $Q_{1next}Q_{0next}$	
	$C = 0$	$C = 1$
00	01	01
01	10	11
10	00	00
11	00	00

(b)

FIGURE 6.7 A next-state truth table with four states and one input signal *C*: (a) original format; (b) new format.

the next state is, given the current state and input values. Both tables show the same information, but we have just rearranged the rows and columns in the truth table to get the next-state table. The rows in the next-state table are labeled with the current states, the columns are labeled with the inputs, and the entries in the table are the next-state values. We will use this new format to show the next-state table.

Having two flip-flops, Q_1 and Q_0 in the state memory of the FSM, there will be four different encodings, 00, 01, 10, and 11, for the current state. There is one input signal, C, with the two possible values, 0 and 1. Thus, with three input variables (Q_1, Q_0, and C), there will be a total of eight (2^3) entries in the table. Each entry in the table is composed of two bits: the leftmost bit, Q_{1next}, is the next value for the Q_1 flip-flop, and the rightmost bit, Q_{0next}, is the next value for the Q_0 flip-flop. Writing these two bits together for each entry is equivalent to having two separate truth tables. It is easier to see the next-state encodings by combining them together. Together, the two bits, $Q_{1next}Q_{0next}$, in each of the entries in the table denote the next-state values for the two flip-flops. These next-state values are obtained from substituting the current state values, Q_1Q_0, and the input value, C, into the next-state equations 6.3 and 6.4.

For example, to get the Q_{1next} value for the top-left entry (the left bit in the color entry), we substitute the current state values, $Q_1 = 0$ and $Q_0 = 0$, and the input value $C = 0$ into Equation 6.3, which gives

$$\begin{aligned} Q_{1next} &= Q_1'Q_0 \\ &= 0' \cdot 0 \\ &= 1 \cdot 0 \\ &= 0 \end{aligned}$$

Similarly, substituting the same values for Q_1, Q_0, and C into Equation 6.4 will give us the Q_{0next} value for that same top-left entry.

$$\begin{aligned} Q_{0next} &= Q_1'Q_0' + CQ_1' \\ &= 0' \cdot 0' + 0 \cdot 0' \\ &= 1 + 0 \\ &= 1 \end{aligned}$$

Therefore, the top-left entry has the next-state value 01 for $Q_{1next}Q_{0next}$. The rest of the entries in the next-state table are obtained in the same manner by substituting the corresponding values for Q_1, Q_0, and C into the two next-state equations 6.3 and 6.4.

The top-left entry tells us that if the current state Q_1Q_0 is 00 and the input signal C is 0, then the next state $Q_{1next}Q_{0next}$ that the FSM will go to is 01. The top-right entry tells us that if the current state is 00 and the input signal C is 1, then the next state is also 01. This means that the transition from state 00 to 01 does not depend on the input condition C, so this is an unconditional transition. From state 01, there are two conditional transitions: the FSM will transition to state 10 if the condition $(C = 0)$ is true; otherwise, if $(C = 1)$ is true, then it will transition to state 11. From either state 10 or 11, the FSM will go to state 00 unconditionally.

6.3.3 Output Equations

The **output equations** are the equations derived from the combinational output logic circuit in the FSM. Depending on the type of FSM (Moore or Mealy), the output equations can be dependent on just the current state, or on both the current state and the inputs.

For the Moore circuit of Figure 6.3(a), the output equation is

$$Y = Q_1'Q_0 \tag{6.5}$$

For the Mealy circuit of Figure 6.3(b), the output equation is

$$Y = CQ_1'Q_0 \tag{6.6}$$

A typical FSM will have many output signals, and every output signal will have one equation for it.

6.3.4 Output Table

The **output table** is the truth table that is derived from the output equations. The output tables for the Moore and Mealy FSMs are slightly different from each other. For the Moore FSM, the output table lists the output values for every combination of the current state. For the Mealy FSM, however, the output table lists the output values for every combination of the current state and input values. These output values are obtained by substituting the current state and input values into the appropriate output equations.

Figures 6.8(a) and (b) show the output tables for the Moore and Mealy FSMs as derived from the output equations 6.5 and 6.6, respectively. For the Moore FSM, the output signal Y is dependent only on the current state value Q_1Q_0; whereas, for the Mealy FSM, the output signal Y is dependent on both the current state and input C.

Current State Q_1Q_0	Output Y
00	0
01	1
10	0
11	0

(a)

Current State Q_1Q_0	Output Y	
	$C = 0$	$C = 1$
00	0	0
01	0	1
10	0	0
11	0	0

(b)

FIGURE 6.8 Output table: (a) for a Moore FSM; (b) for a Mealy FSM.

6.3.5 **State Diagram**

The last step in the analysis of an FSM circuit is to derive the state diagram. The state diagram is obtained directly from interpreting the next-state table and the output table.

The FSM circuit from Figure 6.6 uses two flip-flops for its state memory. Two flip-flops can have four different state encodings; therefore, the state diagram will have four nodes. We start by drawing the four nodes and labeling them with the four different combinations of the state encodings as shown in Figure 6.9(a).

Next, we draw the edges based on the information in the next-state table. For each next-state entry in the next-state table shown in Figure 6.7(b), there is a corresponding directed edge pointing from that current state to that next state. The corresponding input condition for that transition is labeled on that edge. For example, when the current state Q_1Q_0 is 01, the next state $Q_{1next}Q_{0next}$ is 10 when C is 0, therefore, there is a directed edge going from the node labeled 01 to the node labeled 10. Because this

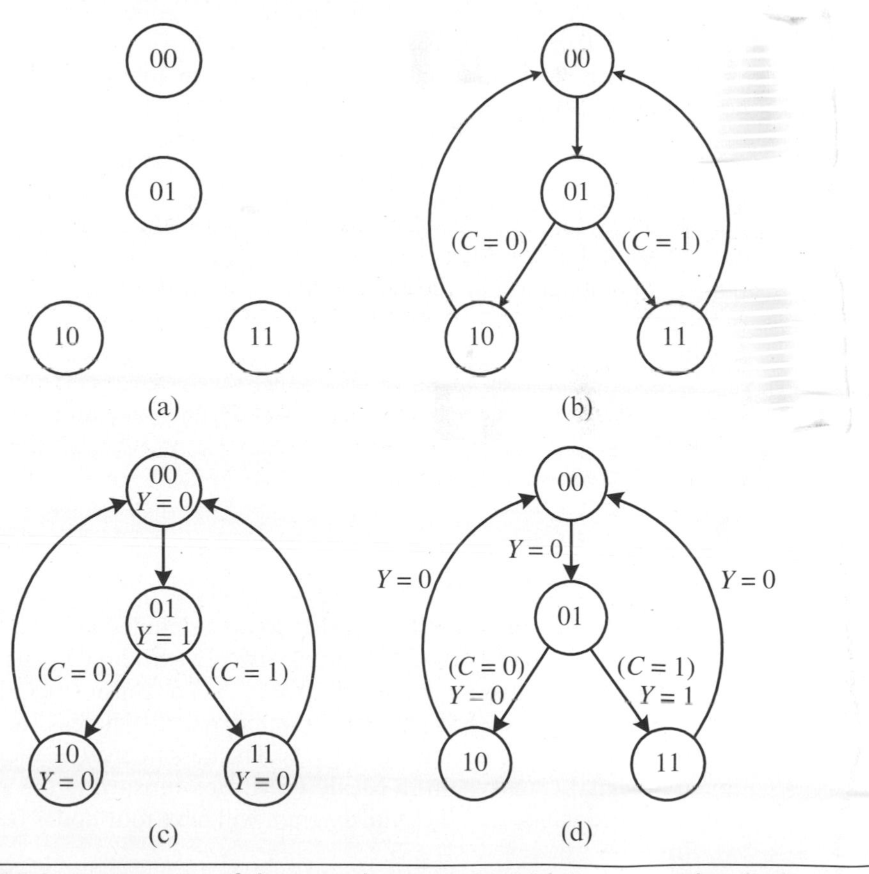

FIGURE 6.9 Construction of the state diagram: (a) initial diagram with only the nodes; (b) diagram with the edges added; (c) final state diagram for the Moore FSM; (d) final state diagram for the Mealy FSM.

transition occurs only when the input signal C is a 0, the edge is labeled with the condition $(C = 0)$ as shown in Figure 6.9(b).

If the same transition from a current state to a next state occurs for all possible input conditions, then it is an unconditional transition. For this, only one edge is drawn and no label is needed on that edge. For example, from the current state 00, the next state is 01 for all possible values of C, making it an unconditional transition. Thus, there is only one edge going from state 00 to state 01, and it has no conditional label on it.

For the Moore FSM, the output signals are labeled inside the node or next to the node based on the output table. For example, looking at the output table from Figure 6.8(a), we see that the output signal Y is a 0 when the FSM is in state 00. Therefore, we add the output signal label $Y = 0$ inside the node 00 as shown in Figure 6.9(c). This is repeated for all of the entries in the output table. For the Mealy FSM, the output signals are labeled on an edge because it is also dependent on the input condition signal. For example, the output signal Y is a 1 when the current state is 01 and the condition C is a 1; therefore, the edge going from 01 to 11 has the output signal $Y = 1$ labeled on it as shown in Figure 6.9(d).

The complete Moore state diagram shown in Figure 6.9(c) is derived from the next-state table from Figure 6.7(b) and the Moore output table from Figure 6.8(a). In this state diagram, the output values are labeled inside or next to a state because the output signals are dependent only on the current state. The Mealy state diagram shown in Figure 6.9(d) is derived from the same next-state table from Figure 6.7(b), but using the Mealy output table from Figure 6.8(b). In this state diagram, the output values are labeled next to the edges because the output signals are dependent on both the current state and the input condition.

6.3.6 **Example**

Example 6.2 illustrates the complete process of analyzing an FSM.

EXAMPLE 6.2

Analysis of an FSM

We will follow the steps described earlier to do a detailed analysis of the FSM circuit shown in Figure 6.10. This FSM contains two D flip-flops for its state memory, one input signal C, and two output signals X and Y. The output signal X is dependent on both the current state Q_1Q_0, and the input signal C, whereas the output signal Y is dependent only on the current state Q_1Q_0. The dependency of the X output signal on the input signal C makes this a Mealy FSM. Before starting the analysis process, we already can conclude that the state diagram will have four nodes because the FSM has two flip-flops.

Step 1 of the analysis is to derive the next-state equations, which are derived from the combinational next-state logic circuit. These equations are dependent on the current state of the flip-flops, Q_1 and Q_0, and the input, C. One equation is needed for

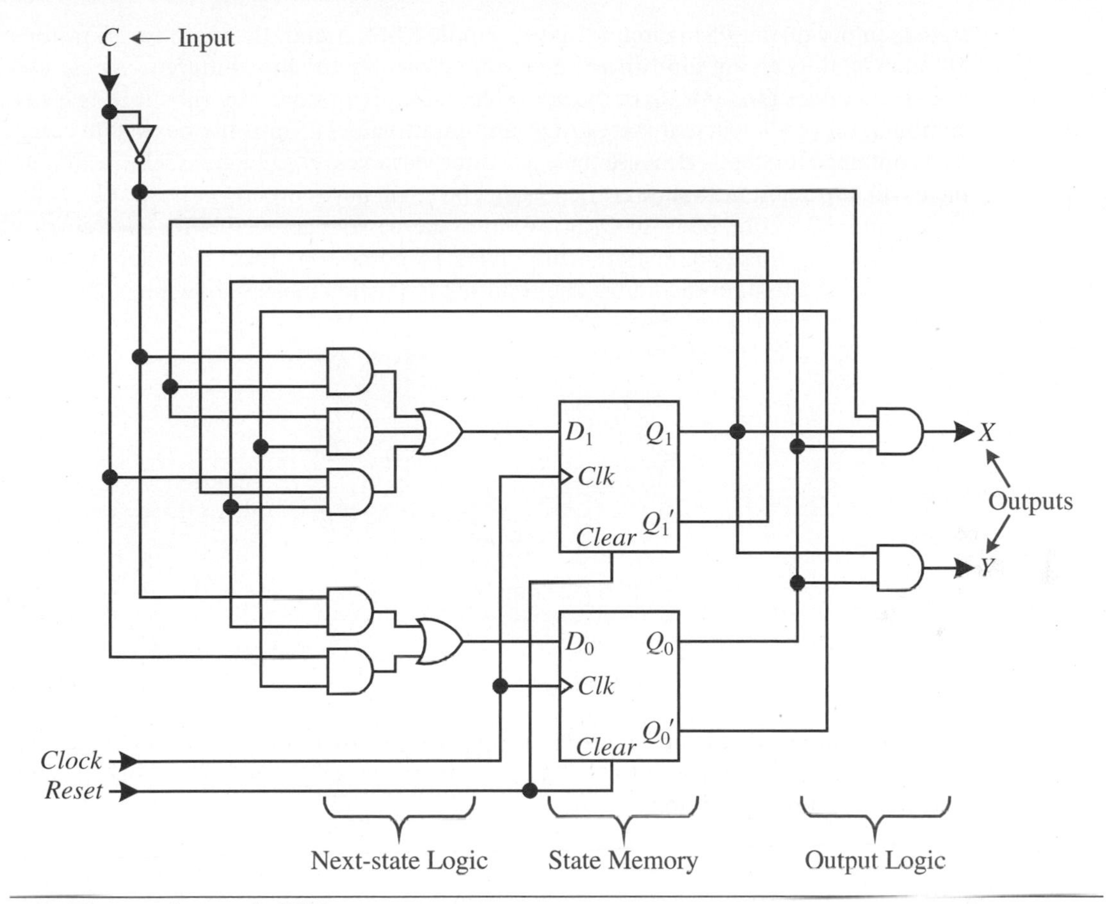

FIGURE 6.10 A sample FSM.

every data input of all of the flip-flops in the state memory. Our sample circuit has two flip-flops having the two data inputs D_1 and D_0, so we will have two next-state equations. Analyzing the top combinational next-state circuit that outputs a value to D_1 gives us the following next-state equation.

$$Q_{1next} = D_1 = C'Q_1 + Q_1Q_0' + CQ_1'Q_0$$

Analyzing the bottom combinational next-state circuit that outputs a value to D_0 gives us the following next-state equation.

$$Q_{0next} = D_0 = C'Q_0 + CQ_0'$$

Step 2 is to derive the next-state table from the next-state equations. The number of current states in the next-state table is deduced from the number of flip-flops in the state memory of the FSM circuit. In our sample FSM circuit, there are two flip-flops Q_1 and Q_0, thus giving four different combinations for the four different states. The next-state values $Q_{1next}Q_{0next}$ in the next-state table are obtained by substituting every combination of the current state Q_1Q_0, and input values C into the next-state equations obtained in Step 1. Because there are three variables, Q_1, Q_0, and C the next-state table will have eight next-state entries. Each entry will have two bits—the first bit is for Q_{1next}, and the second bit is for Q_{0next}. Writing these two bits together for each entry is equivalent to having two separate truth tables. By combining them together, it makes it easier to see the state encoding. The resulting next-state table is shown next.

Current State Q_1Q_0	Next State $Q_{1next}Q_{0next}$	
	$C = 0$	$C = 1$
00	00	01
01	01	10
10	10	11
11	11	00

For example, to find the Q_{1next} value for the current state $Q_1Q_0 = 00$ and $C = 1$ (the color entry in the table), we substitute the values $Q_1 = 0$, $Q_0 = 0$, and $C = 1$ into the equation $Q_{1next} = C'Q_1 + Q_1Q_0' + CQ_1'Q_0 = (1' \cdot 0) + (0 \cdot 0') + (1 \cdot 0' \cdot 0)$ to get the value of 0. Similarly, we get Q_{0next} by substituting the same values for Q_1, Q_0, and C into the equation $Q_{0next} = C'Q_0 + CQ_0' = (1' \cdot 0) + (1 \cdot 0')$ to get the value of 1. Therefore, $Q_{1next}Q_{0next} = 01$.

Step 3 is to derive the output equations from the combinational output logic circuit. One output equation is needed for every output signal. For this example, there are two output signals X and Y. The output signal X is dependent on both the current state Q_1Q_0 and the input signal C. The output signal Y is dependent only on the current state Q_1Q_0. By performing a combinational analysis of the output logic circuit for the output signals X and Y, we obtain the following two output equations.

$$X = C'Q_1Q_0$$

$$Y = Q_1Q_0$$

Step 4 is to derive the output table. Just like the next-state table, the output table is obtained by substituting all possible combinations of the current state and input signal values into the output equations. The output table for this example is shown next. Because output X is dependent on both the current state Q_1Q_0 and the input C, it has two columns, one for $C = 0$ and one for $C = 1$. Output Y is dependent only on the current state Q_1Q_0, so it has only one column. The resulting output table for both X and Y is shown next.

Current State Q_1Q_0	Output X		Output Y
	$C = 0$	$C = 1$	
00	0	0	0
01	0	0	0
10	0	0	0
11	1	0	1

Step 5 is to draw the state diagram, which is derived directly from the next-state and output tables. Every current state in the next-state table will have a corresponding node in the state diagram labeled with that current state's encoding. For this example, our state diagram will have four nodes. For every next-state entry in the next-state table, there will be a corresponding directed edge. This edge originates from the node labeled with the current state and ends at the node labeled with the next-state entry. The edge is labeled with the corresponding input condition.

For example, in the next-state table, when the current state Q_1Q_0 is 00, the next state $Q_{1next}Q_{0next}$ is 01 for the input $C = 1$. Hence, in the state diagram, there is a directed edge from node 00 to node 01 with the label $(C = 1)$. The output signal Y is labeled inside the nodes because it is dependent only on the current state, whereas, the output signal X is labeled on the edges because it is also dependent on the input condition. The complete state diagram for this example is shown next.

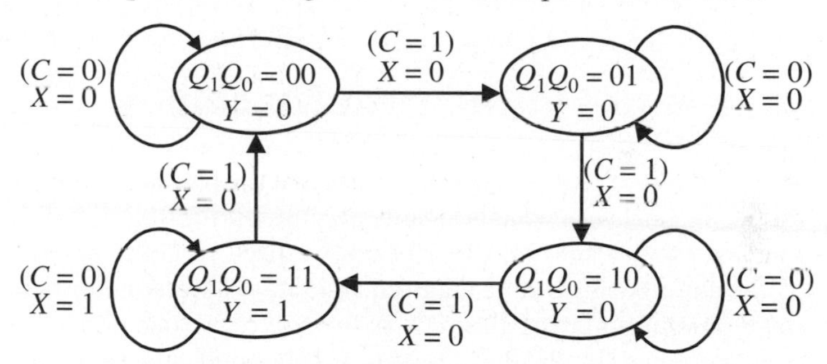

The following is a sample timing diagram for the execution of this FSM circuit.

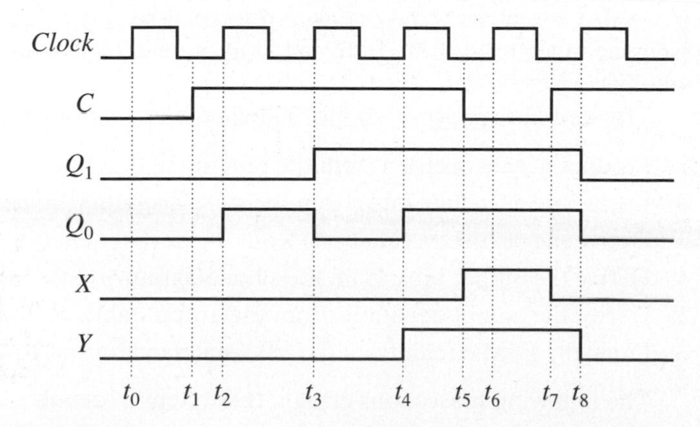

On reset, both flip-flop values are set to 0, so the FSM starts execution from state 00. The two D flip-flops used in the circuit are positive edge-triggered flip-flops, so they will change their states at every rising edge of the clock. The first rising clock edge is at time t_0. Normally, the flip-flops will change state at this time, however, because C is at 0, the flip-flops' values will remain at 0. At time t_1, the input C changes to a 1, so that, at the next rising clock edge at time t_2, the flip-flop value Q_1Q_0 changes to 01. At the next two rising clock edges at t_3 and t_4, with C still at a 1, the value for Q_1Q_0 changes to 10, and then to 11, respectively. At time t_4, when $Q_1Q_0 = 11$, the output Y also changes to a 1 because $Y = Q_1Q_0$. The output X at time t_4, however, remains at a 0 because it is dependent not only on the current state but also on the input C. At time t_5, input C changes to a 0, but the output Y remains at a 1 because Y is not dependent on C. However, output X will change to a 1 because $X = C'Q_1Q_0$. At time t_5, Q_1Q_0 remains the same at 11 through the next rising clock edge at t_6 because C is still at a 0. At time t_7, C changes back to a 1, which results in output X changing back to a 0. Furthermore, it will cause the state memory Q_1Q_0 to change to the next state, which is back to 00 at the next rising clock edge at time t_8, and the cycle repeats.

When $C = 1$, the FSM cycles through the four states in order repeatedly. When $C = 0$, the FSM stops at the current state until C is asserted again. If we interpret the four state encodings as a decimal number, then we can conclude that the circuit of Figure 6.10 is for a modulo-4 up-counter that cycles through the four values 0, 1, 2, and 3. The input C enables or disables the counting.

6.4 Synthesis of Finite-State Machines

The **synthesis of finite-state machines** is just the reverse of the analysis of FSMs. In synthesis, we start with what is usually an ambiguous functional description of the circuit that we want. From this description, we need to come up with the precise operation of the circuit using a state diagram. The state diagram allows us to construct the next-state and output tables. From these two tables, we can derive the next-state and output equations, and finally, the complete FSM circuit.

During the synthesis process, many possible circuit optimizations in terms of the circuit size, speed, and power consumption can be performed. Circuit optimization is discussed in a later section. In this section, we will focus only on synthesizing a functionally correct FSM.

The steps for the synthesis of FSM circuits are as follows:

1. Produce a state diagram from the functional description of the circuit.
2. Derive the next-state table from the state diagram.
3. Derive the next-state equations from the next-state table.
4. Derive the output table from the state diagram.
5. Derive the output equations from the output table.
6. Draw the FSM circuit from the next-state and output equations.

The following subsections explain these steps in detail.

6.4.1 **State Diagram**

The first step in the synthesis of an FSM is to derive the state diagram for it. We might be given an ambiguous or incomplete functional description of a circuit, which occurs when not all possible situations of an event or behavior are specified. In order to translate an ambiguous description into a precise state diagram, the designer must have a full understanding of the functional behavior of the circuit in question. In addition, the designer might need some ingenuity and creativity to fill in the missing gaps. Meaningful assumptions need to be made and stated clearly, and ambiguous situations need to be clarified. This is the one step in the design process that has no clear-cut answer. In this step, we rely on the knowledge and expertise of the designer to come up with a correct and meaningful state diagram.

We will demonstrate the synthesis of an FSM circuit based on the C-style pseudocode shown in Figure 6.11. Do not try to interpret the logical execution of the code, because it does not perform anything meaningful. Furthermore, this section is not about optimizing the code by modifying it to make it shorter, although optimizing the code this way might produce a smaller FSM circuit. In this section, the focus is on learning how to convert any given pseudocode to an FSM circuit that realizes it.

Section 6.2 already has discussed in detail how to create a state diagram based on any given pseudocode. Each unconditional data manipulation is assigned to a node in the state diagram. Conditional data manipulations can be assigned either to a node or to a conditional edge (resulting in either a Moore or a Mealy FSM). If two data manipulations can be performed in parallel at the same time, then both of them can be assigned to the same node to reduce the total number of nodes in the state diagram. If two data manipulations must be performed sequentially (one after the other), then they need to be assigned to two different nodes.

The pseudocode shown in Figure 6.11 contains four signal assignment statements—two $Y = 0$ and two $Y = 1$. We assign one state to each of the four signal assignment statements. The first $Y = 0$ is assigned to state s_0, the second $Y = 0$ is assigned to state s_1, and so on, as shown by the comments in the pseudocode.

After the first $Y = 0$ statement, the IF statement conditionally determines whether to execute the second $Y = 0$ statement or the $Y = 1$ statement. Therefore, from state s_0, one edge goes to state s_1, and one edge goes to state s_2. The labels on these two edges are the conditions for the IF statement. The edge going to state s_1 has the label $(B = 0)$,

```
REPEAT {
  Y = 0                 // s0
  IF (B = 0) THEN
    Y = 0               // s1
  ELSE
    Y = 1               // s2
  END IF
  Y = 1                 // s3
}
```

FIGURE 6.11 Functional description using C-style pseudocode for synthesizing an FSM.

and the edge going to state s_2 has the label $(B = 1)$. Remember that all possible outcomes of all the conditional tests must be labeled on the outgoing edges from the node in order to make it deterministic. In other words, for any condition outcome, you need to know exactly which node to go to next. The operation $Y = 0$ is performed in state s_1 and the operation $Y = 1$ is performed in state s_2. After executing state s_1 or state s_2, state s_3 is executed, so both states s_1 and s_2 have an unconditional edge going from these two states to s_3. Finally, because of the unconditional REPEAT loop, there is an unconditional edge that goes from s_3 back to state s_0. The resulting state diagram is shown next.

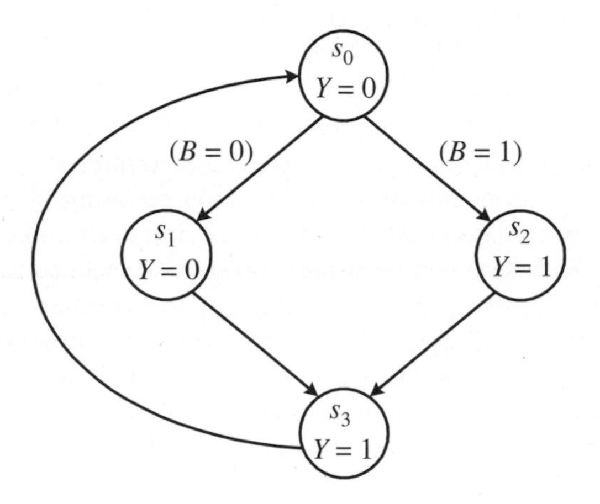

6.4.2 Next-State Table

Having derived the state diagram, it is easy to derive both the next-state and output tables from it. Because the next-state and output tables, and the state diagram portray the same information but depicted in different formats, it requires only a straightforward translation from one to the other.

The resulting next-state table derived from the state diagram is shown next.

Current State Q_1Q_0	Next State $Q_{1next}Q_{0next}$	
	$B = 0$	$B = 1$
s_0 00	s_1 01	s_2 10
s_1 01	s_3 11	s_3 11
s_2 10	s_3 11	s_3 11
s_3 11	s_0 00	s_0 00

The row labels are the current state Q_1Q_0 and the column labels are the input conditions for B. The table entries are the next states $Q_{1next}Q_{0next}$. Translating directly from the state diagram, from the current state s_0, if the input condition $(B = 0)$ is true, then the next state is s_1. Correspondingly, in the next-state table, the entry for the intersection of the current state s_0 and the input $B = 0$ (the color entry in the table) is s_1.

In the next-state table, the actual encodings for the states also are given. To encode the four states, two flip-flops Q_1 and Q_0 are required. In the example, the encodings given to the four states s_0, s_1, s_2, and s_3 are just the four different combinations of the two flip-flop values 00, 01, 10, and 11, respectively. This encoding assignment, however, does not need to be. Using different encoding schemes can give different results in terms of circuit size, speed, and power consumption. This optimization technique is further discussed in Section 6.5.2.

6.4.3 Next-State Equations

The next-state equations are derived from the next-state table. Because we needed two D flip-flops to encode the four states in our state diagram, we will have two next-state equations, one for each of thc D input to the D flip-flop. Each entry in the next-state table has two bits $Q_{1next}Q_{0next}$. When deriving the next-state equations, we need to separate these two bits and look at them individually as two separate truth tables. Extracting the leftmost bit in every entry in the table will give us the truth table for the Q_{1next} equation, and extracting the rightmost bit in every entry in the table will give us the truth table for the Q_{0next} equation. Remember that the characteristic equation for the D flip-flop is $Q_{next} = D$, so the values for Q_{next} and D are the same. The separated truth tables, in the form of a K-map, and their corresponding next-state equations for Q_{1next} and Q_{0next} are shown next.

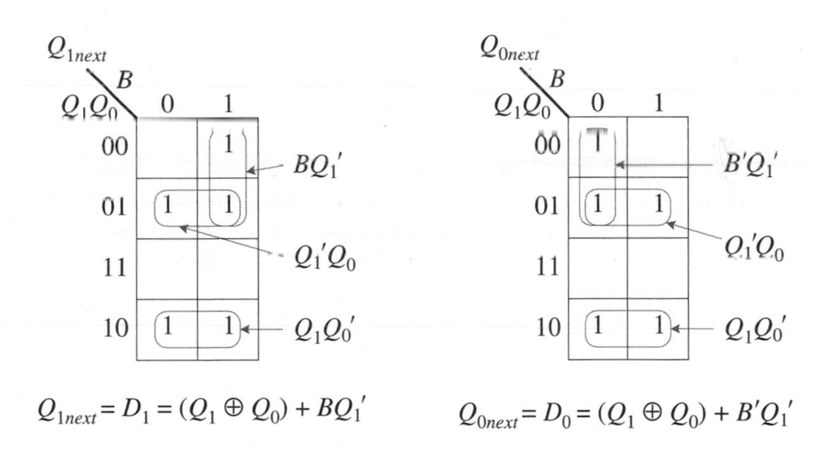

These two equations allow us to construct the combinational next-state logic circuit for the D inputs to the two D flip-flops in our state memory.

6.4.4 Output Table and Output Equations

The output table and output equations are used to derive the output logic circuit in the FSM. The output table is obtained directly from the state diagram. From this state diagram, we see that the output signal Y is dependent only on the state. In states s_0 and s_1, Y is assigned the value 0 and in states s_2 and s_3, Y is assigned the value 1. Hence, we get the resulting output table shown next.

Current State Q_1Q_0	Output Y
s_0 00	0
s_1 01	0
s_2 10	1
s_3 11	1

The output equation as derived from the output truth table is simply

$$Y = Q_1$$

6.4.5 **FSM Circuit**

We will use the sample FSM circuit shown in Figure 6.3 as a template. The number of D flip-flops to use for our state memory was determined when the states were encoded. As noted before, our FSM circuit requires two D flip-flops for its state memory. The two *Clk* signals to the two flip-flops are connected together to the main *Clock* input signal. The two *Clear* signals to the two flip-flops are connected together to the main *Reset* input signal. The next-state logic circuit is drawn from the two next-state equations obtained in Section 6.4.3, and the output logic circuit is drawn from the output equation obtained in Section 6.4.4. Connecting these three parts, next-state circuit, state memory, and output circuit together produce the final FSM circuit shown next.

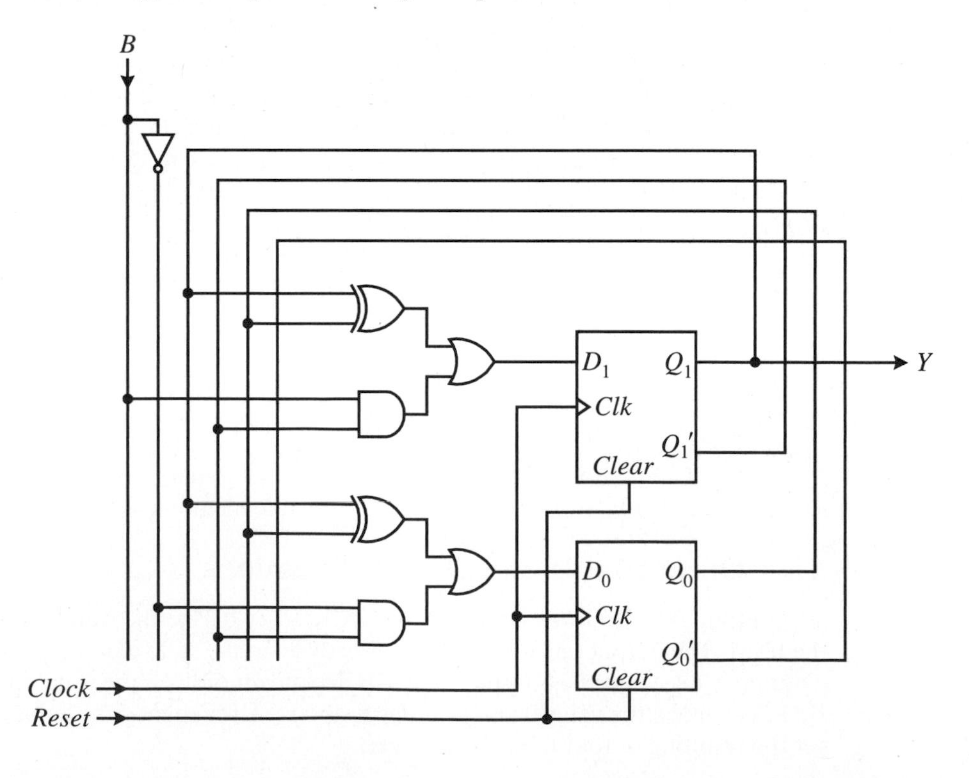

6.5 Optimizations for FSMs

In designing any digital circuit, in addition to getting a functionally correct circuit, we like to optimize it for size, speed, and power consumption. In this section, we will discuss briefly some of the issues involved in optimizing the size of the FSM circuit. In general, reducing the size of a circuit also will make it run faster and use less power.

Because FSM circuits contain the next-state logic and the output logic combinational circuits, we can optimize the FSM by reducing the size of these combinational circuits using the methods as described in Chapter 3. For optimizing the state memory, we can reduce the number of states needed by the FSM or use different state encoding schemes.

6.5.1 State Reduction

FSM circuits with fewer states most likely will result in a smaller circuit, because the number of states directly translates to the number of flip-flops needed. Fewer flip-flops imply a smaller state memory for the FSM, which also means fewer flip-flop inputs. Fewer flip-flop inputs will require fewer next-state equations and so the next-state logic circuit also is reduced.

There are two levels in which we can reduce the number of states. At the pseudocode description level, we can try to optimize the code by shortening it, if possible. We also can reduce the number of states needed by assigning two or more data operations to the same state if these data operations can be performed in parallel. Whether two operations can be performed in parallel is determined by two factors: data dependency and availability of functional units. In the obvious case, if we have two addition instructions, and the second addition is dependent on the result from the first instruction, then these two instructions cannot be performed in parallel. However, if we have two additions that are totally independent of each other, such as

$$x = a + b$$
$$y = c + d$$

we still cannot perform these two instructions in parallel if we only have one adder available to use. The choice of what and how many functional units to include in the microprocessor is discussed in Chapter 7.

After obtaining a state diagram, we still might be able to reduce the number of states by removing equivalent states. Two states are said to be equivalent if the following two conditions are true:

1. Both states produce the same output for every input.
2. Both states have the same next state for every input.

If two states are equivalent, we can remove one of them and use instead the other equivalent state. The resulting FSM still will be functionally equivalent.

Another way to reduce the number of states is to use the Mealy FSM model instead of the Moore FSM model. The reason is that if outputs are dependent on both the current state and input conditions, then we do not need to use a separate state to do the outputs. The tradeoff is that the output logic circuit most likely will be larger. Sections 6.6.5 and 6.6.6 will show examples of this tradeoff.

6.5.2 State Encoding

When initially drawing the state diagram for an FSM circuit, it is preferred to use symbolic state names. However, these state names eventually must be encoded with a unique binary bit string. The process of state encoding is to determine how many flip-flops are required to represent the states in the next-state table or state diagram, and then to assign a unique binary bit string to each of the states. In all of the examples presented so far, we have used the straight binary encoding scheme, where n flip-flops are needed to encode 2^n states. For example, for four states, state s_0 gets the encoding 00, s_1 gets the encoding 01, s_2 gets 10, and s_3 gets the encoding 11. However, there is no reason why we cannot use a different encoding for the states. In fact, we do want to use a different encoding if it will result in a smaller circuit. This straight binary encoding scheme does not always lead to the smallest FSM circuit. Other encoding schemes, such as the minimum bit change, minimize 1-bit, prioritized adjacency, and one-hot encoding, might result in a smaller circuit.

For the **minimum bit change** scheme, binary encodings are assigned to the states in such a way that the total number of bit changes for all state transitions is minimized. In other words, if every edge in the state diagram is assigned a weight that is equal to the number of bit changes between the source encoding and the destination encoding of that edge, this scheme would select the one that minimizes the sum of all of these edge weights.

For example, given the state diagram with four states shown in Figure 6.12(a), the minimum bit change scheme would use the encoding shown in (b) and not the encoding shown in (c). In both Figures 6.12(b) and (c), the number of bit changes between the encodings of two states joined by an edge is labeled on that edge. For example, in Figure 6.12(b), the number of bit changes between state $s_1 = 01$ and $s_2 = 11$ is 1. The encoding used in Figure 6.12(b) has a smaller sum of all of the edge weights than the encoding used in (c).

Notice that, even though the encoding of Figure 6.12(b) produces the smallest total edge weight, there are several other ways to encode these four states that also will produce the same total edge weight (e.g., assigning 00 to s_1 instead of to s_0, 01 to s_2 instead of to s_1, 11 to s_3, and 10 to s_0).

For the **minimize 1-bit** scheme, the state with the most incoming edges is encoded with the least number of 1 bits. In Figure 6.13, (b) uses the straight binary encoding and (c) uses the minimize 1-bit encoding. Their corresponding next-state tables and equations are shown in Figures 6.13(d) and (e), respectively. In most situations, the next-state table with fewer 1 bits will result in a smaller next-state circuit.

For the **prioritized adjacency** scheme, adjacent states to any state s are given certain priorities. Encodings are assigned to these adjacent states such that those with a higher priority will have an encoding that has fewer bit changes from the encoding of state s than those adjacent states with a lower priority. The minimize 1-bit scheme is

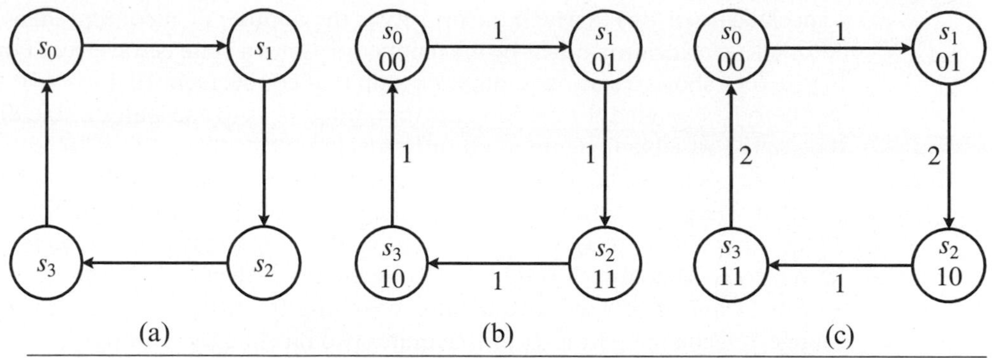

FIGURE 6.12 Minimum bit change encoding: (a) a state diagram with four states; (b) encoding with a total weight of four; (c) encoding with a total weight of six.

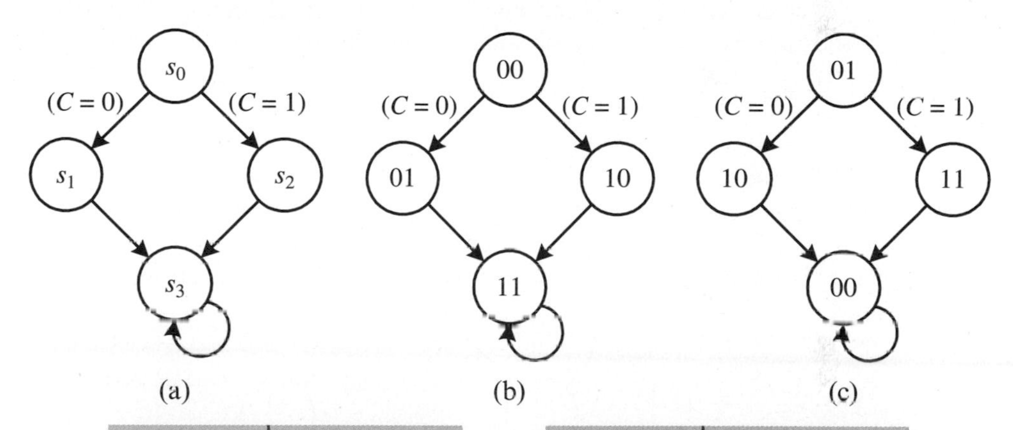

Current State Q_1Q_0	Next State $Q_{1next}Q_{0next}$	
	$C = 0$	$C = 1$
00	01	10
01	11	11
10	11	11
11	11	11

$$Q_{1next} = C + Q_1 + Q_0$$
$$Q_{0next} = C' + Q_1 + Q_0$$

(d)

Current State Q_1Q_0	Next State $Q_{1next}Q_{0next}$	
	$C = 0$	$C = 1$
00	00	00
01	10	11
10	00	00
11	00	00

$$Q_{1next} = Q_1'Q_0$$
$$Q_{0next} = CQ_1'Q_0$$

(e)

FIGURE 6.13 Minimize 1-bit encoding comparison: (a) state diagram with four states; (b) using straight binary encoding; (c) using the minimize 1-bit encoding; (d) next-state table for the straight binary encoding; (e) next-state table for the minimize 1-bit encoding.

a special case of this in which the priority is the number of incoming edges to a state. Another prioritization might be the number of times a state is traversed. For example, Figure 6.14 shows a state diagram for a loop that counts from 10 down to 0. The statements to be executed for the loop are annotated next to the states in the diagram. In state s_0 the variable C is initialized to 10 (decimal). In state s_1, C is decremented by 1. Next, the condition $(C = 0)$ is tested, and if it is true then it will go to state s_3, otherwise it will go to state s_2. Because of the nature of the loop, we know that the edge going from state s_1 to s_2 will be traversed many more times than the edge going from state s_1 to s_3. Therefore, if state s_1 is encoded with 01, then we will want to encode state s_2 with 11 and state s_3 with 10 because changing from 01 to 11 requires only one bit change, but changing from 01 to 10 requires two bit changes. This particular prioritized adjacency scheme might be good for reducing power consumption because the amount of power used in a circuit is related to the number of bit changes.

For the **one-hot encoding** scheme, each state is assigned one flip-flop. A state is encoded with its flip-flop having a 1 value, while all of the other flip-flops have a 0 value. For example, the one-hot encoding for four states would be 0001, 0010, 0100, and 1000, as shown in Figure 6.15. The advantage of this scheme is that we are minimizing the number

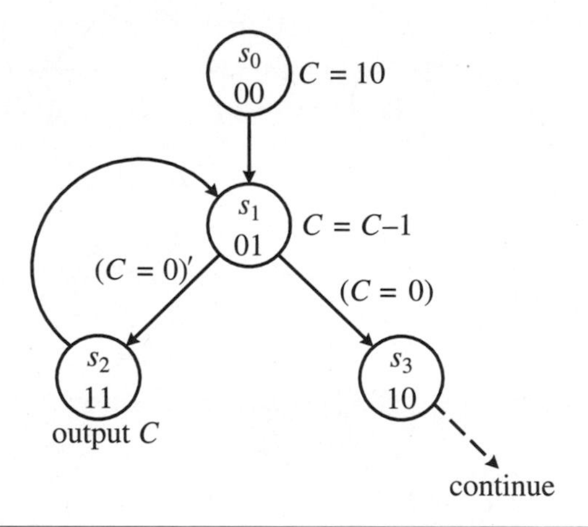

FIGURE 6.14 Prioritized adjacency encoding.

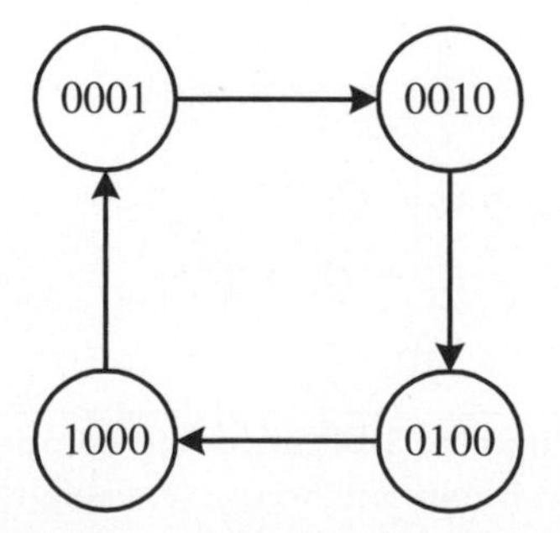

FIGURE 6.15 One-hot encoding.

of 1 bits needed to encode a state, and there will be at most two bit changes needed to go between any two states. The disadvantage is that many more flip-flops are needed.

6.5.3 Unused States

In a real-world situation, the number of states used in the state diagram is most likely not a power of two. For example, a modulo-6 up-counter will have six states. To encode six states, we need at least three flip-flops, because two flip-flops can encode only four different combinations. However, three flip-flops give eight different combinations, and so two combinations are not used. The questions are: What do we do with these unused encodings? In the next-state table, what next state values do we assign to these unused states? Do we just ignore them?

If the FSM can never be in any of the unused states, then it does not matter what their next states are. In this case, we can put don't-care values for their next states. The resulting next-state circuit might be smaller because of the don't-care values.

But what if, by chance, the FSM enters one of these unused states? The operation of the FSM will be unpredictable because we do not know what the next state is. Well, this is not exactly true because, even though we started with the don't-cares, we have mapped them to a fixed next-state equation. So, these unused states do have definite next states. It is just that these next states are not what we wanted. Therefore, the resulting FSM operation will be incorrect if it ever enters one of the unused states. If this FSM is used in a mission-critical control unit, we do not want even this slight chance to occur.

One solution is to use the initialization or starting state as the next state for these unused state encodings. This way, the FSM will restart from the beginning if it ever enters one of these unused states.

6.6 FSM Construction Examples

We will now provide several examples to illustrate the complete process of synthesizing FSMs.

6.6.1 Car Security System—Version 3

Let us revisit the car security system example from Chapters 2 and 6. Recall that in the first version (Chapter 2) the circuit is a combinational circuit. The problem with a combinational circuit is that after the alarm is triggered, by opening the door, for example, the alarm can be turned off immediately by closing the door again. However, what we want is that after the alarm is triggered, it should remain on even after the door is closed again, and the only way to turn it off is to turn off the master switch.

This requirement suggests that we need a sequential circuit in which the output is dependent not only on the current input switch settings, but also on the current state of the alarm. Version 2 of the car security system in Chapter 6 used an ad hoc approach to resolve this issue by adding a SR latch. In this section, we will use a more formal approach by designing an FSM for the car security system.

We start by deriving the state diagram for the system, as shown in Figure 6.16(a). In addition to the three input switches M, D, and V (for *Master*, *Door*, and *Vibration*),

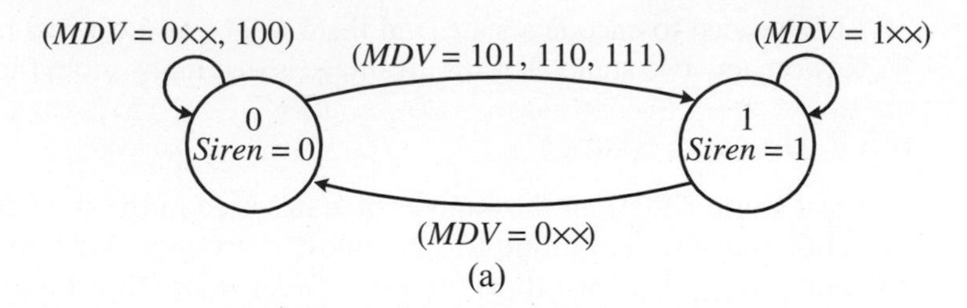

(a)

Current State Q_0	Next State Q_{0next}							
	M, D, V							
	000	001	010	011	100	101	110	111
0	0	0	0	0	0	1	1	1
1	0	0	0	0	1	1	1	1

(b)

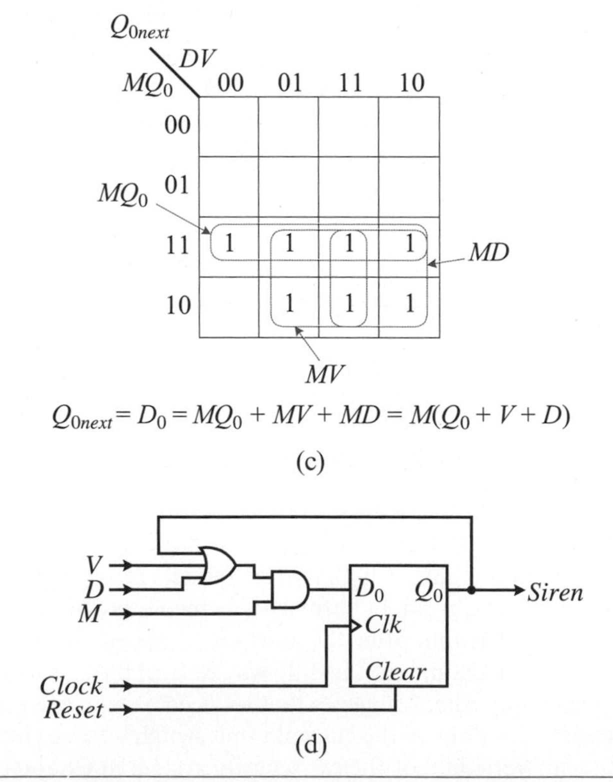

(d)

FIGURE 6.16 Car security system—Version 3: (a) state diagram; (b) next-state table; (c) K-map and next-state equation; (d) circuit.

we need two states 1 and 0 to represent whether the siren is on or off, respectively. If the siren is on (i.e., in state 1), then it will remain in that state as long as the master switch is still on, so it doesn't matter whether the door is now closed or open. This is represented by the edge that goes from state 1 and loops back to state 1 with the label ($MDV = 1\times\times$). From the on (1) state, the only way to turn off the siren is to turn off the master switch. This is represented by the edge going from state 1 to state 0 with the label ($MDV = 0\times\times$). If the siren is off (i.e., in state 0), it is turned on when the master switch is on, and either the door switch or the vibration switch is on. This is represented by the edge going from state 0 to state 1 with the labels ($MDV = 101, 110$, or 111). Finally, from the off state, the siren will remain off when either the master switch remains off or if the master switch is on but none of the other two switches is on. This is represented by the edge from state 0 looping back to state 0 with the labels ($MDV = 0\times\times, 100$).

The state diagram is translated to the corresponding next-state table using one D flip-flop for the state memory, as shown in Figure 6.16(b). Doing a 4-variable K-map on the next-state table gives us the next-state equation shown in Figure 6.16(c). The final circuit for this car security system is shown in Figure 6.16(d). The circuit uses one D flip-flop. The next-state circuit is derived from the next-state equation, which produces the signal for the D_0 input of the flip-flop. No extra output signal is needed because the Q_0 output of the flip-flop directly drives the siren.

6.6.2 Modulo-6 Up-Counter

In this example, we will design a modulo-6 up-counter with a count enable input C, and an output signal Y. The count is to be represented directly by the contents of the state memory flip-flops. When the input C is asserted, the FSM will transition to the next state, once per clock cycle. The output Y is asserted when the count is equal to five. We will follow the steps described earlier to synthesize an FSM circuit for this modulo-6 up-counter.

Step 1 of the synthesis process is to construct the state diagram. From the above functional description, we need to construct a state diagram that will show the precise operation of the circuit. A modulo-6 up-counter counts from zero to five, and then back to zero. Because the count is represented by the state memory flip-flop values and we have six different counts (from zero to five), we will need three flip-flops Q_2, Q_1, and Q_0 that will produce the sequence 000, 001, 010, 011, 100, 101, 000, 001, ... when C is asserted; otherwise, when C is de-asserted, the counting stops. The remaining two states, 110 and 111, are not used. From state 000, which is the count zero, there will be an edge that goes to state 001 with the label ($C = 1$). From state 001, there is an edge that goes to state 010 with the label ($C = 1$), and so on. For the counting to stop at each count, there will be edges at each state that loop back to itself with the label ($C = 0$). Furthermore, we want to assert Y when the count is five, which is state 101, so in this state, we set Y to a 1. For the rest of the states, Y is set to a 0. Hence, we obtain the state diagram shown next for a modulo-6 up-counter.

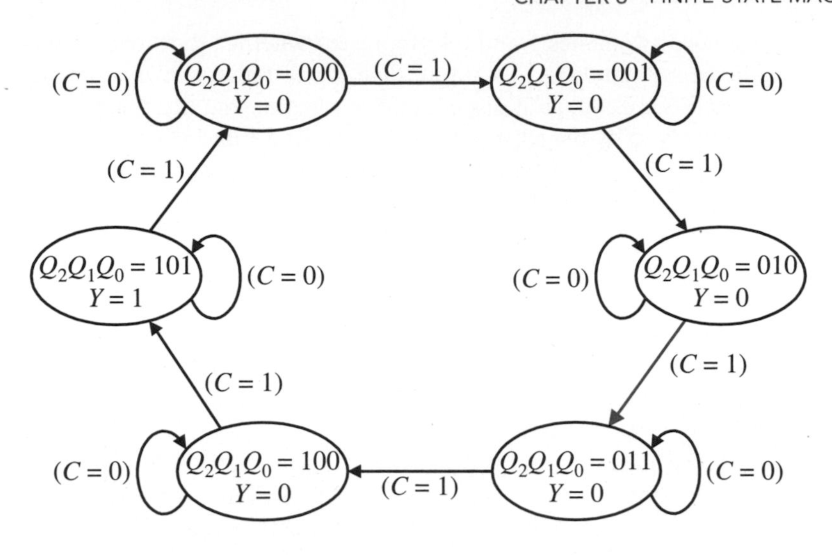

Step 2 is to derive the next-state table, which is a direct translation from the state diagram. We have three flip-flops Q_2, Q_1, and Q_0, and one primary input C. The current states for the flip-flops are listed down the rows of the table, while the input conditions are listed across the columns. The entries in the table are the next-state values. Each next-state value is composed of the three flip-flop values Q_{2next}, Q_{1next}, and Q_{0next}. We follow the edges in the state diagram to determine the three flip-flop values $Q_{2next}Q_{1next}Q_{0next}$, for each entry. For example, if the current state $Q_2Q_1Q_0$ is 010 and the input is $(C = 1)$, then the next state $Q_{2next}Q_{1next}Q_{0next}$ is 011. This transition is highlighted by the color edge in the state diagram and the corresponding color entry in the next-state table shown next.

For the two unused states, we have given the don't-care values for their next states.

Current State $Q_2Q_1Q_0$	**Next State** $Q_{2next}Q_{1next}Q_{0next}$	
	(C = 0)	**(C = 1)**
000	000	001
001	001	010
010	010	**011**
011	011	100
100	100	101
101	101	000
110	×××	×××
111	×××	×××

Step 3 is to derive the next-state equations for all of the flip-flop inputs in terms of the current state and the primary input. These equations are obtained directly from the next-state table. To derive each next-state equation, we need to visualize the three next-state bits $Q_{2next}Q_{1next}Q_{0next}$ as three separate truth tables and look at them individually. If you find it difficult to visualize them as separate tables, then you first might want to actually separate these three bits into three separate truth tables. For the Q_{2next} equation, we consider just the leftmost bit in each entry in the truth table. Looking at all of the leftmost bits, there are four 1-minterms, giving the equation

$$Q_{2next} = C'Q_2Q_1'Q_0' + C'Q_2Q_1'Q_0 + CQ_2'Q_1Q_0 + CQ_2Q_1'Q_0'$$

Note in the equation that we have replaced the condition $(C = 0)$ with C', and $(C = 1)$ with C. This is just another way of writing it, and it is shorter.

The equation for Q_{1next} is derived from considering just the middle bit for all of the entries in the next-state table, and the equation for Q_{0next} is derived from just the rightmost bit, giving us the following two equations:

$$Q_{1next} = C'Q_2'Q_1Q_0' + C'Q_2'Q_1Q_0 + CQ_2'Q_1'Q_0 + CQ_2'Q_1Q_0'$$

$$Q_{0next} = C'Q_2'Q_1'Q_0 + C'Q_2'Q_1Q_0 + C'Q_2Q_1'Q_0 + CQ_2'Q_1'Q_0' + CQ_2'Q_1Q_0' + CQ_2Q_1'Q_0'$$

Because these next-state equations will be used to construct the next-state circuit, they should be simplified to make the circuit smaller. The three K-maps and simplified next-state equations for Q_{2next}, Q_{1next}, and Q_{0next} are shown next. In the K-maps, don't-care values are used for the next-state values for the two unused state encodings 110 and 111. However, in the final simplified next-state equations, these don't-care values are replaced with actual values, so these two unused states do have an actual next state. (See Problem 6.10.)

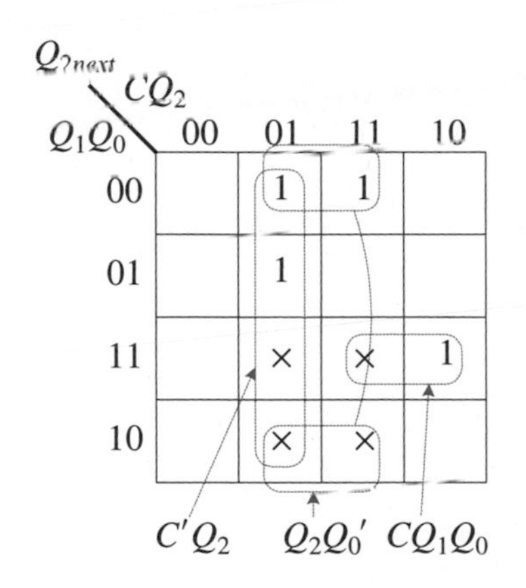

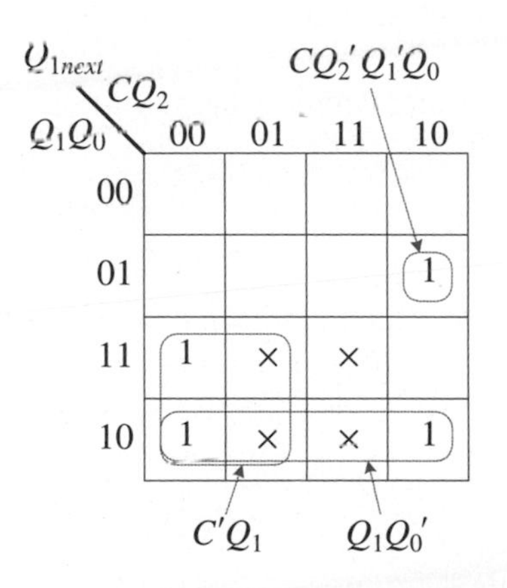

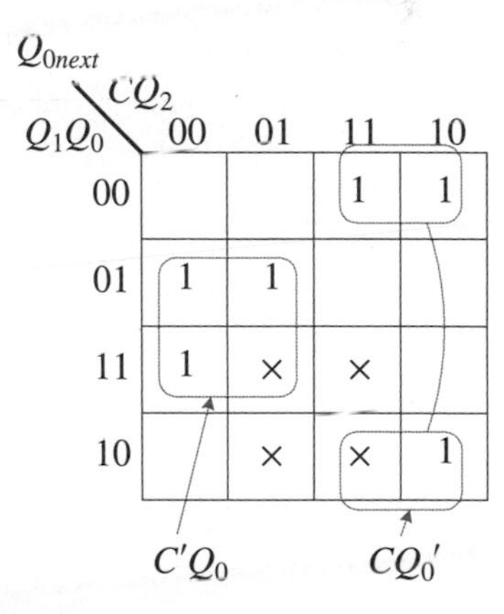

$$Q_{2next} = C'Q_2 + Q_2Q_0' + CQ_1Q_0$$

$$Q_{1next} = C'Q_1 + Q_1Q_0' + CQ_2'Q_1'Q_0$$

$$Q_{0next} = C'Q_0 + CQ_0' = C \oplus Q_0$$

Steps 4 and 5 are to derive the output table and equations. There is one equation for every output signal. Because the value of Y is labeled next to each node, it is dependent only on the current state. From the state diagram, Y is asserted only in state 101, so Y has a 1 only in that current-state entry, while the rest of them are 0s. The output table and equation are shown next.

Current State $Q_2Q_1Q_0$	Output Y
000	0
001	0
010	0
011	0
100	0
101	1

$$Y = Q_2Q_1'Q_0$$

Step 6 is to draw the FSM circuit. Our state memory consists of three D flip-flops with the three inputs D_2, D_1, and D_0. The values given to these three D inputs are directly from the three next-state equations because the characteristic equation for the D flip-flop is $Q_{next} = D$. Thus, we use the three previously derived next-state equations

$$Q_{2next} = D_2 = C'Q_2 + Q_2Q_0' + CQ_1Q_0$$

$$Q_{1next} = D_1 = C'Q_1 + Q_1Q_0' + CQ_2'Q_1'Q_0$$

$$Q_{0next} = D_0 = C'Q_0 + CQ_0' = C \oplus Q_0$$

to construct our next-state circuit. The output circuit is constructed from the output equation

$$Y = Q_2Q_1'Q_0$$

The complete FSM circuit is shown next. Following the template FSM circuit, the *Clk* signals to the two flip-flops are connected together to the main *Clock* input signal, and the *Clear* signals to the two flip-flops are connected together to the main *Reset* input signal.

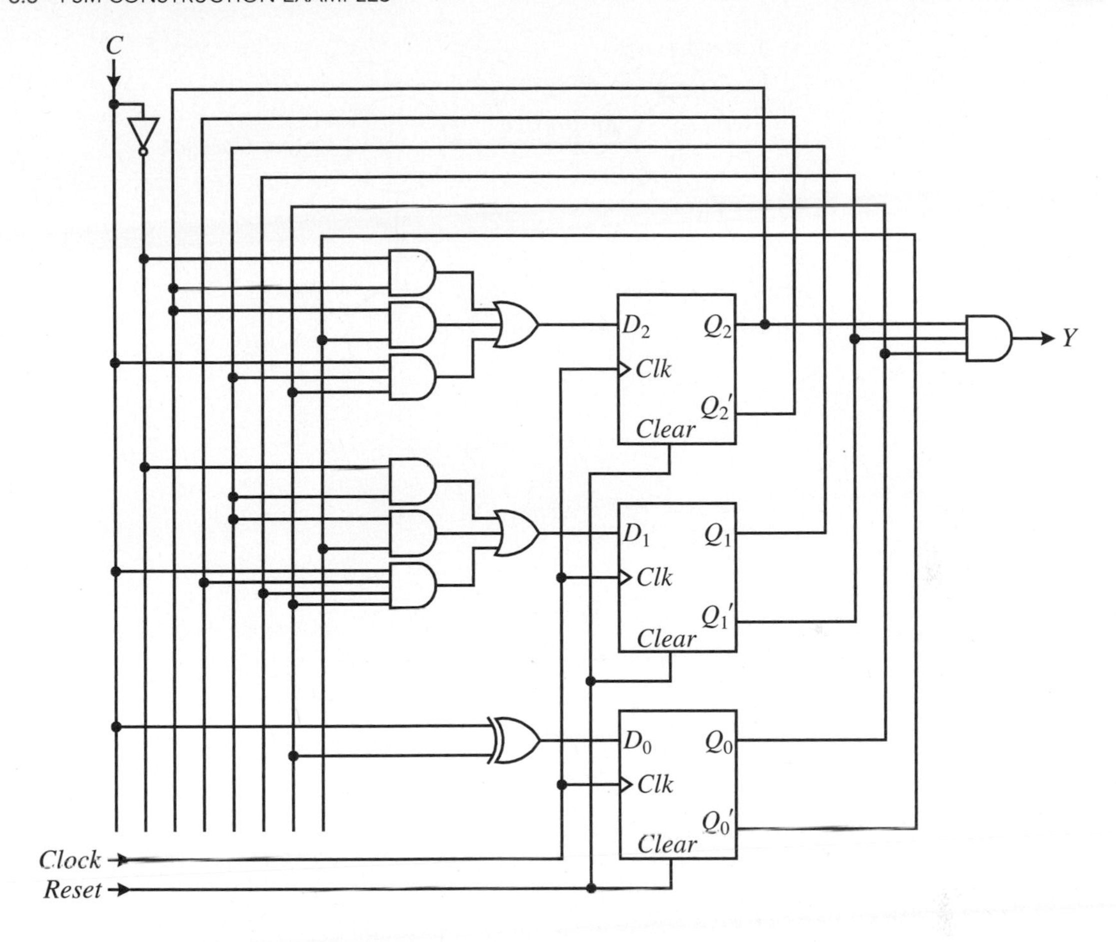

6.6.3 One-Shot Circuit

In this example, we will synthesize an FSM for a one-shot circuit, which outputs a single short pulse when given an input of arbitrary time length. In this FSM circuit, the length of the single short pulse will be one clock cycle. The state diagram for this circuit is shown in Figure 6.17(a).

State s_0, encoded as 00, is the reset state, and the FSM waits for a key press in this state. When a switch is pressed, the FSM goes to state s_1 (encoded as 01), and outputs a single short pulse. From s_1, the FSM unconditionally goes to state s_2 (encoded as 11), and turns off the one-shot pulse. Therefore, the pulse lasts only for one clock cycle, regardless of how long the key is pressed. To break the loop and wait for another key press, the FSM has to wait for the release of the key in state s_2. When the key is released, the FSM goes back to state s_0 to wait for another key press.

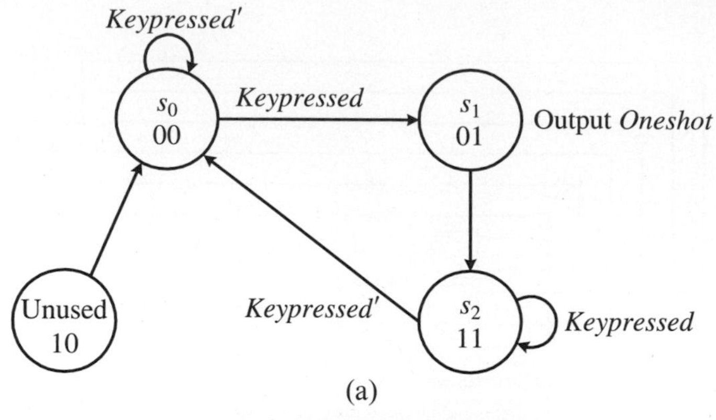

(a)

Current State Q_1Q_0	Next State $Q_{1next}Q_{0next}$	
	Keypressed	
	0	**1**
00	00	01
01	11	11
11	00	11
10 Unused	00	00

(b)

Q_{1next}

Q_1Q_0 \ *Keypressed*	0	1
00		
01	1	1
11		1
10		

$$Q_{1next} = Q_1'Q_0 + Q_0 Keypressed$$

Q_{0next}

Q_1Q_0 \ *Keypressed*	0	1
00		1
01	1	1
11		1
10		

$$Q_{0next} = Q_1' Keypressed + Q_1'Q_0 + Q_0 Keypressed$$

(c)

FIGURE 6.17 FSM for one-shot circuit: (a) state diagram; (b) next-state table; (c) next-state equations and K-maps; (d) output table and output equation; (e) FSM circuit. *(continued on next page)*

This state diagram uses two bits to encode the three states; state encoding 10 is not used. The state diagram shows that, if the FSM somehow gets to state 10, it unconditionally will go to the reset state 00 in the next clock cycle.

Current State Q_1Q_0	Output *Oneshot*
00	0
01	1
11	0
10	0

$$Oneshot = Q_1'Q_0$$

(d)

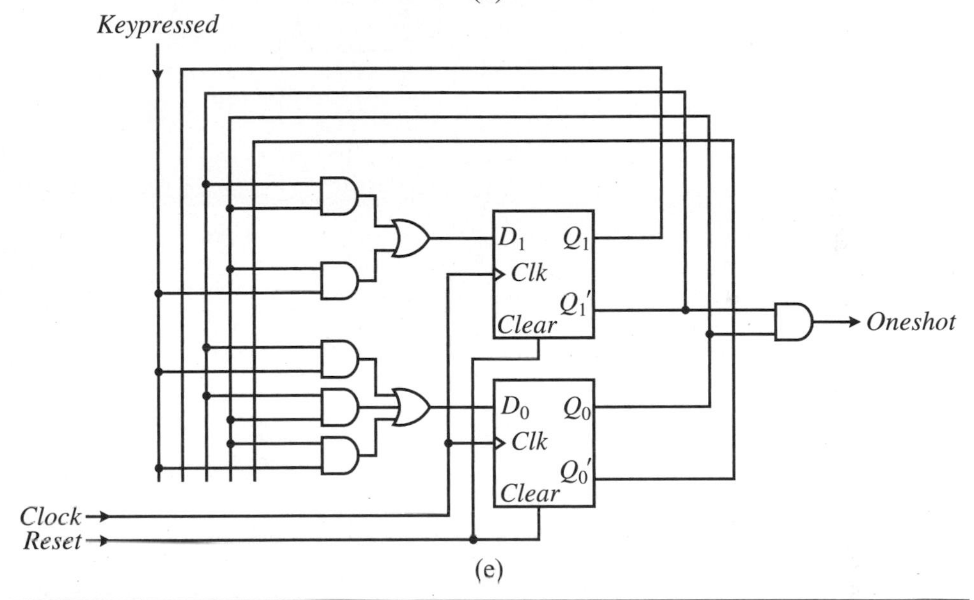

(e)

FIGURE 6.17 FSM for one-shot circuit: (a) state diagram; (b) next-state table; (c) next-state equations and K-maps; (d) output table and output equation; (e) FSM circuit.

The corresponding next-state table and next-state equations are shown in Figures 6.17(b) and (c), respectively. The output table and output equation are shown in Figure 6.17(d) and finally, the complete FSM circuit in Figure 6.17(e).

6.6.4 Simple Microprocessor Control Unit

In this example, we will synthesize an FSM that illustrates what a simple control unit of a microprocessor is like. We start with the state diagram as shown in Figure 6.18(a). Each state is labeled with a state name s_0, s_1, s_2, and s_3, and has two output signals x and y. There are also two conditional status signals, *Start* and $(n = 9)$ labeled on four of the edges, while the rest of the edges do not have any conditions. From state s_0, the conditional edge labeled *Start* is taken when *Start* = 1; otherwise, the edge labeled *Start′* is taken. Similarly, from state s_2, the edge with the label $(n = 9)$ is taken when the condition is true (i.e., when the value of variable n is equal to nine). If n is not equal to nine, then the edge with the label $(n = 9)'$ is taken.

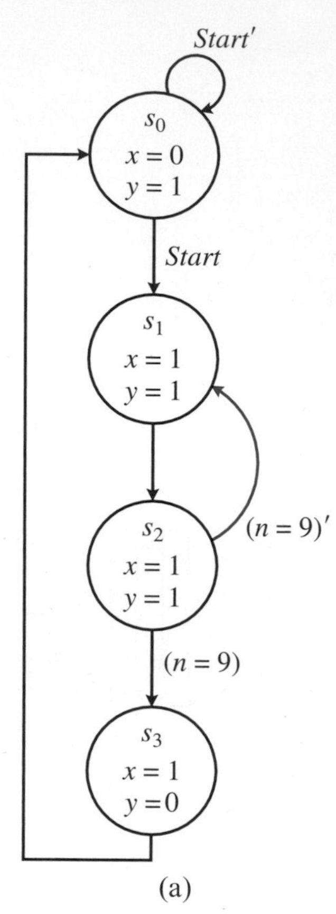

Current State Q_1Q_0	Next State $Q_{1next}Q_{0next}$			
	Start, $(n = 9)$			
	00	**01**	**10**	**11**
s_0 00	s_0 00	s_0 00	s_1 01	s_1 01
s_1 01	s_2 10	s_2 10	s_2 10	s_2 10
s_2 10	s_1 01	s_3 11	s_1 01	s_3 11
s_3 11	s_0 00	s_0 00	s_0 00	s_0 00

(b)

Q_{1next}

Q_1Q_0 \ *Start*, $(n = 9)$	00	01	11	10
00				
01	1	1	1	1
11				
10		1	1	

$Q_1'Q_0$ $\quad Q_1Q_0'(n = 9)$

$$Q_{1next} = D_1 = Q_1'Q_0 + Q_1Q_0'(n = 9)$$

Q_{0next}

Q_1Q_0 \ *Start*, $(n = 9)$	00	01	11	10
00			1	1
01				
11				
10	1	1	1	1

Q_1Q_0' $\quad StartQ_0'$

$$Q_{0next} = D_0 = Q_1Q_0' + StartQ_0'$$

(c)

FIGURE 6.18 **Synthesis of a simple microprocessor control unit FSM: (a) state diagram; (b) next-state table; (c) K-maps and next-state equations for Q_{1next} and Q_{0next}; (d) output table; (e) K-maps and output equations; (f) FSM circuit.** *(continued on next page)*

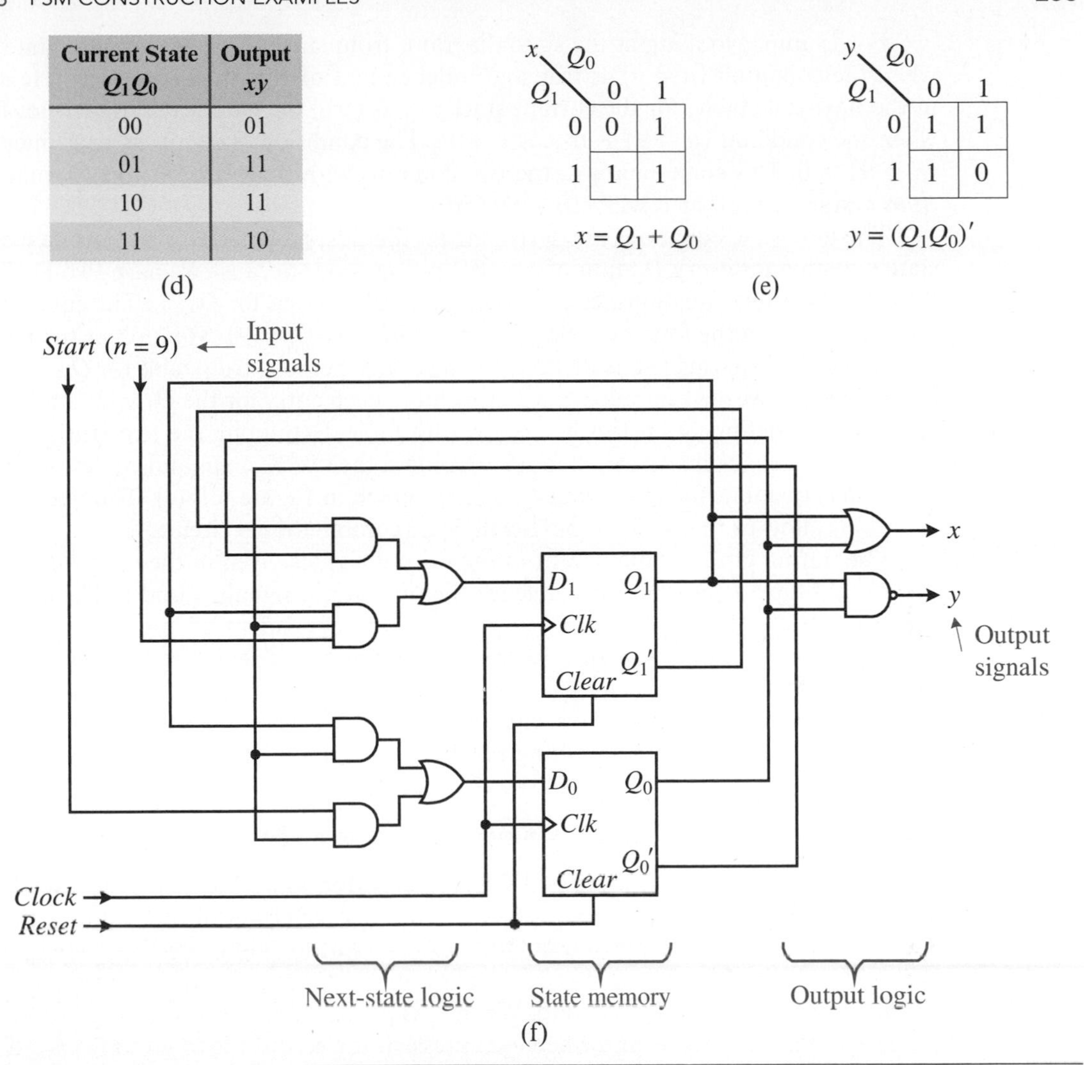

Current State Q_1Q_0	Output xy
00	01
01	11
10	11
11	10

FIGURE 6.18 Synthesis of a simple microprocessor control unit FSM: (a) state diagram; (b) next-state table; (c) K-maps and next-state equations for Q_{1next} and Q_{0next}; (d) output table; (e) K-maps and output equations; (f) FSM circuit.

Two flip-flops Q_0 and Q_1 are needed in order to encode the four states. For simplicity, we will use the binary value of the index of the state name to be the encoding for that state, so the encoding for state s_0 is $Q_1Q_0 = 00$ and the encoding for state s_1 is $Q_1Q_0 = 01$, and so on.

From the above interpretation, we are able to derive the next-state table, as shown in Figure 6.18(b). The four current states for Q_1Q_0 are listed down the four rows. The four columns are for the four combinations of the two conditional input signals *Start* and $(n = 9)$. For example, the column with the value *Start*, $(n = 9) = 10$ means *Start* $= 1$ and $(n = 9) = 0$. The condition $(n = 9) = 0$ means that the condition $(n = 9)$ is false, which means $(n = 9)'$ is true. The entries in the table are the next states $Q_{1next}Q_{0next}$ for the two flip-flops.

For example, looking at the state diagram, from state s_2, we go back to state s_1 when the condition $(n = 9)'$ is true and independent of the *Start* condition. Hence, in the next-state table, for the current state row s_2 (10), the two next-state entries for when the condition $(n = 9)'$ is true is s_1 (01). The condition "$(n = 9)'$ is true" means $(n = 9) = 0$. This corresponds to the two columns with the labels 00 and 10, that is, *Start* can be either 0 or 1, while $(n = 9)$ is 0.

The next-state equations are derived from the next-state table. There is one next-state equation for every D input of every flip-flop used. Because we have two D flip-flops, we have two equations: one for Q_{1next} and the second for Q_{0next}. The equations are dependent on the four variables Q_1, Q_0, *Start*, and $(n = 9)$. We look at the next-state table as one having two truth tables merged together: one truth table for Q_{1next} and one for Q_{0next}. We look at only the leftmost bit in each entry for the Q_{1next} truth table, and only the rightmost bit for the Q_{0next} truth table. Extracting the two truth tables from the next-state table in this manner, we obtain the two K-maps and corresponding next-state equations for Q_{1next} and Q_{0next}, as shown in Figure 6.18(c). The next-state equations allow us to derive the next-state logic combinational circuit.

The output table is obtained from the output signals given in the state diagram. The output table is just the truth table for the two output signals, x and y. The output signal equations derived from the output table are dependent on the current state, Q_1Q_0. The output table, K-maps, and output equations are shown in Figures 6.18(d) and (e).

From the next-state and output equations, we can easily produce the next-state and output logic circuits, and the resulting FSM circuit is shown in Figure 6.18(f).

6.6.5 Elevator Controller Using a Moore FSM

A new building with two floors will have an elevator, and you have been asked to design the FSM controller circuit for the elevator. On each floor of the building, there is a button for a person to press to call the elevator. Button f_1 is located on floor 1, and button f_2 is located on floor 2. Inside the elevator, there are two buttons (e_1 and e_2) to tell the elevator which floor to go to. When e_1 is pressed, the elevator is to go to floor 1 if it is not already there, and when e_2 is pressed, the elevator is to go to floor 2 if it is not already there. Finally, there are two input signals at_1 and at_2 that are asserted automatically by the elevator mechanism when it reaches that particular floor. at_1 is asserted when the elevator is at floor 1, and at_2 is asserted when the elevator is at floor 2.

A 2-bit output signal $go_{1\text{-}0}$ controls the elevator motor to turn on and off, and which floor to go to. The elevator motor is turned on when go_1 is asserted with a 1, and turned off when go_1 is de-asserted. The output signal go_0 specifies which floor to go to: the elevator goes to floor 1 if $go_0 = 0$, and floor 2 if $go_0 = 1$. There are also two LEDs on each of the two floors to show the elevator's location. The output signal led_1 is turned on if the elevator is at floor 1, and led_2 is turned on if the elevator is at floor 2. Both LEDs are turned off when the elevator is moving between floors. A picture of the elevator setup and a summary of the I/O signals are shown next.

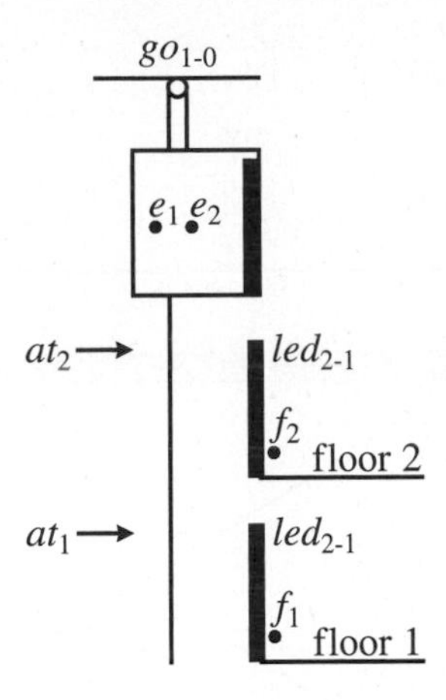

Inputs:

- f_1, f_2: Buttons at each floor to call the elevator. f_1 is on floor 1, f_2 is on floor 2.
- e_1, e_2: Buttons inside the elevator to tell the elevator which floor to go to.
- at_1, at_2: Signals from the elevator mechanism that get asserted depending on the elevator's current location.

Outputs:

- go_1: 0 to turn off the elevator motor, and 1 to turn on the motor.
- go_0: 0 to go to floor 1, and 1 to go to floor 2.
- led_1, led_2: LEDs on each of the two floors to show the elevator's current location.

Although there are two sets of button inputs, f_i and e_i, they serve the same function, in that when pressed, they will call the elevator to go to a particular floor. So we can reduce the number of inputs by combining the corresponding f_i and e_i signals into one signal f_i. We will assume that if both buttons (f_2 and f_1) are pressed at the same time, then nothing will change and the FSM will remain in the current state.

From the description of the elevator system, we can derive the state diagram for a Moore FSM as shown in Figure 6.19(a). There are two states for when the elevator is at a particular floor, and two intermediate states for when the elevator is moving to the other floor. Starting at state 00 (floor 1), we wait for f_2 to be pressed. Any other combinations of button presses will not change states. When f_2 is asserted, we go to the intermediate state 11 in which the elevator is going to floor 2. In this state, we set $go_{1\text{-}0}$ to 11 to turn on the elevator motor to go to floor 2, and turn off both LEDs. We continue in state 11 until at_2 is asserted, telling us that the elevator has arrived at floor 2, and we transition to state 10 (floor 2). In state 10, we turn off the motor by setting $go_{1\text{-}0}$ to 0x, and turning on led_2. Going down from floor 2 to floor 1 is similar with the corresponding changes.

There are a total of four inputs f_1, f_2, at_1, and at_2, and if we enumerate all possible combinations of these four inputs, our next-state table would have 16 columns. However, notice that the f_i's and the at_i's are mutually exclusive, in that the f_i's are used only in

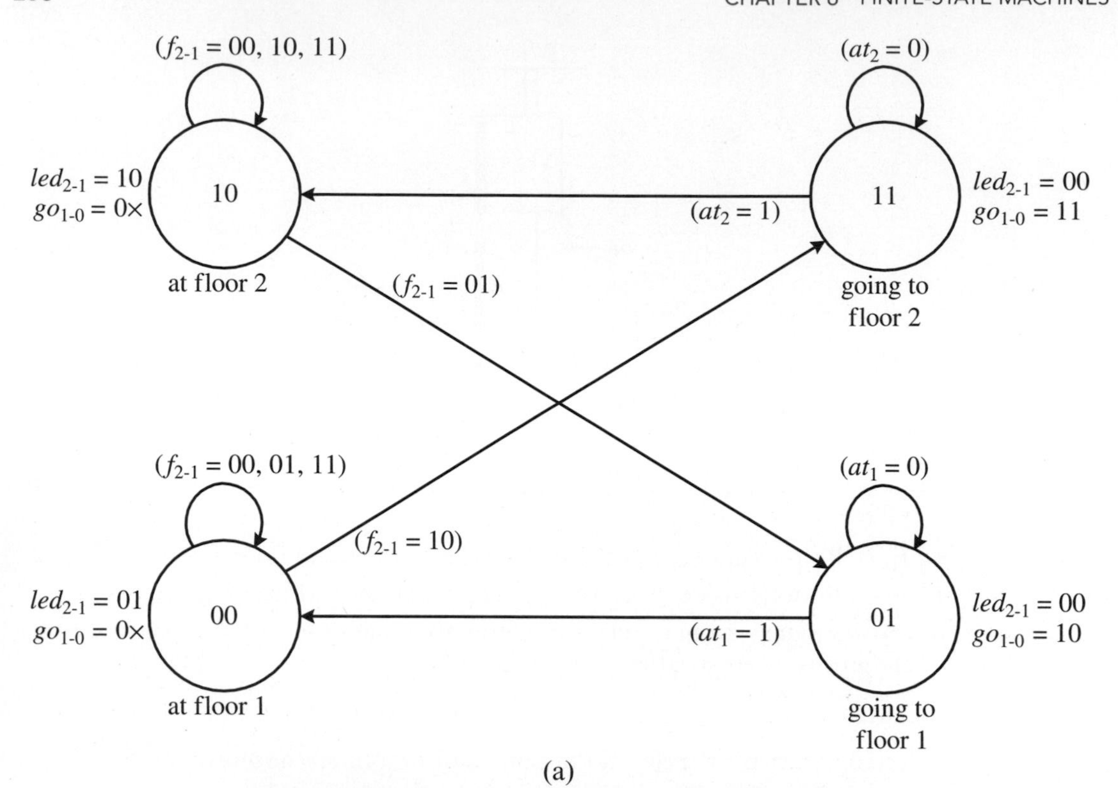

(a)

Current State Q_1Q_0	Next State $Q_{1next}Q_{0next}$					
	$f_{2\text{-}1}$ =				at_i =	
	00	**01**	**10**	**11**	**0**	**1**
00	00	00	11	00		
01					01(at_1')	00(at_1)
10	10	01	10	10		
11					11(at_2')	10(at_2)

(b)

$$Q_{1next} = D_1 = Q_1'Q_0'(f_2 f_1') + Q_1Q_0'(f_1' + f_2) + Q_1Q_0$$

$$Q_{0next} = D_0 = Q_1'Q_0'(f_2 f_1') + Q_1'Q_0(at_1') + Q_1Q_0'(f_2' f_1) + Q_1Q_0(at_2')$$

(c)

FIGURE 6.19 Elevator controller using Moore model: (a) state diagram; (b) next-state table; (c) next-state equations; (d) output table; (e) output equations; (f) FSM circuit.
(continued on next page)

Current State Q_1Q_0	Output	
	$led_{2\text{-}1}$	$go_{1\text{-}0}$
00	01	0×
01	00	10
10	10	0×
11	00	11

(d)

$$led_2 = Q_1Q_0'$$
$$led_1 = Q_1'Q_0'$$
$$go_1 = Q_0$$
$$go_0 = Q_1$$

(e)

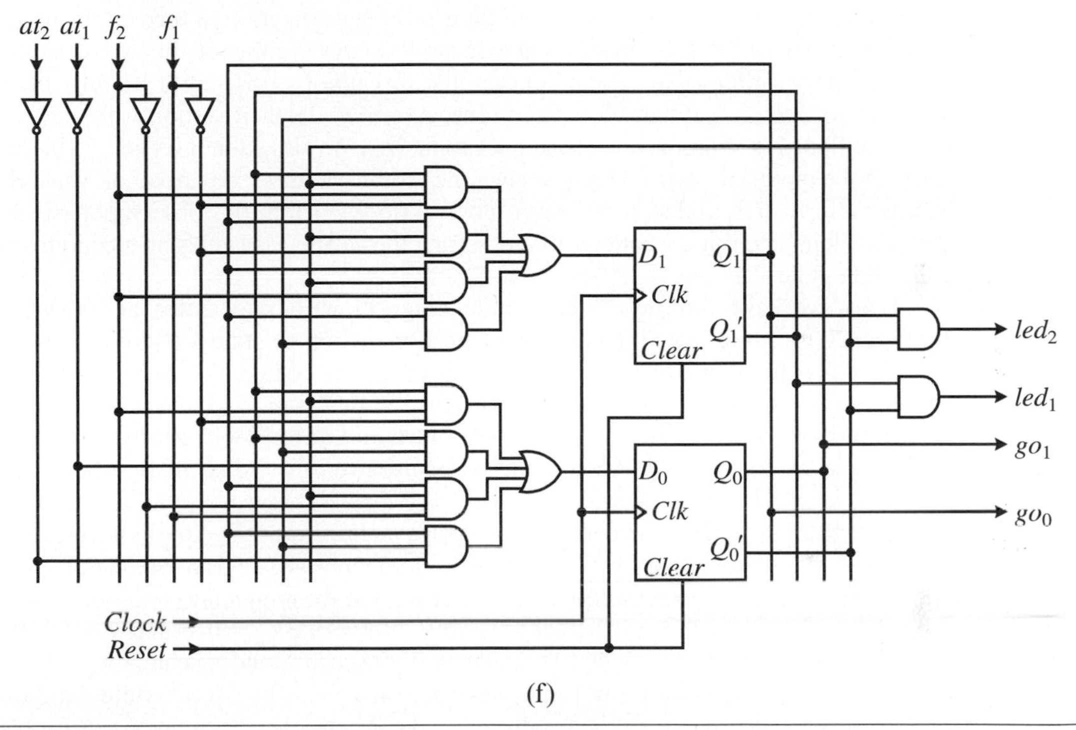

(f)

FIGURE 6.19 Elevator controller using Moore model: (a) state diagram; (b) next-state table; (c) next-state equations; (d) output table; (e) output equations; (f) FSM circuit.

states 00 and 10, and the at_i's are used only in states 01 and 11. Therefore, they can be separated to reduce the number of columns needed. The two at_i's are also mutually exclusive, so in the table we just have one column for at_i, but we need to remember to replace it with either at_1 or at_2 depending on which state we are checking. All of these changes are just to make the next-state table smaller and more manageable, but this can cause confusion if we are not careful when deriving the next-state equations. The next-state table is shown in Figure 6.19(b), and the next-state equations are shown in Figure 6.19(c).

The output table and output equations, shown in Figures 6.19(d) and (e), respectively, can easily be obtained from the state diagram. The final FSM circuit, as derived from the next-state and output equations, is shown in Figure 6.19(f).

6.6.6 Elevator Controller Using a Mealy FSM

We noted in Section 6.1 that the Moore and Mealy FSM models provide a tradeoff between the size of the state memory and next-state logic circuit with that of the output logic circuit. To illustrate this difference, we will design the same elevator controller from Section 6.6.5, but using a Mealy FSM instead.

First, we note that we can remove the two intermediate states (01 and 11) from the Moore state diagram shown in Figure 6.19(a), and still get the same functionality based on the description of the problem. The original two intermediate states in the Moore state diagram were used to turn on the motor and turn off the LEDs. Without the two intermediate states, we need to decide when and where in the state diagram to turn on the motor and turn off the LEDs. It is obvious that we cannot do this in either of the two remaining states when the elevator is at a floor. For example, we cannot set $go_{1\text{-}0} = 11$ (to turn on the motor to go to floor 2) in state 00 (at floor 1) because, if we do, then as soon as the elevator arrives at floor 1, it will immediately go back to floor 2. We also cannot set $go_{1\text{-}0} = 0\times$ (to turn off the motor) in state 00 because then the motor will never be turned on when the elevator is at floor 1, and so it will never move to floor 2. From this observation, we see that we will have to put the actions, that is, setting the *go* and *led* signals on a conditional edge in the state diagram.

Thus, we have our new Mealy state diagram with two states as shown in Figure 6.20(a). From state 0 (floor 1), the only transition to state 1 (floor 2) is when $f_{2\text{-}1}at_2 = 101$. In other words, the transition from state 0 to state 1 occurs only when f_2 is pressed ($f_{2\text{-}1} = 10$) and the elevator has reached floor 2 ($at_2 = 1$). On this transition, we will turn off the motor (set $go_{1\text{-}0} = 0\times$) and turn on led_2 (set $led_{2\text{-}1} = 10$) because the elevator has reached floor 2. Additionally, there are two other edges that go from state 0 back to itself. The reason for needing these two edges is that we have different output signals depending on the input conditions. We turn on the motor to go to floor 2 and turn off all the LEDs, that is, set $go_{1\text{-}0} = 11$ and $led_{2\text{-}1} = 00$, only when f_2 is pressed ($f_{2\text{-}1} = 10$) and the elevator has not reached floor 2 ($at_2 = 0$) yet. For the remaining conditions, we do not want to change anything, so we keep the motor off with $go_{1\text{-}0} = 0\times$, and $led_{2\text{-}1} = 01$. Going down from floor 2 to floor 1 is similar, with the corresponding changes.

The f_i and at_i signals are not mutually exclusive as in the Moore state diagram, but the two at_i signals are still mutually exclusive. Therefore, for the next-state table, it will be clearer to enumerate all possible combinations of these three bits, f_2, f_1, and at_i. We still use just at_i to denote both of the *at* signals, and then replace them with the appropriate index when we derive the next-state equations. The next-state table and next-state equations are shown in Figures 6.20(b) and (c), respectively.

All of the output signals are now dependent not only on the current state, but also on the input signals. In the output table shown in Figure 6.20(d), we have enumerated all possible input conditions. The 4-bit output signal entries in the table are listed in the order led_2, led_1, go_1, and go_0. The values for each entry are obtained directly from the state diagram. For example, for the edge going from state 0 to state 1 labeled with the condition ($f_{2\text{-}1}at_2 = 101$), the output signals assigned to this edge are $led_{2\text{-}1} = 10$ and $go_{1\text{-}0} = 0\times$, so the entry in the output table for when $Q = 0$ and $f_{2\text{-}1}at_2 = 101$ is $100\times$.

From the output table, we derive the output equations shown in Figure 6.20(e). The final FSM circuit is shown in Figure 6.20(f).

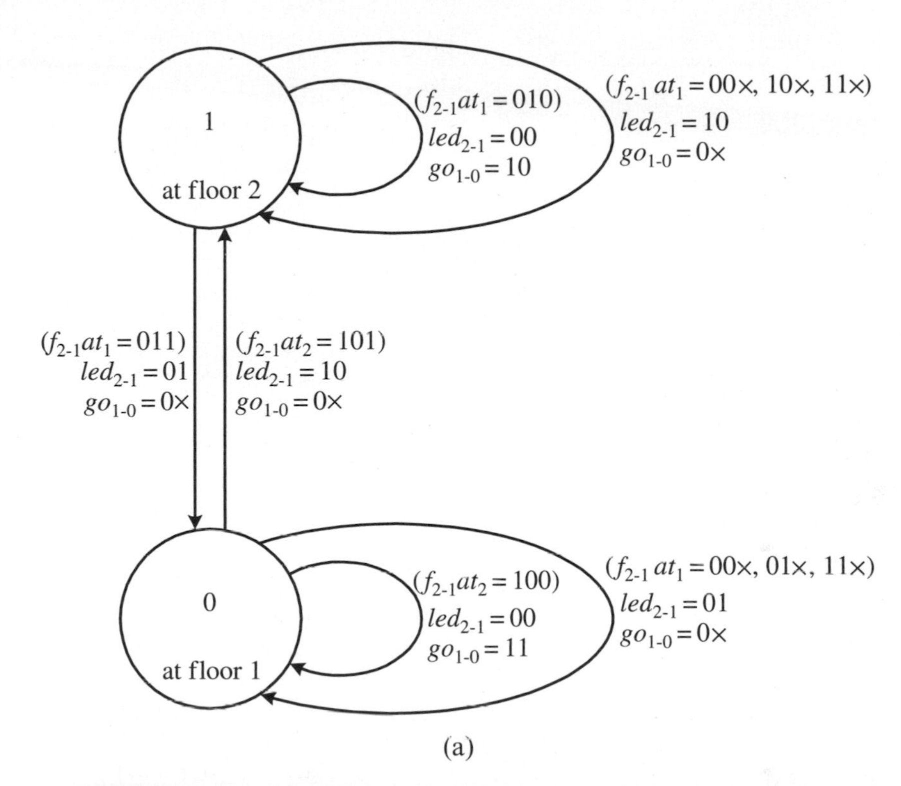

(a)

Current State Q_0	**Next State** Q_{0next} $f_{2\text{-}1}at_i =$							
	000	**001**	**010**	**011**	**100**	**101**	**110**	**111**
0	0	0	0	0	0	1	0	0
1	1	1	1	0	1	1	1	1

(b)

$$Q_{0next} = D_0 = Q_0'(f_2 f_1' at_2) + Q_0(f_2' f_1 at_1)'$$

(c)

FIGURE 6.20 Elevator controller using Mealy model: (a) state diagram; (b) next-state table; (c) next-state equations; (d) output table; (e) output equations; (f) FSM circuit.
(continued on next page)

Current State Q_0	Outputs $led_{2\text{-}1}go_{1\text{-}0}$							
	$f_{2\text{-}1}at_i$ =							
	000	**001**	**010**	**011**	**100**	**101**	**110**	**111**
0	010×	010×	010×	010×	0011	100×	010×	010×
1	100×	100×	0010	010×	100×	100×	100×	100×

(d)

$$led_2 = Q_0'(f_2 f_1' at_2) + Q_0(f_2' f_1' + f_2)$$
$$led_1 = Q_0'(f_2' + f_2 f_1) + Q_0(f_2' f_1 at_1)$$
$$go_1 = Q_0'(f_2 f_1' at_2') + Q_0(f_2' f_1 at_1')$$
$$go_0 = Q_0'(f_2 f_1' at_2')$$

(e)

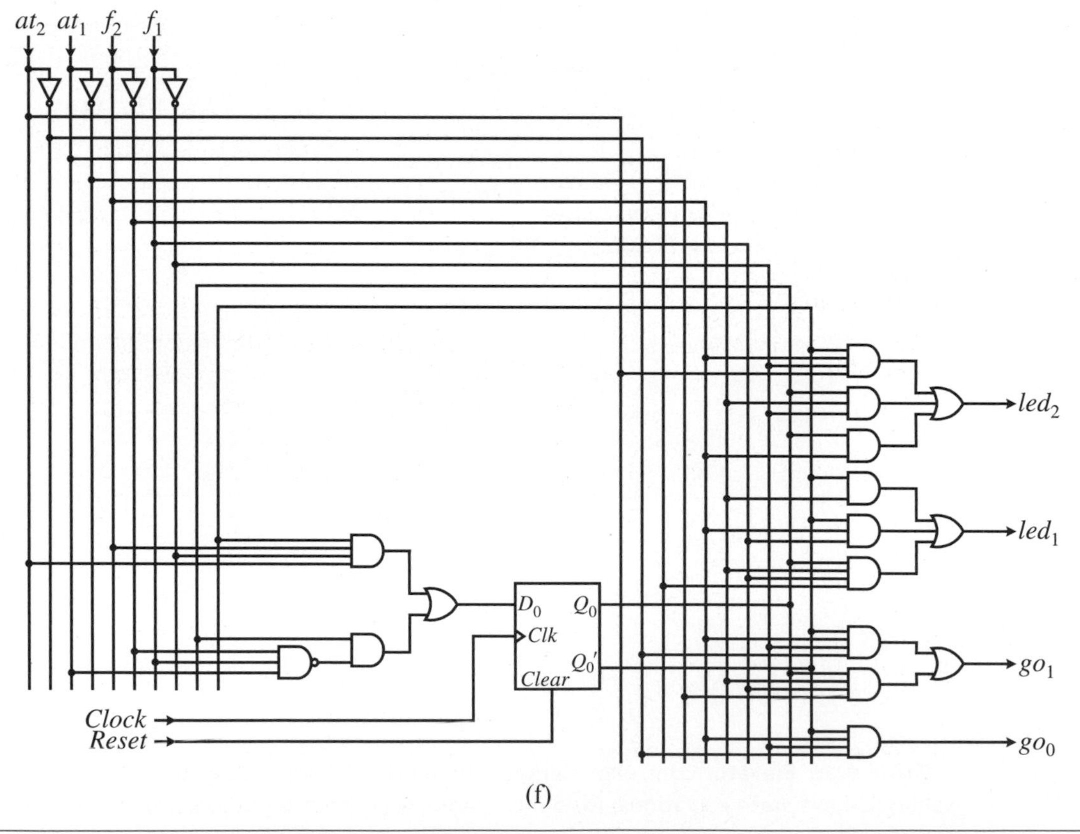

(f)

FIGURE 6.20 Elevator controller using Mealy model: (a) state diagram; (b) next-state table; (c) next-state equations; (d) output table; (e) output equations; (f) FSM circuit.

Comparing the Moore and Mealy FSM circuits for the elevator controller shown in Figures 6.19(f) and 6.20(f), respectively, it is obvious that the state memory and the next-state logic circuit for the Mealy FSM is much smaller, whereas the output logic circuit for the Moore FSM is much smaller. In general, we can conclude that Mealy FSMs typically will have a smaller state memory and next-state logic circuit, but a larger output logic circuit; whereas, Moore FSMs are just the opposite.

The observant reader might have noticed, however, that there is a slight difference in the actual operation of the two elevator controllers. For the Moore version, if a person is on floor 1 (state 00) and presses f_2, the FSM will immediately go to the intermediate state 11. In this intermediate state, the FSM remembers the fact that f_2 has been pressed, and changing any of the f_i signals while in this state will not do anything. Only asserting at_2 while in this state will cause the FSM to go to floor 2 (state 10). On the other hand, for the Mealy version, if a person is on floor 1 (state 0) and presses f_2, the FSM does not change state, therefore, it does not remember the fact that f_2 has been pressed. Therefore, if f_2 is de-asserted before asserting at_2, it will not cause the FSM to go to floor 2 (state 1). In order for the FSM to go to floor 2 (state 1), both f_2 and at_2 must be asserted, and f_1 also must be de-asserted.

6.7 Verilog and VHDL Code for FSM Circuits

Writing HDL code for FSM circuits usually is done at the behavioral level. The advantage of writing behavioral HDL code is that we do not need to synthesize the circuit manually. The synthesizer automatically will produce the netlist for the complete circuit from the behavioral code. The only information that we need to know is the state diagram for the FSM.

The states in the state diagram are implemented using a `case` statement. Each case in the `case` statement corresponds to a state in the state diagram. Because the HDL synthesizer automatically takes care of the state encoding, the states need only to be labeled with their logical names. A state variable is used to remember the current state of the FSM. Unconditional edges are implemented simply as an assignment of a new state value to the state variable. Conditional edges will have a conditional `if` statement to test for the condition before the state variable assignment. The output signal information in the state diagram is used to derive the output logic, and is implemented with assignment statements to assign the given values to the output signals.

The remaining subsections illustrate the behavioral Verilog and VHDL code for both the Moore and Mealy FSM state diagrams based on the two elevator controller examples discussed in Sections 6.6.5 and 6.6.6. The two state diagrams are repeated here in Figure 6.21 for convenience.

6.7.1 Behavioral Verilog Code for a Moore FSM

The Moore FSM Verilog code shown in Figure 6.22 is based on the Moore FSM state diagram shown in Figure 6.21(a). The main portion of the Verilog code contains two `always` blocks: one for the next-state logic and the other for the output logic.

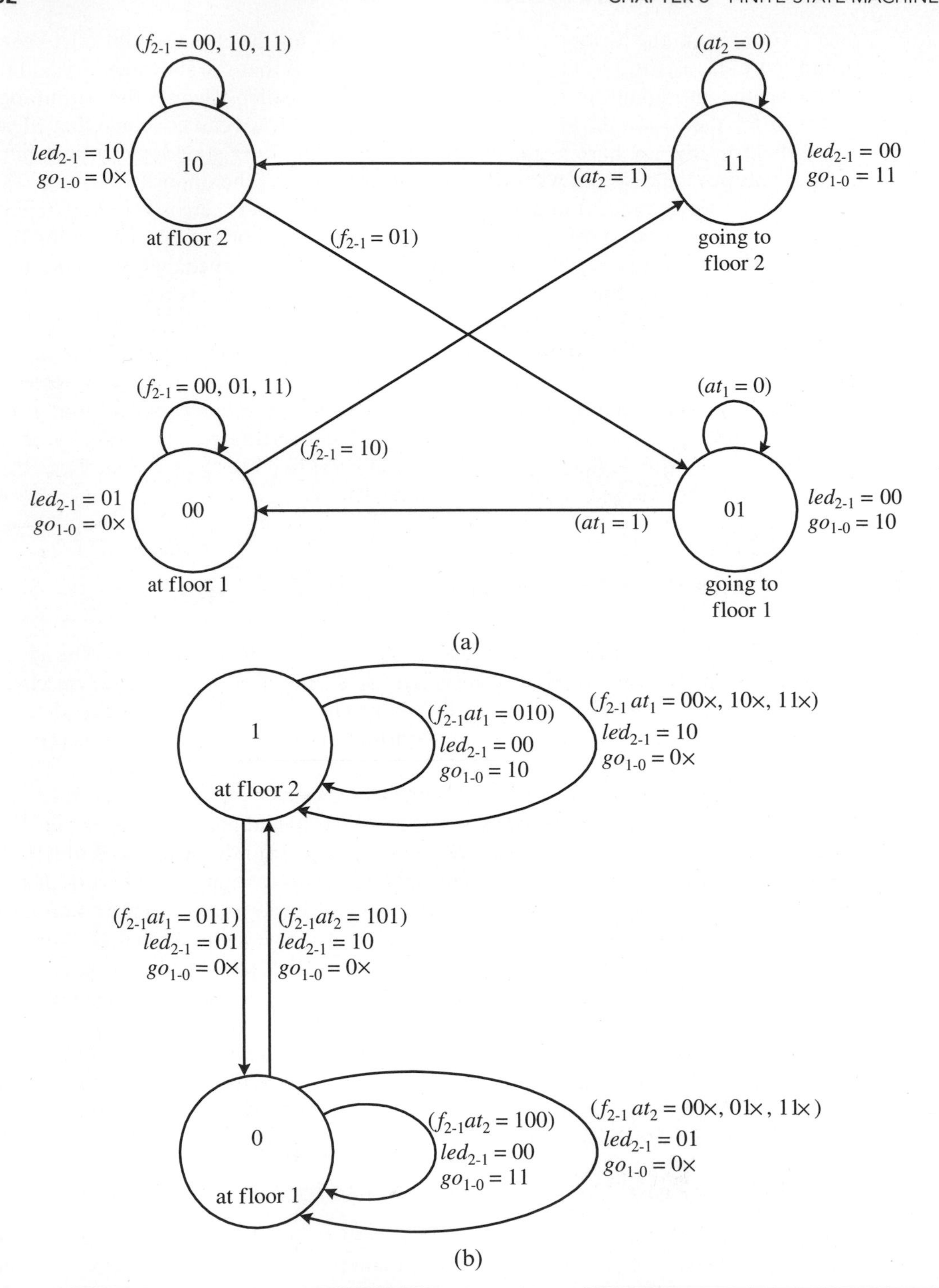

FIGURE 6.21 State diagrams: (a) for a Moore FSM; (b) for a Mealy FSM.

```
module MooreFSM (
  input Clock, Reset,
  input [2:1] f, at,
  output reg [1:0] go,
  output reg [2:1] led,
  output [1:0] debug_state
);
  reg [1:0] state;
  parameter s00 = 2'b00, s01 = 2'b01, s10 = 2'b10, s11 = 2'b11;

  assign debug_state = state;

  // next-state logic
  always @ (posedge Clock or posedge Reset) begin
    if (Reset) begin
      state <= s00;
    end else
      case (state)
      s00: begin   // floor 1
        if (f == 2'b10)
          state <= s11;
        else
          state <= s00;
        end
      s01: begin
        if (at[1])
          state <= s00;
        else
          state <= s01;
        end
      s10: begin   // floor 2
        if (f == 2'b01)
          state <= s01;
        else
          state <= s10;
        end
      s11: begin
        if (at[2])
          state <= s10;
        else
          state <= s11;
        end
      endcase
  end

  // output logic - depends only on the state
  always @ (state) begin
    case (state)
      s00: begin // floor 1
```

FIGURE 6.22 Behavioral Verilog code for a Moore FSM. *(continued on next page)*

```
          led = 2'b01;
          go = 2'b00;
          end
        s01: begin
          led = 2'b00;
          go = 2'b10;
          end
        s10: begin   // floor 2
          led = 2'b10;
          go = 2'b00;
          end
        s11: begin
          led = 2'b00;
          go = 2'b11;
          end
      endcase
    end
  endmodule
```

FIGURE 6.22 Behavioral Verilog code for a Moore FSM.

The module first declares all of the input and output signals for the circuit. The *Clock* input signal determines the speed in which the FSM will transition from one state to the next, and the *Reset* signal will be used to initialize all of the state memory flip-flops to zero. Based on the state diagram, two 2-bit input signals *f* and *at*, and two 2-bit output signals *go* and *led* are also declared. For debugging purposes, we have added an extra output signal *debug_state* to show the current state of the FSM.

The state variable and the symbolic names for the four states *s00*, *s01*, *s10*, and *s11* are declared next. The symbolic state names also are assigned with their actual encodings. Alternatively, we do not need to declare and use the symbolic state names. Instead, we can just use the binary encodings in the state assignments and `case` statement as shown later in the Mealy FSM code.

The two `always` blocks are executed concurrently: the first block defines the next-state logic circuit inside the control unit, and the second defines the output logic circuit inside the control unit. The main statement inside these two blocks is the `case` statement that determines the current state of the FSM.

In the next-state logic block, the current state of the FSM is initialized to *s00* on reset. The `case` statement is executed only at the rising clock edge because of the `posedge` *Clock* in the sensitivity list. Therefore, the *state* signal is assigned a new state value at every rising clock edge. The new state value is, of course, dependent on the current state and input signals. For example, if the current state is *s00*, then the case for *s00* is selected. From the state diagram, we see that when in state *s00*, the next state is dependent on the 2-bit input signal $f_{2\text{-}1}$. Therefore, an `if` statement is used in the code. If $f_{2\text{-}1}$ is 2'b10 (i.e., 10) then the new state *s11* is assigned to the *state* variable, otherwise, the current state *s00* is assigned to *state*. For the latter case, even though we are not changing the state value *s00*, we still need to make that assignment to prevent the synthesizer from using a memory element for the *state* signal. Recall from Section 5.7 that a memory element is used for a signal if the signal is not assigned a value for all possible cases. The rest of the cases in the `case` statement are written similarly based on the remaining edges in the state diagram.

In the output logic `always` block, all of the output signals must be assigned a value in every case because the output circuit is a combinational circuit, and so we do not want memory elements to be created for these output signals. For each state in the `case` statement, the values assigned to each of the output signals are obtained directly from the state diagram. For this example, we have two 2-bit output signals $led_{2\text{-}1}$ and $go_{1\text{-}0}$. For debugging purposes, the *state* variable is assigned to the *debug_state* output signal so that we can see the current state of the FSM.

6.7.2 Behavioral Verilog Code for a Mealy FSM

The Mealy FSM Verilog code shown in Figure 6.23 is based on the Mealy FSM state diagram shown in Figure 6.21(b). This Mealy FSM code basically follows the same format as the Moore FSM code. The main difference is in the output logic block, where

```
module MealyFSM (
  input Clock, Reset,
  input [2:1] f, at,
  output reg [1:0] go,
  output reg [2:1] led,
  output debug_state
);
  reg  state;
  assign debug_state = state;

  // next-state logic
  always @ (posedge Clock or posedge Reset) begin
    if (Reset) begin
      state <= 1'b0;
    end else
      case (state)
      1'b0: // floor 1
        if ({f,at[2]} == 3'b101)
          state <= 1'b1;
      1'b1:  // floor 2
        if ({f,at[1]} == 3'b011)
          state <= 1'b0;
      endcase
  end
```

FIGURE 6.23 Behavioral Verilog code for a Mealy FSM. *(continued on next page)*

```
    // output logic - depends on the state and inputs
    always @ (state) begin
      case (state)
        1'b0: begin // floor 1
          if ((f == 2'b00) | (f == 2'b01) | (f == 2'b11)) begin
            led = 2'b01;
            go =  2'b00;
          end else if ({f,at[2]} == 3'b100) begin
            led = 2'b00;
            go =  2'b11;
          end else if ({f,at[2]} == 3'b101) begin
            led = 2'b10;
            go =  2'b00;
            end
          end
        1'b1: begin // floor 2
          if ((f == 2'b00) | (f == 2'b10) | (f == 2'b11)) begin
            led = 2'b10;
            go =  2'b00;
          end else if ({f,at[1]} == 3'b010) begin
            led = 2'b00;
            go =  2'b10;
          end else if ({f,at[1]} == 3'b011) begin
            led = 2'b01;
            go =  2'b00;
            end
          end
      endcase
    end // always
  endmodule
```

FIGURE 6.23 Behavioral Verilog code for a Mealy FSM.

`if` statements are used to test for the input conditions before assigning appropriate values to the output signals. The transitioning of the states and the assignment of output signals follow the state diagram directly.

6.7.3 Behavioral VHDL Code for a Moore FSM

The behavioral VHDL code for a Moore FSM shown in Figure 6.24 is based on the Moore FSM state diagram shown in Figure 6.21(a). Except for the syntactical differences, the structure of this VHDL code is similar to the Verilog code. The main portion of the VHDL code contains two processes: a next-state logic process and an output logic process. The ENTITY section declares all of the input and output signals for the circuit. The *Clock* input signal determines the speed in which the FSM will transition from one state to the next, and the *Reset* input signal will be used to initialize all of the state memory flip-flops to zero. Based on the state diagram, two 2-bit input signals *f* and *at,* and two 2-bit output signals *go* and *led* are also declared.

```
LIBRARY IEEE;
USE IEEE.STD_LOGIC_1164.ALL;

ENTITY MooreFSM IS PORT(
  Clock, Reset: IN STD_LOGIC;
  f, at: IN STD_LOGIC_VECTOR(2 DOWNTO 1);
  go: OUT STD_LOGIC_VECTOR(1 DOWNTO 0);
  led: OUT STD_LOGIC_VECTOR(2 DOWNTO 1)
  );
END MooreFSM;

ARCHITECTURE Behavioral OF MooreFSM IS
  TYPE state_type IS (s00, s01, s10, s11);
  SIGNAL state: state_type;
BEGIN

  -- next-state logic
  next_state_logic: PROCESS (Clock, Reset)
  BEGIN
    IF (Reset = '1') THEN
      state <= s00;
    ELSIF (RISING_EDGE(Clock)) THEN
      CASE state IS
        WHEN s00 =>
          IF (f = "10") THEN
            state <= s11;
          ELSE
            state <= s00;
          END IF;
        WHEN s01 =>
          IF (at(1) = '1') THEN
            state <= s00;
          ELSE
            state <= s01;
          END IF;
        WHEN s10 =>
          IF (f = "01") THEN
            state <= s01;
          ELSE
            state <= s0;
          END IF;
        WHEN s11 =>
          IF (at(2) = '1') THEN
            state <= s10;
          ELSE
            state <= s11;
          END IF;
      END CASE;
    END IF;
```

FIGURE 6.24 Behavioral VHDL code for a Moore FSM. *(continued on next page)*

```
  END PROCESS;
  -- output logic - depends only on the state
  output_logic: PROCESS (state)
  BEGIN
    CASE state is
      WHEN s00 =>
        led <= "01";
        go <= "00";
      WHEN s01 =>
        led <= "00";
        go <= "10";
      WHEN s10 =>
        led <= "10";
        go <= "00";
      WHEN s11 =>
        led <= "00";
        go <= "11";
    END CASE;
  END PROCESS;
END Behavioral;
```

FIGURE 6.24 **Behavioral VHDL code for a Moore FSM.**

The ARCHITECTURE section starts out with using the TYPE statement to define the symbolic names for the four states *s00*, *s01*, *s10*, and *s11*, as used in the state diagram. The actual encoding of the states is done automatically by the synthesizer. The SIGNAL statement declares the signal *state* to store the current state of the FSM. Two processes in the ARCHITECTURE section execute concurrently: the *next_state_logic* PROCESS, and the *output_logic* PROCESS. As the name suggests, the *next_state_logic* PROCESS defines the next-state logic circuit inside the control unit, and the *output_logic* PROCESS defines the output logic circuit inside the control unit. The main statement within these two processes is the CASE statement that determines the current state of the FSM.

In the *next_state_logic* PROCESS, the current state of the FSM is initialized to *s00* on reset. The CASE statement is executed only at the rising clock edge because of the conditional test (RISING_EDGE(*Clock*)) in the IF statement. Therefore, the *state* signal is assigned a new state value at every rising clock edge. The new state value is, of course, dependent on the current state and input signals, if any. For example, if the current state is *s00*, the case for *s00* is selected. From the state diagram, we see that when in state *s00*, the next state is dependent on the 2-bit input signal $f_{2\text{-}1}$. Therefore, an IF statement is used in the code. If $f_{2\text{-}1}$ is "10" then the new state *s11* is assigned to the *state* variable, otherwise, the current state *s00* is assigned to *state*. For the latter case, even though we are not changing the state value *s00*, we still need to make that assignment to prevent the synthesizer from using a memory element for the *state* signal. Recall from Section 5.7 that a memory element is used for a signal if the signal is not assigned a value for all possible cases. The rest of the cases in the CASE statement are written similarly based on the remaining edges in the state diagram.

In the *output_logic* PROCESS, all of the output signals must be assigned a value in every case because the output circuit is a combinational circuit, and so we do not want memory elements to be created for these output signals. For each state in the CASE statement in the *output_logic* PROCESS, the values assigned to each of the output signals are obtained directly from the state diagram. For this example, we have two 2-bit output signals $led_{2\text{-}1}$ and $go_{1\text{-}0}$.

6.7.4 Behavioral VHDL Code for a Mealy FSM

The behavioral VHDL code for a Mealy FSM shown in Figure 6.25 is based on the Mealy FSM state diagram shown in Figure 6.21(b). This Mealy FSM code basically follows the same format as the Moore FSM code. The transitioning of the states and the assignment of output signals follow the state diagram directly. The main difference is in the output logic block, where IF statements are used to test for the input conditions before assigning appropriate values to the output signals.

```
LIBRARY IEEE;
USE IEEE.STD_LOGIC_1164.ALL;

ENTITY MealyFSM IS PORT(
  Clock, Reset: IN STD_LOGIC;
  f, at: IN STD_LOGIC_VECTOR(2 DOWNTO 1);
  go: OUT STD_LOGIC_VECTOR(1 DOWNTO 0);
  led: OUT STD_LOGIC_VECTOR(2 DOWNTO 1);
  debug_state: OUT STD_LOGIC
  );
END MealyFSM;

ARCHITECTURE Behavioral OF MealyFSM IS
  SIGNAL state: STD_LOGIC;
BEGIN
  debug_state <= state;

  -- next-state logic
  next_state_logic: PROCESS (Clock, Reset)
  BEGIN
    IF (Reset = '1') THEN
      state <= '0';
    ELSIF (RISING_EDGE(Clock)) THEN
      CASE state IS
        WHEN '0' =>
          IF ((f = "10") AND (at(2) = '1')) THEN
            state <= '1';
          END IF;
        WHEN '1' =>
          IF ((f = "01") AND (at(1) = '1')) THEN
            state <= '0';
          END IF;
```

FIGURE 6.25 Behavioral VHDL code for a Mealy FSM. *(continued on next page)*

```
      END CASE;
    END IF;
  END PROCESS;

  -- output logic - depends on the state and inputs
  output_logic: PROCESS (state)
  BEGIN
    CASE state is
      WHEN '0' =>
        IF ((f = "00") OR (f = "01") OR (f = "11")) THEN
          led <= "01";
          go <= "00";
        ELSIF ((f = "10") AND (at(2) = '0')) THEN
          led <= "00";
          go <= "11";
        ELSIF ((f = "10") AND (at(2) = '1')) THEN
          led <= "10";
          go <= "00";
        END IF;
      WHEN '1' =>
        IF ((f = "00") OR (f = "10") OR (f = "11")) THEN
          led <= "10";
          go <= "00";
        ELSIF ((f = "01") AND (at(1) = '0')) THEN
          led <= "00";
          go <= "10";
        ELSIF ((f = "01") AND (at(1) = '1')) THEN
          led <= "01";
          go <= "00";
        END IF;
    END CASE;
  END PROCESS;
END Behavioral;
```

FIGURE 6.25 Behavioral VHDL code for a Mealy FSM.

6.8 PROBLEMS

6.1. Analyze and derive the state diagram for each of the following FSMs:

a) C is an input, and a and b are outputs.

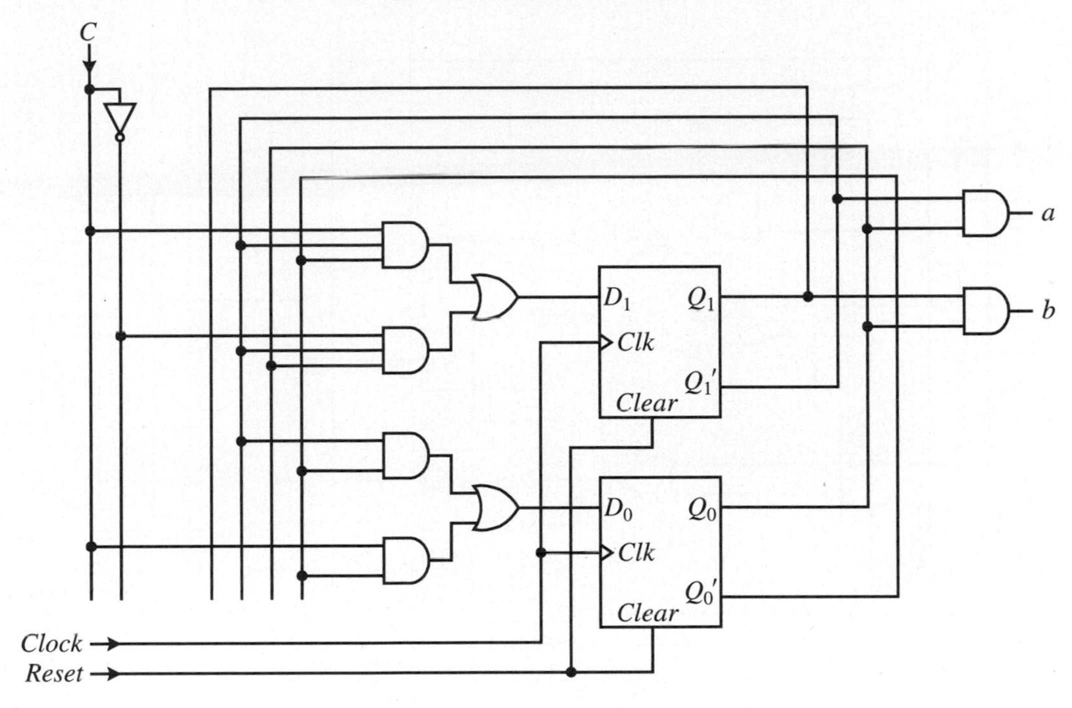

b) C is an input, and a and b are outputs.

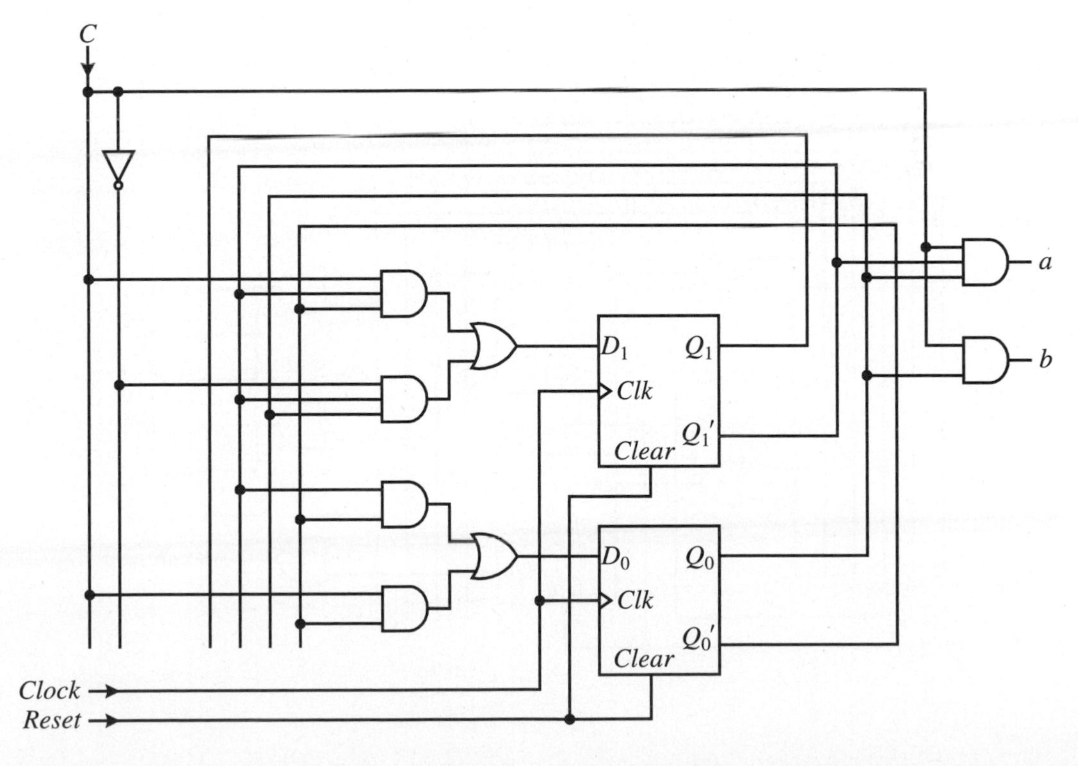

c) *A* and *B* are inputs, and *X* and *Y* are outputs.

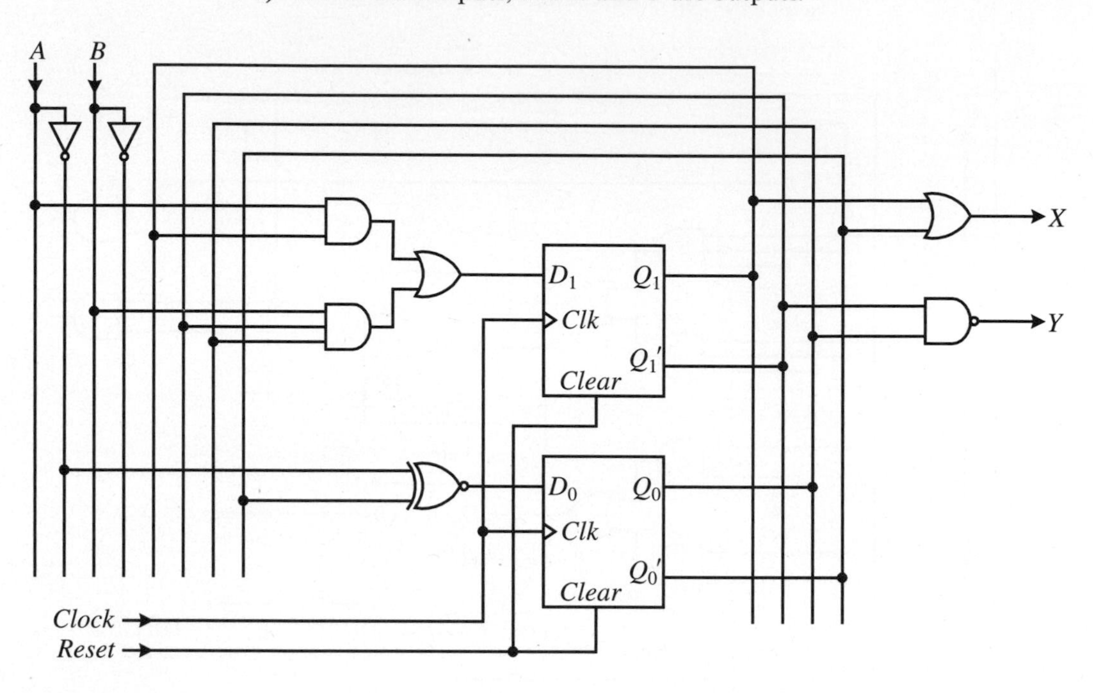

d) $(Z \neq 0)$ is an input, and *ClrX*, *LoadY*, *inZ*, *LoadX*, *stat1*, *LoadZ*, and *subtract* are outputs.

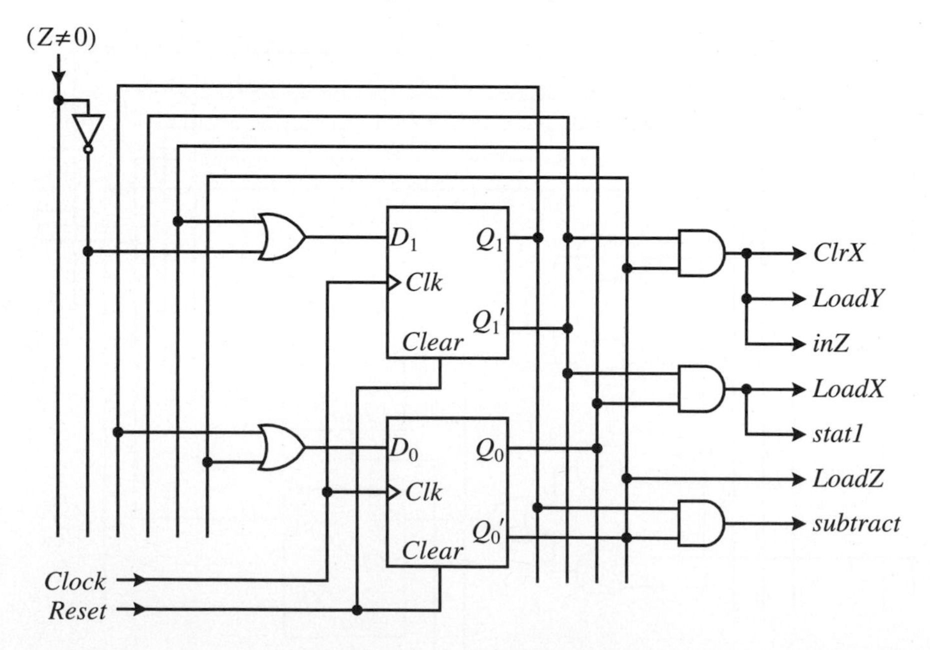

e) *Start* is an input, and *LoadN* and *LoadM* are outputs.

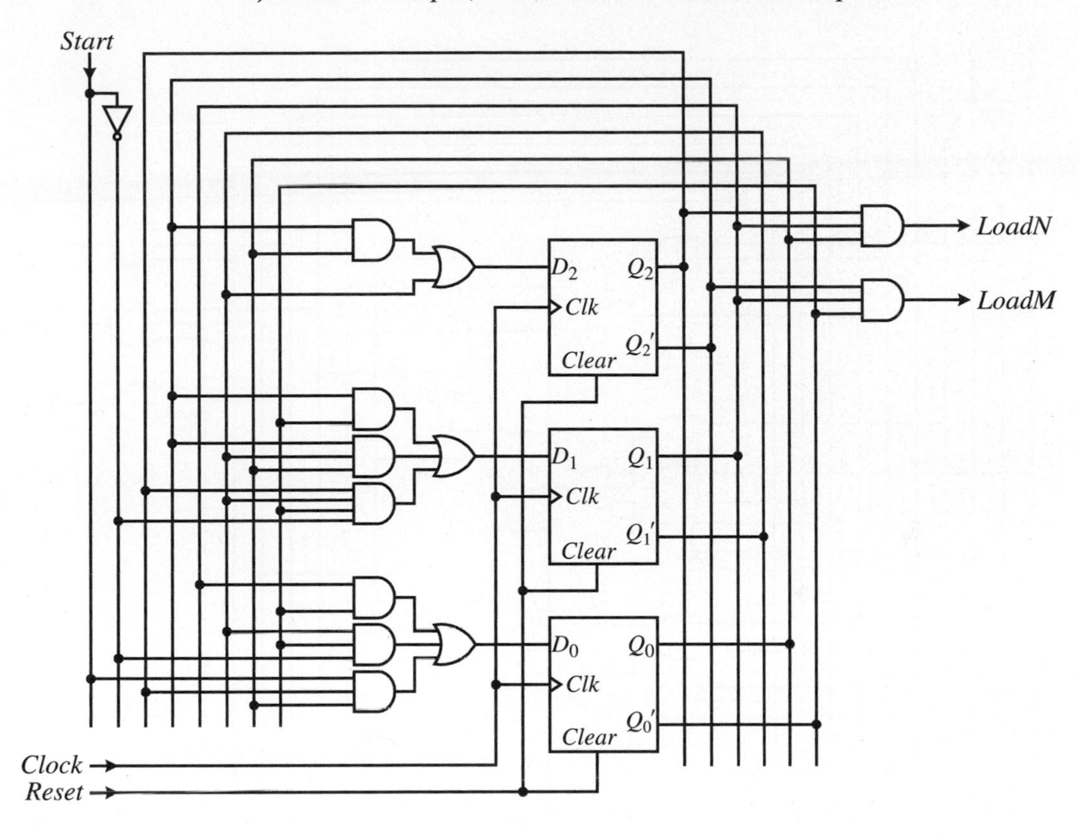

f) *Yes* and *No* are inputs, and *a*, *b*, and *c* are outputs.

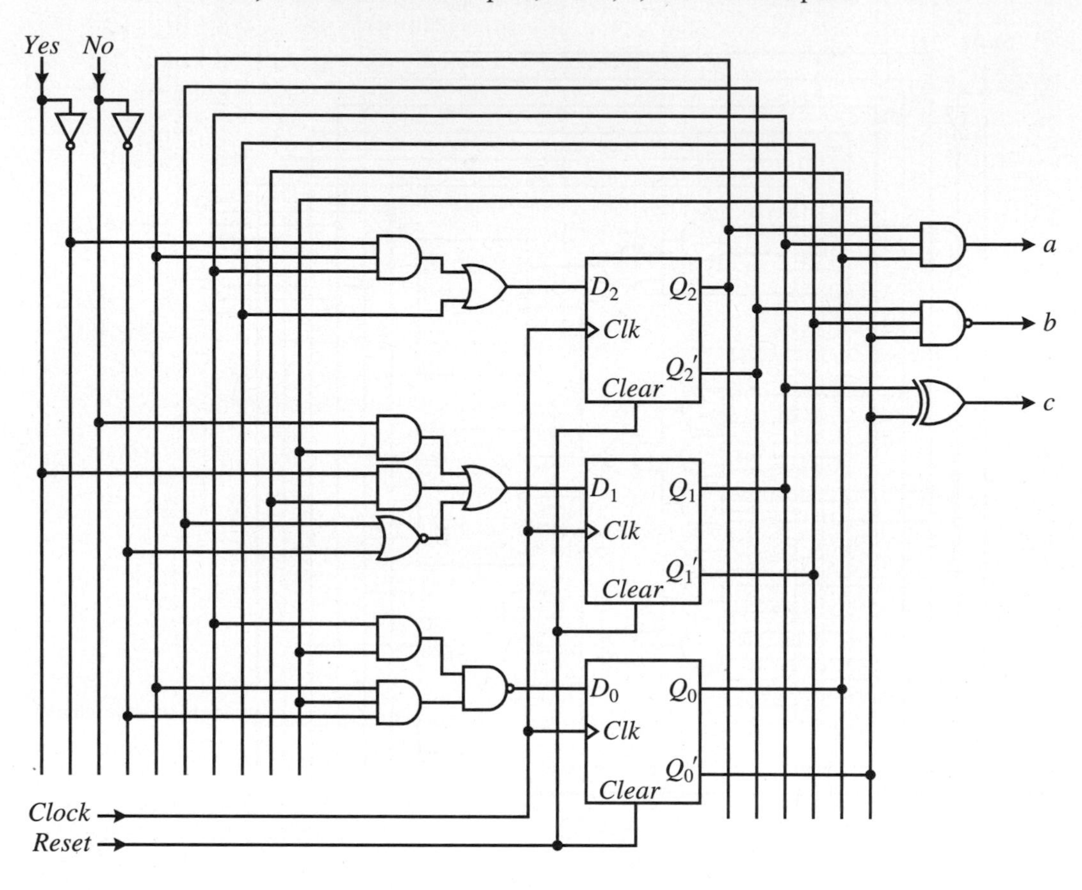

6.2. Analyze and derive the state diagram for each of the following FSMs:

a) C is an input, and a and b are outputs.

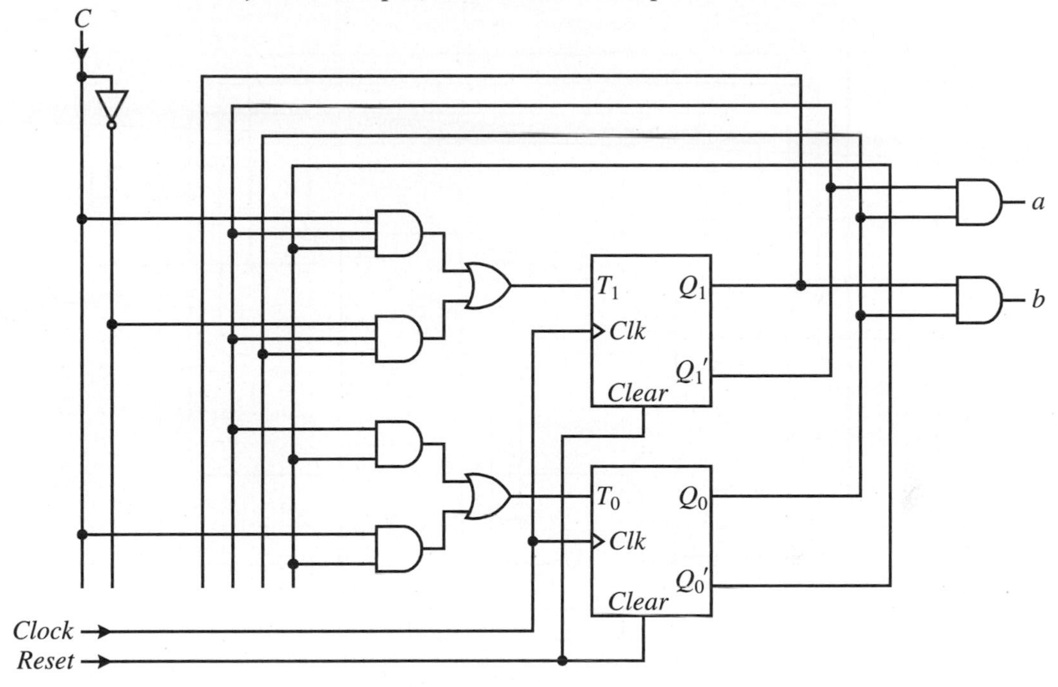

b) C is an input, and a and b are outputs.

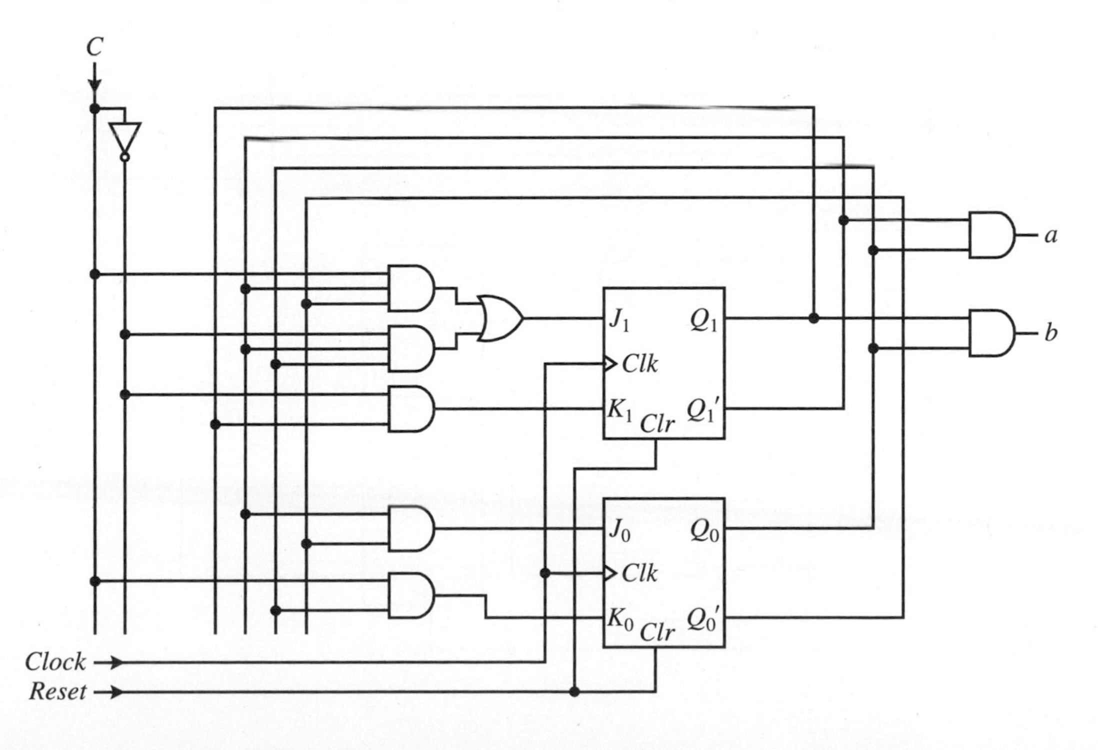

c) *C* is an input, and *a* and *b* are outputs.

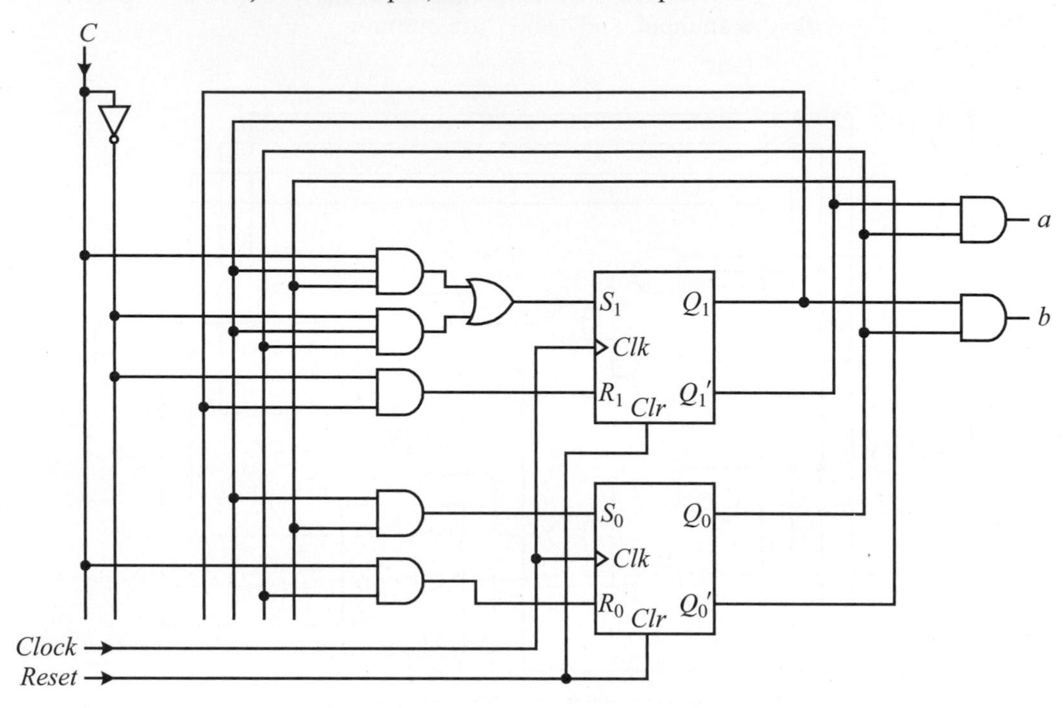

d) *C* is an input, and *a* and *b* are outputs.

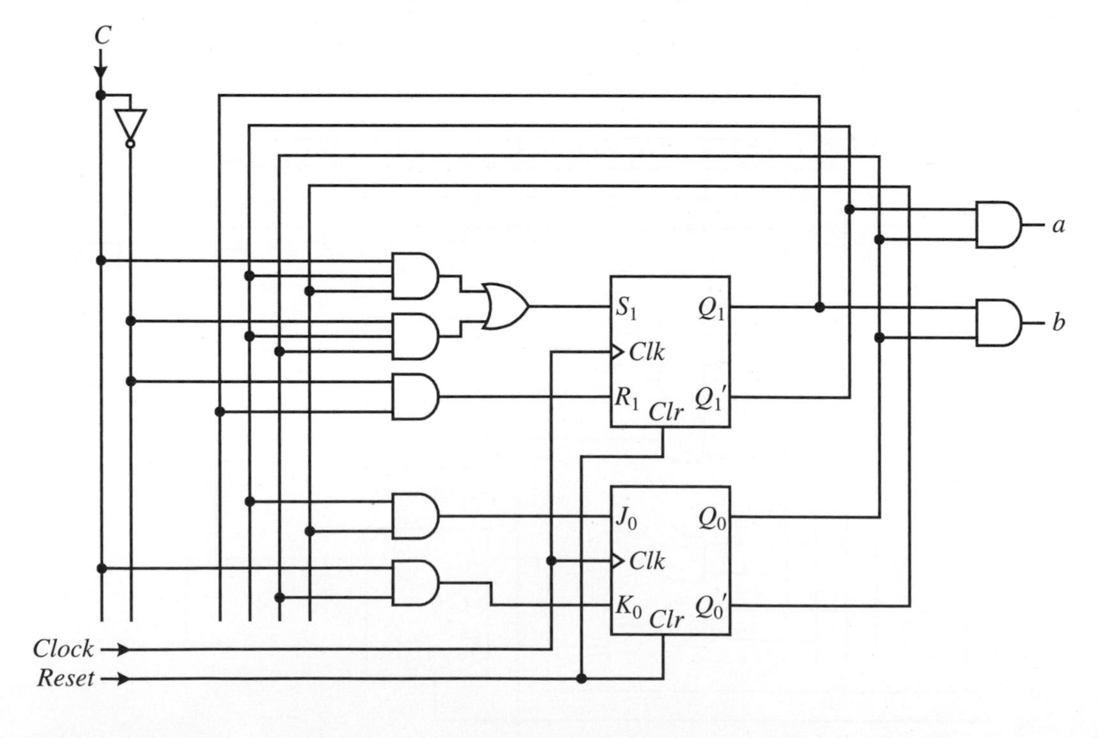

6.3. Synthesize an FSM circuit using D flip-flops for the following state diagrams:

a) A is an input.

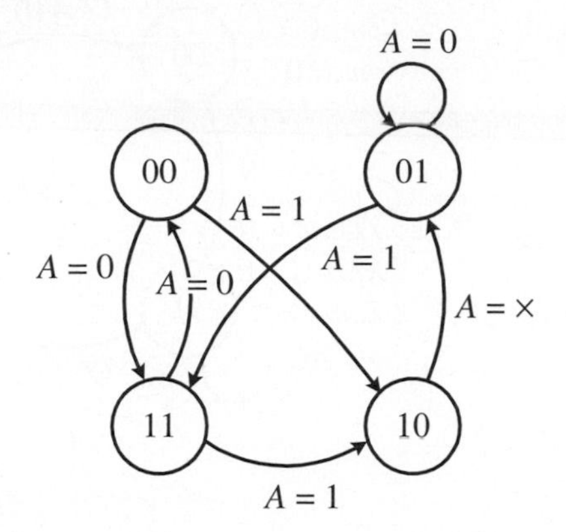

b) J and K are inputs, and Q is an output. (This is the JK flip-flop.)

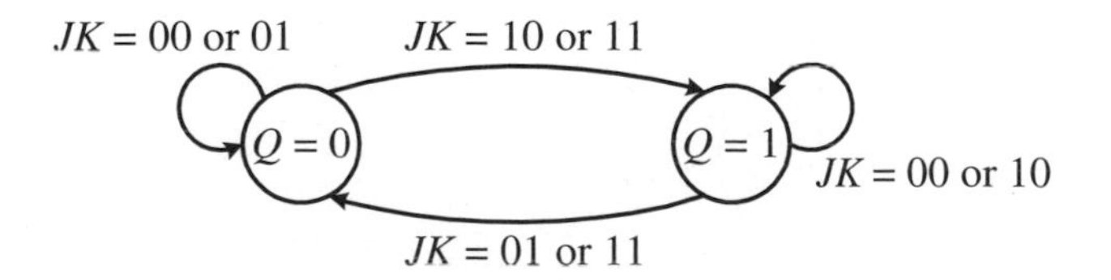

c) T is an input, and Q is an output. (This is the T flip-flop.)

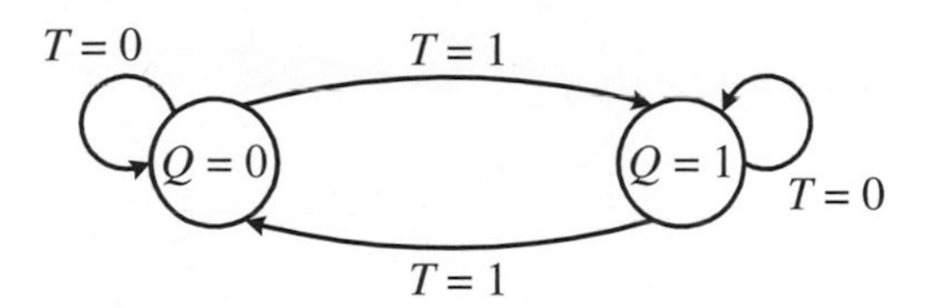

d) S and R are inputs, and Q is an output. (This is the SR flip-flop.)

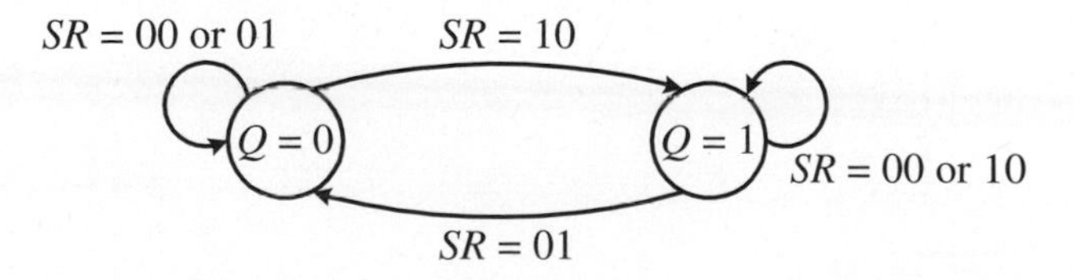

e) $(Z \neq 0)$ is an input, and *YLoad*, *Xload*, *Zmux*, and *out* are outputs.

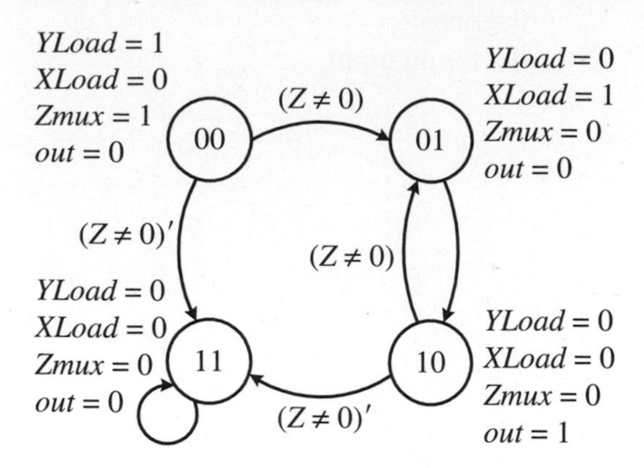

f) *C* is an input, and *X* is an output.

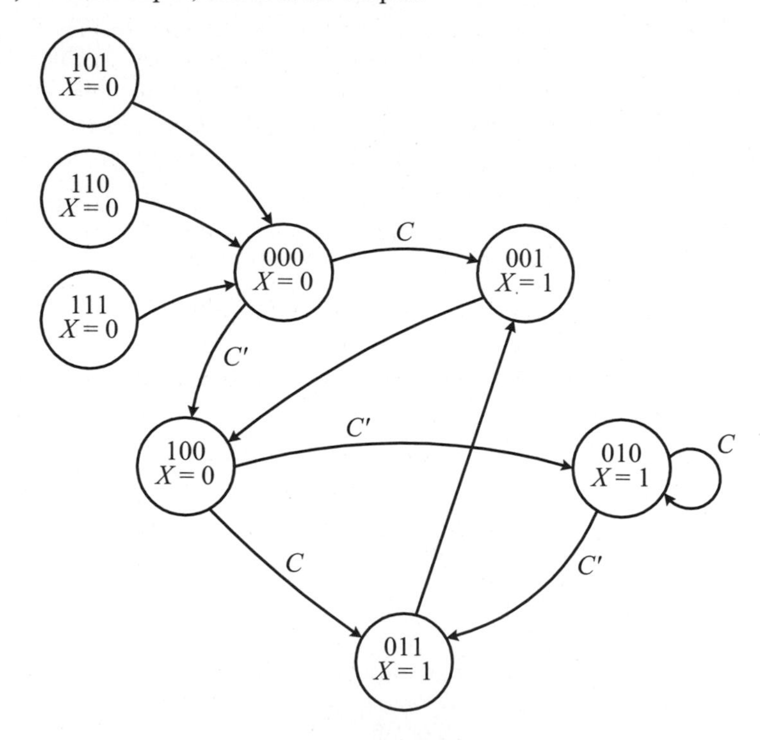

g) $(x = 0)$ and $(x = y)$ are inputs, and A is an output.

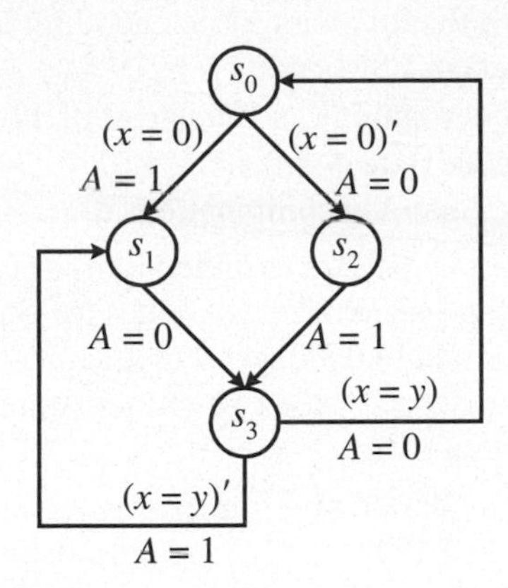

6.4. Design an FSM for a modulo-4 up/down-counter using D flip-flops. The count is represented by the content of the flip-flops. The circuit has a *Count* signal and an *Up* signal. The counter counts when *Count* is asserted, and stops when *Count* is de-asserted. The *Up* signal determines the direction of the count. When *Up* is asserted, the count increments by one at each clock cycle. When *Up* is de-asserted, the count decrements by one at each clock cycle.

6.5. Design an FSM for a modulo-5 up-counter using D flip-flops similar to Problem 6.4, but without the *Up* signal.

6.6. Design an FSM for a modulo-5 up/down-counter using D flip-flops similar to Problem 6.4.

6.7. Design an FSM counter that counts the following decimal sequence.

3, 7, 2, 6, 3, 7, 2, 6, …

The count is to be represented directly by the contents of the D flip-flops. The counting starts when the control input C is asserted and stops whenever C is de-asserted. Assume that the next state from all unused states is the state for the first count in the sequence (i.e., the state for 3).

6.8. Design an FSM counter that counts the following decimal sequence.

1, 4, 6, 7, 1, 4, 6, 7, ….

The count is to be represented directly by the contents of three D flip-flops. The counter is enabled by the input C. The count stops when $C = 0$. The next state from all unused states are undefined.

6.9. Repeat the construction of the One-Shot FSM circuit from Section 6.6.3 but with the following state encoding changes. Which encoding results in the smallest FSM circuit?

a) Encode state s_2 as 10 instead of 11, and encode the unused state as 11.

b) Encode state s_0 as 0001, s_1 as 0010, s_2 as 0100, and the unused state as 1000. All remaining combinations are unused and their next state is undefined.

6.10. In Section 6.6.2, we designed a modulo-6 counter where two of the state encodings have unknown values for their next states. In the final simplified equations, actual values were assigned to the don't-care values. Determine the next states for the two unused states 110 and 111 for that modulo-6 FSM circuit.

6.11. Manually design and implement on an FPGA the following FSM circuit. Make the LEDs in the 7-segment display move in a clockwise direction around in a circle (i.e., turn on and off the LED segments in this order: segment a, b, c, d, e, f, a, b, and etc).

6.12. Manually design and implement on an FPGA the following FSM circuit. This is similar to Problem 6.11, but make one 7-segment LED display in a clockwise direction and the other in a counterclockwise direction.

6.13. Manually design and implement on an FPGA the following FSM circuit. This is similar to Problem 6.11, but make it so that each time a push-button switch is pressed, the display changes directions.

6.14. Manually design and implement on an FPGA the following FSM circuit. Input from the eight DIP switches. Output on the 7-segment the decimal number that represents the number of DIP switches that are in the on position.

6.15. Manually design and implement on an FPGA an FSM circuit for controlling three switches T_1, T_2, and T_3, and three lights L_1, L_2, and L_3. Each light is turned on by the corresponding switch (e.g., T_1 turns on L_1). Initially, all switches are off. The first switch that is pressed will turn on its corresponding light. When the first light is turned on, it will remain on, while the other two lights remain off, and they are unaffected by subsequent switch presses until reset.

6.16. In Section 6.6.5, we created an elevator controller for two floors. Modify the controller to work for four floors. Use the same inputs and outputs as in the two-floor building but add more to extend it for four floors. The I/O signals are summarized next.

Inputs:

- f_1, f_2, f_3, and f_4: Buttons at each floor to call the elevator. f_1 is on floor 1, f_2 is on floor 2, and so on.
- e_1, e_2, e_3, and e_4: Buttons inside the elevator to tell the elevator which floor to go to.
- at_1, at_2, at_3, and at_4: Signals from the elevator mechanism to say which floor the elevator is currently at.

Outputs:

- go_2: 1 to turn on the elevator motor, and 0 to turn off the motor.
- $go_{1\text{-}0}$: 00 to go to floor 1, 01 to go to floor 2, 10 to go to floor 3, and 11 to go to floor 4.
- led_1, led_2, led_3, and led_4: LEDs on each of the four floors to show which floor the elevator is currently at.

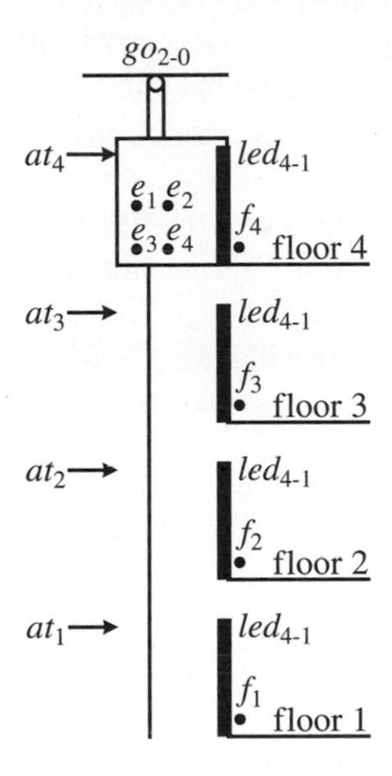

6.17. Design an FSM circuit for controlling a simple home security system. The operation of the system is as follows.

Inputs:

- Front gate switch (*FS*)
- Motion detector switch (*MS*)
- Asynchronous Reset switch (*R*)
- Clear switch (*C*)

Outputs:

- Front gate melody (*FM*)
- Motion detector melody (*MM*)
- When the reset switch (*R*) is asserted, the FSM goes to the initialization state (*S_init*) immediately. The encoding for the initialization state is zeros for all the flip-flops.
- From state *S_init*, the FSM unconditionally goes to the wait state (*S_wait*).
- From state *S_wait*, the FSM waits for one of the four switches to be activated. All the switches are active-high, so when a switch is pressed

or activated, it sends out a 1. The following actions are taken when a switch is pressed:

- When *FS* is pressed, the FSM goes to state *S_front*. In state *S_front*, the front gate melody is turned on by setting $FM = 1$. The FSM remains in state *S_front* until the clear switch is pressed. Once the clear switch is pressed, the FSM goes back to *S_wait*.
- When *MS* is activated, the FSM goes to state *S_motion*. In state *S_motion*, *MM* is turned on with a 1. *MM* will remain on for two more clock periods and then it will go back to *S_wait*.
- From any state, as soon as *R* is pressed, the FSM immediately goes back to state *S_init*.
- Pressing the *C* switch only affects the FSM when it is in state *S_front*. The *C* switch has no effect on the FSM when it is in any other states.
- Any unused state encoding will have *S_init* as their next state.

CHAPTER 7

Dedicated Microprocessors

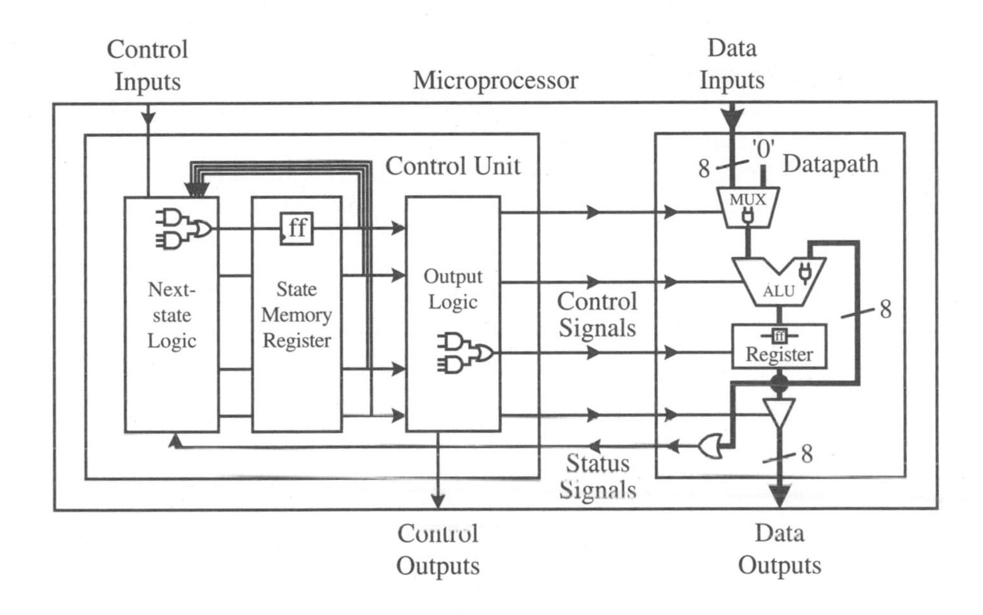

All microprocessors can be divided into two main categories: **dedicated microprocessors** and **general-purpose microprocessors**. General-purpose microprocessors are capable of performing a variety of computations. In order to achieve this goal, each computation is not hardwired into the processor, but rather it is represented by a sequence of instructions in the form of a program that is stored in the memory and executed by the microprocessor. The program in the memory can be changed easily so that another computation can be performed. Because of the general nature of the processor, it is likely that in performing a specific computation not all of the resources available inside the general-purpose microprocessor are used.

Dedicated microprocessors, also known as **application-specific integrated circuits (ASICs)**, on the other hand, are dedicated to performing only one task. The instructions for performing that one task are hardwired into the processor. In other words, no memory is required to store the program because the program is built into the microprocessor circuit itself. If the dedicated microprocessor is customized completely, then only those resources that are required by the computation are included in the microprocessor, so no resources are wasted. Another advantage of building the program instructions directly into the microprocessor circuit itself is that the execution speed of the program is many times faster than if the instructions are stored in memory, because memory access is typically many times slower than the microprocessor operation speed.

The design of a microprocessor, whether it is a dedicated microprocessor or a general-purpose microprocessor, can be divided into two main parts: the datapath and the control unit, as shown in Figure 7.1.

The **datapath** is responsible for all of the operations to be performed on the data by the microprocessor. It includes: (1) functional units such as adders, shifters, multipliers, ALUs, and comparators for the actual manipulation of the data; (2) registers and other

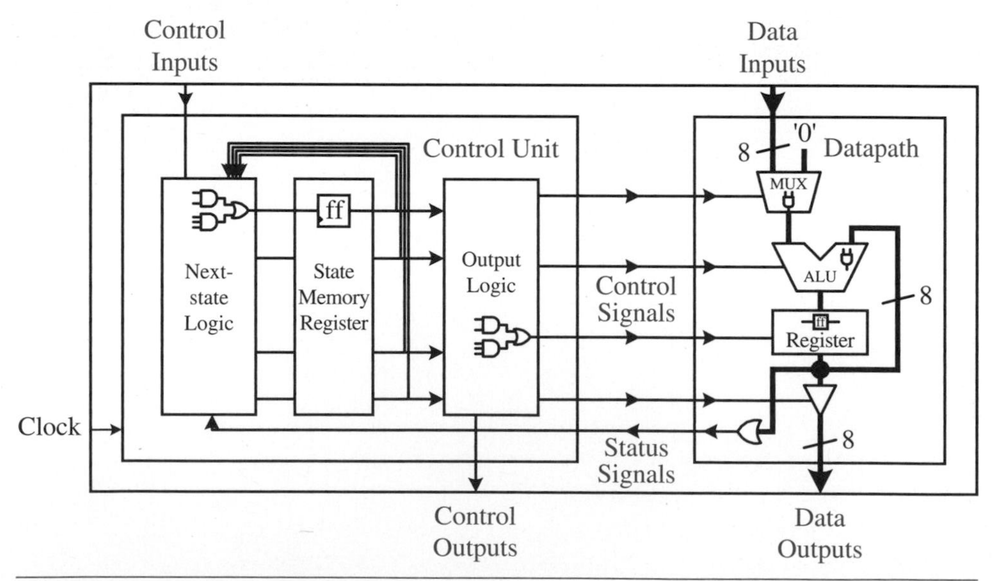

FIGURE 7.1 Block diagram of a microprocessor.

memory elements for the temporary storage of data; and (3) buses, multiplexers, and tri-state buffers for the transfer of data among the different components in the datapath. External data can enter the datapath through the **data input** lines, and results from the computation can be returned through the **data output** lines. These signals serve as the primary input and output data ports for the microprocessor.

In order for the datapath to function correctly, appropriate **control signals** must be asserted at the right time. Control signals are needed for all of the select and control lines for all of the components used in the datapath. These include all of the select lines for multiplexers, ALUs, and other functional units having multiple operations; all of the read and write enable signals for registers and register files; address lines for register files; and enable signals for tri-state buffers. Thus, the operation of the datapath is determined by which control signals are asserted or de-asserted and at what time. For the datapath to operate automatically, these control signals must be generated by the control unit.

The **control unit** (or **controller**) is responsible for controlling all of the operations of the datapath by providing appropriate control signals to the datapath at the appropriate times. Sometimes, the control unit requires the datapath to provide information in the form of **status signals** in order to determine what to do next. These status signals are usually from the output of comparators. The comparator tests for a given logical condition between two data values in the datapath. These values are obtained either from memory elements or directly from the output of functional units, or are hardwired as constants. For example, in a conditional loop situation, the status signal provides the result of the condition being tested, and using this information, the control unit then can decide whether to repeat or exit the loop.

The **control inputs** are the primary external input signals to the control unit. These are the external signals for controlling the operation of the microprocessor. For example, a *Start* signal will tell the microprocessor to start executing, or a *Reset* signal will reset the state memory to the initialization state. The **control outputs** are the primary output signals from the microprocessor to the external world. For example, when a microprocessor is finished executing an algorithm, it can output a *Done* signal to let the user know that it is done, and that the data being output by the datapath is valid.

The control unit itself is a finite-state machine (FSM), and the circuit for it is derived exactly as discussed in Chapter 6 having the next-state logic circuit, the state memory register, and the output logic circuit. As we saw in Chapter 6, the FSM operates by transitioning from one state to another at the rate of one state per clock cycle. By stepping through a sequence of states, the control unit automatically controls the operations of the datapath. The state that the FSM is in is determined by the content of the state memory. In every state, the output logic in the FSM will generate all of the appropriate control signals for controlling the datapath. The datapath, in return, provides status signals for the next-state logic.

The FSM circuits from Chapter 6 simply show that the FSM has input and output signals. For an FSM to be a control unit for a microprocessor, we need to connect the corresponding control and status signals together between the control unit and the datapath. After we have made the connections between the control unit and the datapath, the resulting circuit is a complete microprocessor.

This method of manually constructing a dedicated microprocessor is referred to as the **FSM+D** (FSM *plus* datapath) model because the control unit and the datapath are designed and constructed separately, and then they are connected together using the control and status signals. In this chapter, we start with a detailed discussion about the manual construction of dedicated microprocessors using the FSM+D model. Then in Section 7.6, we will show how dedicated microprocessors can be constructed automatically with HDL code using the **FSMD** (FSM *with* datapath) and **behavioral** models.

There are situations in which a dedicated microprocessor is used only to control external devices and does not need to perform any data manipulations at all. If this is the case, then the datapath is not needed and the microprocessor will only have a control unit. The output signals from the control unit will be used directly to control the external devices. Normally we would not call this a microprocessor, but a **microcontroller** or just **controller**. And because a control unit is just a FSM, these microcontrollers are just FSMs as discussed in Chapter 6. We saw an example of this in the elevator controller discussed in Sections 6.6.5 and 6.6.6.

7.1 Need for a Datapath

In Chapter 4, we learned how to design functional units to perform single, simple data operations, such as the adder for adding two numbers or the comparator for comparing two values. The next logical question to ask is how do we design a circuit to perform more complex data operations or operations that involve multiple steps? For example, how do we design a circuit for adding four numbers or a circuit for adding a million numbers? For adding four numbers, we can connect three adders together, as shown in Figure 7.2(a). However, for adding a million numbers, we really don't want to connect a million minus one adders together in a similar fashion. Instead, we want a circuit with just one adder and to use it a million times. A datapath circuit allows us to do just that, and that is to perform many operations involving multiple steps. Figure 7.2(b) shows a simple datapath using one adder to add as many numbers as we want. In order for this to be possible, a register is needed to store the temporary result after each addition. The temporary result from the register is fed back to the input of the adder so that the next number can be added to the current sum.

Because the datapath is responsible for performing all of the functional operations of a microprocessor and the microprocessor is for solving problems, therefore, the datapath must be able to perform all of the operations that are required to solve the given problem. For example, if the problem requires the addition of two numbers, then the datapath must contain an adder. If the problem requires the storage of three temporary variables, then the datapath must have three registers. However, even with these requirements, there are still many options as to what actually is implemented in the datapath. For example, an adder can be implemented as a single adder circuit, or as part of the ALU. These functional units can be used many times. Registers can be separate register units or combined in a register file. Furthermore, two temporary variables can share the same register if they are not needed at the same time.

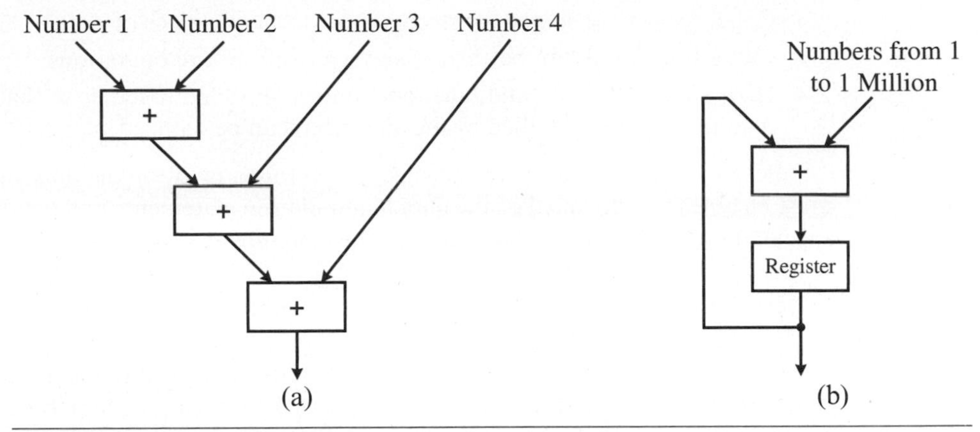

FIGURE 7.2 Circuits to add several numbers: (a) combinational circuit to add four numbers; (b) datapath to add one million numbers.

7.2 Constructing the Datapath

The datapath is responsible for performing all of the data operations specified by a given problem. The goal for designing a dedicated datapath, therefore, is to build a circuit that is able to perform all of these data operations. Datapath design also is referred to as **register-transfer level** (RTL) design. In a RTL design, we look at how data are transferred from one register to another, or back to the same register. If the same data are written back to a register without any modifications, then no meaningful work has been done. Therefore, before writing the data to a register, the data usually pass through one or more functional units, and get modified.

The sequence of RTL operations—read data from a register, modify data by functional units, and write result to a register—is referred to as a **register-transfer operation**. Every register-transfer operation must complete within one clock cycle (which is equivalent to one state of the FSM, because the FSM changes state at every clock cycle). Furthermore, in a single register-transfer operation, a functional unit cannot be used more than once unless it is used by different register-transfer operations in different clock cycles. In other words, a functional unit can be used only once in the same clock cycle, but can be used again in a different clock cycle.

When designing a datapath for a problem, we will specify the problem in the form of an algorithm written in C-style pseudocodes. The logical interpretation of the algorithm is irrelevant in what we are trying to do at this point, so when given a certain segment of code, we will just take the code as is and will not optimize it in any manner.

In a RTL design, we focus on how data move from register to register via some functional units where they are modified. In the design process, we need to decide on the following issues:

- What kind of registers to use, and how many are needed?
- What kind of functional units to use, and how many are needed?

- Can a register be shared for storing more than one piece of data?
- Can a functional unit be shared between two or more operations?
- How are the registers and functional units connected together so that all of the data movements specified by the algorithm can be realized?

Because the datapath is responsible for performing all of the data operations, it must be able to perform all of the data manipulation statements and conditional tests specified by the algorithm. For example, the assignment statement

$$A = A + 3$$

takes the value that is stored in the variable A, adds the constant 3 to it, and stores the result back into A. Note that the initial value of A is irrelevant here because that is a logical issue. In order for the datapath to perform the data operation specified by this statement, the datapath must have a register for storing the value A. Furthermore, there must be an adder for performing the addition. The constant 3 can be hardwired into the circuit as a binary value.

The next question to ask is how do we connect the register, the adder, and the constant 3 together so that the execution of the assignment statement can be realized. Recall from Section 5.11 that the operation of a register is such that a value stored in the register is available at the Q output of the register. Because we want to add $A + 3$, we connect the Q output of the register to the first operand input of the adder, and connect the constant 3 to the second operand input of the adder. We want to store the result of the addition back into A (i.e., back into the same register), therefore, we connect the output of the adder to the D input of the same register, as shown in Figure 7.3(a).

The storing of the adder result into the register is accomplished by asserting the *Load* signal of the register, which is connected to the external *Aload* signal in the circuit diagram. This *ALoad* signal is an example of what we have been referring to as the datapath control signal. This control signal controls the operation of this datapath. The control unit will control this signal by either asserting or de-asserting it.

The actual storing of the value into the register, however, does not occur immediately when *ALoad* is asserted. Because the register is synchronous to the clock signal, the actual storing of the value occurs at the next active clock edge. As a result, the new value of A is not available at the Q output of the register during the current clock cycle, but is available at the beginning of the next clock cycle.

As another example, the datapath shown in Figure 7.3(b) can perform the execution of the statement:

$$A = B + C$$

where B and C are two different variables stored in two separate registers, thus providing the two operand inputs to the adder. The output of the adder is connected to the D input of the A register for storing the result of the adder.

The execution of the statement is realized simply by asserting the *ALoad* signal, and the actual storing of the value for A occurs at the next active edge of the clock. During

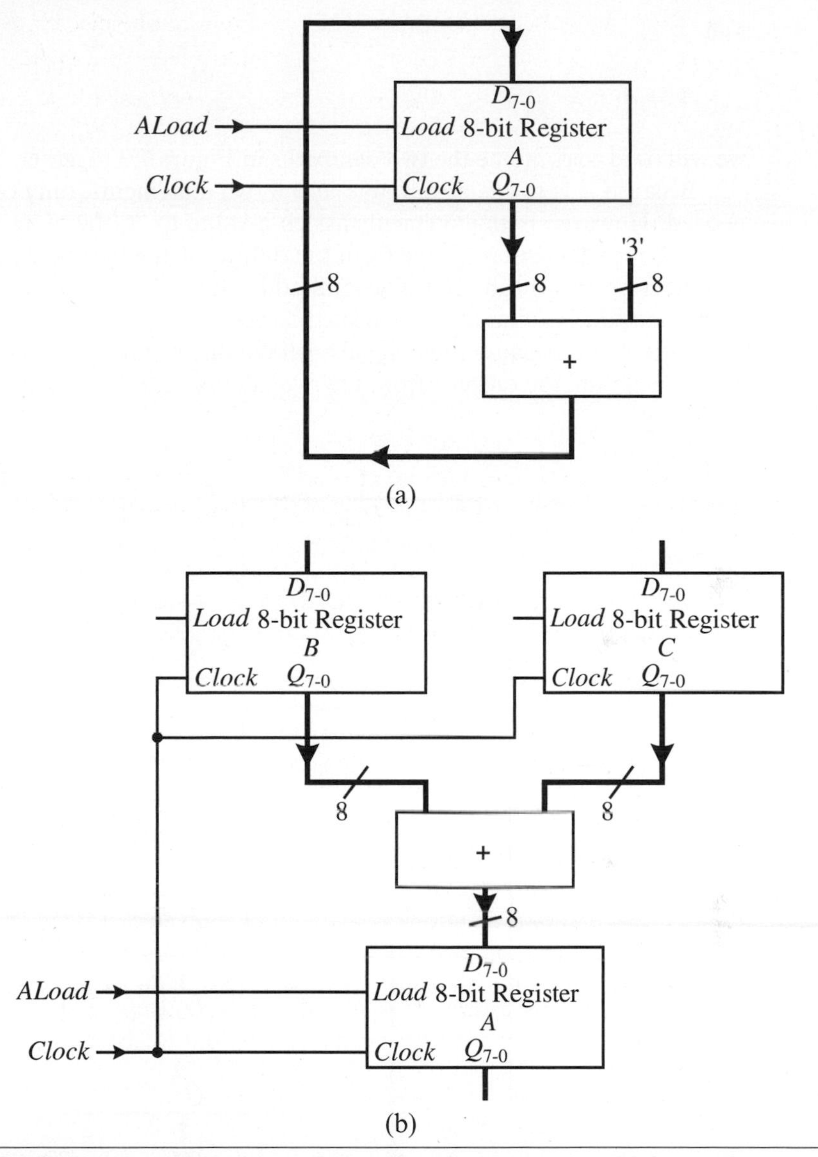

FIGURE 7.3 Sample datapaths: (a) for performing $A = A + 3$ (b) for performing $A = B + C$.

the current clock cycle, the adder will perform the addition of B and C, and the result from the adder must be ready and available at its output before the end of the current clock cycle so that, at the beginning of the next clock cycle (i.e., the next active clock edge), the correct value will be written into A. Because we are not writing any values to register B or C, we do not need to control the two *Load* signals for these two registers.

If we want a single datapath that can perform both of the statements:

$$A = B + C$$

and

$$A = A + 3$$

we will need to combine the two datapaths in Figure 7.3 together.

Because A is the same variable in the two statements, only one register for A is needed. However, both statements assign a value to A, therefore register A now has two different data sources: one from the output of the first adder for $B + C$, and the second from the output of the second adder for $A + 3$. The problem is that two or more data sources cannot be connected directly together to one destination, as shown in Figure 7.4(a) because their signals will collide, resulting in incorrect values. In the circuit diagram, the outputs from the two adders cannot be connected together to go

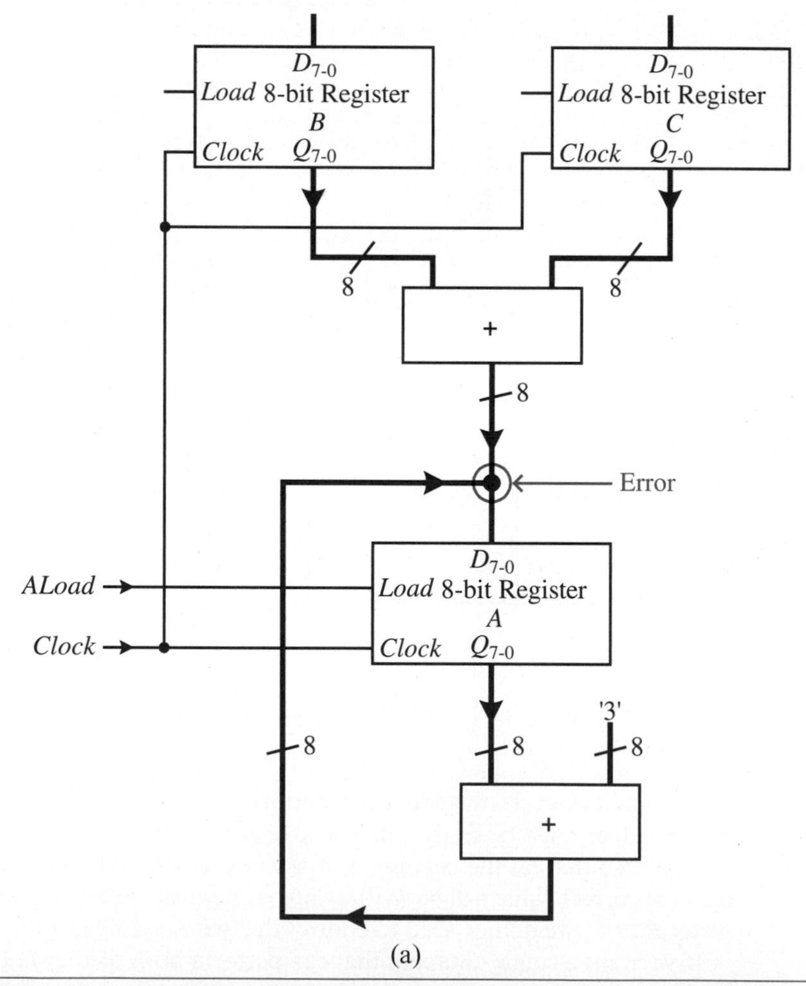

FIGURE 7.4 Datapath for performing $A = A + 3$ and $A = B + C$: (a) without multiplexer—wrong; (b) with multiplexer—correct. *(continued on next page)*

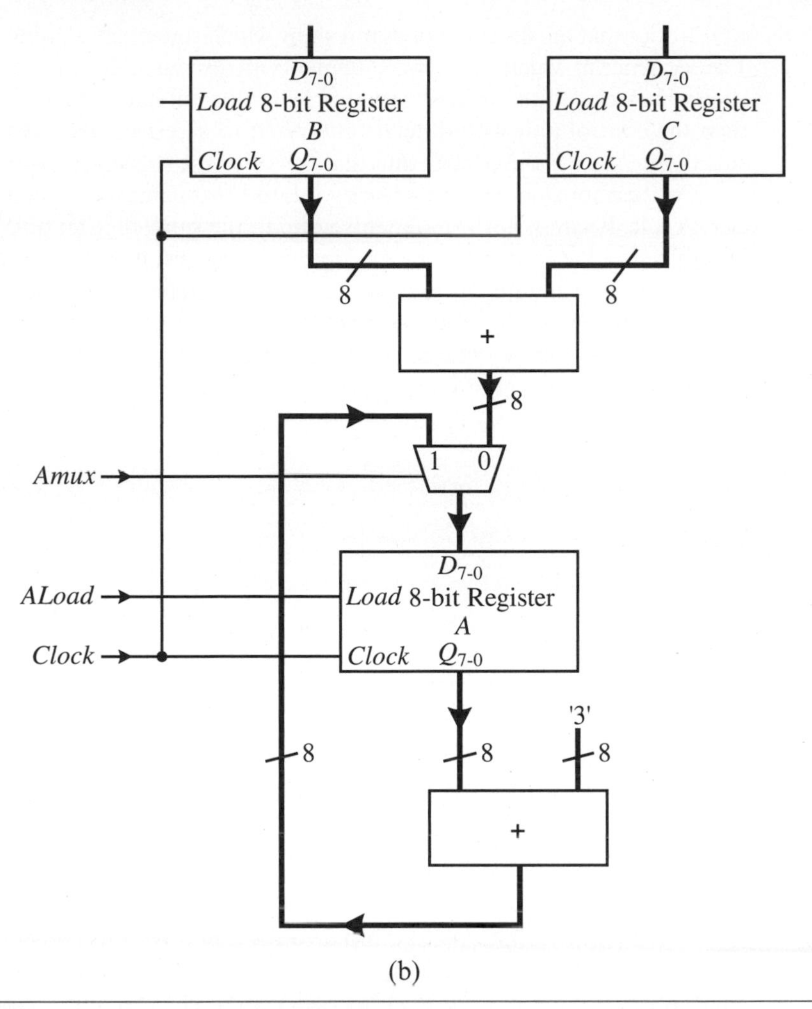

FIGURE 7.4 Datapath for performing $A = A + 3$ and $A = B + C$: (a) without multiplexer—wrong; (b) with multiplexer—correct.

to the D input of register A at the blue highlighted point. The solution is to use a multiplexer to select which of the two sources to pass to register A. The correct datapath using the multiplexer is shown in Figure 7.4(b).

Both statements assign a value to A, so *ALoad* must be asserted for the execution of both statements. The actual value that is written into A, however, depends on the selection of the multiplexer. If *Amux* is asserted, then input 1 of the mux is selected and so the result from the bottom adder (i.e., the result from $A + 3$) passes through the mux and is stored into A; otherwise, input 0 of the mux is selected and the result from the top adder is stored into A. Because the two adders are combinational circuits and the value from a register is always available at its output Q, the results from the two additions are always available at the two inputs of the multiplexer. But depending on the *Amux* control signal, only one value will pass through to the D input of the A register.

Notice that the datapath does not show which statement is going to be executed first. The sequence in which these two statements are executed depends on whether the signal *Amux* is asserted first or de-asserted first. If this datapath is part of a microprocessor, then the control unit would determine when to assert or de-assert this *Amux* control signal, because it is the control unit that performs the sequencing of datapath operations.

Furthermore, notice that these two statements cannot be executed within the same clock cycle. Because both statements write to the same register and a register can latch in only one value at an active clock edge, only one result from one adder can be written into the register in one clock cycle. The other statement will have to be performed in another clock cycle, but not necessarily the next cycle.

Example 7.1 shows the construction of a datapath for the same two instructions shown above but using only one adder.

EXAMPLE 7.1

Designing a dedicated datapath

Design a datapath that can execute the two statements:

$$A = B + C$$

and

$$A = A + 3$$

using only one adder.

The only difference between this datapath and the one in Figure 7.4(b) is that it should use only one adder. So starting with this one adder, in order to execute the first statement, the first operand input to the adder is from register B, and the second operand input to the adder is from register C. However, to execute the second statement, the two input operands to the adder are register A and the constant 3. Because both input operands to the adder have two different sources, again we must use a multiplexer for each of them. The output of the two multiplexers will connect to the two adder input operands, as shown in Figure 7.5. For both statements, the result of the addition is stored in register A, therefore, the output of the adder connects to the input of register A.

Note that the two select lines for the two multiplexers can be connected together. This is possible because the two operands B and C for the first statement are connected to input 0 of the two multiplexers, respectively, and the two operands A and 3 for the second statement are connected to input 1 of the two multiplexers, respectively. Thus, de-asserting the *Mux* select signal will pass the two correct operands for the first statement, and likewise, asserting the *Mux* select signal will pass the two correct operands for the second statement. We want to reduce the number of control signals for the datapath as much as possible, because minimizing the number of control signals will minimize the size of the output circuit in the control unit.

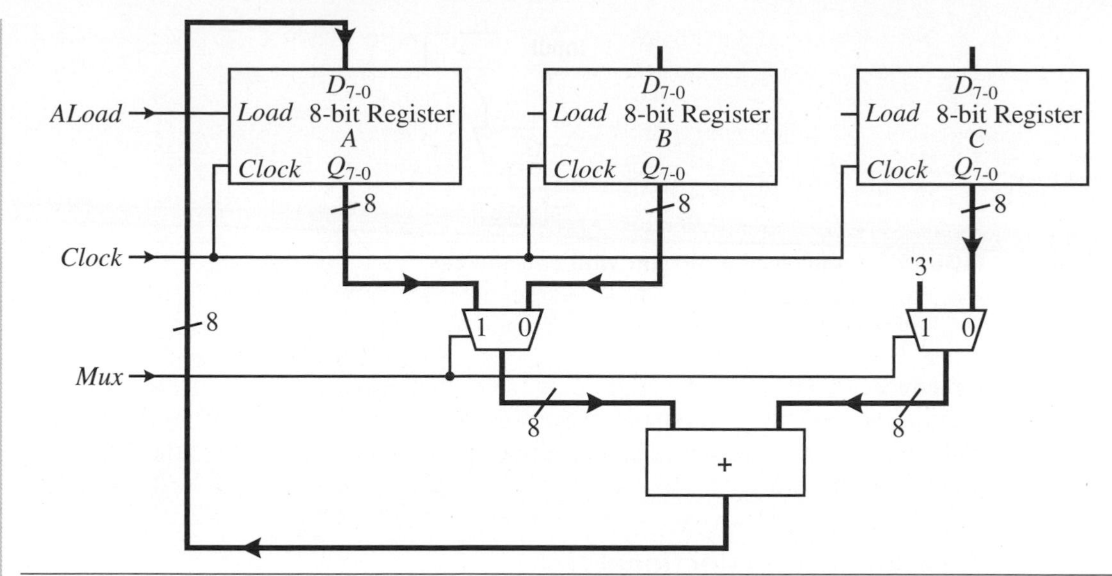

FIGURE 7.5 Datapath for performing $A = A + 3$ and $A = B + C$ using only one adder.

7.2.1 Selecting Registers

In most situations, one register is needed for each variable used by the algorithm. However, if two variables are not used at the same time, then they can share the same register. If two or more variables share the same register, then the data transfer connections leading to the register and out from the register usually are made more complex, because the register now has more than one source and destination. Having multiple destinations is not too big of a problem, because we can connect all of the destinations to the same source.[1] However, having multiple sources will require a multiplexer to select one of the several sources to transfer to the destination. Figure 7.6 shows a circuit with a register having two sources—one from an external input and one from the output of an adder. A multiplexer is needed in order to select which of these two sources is to be the input to the register.

After deciding how many registers are needed, we still need to determine whether to use a single register file containing enough register locations, separate individual registers, or a combination of both for storing the variables in. Furthermore, registers with built-in special functions, such as shift registers and counters, also can be used. For example, if the algorithm has a FOR loop statement, a single counter register can be used not only to store the count variable but also to increment the count. This way,

[1] This is true only theoretically. In practice, there are fan-in (multiple sources with one destination) and fan-out (one source with multiple destinations) maximum limits that must be observed.

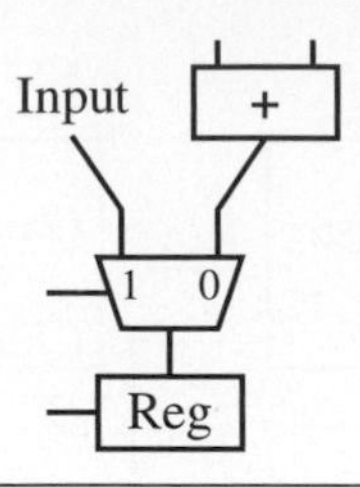

FIGURE 7.6 Circuit of a register with two sources.

we reduce not only the number of components, but also the number of datapath connections between the components. Decisions for selecting the type of registers to use will affect how the data transfer connections between the registers and the functional units are connected.

7.2.2 Selecting Functional Units

It is fairly straightforward to decide what kind of functional units is required. For example, if the algorithm requires the addition of two numbers, then the datapath must include an adder. However, we still need to decide whether to use a dedicated adder, an adder–subtractor combination, or an ALU (which has the addition operation implemented). Of course, these questions can be answered by knowing what other data operations the algorithm needs. If the algorithm has only an addition and a subtraction, then you might want to use the adder–subtractor combination unit. On the other hand, if the algorithm requires several addition operations, do we use just one adder or several adders?

Using one adder might decrease the datapath size in terms of the number of functional units, but it also might increase the datapath size because more complex data transfer paths are needed. For example, if the algorithm contains the following two addition operations

$$a = b + c$$
$$d = e + f$$

Using two separate adders will result in the datapath shown in Figure 7.7(a); whereas, using one adder will require the use of two extra 2–to–1 multiplexers to select which register will supply the input to the adder operands, as shown in Figure 7.7(b). Furthermore, this second datapath requires two extra control signals for the two multiplexers. In terms of execution speed, the datapath in Figure 7.7(a) can execute both addition statements simultaneously within the same clock cycle because they are independent of each other. However, the datapath in Figure 7.7(b) will have to execute these two additions sequentially in two different clock cycles because only one adder is available. This is a tradeoff issue between speed versus size, and the final decision on which datapath to use is up to the designer.

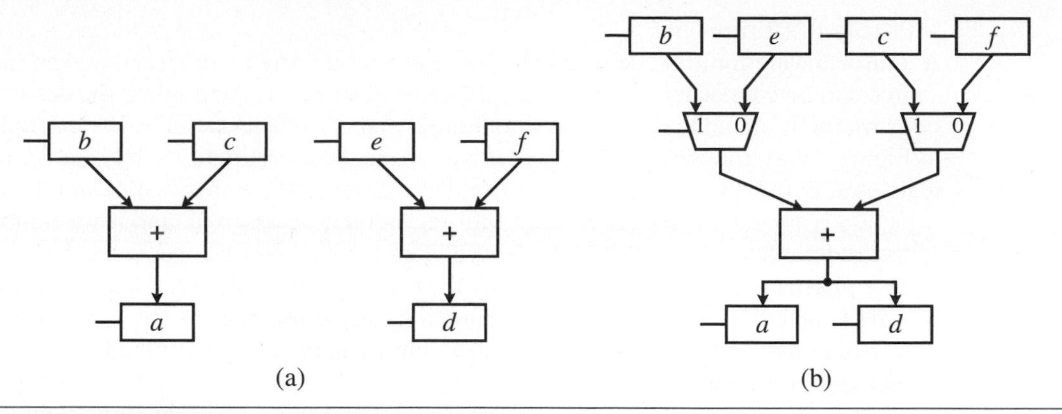

FIGURE 7.7 Datapaths for realizing two addition operations: (a) using two separate adders; (b) using one adder.

7.2.3 Data Transfer Methods

There are several methods in which the registers and the functional units can be connected together so that the correct data transfers between the different units can be made.

Multiple Sources

If the input to a unit has more than one source, then a multiplexer must be used to select which one of the multiple sources to use. The sources can be from registers, constant values, or outputs from other functional units. Figure 7.8 shows two such examples. In Figure 7.8(a), the left operand of the adder has four sources: two from two different registers, one from the constant 8, and one from the output of an ALU. In Figure 7.8(b), register *a* has two sources: one from the constant 8 and one from the output of an adder. The multiplexer select lines (*Muxselect*), will be controlled by the control unit to determine which of the sources will be passed to the destination.

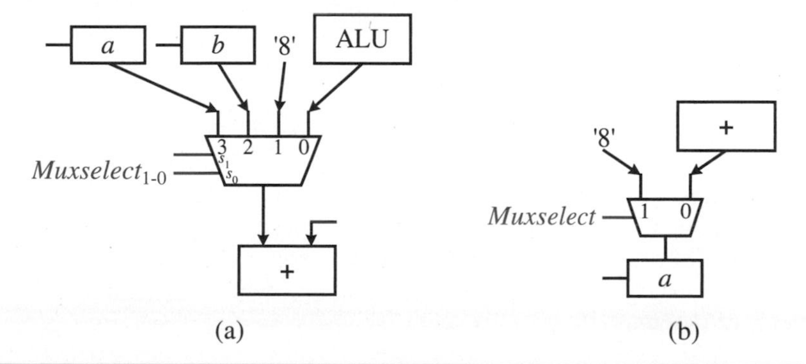

FIGURE 7.8 Examples of multiple sources using multiplexers: (a) an adder operand having four sources; (b) a register having two sources.

Multiple Destinations

A source having multiple destinations does not require any extra circuitry. The one source can be connected directly to the different destinations, and all of the destinations where the data are not needed would simply ignore the data source. For example, in Figure 7.9(a), the output of the adder has two register destinations: register *a*, and register *d*. If the output of the adder is intended for register *a* then the *Load* line for register *a* is asserted, while the *Load* line for register *d* is de-asserted; and if the output of the adder is for register *d*, then the *Load* line for register *d* is asserted, while the *Load* line for register *a* is not. In either case, only the register that the data is intended for will have its *Load* line asserted while the other units simply ignore the data by not asserting their *Load* lines. Again it is up to the control unit to assert the correct load signal in a particular clock cycle.

Connecting two or more combinational functional units as destinations from the same source as shown in Figure 7.9(b) also is correct. In this case, register *a* provides the source to both the adder and the subtractor, from which they will take the data, manipulate it, and output their respective results. However, only the register that is connected to the output of the needed functional unit will have its *Load* signal asserted. The output from the other functional unit will not be stored and so its result is ignored. For example, if we need to perform the addition and not the subtraction, then only the *Load* signal for register *c*, which is connected to the output of the adder, is asserted. Note that the outputs from both the adder and the subtractor should be connected to a register, otherwise, the result is not saved and so there is no need for that functional unit.

Functionally, it does not matter that both functional units operate on the source, because only the needed result is stored. However, it does require power for the functional units to manipulate the data, so if we want to reduce the power consumption, we would want only the functional unit that is needed to manipulate the data. This, however, is a power optimization issue that is beyond the scope of this book.

Tri-state Bus

Another scheme where multiple sources and destinations can be connected to the same data bus is through the use of tri-state buffers. The point to note here is that, when multiple sources are connected to the same bus, only one source can output at any one

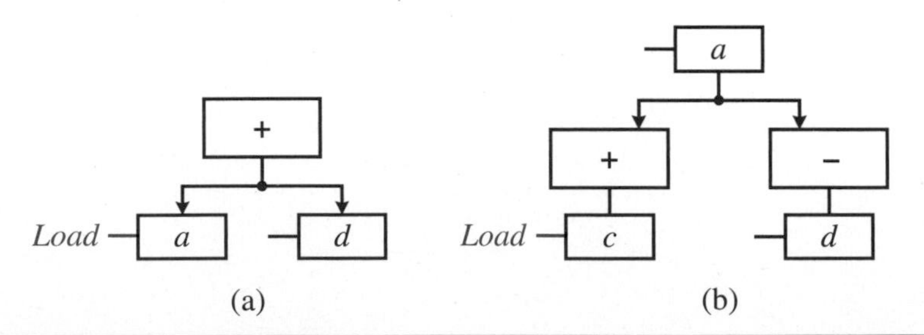

FIGURE 7.9 Examples of multiple destinations: (a) two register destinations; (b) two combinational circuit destinations.

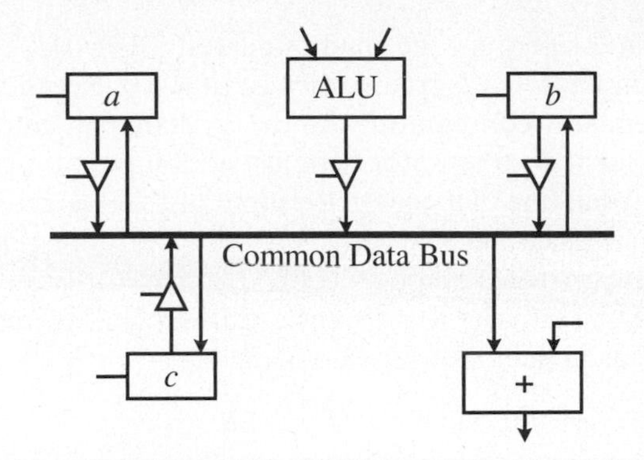

FIGURE 7.10 Multiple sources using tri-state buffers to share a common data bus.

time. If two or more sources output to the same bus at the same time, then there will be data conflicts. This occurs when one source outputs a 0 while another source outputs a 1. By using tri-state buffers to connect between the various sources and the common data bus, we want to make sure that only one tri-state buffer is enabled at any one time, while the rest are all disabled. Tri-state buffers that are disabled output high-impedance Z values, so no data conflicts can occur.

Figure 7.10 shows a common tri-state data bus with five components (three registers, an ALU, and an adder) connected to it. An advantage of using a common tri-state data bus is that the bus is bidirectional, so that data can travel in both directions on the bus. Connections for data going from a component to the bus need to be tri-stated by using a tri-state buffer in between (such as from registers a, b, and c, and the ALU), while connections for data going from the bus to a component need not be (such as data from the bus going to the registers and the adder). Note also that data input and output of a register both can be connected to the same tri-state bus; whereas, the input and output from the same functional unit (such as the adder and the ALU) cannot be connected to the same tri-state bus.

7.2.4 Generating Status Signals

Although the control unit is responsible for the sequencing of statement execution, but in situations where a conditional test is involved, the datapath must supply the result of the conditional test for the control unit so that the control unit can determine what statement to execute next. Status signals are the results of the conditional tests that the datapath supplies to the control unit. Every conditional test that the algorithm has requires a corresponding status signal. These status signals usually are generated by comparators.

For example, if the algorithm has the following IF statement

IF $(A = 0)$ THEN ...

the datapath must have an equality comparator that compares the value in the A register with the constant 0, as shown in Figure 7.11(a). The output of the comparator is the status signal for the conditional test $(A = 0)$. This status signal is a 1 when the condition $(A = 0)$ is true; otherwise, it is a 0. Recall from Section 4.9 that the circuit for the equality comparator with the constant 0 is simply a NOR gate, so we can replace the box for the comparator with just an 8-input NOR gate as shown in Figure 7.11(b). The thick bus line that goes to the input of the NOR gate is another way to show a multi-input gate. Since the bus width is 8 bits wide, it must be an 8-input NOR gate.

There are times when an actual comparator is not needed for generating a status signal. For example, if we want a status signal to test whether a number is an odd number, as in the following IF statement

`IF` (*A* is an odd number) `THEN ...`

we can simply use the A_0 bit of the 8-bit number from register A as the status signal for this condition, because all odd numbers have a 1 in the zero bit position. The generation of this status signal is shown in Figure 7.12.

We will now show the complete process on how to construct a datapath with two examples. Example 7.2 shows the construction of a datapath for solving a simple IF-THEN-ELSE problem. Example 7.3 shows the construction of a datapath for a summation problem to generate and sum the numbers from 1 to 10.

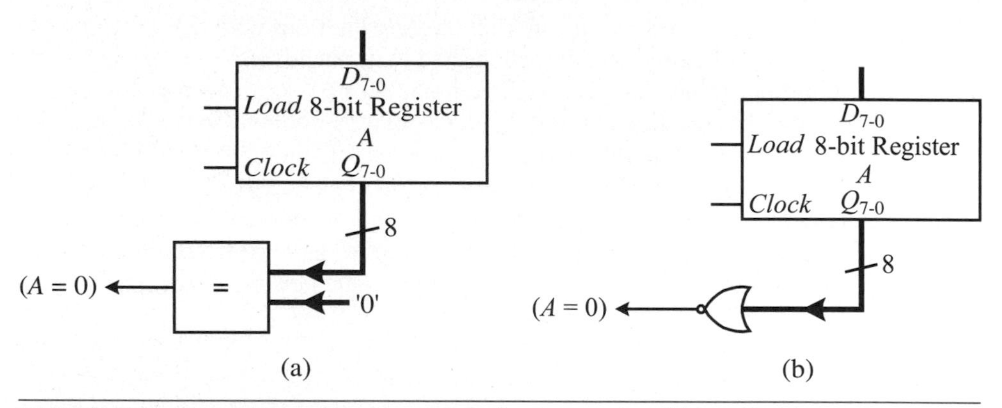

FIGURE 7.11 **Comparator for generating the status signal ($A = 0$): (a) using an "equal to zero" circuit; (b) using a NOR gate.**

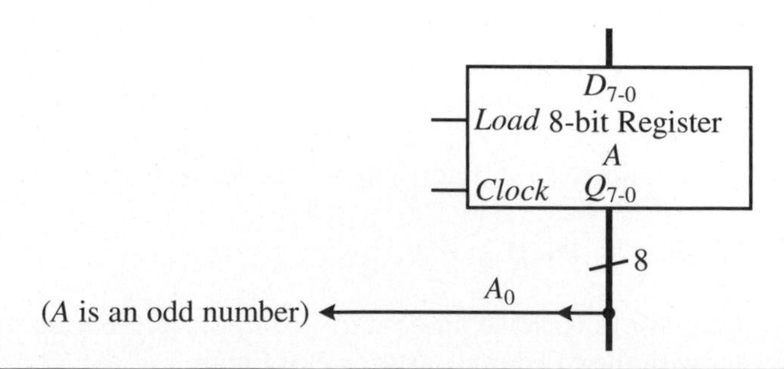

FIGURE 7.12 **Comparator for generating the status signal (A is an odd number).**

EXAMPLE 7.2

Datapath for a simple IF-THEN-ELSE problem

In this example, we want to construct a 4-bit-wide dedicated datapath to solve the simple IF-THEN-ELSE problem shown in Figure 7.13(a). To create a datapath for this algorithm, we need to look at all of the data manipulation statements in the algorithm, because the datapath is responsible for manipulating the data. These data manipulation instructions are the register-transfer operations. In most cases, one data manipulation instruction is equivalent to one register-transfer operation. However, some data manipulation instructions might require two or more register-transfer operations to realize.

The algorithm uses two variables, A and B; therefore, the datapath should have two 4-bit registers—one for each variable. Line 1 of the algorithm inputs a value into A. In order to realize this operation, we need to connect the data input signal *Input* to the D input of register A, as shown in Figure 7.13(b). By asserting the *ALoad* signal, the data input value will be loaded into register A at the next active clock edge.

Line 2 of the algorithm tests the value of A with the constant 5. The datapath in Figure 7.13(b) uses a 4-input AND gate for the equality comparator with the four input bits connected as 0101 to the four output bits of register A. Because 5 in decimal is 0101 in binary, bits 0 and 2 are not inverted for the two 1s in the bit string, while bits 1 and 3 are inverted for the two 0s. With this connection, the AND gate will output a 1 when the input is a 5. The output of this comparator is the 1-bit status signal for the condition $(A = 5)$ that the datapath sends to the control unit.

Given the status signal for the comparison $(A = 5)$, the control unit will decide whether to execute line 3 or line 5 of the algorithm. This sequencing decision is done by the control unit and not by the datapath. The datapath is responsible only for the register-transfer operations. Lines 3 and 5 require loading either an 8 or a 13 into register B. In order to select which one gets loaded, a 2-to-1 multiplexer is needed. One input of the multiplexer is connected to the constant 8 and the other to the constant 13. The output of the multiplexer is connected to the D input of register B, so that one of the two constants can be loaded into the register. Again, which constant is to be loaded into the register is dependent on the condition in line 2. Knowing the result of the test from the status signal, the control unit will generate the correct signal for the multiplexer select line, *Muxsel*. The actual loading of the value into register B is accomplished by asserting the *BLoad* signal.

Finally, the algorithm outputs the value from register B in line 7. This is accomplished by connecting a tri-state buffer to the output of the B register. To output the value, the control unit asserts the enable line, *Out*, connected to the tri-state buffer, and the value from the B register will be passed to the data output lines.

Note that the complete datapath shown in Figure 7.13(b) consists of two separate circuits. This is because the algorithm does not require the values of A and B to be used together. A question we might ask is whether we can connect the output of the comparator to the multiplexer select signal so that the status signal $(A = 5)$ directly controls *Muxsel*. Logically, this is all right, because if the condition $(A = 5)$ is true, then the status signal is a 1. Assigning a 1 to *Muxsel* will select the 1 input of the multiplexer,

```
1      INPUT A
2      IF (A = 5) THEN
3          B = 8
4      ELSE
5          B = 13
6      END IF
7      OUTPUT B
```

(a)

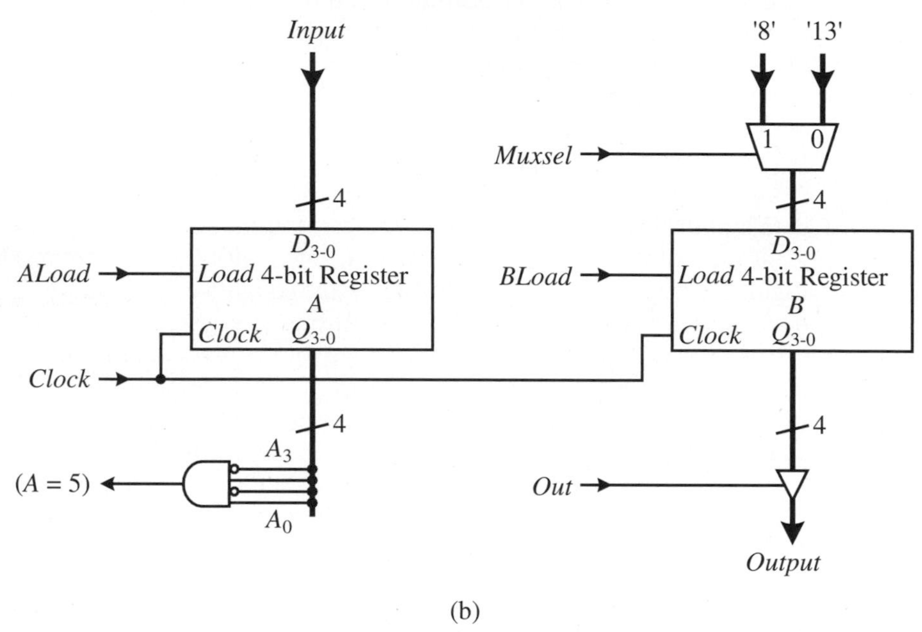

(b)

FIGURE 7.13 A simple IF-THEN-ELSE problem: (a) algorithm; (b) datapath.

thus passing the constant 8 to register B. Otherwise, if the condition $(A = 5)$ is false, then *Muxsel* will get a 0 from the comparator, and the constant 13 will pass through the multiplexer. The advantage of doing this is that the datapath will generate one less status signal and requires one less control signal from the control unit. However, in some situations, we need to be careful with the timing when we use status signals from the datapath to directly control the control signals. So it is best to have the control unit take the $(A = 5)$ status signal from the datapath and then decide how to generate the *Muxsel* control signal.

EXAMPLE 7.3

Datapath for a summation problem to generate and sum the numbers from 1 to 10

In this example, we want to construct an 8-bit-wide dedicated datapath to solve a summation problem to generate and sum the numbers from 1 to 10. The algorithm shown in Figure 7.14(a) for solving this summation problem has five data

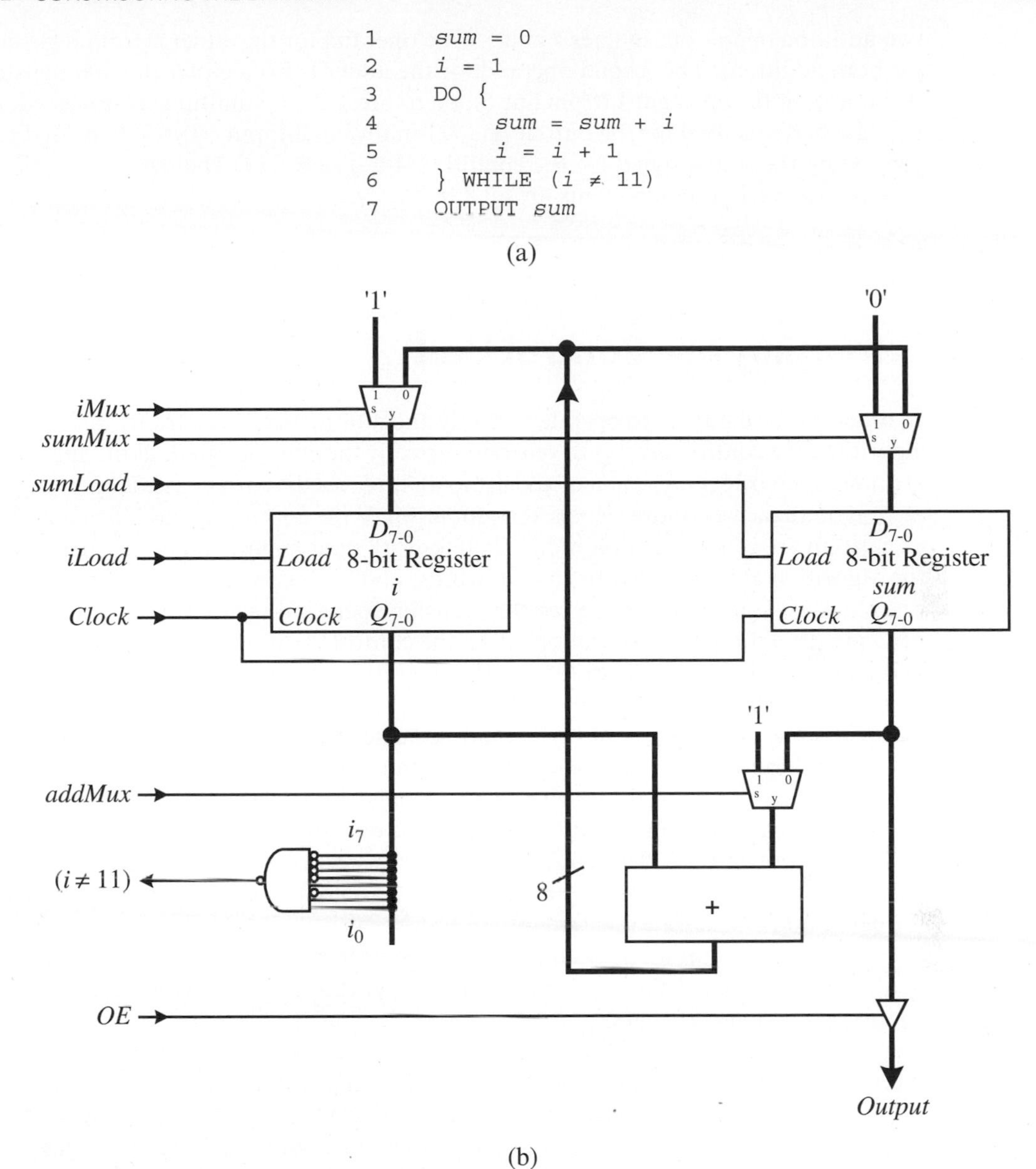

FIGURE 7.14 Summation problem to generate and sum the numbers from 1 to 10: (a) algorithm; (b) datapath.

manipulation statements (1, 2, 4, 5, and 7). The datapath for solving this problem is shown in Figure 7.14(b). Two 8-bit registers are needed for the two variables *sum* and *i*. Two separate multiplexers are needed for the input to the two registers since both variables have two different sources. For register *i*, line 2 assigns the constant 1, and line 5 assigns the result of the addition to it. For register *sum*, line 1 assigns the constant 0, and line 4 assigns the result of the addition to it. One adder is used for the two addition operations in lines 4 and 5. One operand for the adder is from register *i* for both additions. The second operand for the adder is from either the *sum* register (for line 4) or the constant 1 (from line 5), therefore, a 2-to-1 multiplexer is needed. A tri-state buffer is used for the output line 7. Finally, an 8-input NAND gate is used for generating the status signal for the conditional test $(i \neq 11)$. The data connections between the various components are all 8 bits wide.

7.3 Constructing the Control Unit

In order for the datapath to operate correctly and automatically according to the given algorithm, the control unit has to generate correctly the control signals at the appropriate time. The correct operation of the datapath involves the correct assertion or de-assertion of all of the control signals as a group. All of the control signals taken together as a group are called a **control word**. Therefore, all data manipulation instructions in the algorithm are converted to control words, and each control word is executed in one clock cycle to perform one register-transfer operation. A control unit is used to generate the appropriate control signals in the control words so that the datapath can perform all of the required register-transfer operations automatically according to the sequence specified in the algorithm. The control unit is just an FSM and the control signals are the output signals from the output logic circuit that is inside the FSM.

In addition to generating the control signals, the control unit also is needed to control the sequencing of the instructions in the algorithm. The datapath is responsible only for the manipulation of the data; it performs only the register-transfer operations. It is the control unit that determines when each register-transfer operation is to be executed and in what order. The sequencing done by the control unit is established during the derivation of the state diagram for the FSM.

The state diagram shows what register-transfer operation is executed in what state and the sequencing of the execution of these operations. A state is created for each control word, and each state is executed in one clock cycle. The edges in the state diagram are determined by the sequence in which the instructions in the algorithm are executed. The sequential execution of instructions is represented by unconditional transitions between states (i.e., edges with no labels). Execution branches in the algorithm are represented by conditional transitions from a state with two outgoing edges: one with the label for when the condition is true and the other with the label for when the condition is false. If a particular state has more than one condition then all possible combinations of these conditions must be labeled on the outgoing edges from that state. These conditions are the status signals generated by the datapath, and passed to the next-state logic in the FSM.

Once the state diagram is derived, the actual construction of the control unit is accomplished by following the same procedure for constructing an FSM as discussed in Chapter 6.

7.3.1 Deriving the Control Signals

Any given datapath will have a number of control signals for controlling its operations. By asserting or de-asserting these control signals at different times, the datapath can perform different register-transfer operations. Because the execution of an operation requires the correct assertion or de-assertion of all of the control signals together, we would like to think of them as a group rather than as individual signals. All of the control signals for a datapath, when grouped together, are referred to as a control word. Hence, a control word will have one bit for each control signal in the datapath. One register-transfer operation of a datapath, therefore, is determined by the values set in one control word, and so, we can specify the datapath operation simply by specifying the bit string for the control word. Each control word operation will take one clock cycle to perform. By combining multiple control words together in a certain sequence, the datapath will perform the specified operations in the order given.

We will now show the derivation of control words with three examples. Example 7.4 shows the derivation of the control words for executing the two statements $A = A + 3$ and $A = B + C$. Example 7.5 shows the derivation of the control words for executing the statements for the simple IF-THEN-ELSE problem. Example 7.6 shows the derivation of the control words for the summation problem to generate and sum the numbers from 1 to 10.

EXAMPLE 7.4

Deriving the control words for a datapath

The datapath for Example 7.1 and repeated here in Figure 7.15 was designed to execute the two statements $A = A + 3$ and $A = B + C$ using only one adder. This datapath has two control signals, *ALoad* and *Mux*. The control word for this datapath,

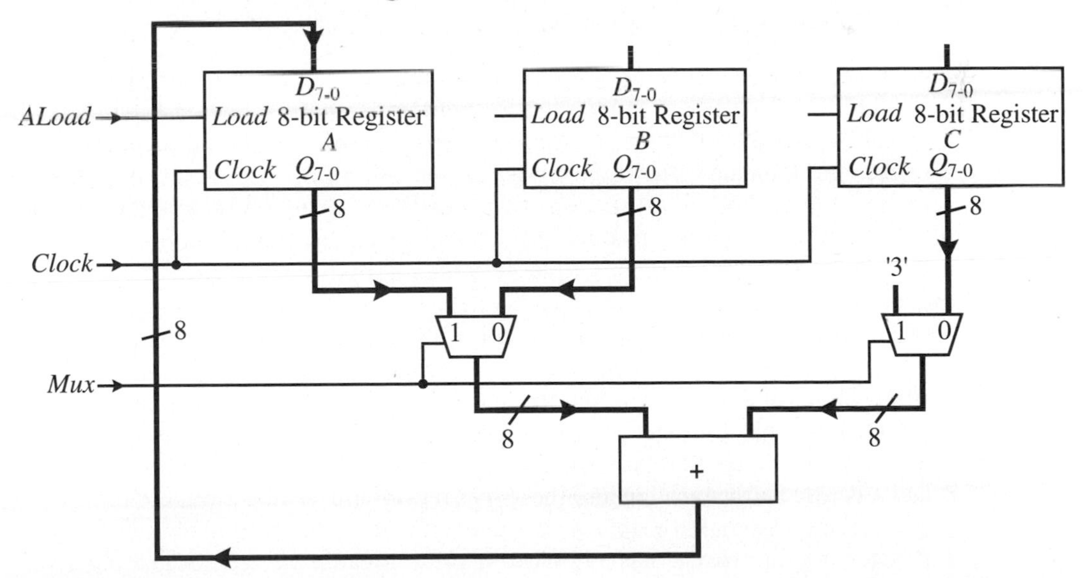

FIGURE 7.15 Datapath for performing $A = A + 3$ and $A = B + C$ using only one adder.

therefore, has two bits—one for each of the two control signals. The ordering of these two bits at this point is arbitrary; however, once decided, we should be consistent with the order. The two control words for performing the two statements are shown in Figure 7.16.

Control word 1 specifies the control word bit string for executing the statement, $A = A + 3$. This is accomplished by asserting both the *ALoad* and the *Mux* signals. When *Mux* is asserted, the output from register A and the constant 3 are passed through the two multiplexers to the adder. By asserting *ALoad*, the result from the adder is stored into register A. Similarly, control word 2 is for executing the statement, $A = B + C$, by asserting *ALoad* but de-asserting *Mux*.

Control Word	Instruction	*ALoad*	*Mux*
1	$A = A + 3$	1	1
2	$A = B + C$	1	0

FIGURE 7.16 Control words for the datapath in Figure 7.15 for performing the two statements: $A = A + 3$ and $A = B + C$.

EXAMPLE 7.5

Control words for the simple IF-THEN-ELSE problem

Figure 7.17 shows the control words for performing the statements in the IF-THEN-ELSE algorithm shown in Figure 7.13(a), and using the datapath shown in Figure 7.13(b). Control word 1 executes the instruction INPUT A. To do this, the *ALoad* signal is asserted, and the data value at the input port will be loaded into register A at the next active clock edge. For this instruction, we do not need to load a value into the B register; therefore, *BLoad* is de-asserted for this control word. Furthermore, because of this, it does not matter what the multiplexer, which supplies a value for the B register, outputs, so *Muxsel* can be a don't-care value. *Out* is de-asserted because we are not doing outputs in this control word. For control words 2 and 3, we want to load one of the two constants into B; therefore, *BLoad* is asserted for both of these control words, and the value for *Muxsel* determines which constant is loaded into B. When *Muxsel* is asserted, the constant 8 is passed to the input of the B register, and when it is de-asserted, the constant 13 is passed to the register. Both *ALoad* and *Out* are de-asserted because we are neither writing into the A register nor outputting a value. Control word 4 asserts the *Out* signal to enable the tri-state buffer, thus outputting the value from the B register. Again, *Muxsel* has a don't-care value because we are not loading a value into the B register. The status signal ($A = 5$) is not used anywhere in the control words.

Control Word	Instruction	*ALoad*	*Muxsel*	*BLoad*	*Out*
1	INPUT *A*	1	×	0	0
2	$B = 8$	0	1	1	0
3	$B = 13$	0	0	1	0
4	OUTPUT *B*	0	×	0	1

FIGURE 7.17 Control words for solving the simple IF-THEN-ELSE problem.

EXAMPLE 7.6

Control words for the summation problem to generate and sum the numbers from 1 to 10

Figure 7.18 shows the control words for performing the data manipulation statements in the algorithm to generate and sum the numbers from 1 to 10 shown in Figure 7.14(a), and using the datapath shown in Figure 7.14(b). Control words 1 and 3 load a new value into *sum* by asserting *sumLoad*; the actual value that is loaded in is determined by the select line *sumMux* for the multiplexer. For these two control words, both *iLoad* and *OE* are disabled because they are not used, and *iMux* can have a don't-care value. For control word 3, *addMux* needs to be de-asserted for the adder to use *sum* as the second operand.

Control words 2 and 4 load a new value into *i* by asserting *iLoad* and selecting the corresponding source with the *iMux* enable line on the multiplexer. For control word 4, *iMux* is asserted for the correct addition operation. Finally, for control word 5, *OE* is asserted to output the *sum* value.

Control Word	Instruction	*sumMux*	*sumLoad*	*iMux*	*iLoad*	*addMux*	*OE*
1	$sum = 0$	1	1	×	0	×	0
2	$i = 1$	×	0	1	1	×	0
3	$sum = sum + i$	0	1	×	0	0	0
4	$i = i + 1$	×	0	0	1	1	0
5	OUTPUT *sum*	×	0	×	0	×	1

FIGURE 7.18 Control words for the summation problem to generate and sum the numbers from 1 to 10.

7.3.2 Deriving the State Diagram

In constructing an FSM, the first step is to derive its state diagram. The state diagram shows what control word is executed in what state, and the sequencing of these states. A state is created for each control word. The edges in the state diagram are determined by the sequence in which the instructions in the algorithm are executed. The sequential execution of instructions is represented by unconditional transitions between states (i.e., edges with no labels). Execution branches in the algorithm are represented by conditional transitions from a state with two outgoing edges: one with the label for when the condition is true and the other with the label for when the condition is false. If there is more than one condition from a particular state, then all possible combinations of these conditions must be labeled on the outgoing edges from that state.

When deriving the state diagram for the control unit, we have to be careful with the timings of the register-transfer operations. The issue here is that when we write a value into a register, this new value is not available at the Q output of the register until the beginning of the next clock cycle. Therefore, if we read from the register in the current clock cycle, we would be reading the old value rather than the new value. When we were designing the FSMs in Chapter 6, this timing issue was not so much of a problem, because we were not using the FSMs to control register-transfer operations in a datapath. They were stand-alone FSMs, and so their input and output signals are independent of each other. However, the FSMs that we are designing here are for controlling the register-transfer operations in a datapath. The output signals from these FSMs are control signals for the datapath, and some of them are used to load registers with new values. These new register values might be used by comparators for testing conditions. The results of these conditional tests are the status signals used by the control unit to determine what next state to go to. Finally, from the different states, different control signals are generated. Therefore, the status (input) signals and the control (output) signals of a control unit are dependent on each other. Thus, when status signals are generated, we need to make sure that they are from tests of the intended register value. Because of these timing issues, extra states might be needed in order to get the correct timing.

We will now construct the state diagram for controlling the dedicated datapath for the simple IF-THEN-ELSE problem from Example 7.2. The algorithm, dedicated datapath, and control words from Examples 7.2 and 7.5 are repeated here in Figure 7.19 for convenience. Example 7.7 shows the naive way of creating the state diagram by simply assigning one state for each control word. However, we will see from the example that the state diagram created this way is incorrect for this problem, because the status signal generated is wrong. Example 7.8 shows the derivation of the corrected state diagram for the IF-THEN-ELSE problem.

EXAMPLE 7.7

Deriving an incorrect state diagram for the simple IF-THEN-ELSE problem

In this example, we will derive the state diagram for the control unit to control the dedicated datapath from Example 7.2 for solving the IF-THEN-ELSE problem. The algorithm for the problem is shown in Figure 7.19(a). The dedicated datapath for this algorithm already has been derived and is shown again in Figure 7.19(b). The algorithm shows that there are four data manipulation instructions: lines 1, 3, 5, and 7. Line 2 is not a

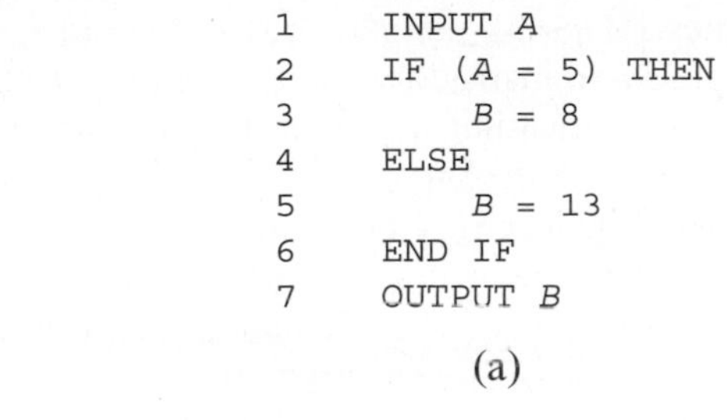

```
1    INPUT A
2    IF (A = 5) THEN
3        B = 8
4    ELSE
5        B = 13
6    END IF
7    OUTPUT B
```

(a)

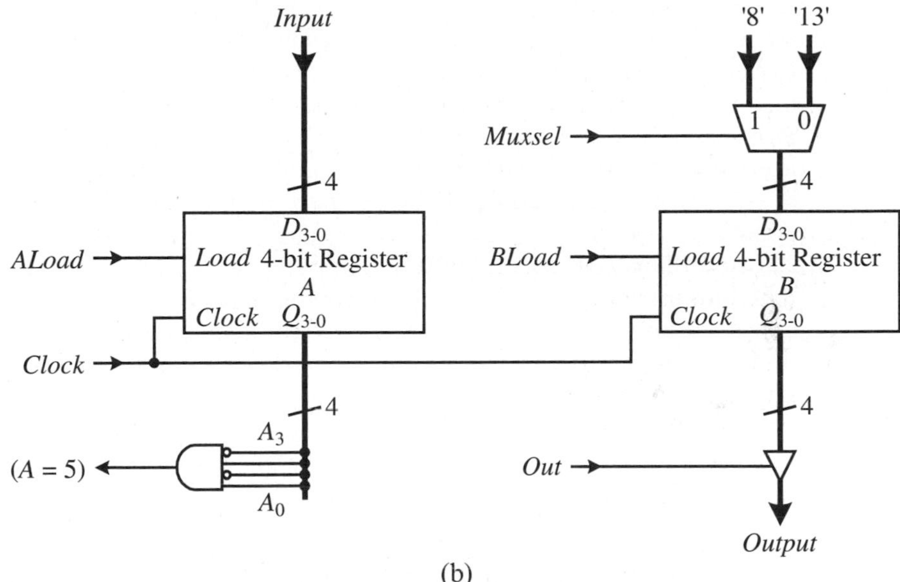

(b)

Control Word	Instruction	*ALoad*	*Muxsel*	*BLoad*	*Out*
1	INPUT *A*	1	×	0	0
2	*B* = 8	0	1	1	0
3	*B* = 13	0	0	1	0
4	OUTPUT *B*	0	×	0	1

(c)

FIGURE 7.19 The simple IF-THEN-ELSE problem: (a) algorithm; (b) dedicated datapath; (c) control words.

data manipulation instruction, but rather, it is a control statement. From these four data manipulation instructions, we already have derived the four corresponding control words shown in Figure 7.19(c) to control this dedicated datapath.

The next step in the construction of the control unit is to derive the state diagram for the FSM. We start by assigning these four control words to four separate states in the state diagram, as shown in Figure 7.20(a). These four states are given the symbolic names *s_input*, *s_equal*, *s_notequal*, and *s_output*, and are annotated with the control word and instruction that is assigned to them. For example, in state *s_input*, we want the control unit to generate the control signals for control word 1 to execute the INPUT A instruction.

After the INPUT A instruction, the execution of the following two instructions, $B = 8$ and $B = 13$, in states *s_equal* and *s_notequal*, respectively, depends on the condition $(A = 5)$ of the IF statement. The outcome of the condition $(A = 5)$ will determine which of the two states the FSM will transition to next. This conditional execution is represented by the two outgoing edges from state *s_input*: one edge going to state *s_equal* with the label $(A = 5)$ for when the condition is true, and the second edge going to state *s_notequal* with the label $(A = 5)'$ for when the condition is false. Finally, the instruction, OUTPUT B, is executed unconditionally after executing either of the two instructions, $B = 8$ or $B = 13$; therefore, from either state *s_equal* or *s_notequal*, there is an unconditional edge going to state *s_output*. The algorithm halts after executing OUTPUT B, so we make the FSM halt in state *s_output* by having an unconditional edge going back to itself.

According to the algorithm, after inputting a value for A in state *s_input*, we need to test for the condition $(A = 5)$. If the condition is true, we go to state *s_equal* to execute the instruction $B = 8$; otherwise, we go to state *s_notequal* to execute the instruction $B = 13$. From either state *s_equal* or *s_notequal*, the next and final state is *s_output*. Let us assume that state *s_input* is executed in clock cycle 1. In clock cycle 2, either state *s_equal* or *s_notequal* is executed. State *s_output* is then executed in clock cycle 3.

There are two important points to understand and remember here:

1. At a rising clock edge, a register is loaded with a new value if its load signal is asserted.
2. At every rising clock edge, the FSM enters a new state—the next state.

As with all of our other examples, we use the rising clock edge as the active edge. The reason for point 1 above is because we are using positive edge-triggered D flip-flops with enable (see Section 5.10) in our registers. So if the flip-flop is enabled by asserting the load signal, then the input data will be stored into the flip-flop at the next rising clock edge. Point 2 is because we also are using positive edge-triggered D flip-flops (see Section 5.9) in the state memory register inside the FSM. However, these flip-flops are always enabled without the need of an enable signal. Therefore, at every rising clock edge, a new value from the next-state logic circuit will be stored into the state memory register, and so, the FSM enters a new state at every rising clock edge.

If we construct the control unit based on the state diagram shown in Figure 7.20(a), then the following scenario can occur. At the first rising clock edge, the FSM enters state *s_input*. Shortly after entering state *s_input*, the FSM asserts the control signal *ALoad* to load in a value for variable A. Because a register is loaded at a rising clock edge, and the first rising clock edge has passed, therefore, the value for A will be stored into the register at the next rising clock edge (i.e., at the beginning of clock cycle 2). However, the FSM also needs to go to the next state (either *s_equal* or *s_notequal*) at the beginning of clock cycle 2. In order for the FSM to know which of the two states to go to, the FSM must know the result of the test condition $(A = 5)$ while it is still in state *s_input*. The dilemma here is that the status signal generated by the comparator for the test $(A = 5)$ is needed in state *s_input*, and the test uses the input value of A. However, this new input value of A is not available at the Q output of the register until the beginning of the next clock cycle. Therefore, what the comparator is reading from the Q output of the register in clock cycle 1 is the old (or current) value of A and not the new input value.

Figure 7.20(b) shows the timing diagram for this state diagram with the incorrect result. At time 0, the user inputs a 5. However, the diagram shows that the value 5 is not loaded into register A until at time 200 ns (the beginning of clock cycle 2). Because the input is a 5, the test for $(A = 5)$ should be true, and the next state for the FSM to go to should be *s_equal*. However, in the timing diagram, we see that the state changes to *s_notequal* at time 200 ns. This is because the conditional test being performed in clock cycle 1 is reading A with the old value of 0 rather than the new value of 5. Hence, at time 400 ns, the output of 13 is incorrect.

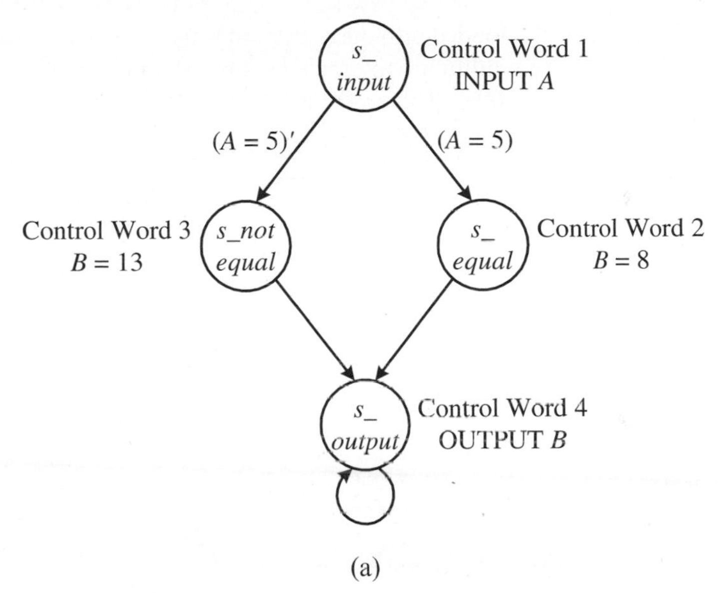

(b)

FIGURE 7.20 Example 7.7: (a) incorrect state diagram; (b) incorrect timing diagram.

EXAMPLE 7.8

Deriving a correct state diagram and the control unit for the simple IF-THEN-ELSE problem

In Example 7.7, even though the comparator is testing for the condition $(A = 5)$, it is not getting the correct value for A in the clock cycle that the test result is needed. In order for the comparator to get the correct value for A, it needs to wait until the value is loaded into the register at the next clock cycle. One simple way to resolve this timing error is to add an extra state after inputting the value for A so that the value can be written into the register before it is read back out for the test. This extra new state, called *s_extra*, follows immediately after state *s_input*, and has no control word assigned to it. Figure 7.21(a) shows this modified state diagram. At the beginning of this *s_extra* state, the input value will have been stored into register A, so when the condition $(A = 5)$ is performed, it will get the correct value for A.

The timing diagram for this new state diagram is shown in Figure 7.21(b). The same input value of 5 is loaded into the register at time 200 ns at the beginning of clock cycle 2. However, the reading of the register for the conditional test does not occur until shortly after time 200 ns. Hence, the test result is equal, and the FSM goes to state *s_equal* and outputs the correct value of 8.

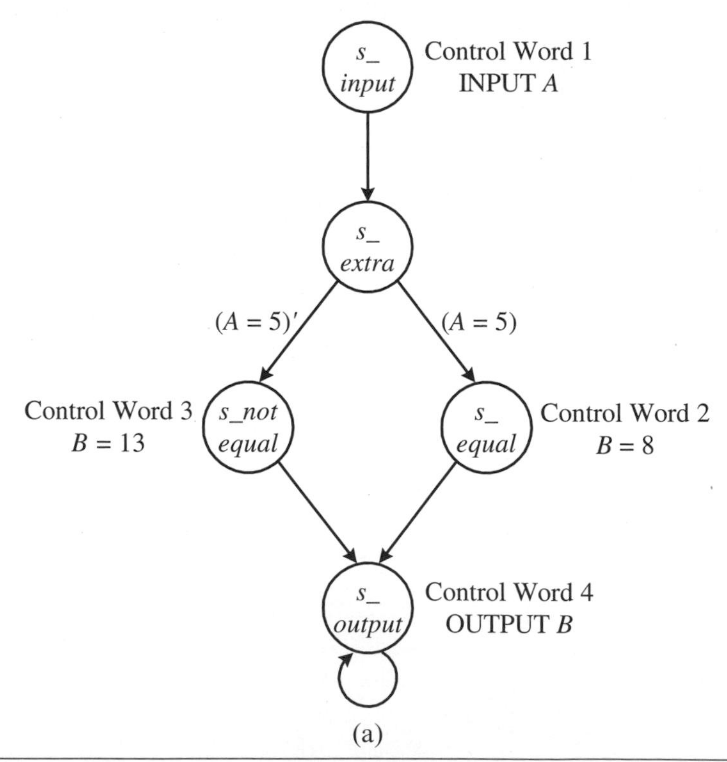

FIGURE 7.21 Example 7.8: (a) correct state diagram; (b) correct timing diagram.
(continued on next page)

(b)

FIGURE 7.21 Example 7.8: (a) correct state diagram; (b) correct timing diagram.

Adding this one extra state is not the only solution to this timing problem. Another way to solve the problem is to connect the comparator for $(A = 5)$ directly to the input signal rather than to the output of the A register, as shown in Figure 7.22. This way, we won't have to wait one extra clock cycle for the input value to be latched into the register before we can test it. We will get the same functional result for both cases, however, in the second case, we will actually be performing the conditional test $(Input = 5)$ instead of $(A = 5)$ as specified in the algorithm. In other words, the test is performed directly with the input value rather than with the value from the A register.

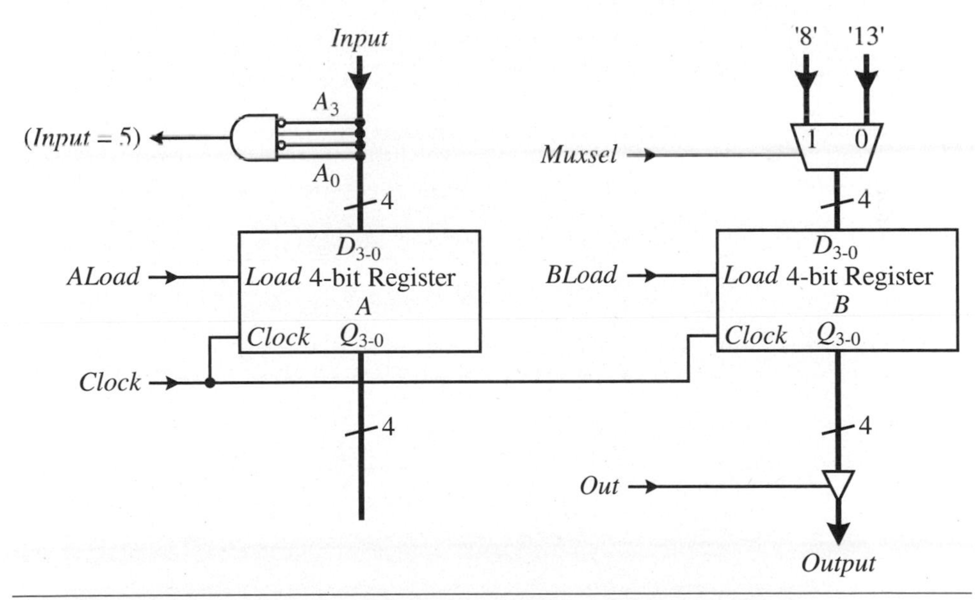

FIGURE 7.22 Alternative datapath for the simple IF-THEN-ELSE problem.

7.3.3 **Timing Issues**

From the last section, we saw that to get a correct state diagram for the microprocessor control unit, we need to be careful with the timing of the register-transfer operations in the datapath. There are two important points that we need to understand regarding the timing of register-transfer operations:

1. Read before write. If we first perform a read operation and then follow it by a write operation on the same register in the same clock cycle, then it is all right. An example of this is executing the statement $i = i + 1$ in a state.
2. Write before read. If we first perform a write operation and then follow it by a read operation on the same register in the same clock cycle, then we will be reading the value before the write. An example of this is updating a register and then testing the value in the same register.

To better understand these two register-transfer timing issues, we will illustrate with the following example to derive the state diagram for the summation problem to generate and sum the numbers from 1 to 10.

EXAMPLE 7.9

State diagram for the summation problem

Figure 7.23(a) shows the algorithm for the summation problem to generate and sum the numbers from 1 to 10. The state diagram for this algorithm is shown in Figure 7.23(b). The states in the state diagram are annotated with the instruction that is executed in it.

```
1    sum = 0
2    i = 1
3    DO {
4         sum = sum + i
5         i = i + 1
6    } WHILE (i ≠ 11)
7    OUTPUT sum
```

(a)

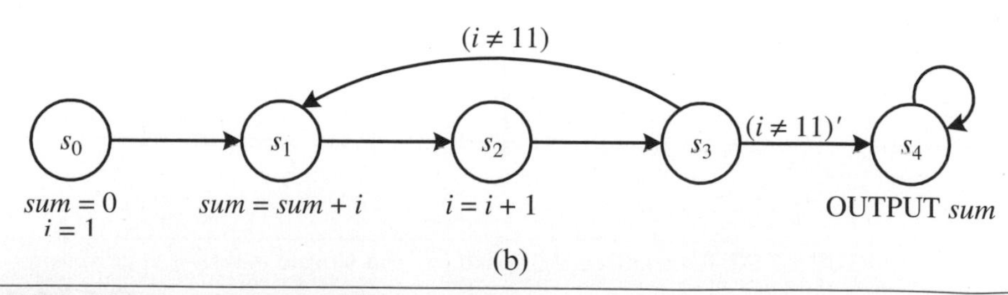

(b)

FIGURE 7.23 Summation problem to generate and sum the numbers from 1 to 10: (a) algorithm; (b) state diagram.

The two initialization statements, 1 and 2, are independent of each other, therefore we can assign both of them to be executed in the same initialization state s_0. State s_1 performs statement 4 to sum i, while state s_2 performs statement 5 to increment i. An extra state s_3 is added that does not perform any operation. This extra state is needed because of timing reasons for testing the condition $(i \neq 11)$ as explained below. Depending on the result of the conditional test, the FSM either loops back to state s_1 or goes to state s_4 to output the sum. The FSM halts in state s_4 by looping back to itself unconditionally.

For the first read-before-write timing issue, let us look at the execution of the instruction $i = i + 1$ in state s_2. The instruction $i = i + 1$ requires both a read and a write of the same register i. The read is for the i that is on the right side of the equal sign, and the write is for the same i that is on the left side of the equal sign. To execute the instruction $i = i + 1$ in the datapath, the *iLoad* signal for register i is asserted in state s_2.

The FSM enters state s_2 at the active (rising) edge of the clock in the current clock cycle as shown in the timing diagram in Figure 7.24. The execution of the instruction $i = i + 1$ begins with first reading from register i and then performing the addition $i + 1$. The current value of i is available for reading from the Q output of the register at the beginning of the current clock cycle for state s_2 because the value of the register is always available at its output. The *iLoad* signal also is asserted shortly after the rising clock edge when the FSM enters state s_2. However, the actual writing of the register does not occur until the next rising clock edge at the start of the next clock cycle in state s_3 because values are written into registers only at the active (rising) clock edge, and the current rising clock edge already has passed. Therefore, even though the *iLoad* signal is asserted in the current clock cycle, it has to wait until the next rising clock edge before a value gets written in. Meanwhile, the adder performs the addition using the current value of i, and the result from the addition will be available shortly before the beginning of the next rising clock edge. Therefore, it is the result of the addition that is written back into register i at the beginning of the next clock cycle when the FSM enters state s_3.

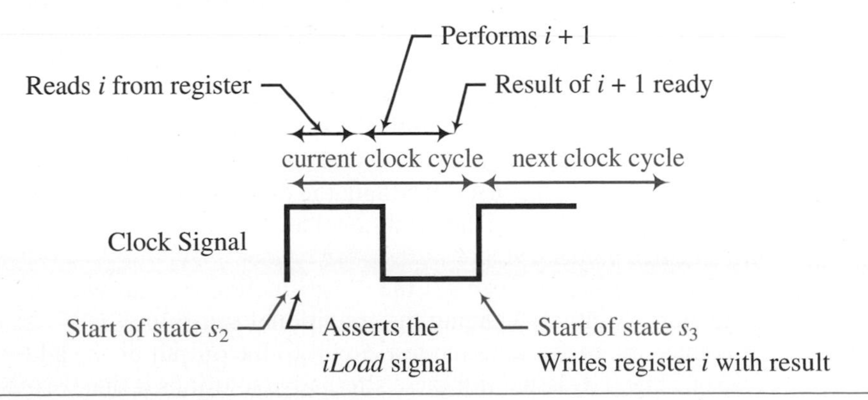

FIGURE 7.24 Read-before-write timings.

A question to ask is what happens if the adder takes longer than a clock period to output the valid result? In other words, what happens if, at the next rising clock edge at the beginning of state s_3, the adder is not yet finished with its calculations and still outputting invalid results? The answer is that the datapath will write the wrong result from the adder into the register. The solution to this problem is to either make a faster adder that will finish in time, or make the clock period longer by slowing down the clock.

From the above analysis, we see that performing a read before a write to the same register in the same clock cycle does not create any data conflict because the reading occurs at the beginning of the current clock cycle. The value that is available at the output of the register in the current clock cycle is still the value before the write back, which is the value before the addition of $i + 1$. Although the *iLoad* signal is asserted in the current clock cycle, the actual writing, however, does not occur until the beginning of the next clock cycle, which happens after the reading and the addition operation has completed.

For the second write-before-read timing issue, let us look at the execution of the same instruction $i = i + 1$ in state s_2. According to the algorithm in Figure 7.23(a), we want to test the condition `(i ≠ 11)` immediately after executing the instruction $i = i + 1$. In other words, after writing the result of the addition into register i, we want to read the new value of i for the conditional test. Because the conditional test is not a datapath operation, it does not require a state in the state diagram where the test is performed, but rather just a label on a conditional edge. Therefore, we incorrectly might draw the following state diagram with the two conditional edges going out from state s_2.

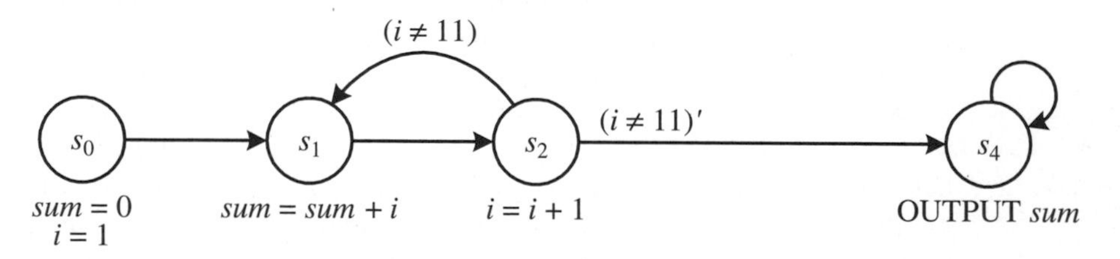

The problem with this is that the conditional test requires the reading of the register i in the same clock cycle that the instruction $i = i + 1$ is performed. Further, from the natural sequential execution of the algorithm, we expect the conditional test to be performed is for the incremented value of i, that is, after the writing back of the result from the addition. However, from the previous timing issue discussion, we know that i is not going to be updated until the next rising clock edge in the next clock cycle, so reading the register in the current clock cycle will get the old value of i, and not the incremented value of i. By adding the extra state s_3 in the state diagram shown in Figure 7.23(b) and doing the conditional test in that state, we will be testing the correct incremented value of i.

Note that we also can get the same functional result without having to add the extra state by either changing the conditional test from `(i ≠ 11)` to `(i ≠ 10)`, or connecting the conditional test circuit to the output of the adder instead of to the register. The only issue with these alternative solutions is that we are not following the original algorithm exactly.

7.3.4 **Deriving the FSM Circuit**

Constructing the FSM circuit here for the microprocessor follows exactly the procedure discussed in Chapter 6 for synthesizing FSMs. After we have derived the correct state diagram, the next step is to convert the state diagram into the next-state table, and then from the next-state table we can construct the next-state logic circuit for the FSM. The total number of states in the state diagram will determine the number of D flip-flops needed for the state memory to give a unique encoding for each of the states. The output logic circuit for the FSM is derived from the control signals in the control words and the states to which the control words are assigned. Finally, combining the next-state logic circuit, the state memory, and the output logic circuit together produces the complete FSM control unit circuit.

We will now show the complete process on how to derive an FSM circuit with two examples. Example 7.10 shows the construction of the FSM circuit for the simple IF-THEN-ELSE problem. Example 7.11 shows the construction of the FSM circuit for the summation problem to generate and sum the numbers from 1 to 10.

EXAMPLE 7.10

Construction of the FSM for the simple IF-THEN-ELSE problem

In this example, we start with the given state diagram shown in Figure 7.25(a). The next-state table shown in Figure 7.25(b) is obtained directly from the state diagram. Since there is a total of five states, three flip-flops are needed to encode them using the straight binary encoding scheme. State *s_input* is encoded as $Q_2Q_1Q_0 = 000$, state *s_extra* is encoded as $Q_2Q_1Q_0 = 001$, and so on. The three remaining encodings (101, 110, and 111) are not used. In normal circumstances, the control unit should never get to one of the unused states. However, because of noise or glitches in the circuit, the FSM might end up in one of these unused states, so it is a good idea to set the next state for all of the unused states to the reset state. In our next-state table, we have added three more rows for these three unused states. Their next state for all input conditions is the reset state, 000. The next-state table is a direct translation from the state diagram but written in a table format. From state 000 (*s_input*) we go to state 001 (*s_extra*) unconditionally. From state 001 (*s_extra*) we go to either 010 (*s_notequal*) or 011 (*s_equal*) depending on the outcome of the condition $(A = 5)$.

The next-state equations shown in Figure 7.25(c) are derived from the next-state table. There is one equation for each of the three flip-flops, and because we are using D flip-flops, $Q_{next} = D$. In the next-state table, the values for the three flip-flops are grouped together. When deriving the next-state equations, we need to look at them separately as three truth tables—one for Q_{2next}, one for Q_{1next}, and one for Q_{0next}. These equations are dependent on the four variables Q_2, Q_1, Q_0, and $(A = 5)$.

The control words also serve as the output table. The output logic circuit for the FSM is derived from the control signals in the control words and the states to which the control words are assigned. Each control signal has one output equation, and these equations are dependent only on the states Q_2, Q_1, and Q_0 of the FSM. We derive the truth tables for these output equations by taking the control word table and replacing all of the control word numbers with the actual encoding of the state to which that control word is assigned. Having the output table, the output equations can be derived easily, as shown in Figures 7.25(d) and (e), respectively.

After deriving the next-state and output equations, we can draw the control unit circuit as shown in Figure 7.25(f). The state memory simply consists of the three D flip-flops. Both the next-state logic circuit and the output logic circuit are combinational circuits, and are constructed from the next-state equations and output equations, respectively. Combining the next-state logic circuit, the state memory, and the output logic circuit together produces the final FSM control unit circuit.

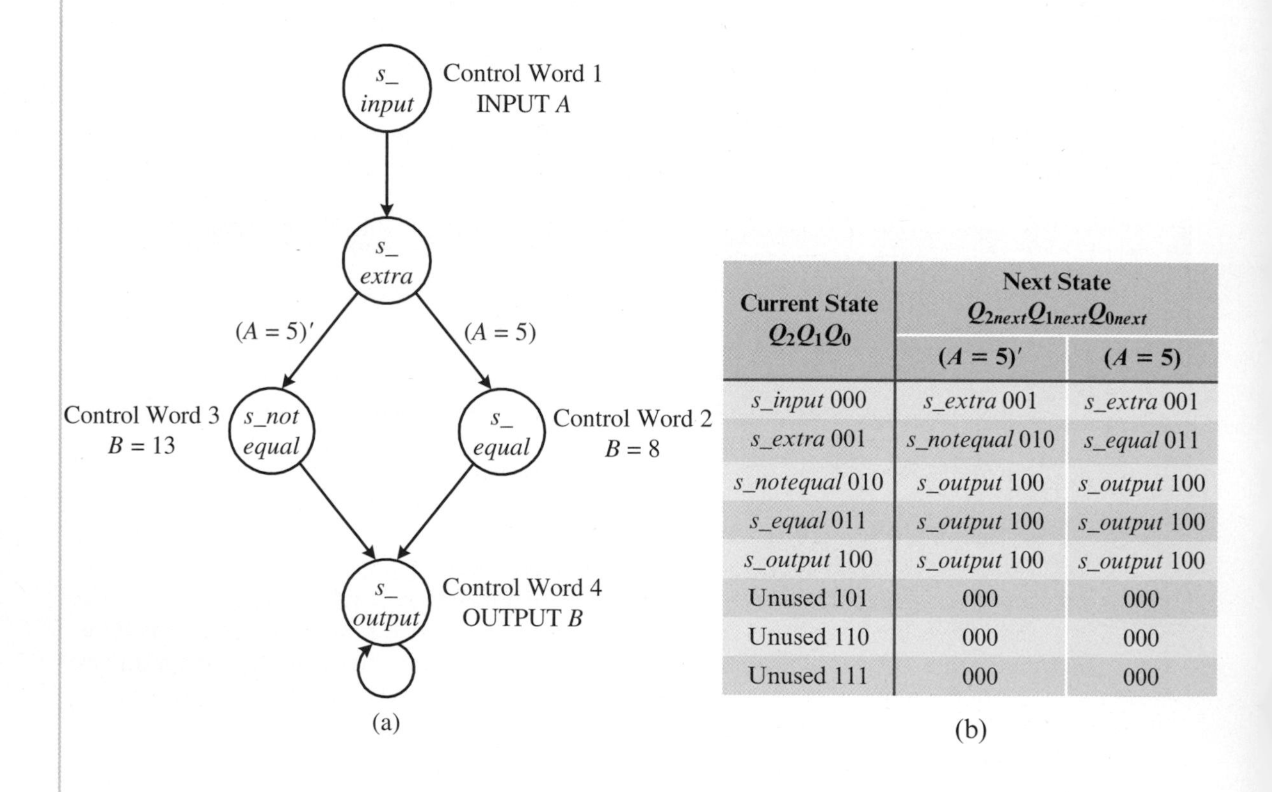

Current State $Q_2Q_1Q_0$	Next State $Q_{2next}Q_{1next}Q_{0next}$	
	$(A = 5)'$	$(A = 5)$
s_input 000	*s_extra* 001	*s_extra* 001
s_extra 001	*s_notequal* 010	*s_equal* 011
s_notequal 010	*s_output* 100	*s_output* 100
s_equal 011	*s_output* 100	*s_output* 100
s_output 100	*s_output* 100	*s_output* 100
Unused 101	000	000
Unused 110	000	000
Unused 111	000	000

(b)

$$Q_{2next} = D_2 = Q_2'Q_1 + Q_2Q_1'Q_0'$$
$$Q_{1next} = D_1 = Q_2'Q_1'Q_0$$
$$Q_{0next} = D_0 = Q_2'Q_1'Q_0' + Q_2'Q_1'(A = 5)$$

(c)

FIGURE 7.25 Construction of the FSM for the simple IF-THEN-ELSE problem: (a) state diagram; (b) next-state table; (c) next-state equations; (d) output table; (e) output equations for the four control signals; (f) circuit. *(continued on next page)*

$Q_2Q_1Q_0$	Instruction	*ALoad*	*Muxsel*	*BLoad*	*Out*
000	INPUT *A*	1	×	0	0
001	No operation	0	×	0	0
010	$B = 8$	0	1	1	0
011	$B = 13$	0	0	1	0
100	OUTPUT *B*	0	×	0	1
101	No operation	0	×	0	0
110	No operation	0	×	0	0
111	No operation	0	×	0	0

(d)

$$ALoad = Q_2'Q_1'Q_0'$$
$$Muxsel = Q_2'Q_1Q_0'$$
$$BLoad = Q_2'Q_1$$
$$Out = Q_2Q_1'Q_0'$$

(e)

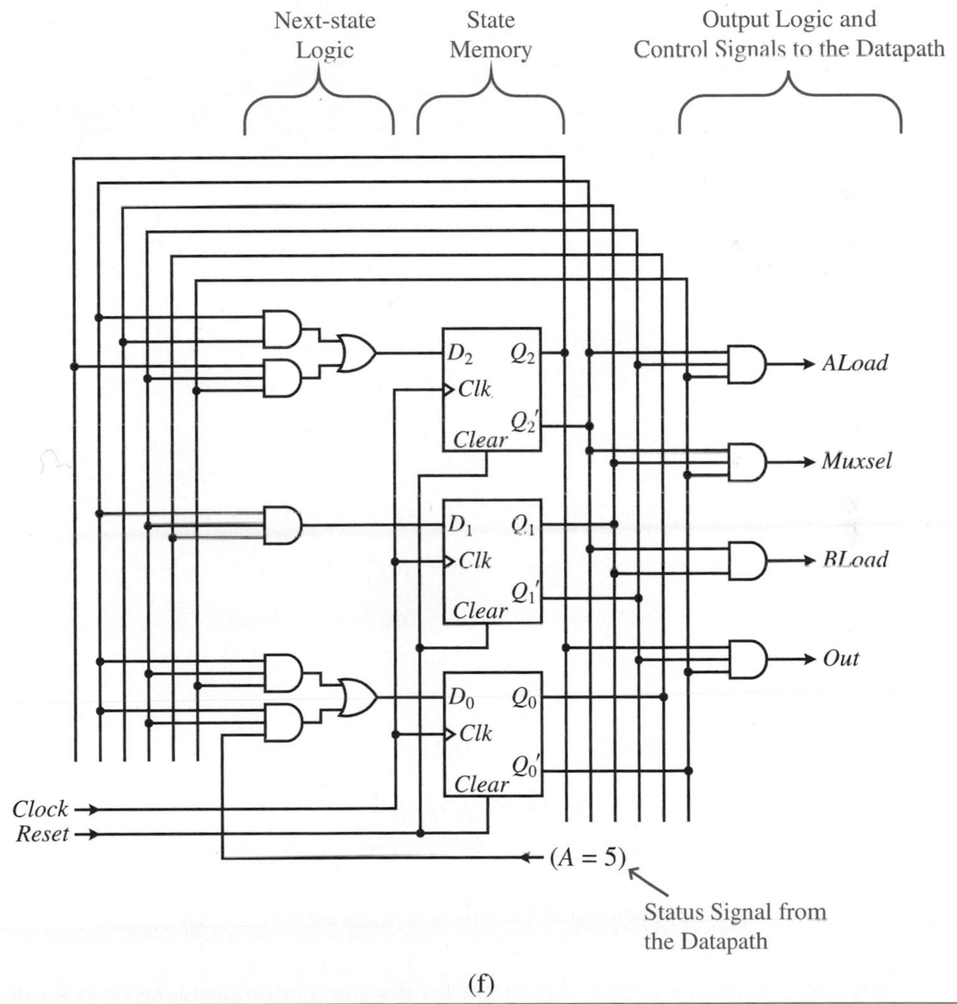

FIGURE 7.25 **Construction of the FSM for the simple IF-THEN-ELSE problem: (a) state diagram; (b) next-state table; (c) next-state equations; (d) output table; (e) output equations for the four control signals; (f) circuit.**

EXAMPLE 7.11

Construction of the FSM for the summation problem

In this example, we will construct the FSM for the summation problem to generate and sum the numbers from 1 to 10. We already have derived the state diagram for this problem in Example 7.9. Starting with this state diagram shown again here in Figure 7.26(a), we obtain the next-state table and next-state equations shown in Figures 7.26(b) and (c), respectively. The output table, obtained from the control words shown in Figure 7.18, and the output equations are shown in Figures 7.26(d) and (e), respectively. Finally, the complete FSM circuit is shown in Figure 7.26(f).

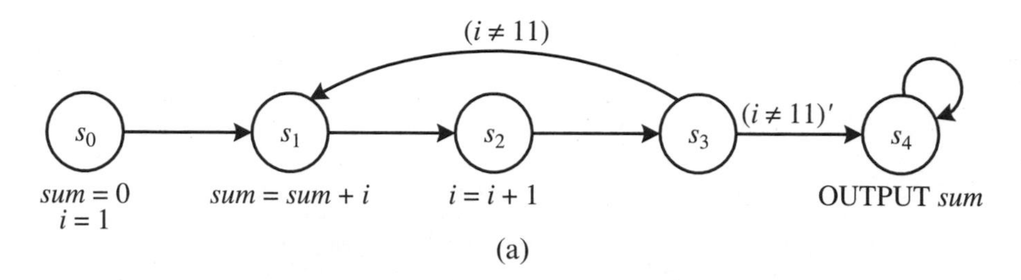

(a)

Current State $Q_2Q_1Q_0$	Next State $Q_{2next}Q_{1next}Q_{0next}$	
	$(i \neq 11)'$	$(i \neq 11)$
s_0 000	s_1 001	s_1 001
s_1 001	s_2 010	s_2 010
s_2 010	s_3 011	s_3 011
s_3 011	s_4 100	s_1 001
s_4 100	s_4 100	s_4 100
Unused 101	000	000
Unused 110	000	000
Unused 111	000	000

(b)

$$Q_{2next} = D_2 = Q_2'Q_1Q_0(i \neq 11)' + Q_2Q_1'Q_0'$$
$$Q_{1next} = D_1 = Q_2'Q_1'Q_0 + Q_2'Q_1Q_0'$$
$$Q_{0next} = D_0 = Q_2'Q_0' + Q_2'Q_1Q_0(i \neq 11)$$

(c)

FIGURE 7.26 Construction of the FSM for the summation problem: (a) state diagram; (b) next-state table; (c) next-state equations; (d) output table; (e) output equations; (f) circuit. *(continued on next page)*

Current State $Q_2Q_1Q_0$	Instruction	*iMux*	*sumMux*	*sumLoad*	*iLoad*	*addMux*	*OE*
s_0 000	$sum = 0; i = 1$	1	1	1	1	×	0
s_1 001	$sum = sum + i$	×	0	1	0	0	0
s_2 010	$i = i + 1$	0	×	0	1	1	0
s_4 100	OUTPUT *sum*	×	×	0	0	×	1

(d)

$$iMux = Q_2'Q_1'Q_0'$$
$$sumMux = Q_2'Q_1'Q_0'$$
$$sumLoad = Q_2'Q_1'$$
$$iLoad = Q_2'Q_0'$$
$$addMux = Q_2'Q_1Q_0'$$
$$OE = Q_2Q_1'Q_0'$$

(e)

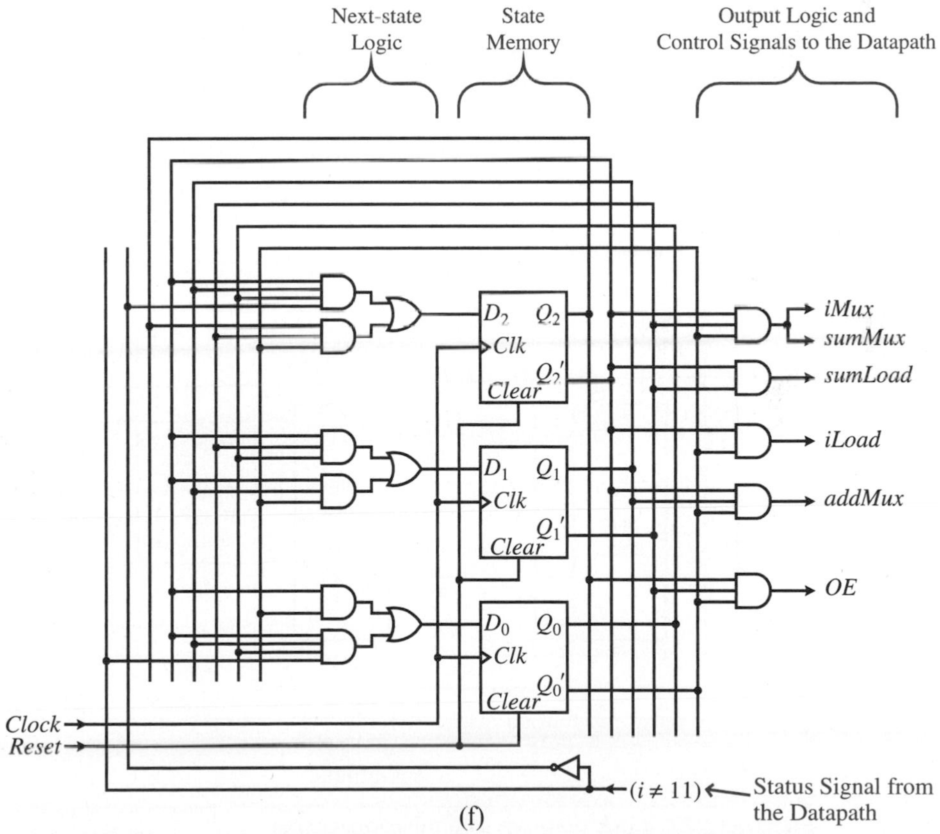

(f)

FIGURE 7.26 Construction of the FSM for the summation problem: (a) state diagram; (b) next-state table; (c) next-state equations; (d) output table; (e) output equations; (f) circuit.

7.4 Constructing the Complete Microprocessor

Having constructed both the datapath and the control unit circuits, we now can connect these two components together to produce the complete dedicated microprocessor. Figure 7.27 shows how these two components are connected together using the corresponding control signals and status signals. Recall that the control signals are generated by the control unit to control the operations of the datapath, while the status signals are generated by the datapath to inform the next-state logic in the control unit as to what the next state should be in the execution of the algorithm. The remaining four sets of input/output signals (control inputs, control outputs, data inputs, and data outputs) that interface with external devices are hardware dependant. An example of a control input is the reset signal to reset the microprocessor. A halt signal to notify the external world that the microprocessor has halted execution is an example of a control output. For testing purposes, we will simply connect them to LEDs and switches.

This method of manually constructing a dedicated microprocessor is referred to as the FSM+D (FSM *plus* datapath) model because the control unit and the datapath are constructed separately, and then they are connected together using the control and status signals. Example 7.12 shows this manual construction of a dedicated microprocessor for the simple IF-THEN-ELSE problem. Example 7.13 shows this manual construction of a dedicated microprocessor for the problem to generate and sum the numbers from 1 to 10.

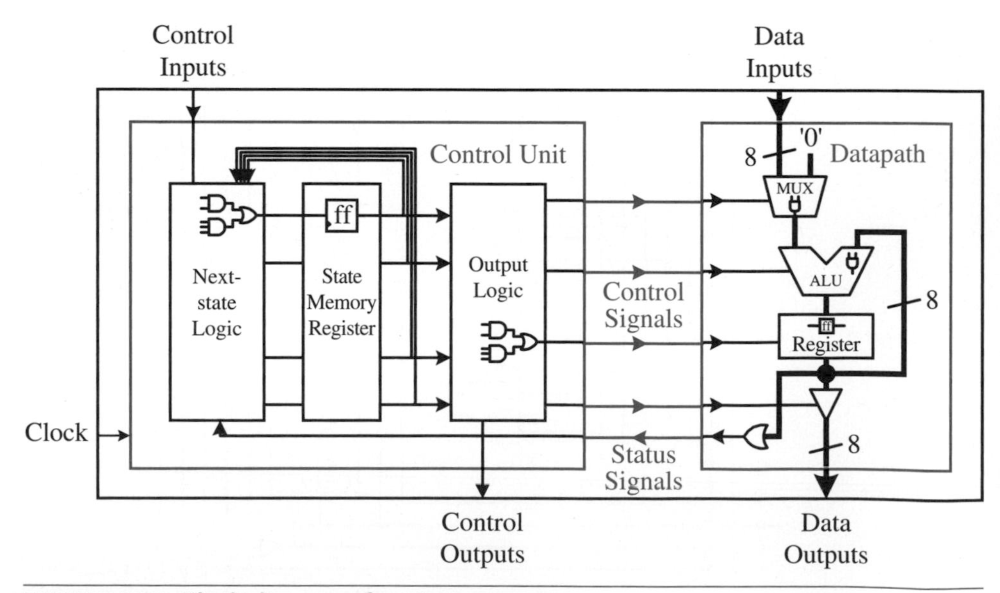

FIGURE 7.27 Block diagram of a microprocessor.

EXAMPLE 7.12

Construction of the complete dedicated microprocessor for the simple IF-THEN-ELSE problem

In the last two sections, we constructed the datapath and the control unit for the simple IF-THEN-ELSE problem as separate circuits. We will now connect them together to form the dedicated microprocessor for executing the given algorithm. The control unit circuit constructed for this microprocessor from Example 7.10 and shown in Figure 7.25(f) is represented by the symbol shown in Figure 7.28(a). The datapath circuit constructed for this microprocessor from Example 7.2 and shown in Figure 7.13(b) is represented by the symbol shown in Figure 7.28(b).

Connecting the datapath and the control unit together forms the complete dedicated microprocessor circuit shown in Figure 7.28(c). The four control signals and the one status signal between the two components are connected together. The data inputs are connected to external switches; the data outputs are connected to external LEDs; the reset signal is connected to an external push button; and the clock signal is connected to an

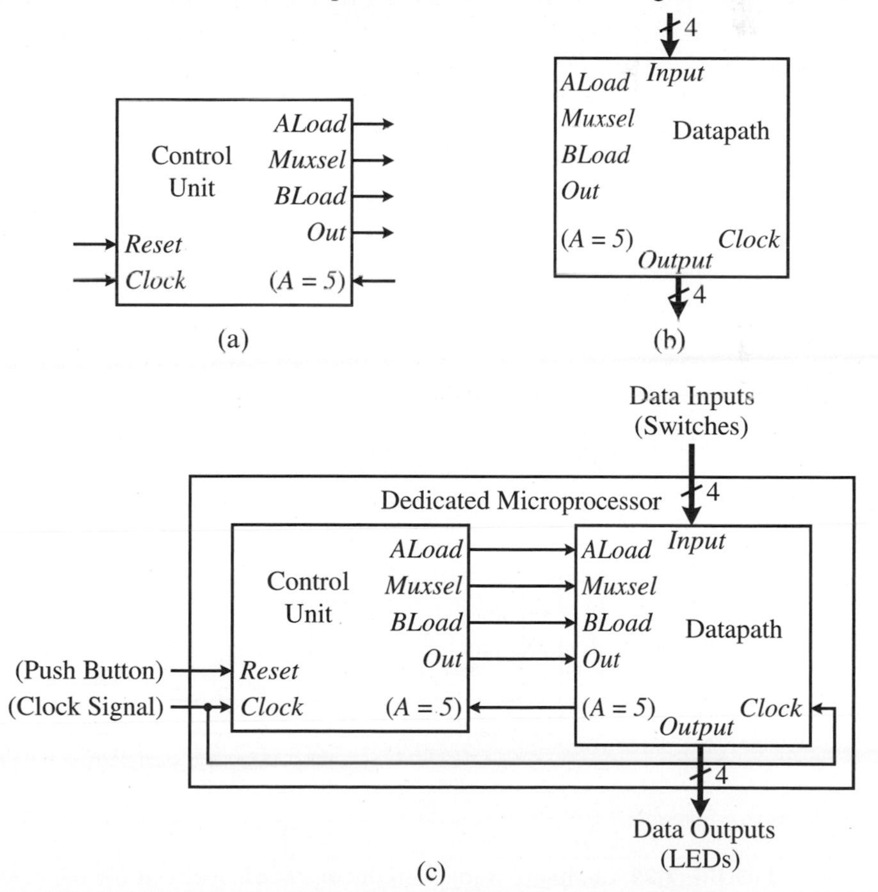

FIGURE 7.28 Complete dedicated microprocessor circuit for the simple IF-THEN-ELSE problem: (a) symbol for the control unit circuit; (b) symbol for the datapath circuit; (c) complete dedicated microprocessor.

external clock source. This dedicated microprocessor will operate according to the simple IF-THEN-ELSE algorithm shown in Figure 7.13(a). The LEDs will show either the number 8 or 13 in binary depending on whether the input from the switches is equal to 5 in binary.

EXAMPLE 7.13

Construction of the complete dedicated microprocessor for the summation problem

The control unit and the datapath for the summation problem to generate and sum the numbers from 1 to 10 already have been derived in Figures 7.26(f) and 7.14(b), respectively. The two symbols representing these two circuits are shown in Figures 7.29(a) and (b), respectively. Connecting the datapath and the control unit together through their corresponding control and status signals forms the complete dedicated microprocessor circuit shown in Figure 7.29(c).

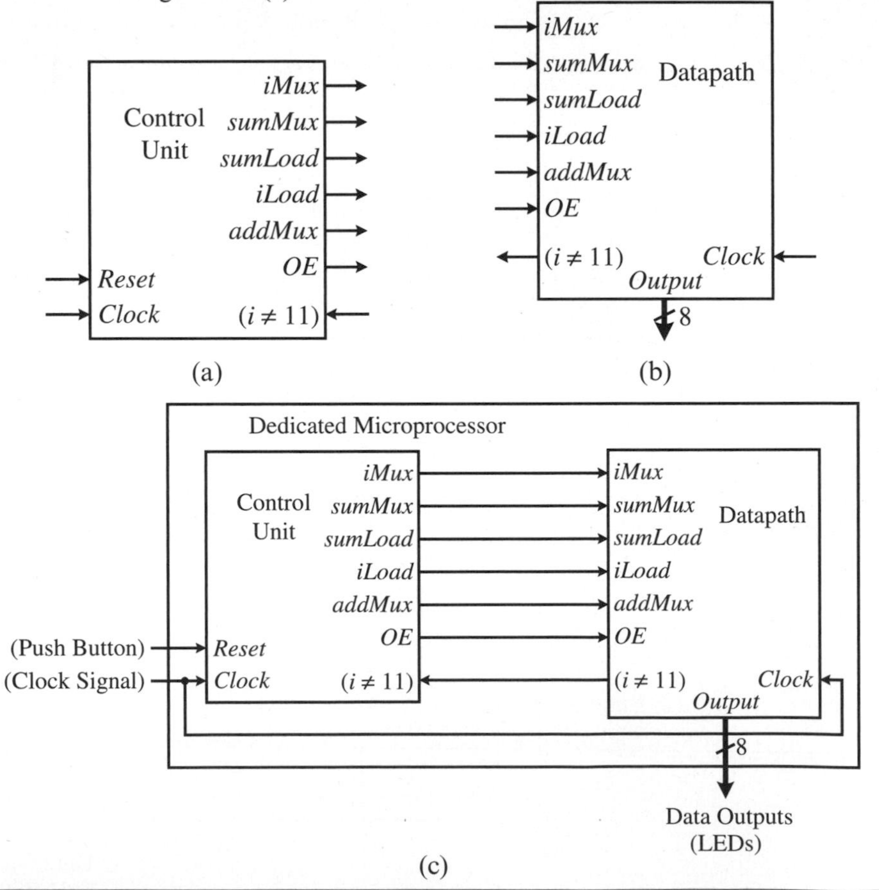

FIGURE 7.29 Complete dedicated microprocessor circuit for the summation problem: (a) symbol for the control unit circuit; (b) symbol for the datapath circuit; (c) complete dedicated microprocessor.

7.5 Dedicated Microprocessor Construction Examples

We now will illustrate the manual design of dedicated microprocessors using the FSM+D method with several examples. The microprocessor circuits produced in these examples are by no means the only correct circuits for solving each of the problems. Just like writing a computer program, there are many ways of doing it. In Section 7.5.1, we show an example of designing a dedicated microprocessor for finding the greatest common divisor (GCD) of two numbers. In Section 7.5.2, we create a dedicated microprocessor for playing the high-low number guessing game. Finally, in Section 7.5.3, we will design a dedicated traffic light controller.

7.5.1 Greatest Common Divisor

Example 7.14 shows the manual construction of a dedicated microprocessor for finding the GCD of two numbers.

EXAMPLE 7.14

Designing a dedicated microprocessor to evaluate the GCD

In this example, we will design a complete dedicated microprocessor to evaluate the GCD of two 8-bit positive numbers *X* and *Y*, using the Euclidean algorithm. The GCD of two positive integers is defined as the largest integer that divides both of them without leaving a remainder. For example, the GCD of 8 and 12 is 4, and the GCD of 3 and 5 is 1. The Euclidean algorithm for solving the GCD problem is shown in Figure 7.30. We first will design a dedicated datapath for the algorithm. Next, we will design the control unit for the datapath. Finally, we will combine these two components together to produce our complete dedicated microprocessor.

The algorithm shown in Figure 7.30 has five data manipulation statements in lines 1, 2, 5, 7, and 10. There are two conditional tests in lines 3 and 4. The dedicated datapath derived for this algorithm is shown in Figure 7.31. Two 8-bit registers are needed to store the two variables *X* and *Y*. One subtractor is used to perform the two subtractions in lines 5 and 7. Two 2-to-1 multiplexers are needed for the input to each of the

```
1   INPUT X
2   INPUT Y
3   WHILE (X ≠ Y){
4      IF (X > Y) THEN
5         X = X - Y
6      ELSE
7         Y = Y - X
8      END IF
9      }
10  OUTPUT X
```

FIGURE 7.30 Euclidean algorithm for solving the GCD problem.

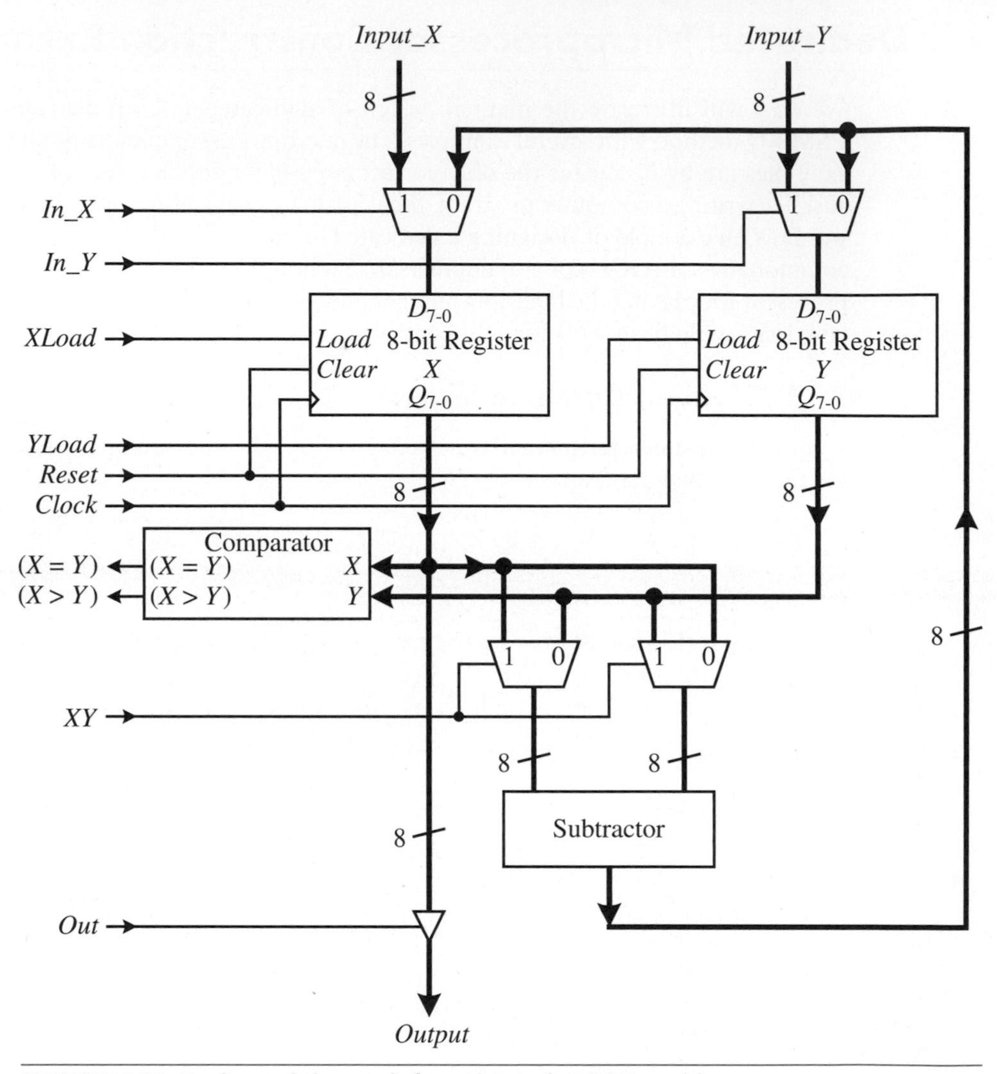

FIGURE 7.31 Dedicated datapath for solving the GCD problem.

two registers, because we initially need to load each register with an input number and subsequently with the result from the subtractor. The two multiplexer select control signals *In_X* and *In_Y* select which of the two sources is to be loaded into the registers *X* and *Y*, respectively. The two register load control signals *XLoad* and *YLoad* load a value into the respective register.

The bottom two 2-to-1 multiplexers, selected by the same *XY* signal, determine the source to the two operands for the subtractor. When *XY* is asserted, the value from register *X* will go to the left operand of the subtractor, and the value from register *Y* will go to the right operand. On the other hand, when *XY* is de-asserted, *Y* will go to the left operand, and *X* will go to the right operand. Thus, this allows the selection of one of the two subtraction operations, $X - Y$ or $Y - X$, to be performed. The *Out* control signal is used to enable the tri-state buffer for outputting the result from register X.

One comparator is used to generate the two conditional status signals, equal to and greater than. The inputs to the comparator are directly from the two X and Y registers. There are two output signals $(X = Y)$ and $(X > Y)$ from the comparator. $(X = Y)$ is asserted if X is equal to Y, and $(X > Y)$ is asserted if X is greater than Y.

This dedicated datapath for solving the GCD problem requires six control signals, *In_X*, *In_Y*, *XLoad*, *YLoad*, *XY*, and *Out*, and generates two status signals, $(X = Y)$ and $(X > Y)$. The control words for this datapath are shown in Figure 7.32(d).

The state diagram for the GCD algorithm requires five states, as shown in Figure 7.32(a). The two input statements are independent of each other, and therefore, can be executed at the same time in the same state (000). An extra "no-operation" state is needed for the correct conditional testing of the updated values of X and Y. If we were to read the result from the conditional test in state 000 (i.e., with the two outgoing conditional edges in 000), we would be reading the old values of X and Y, and not the new updated values because the new values are not written into the registers until the next rising clock edge, which also is when the FSM goes to state 001.

In state 001, we test for the two conditions, $(X = Y)$ and $(X > Y)$. If $(X = Y)$ is true, then the next state is 100, and the FSM stops in this state by looping endlessly back to itself. If $(X = Y)$ is false, then the next state is either 010 or 011, depending on whether the condition $(X > Y)$ is true or false, respectively. In states 010 and 011, the respective subtraction is performed, and then unconditionally they both go back to state 001 to test the conditions again.

This state diagram does not have a *Start* signal, so in order for the resulting microprocessor to read the inputs correctly, we first must set up the input numbers and then assert the *Reset* signal to clear the state memory flip-flops to 0. This way, when the FSM starts executing from state 000, the two input numbers are ready to be read in.

The next-state table, as derived from the state diagram, is shown in Figure 7.32(b). The table requires five variables: three to encode the five states, Q_2, Q_1, and Q_0, and two for the status signals, $(X = Y)$ and $(X > Y)$. There are three unused state encodings, 101, 110, and 111, which will unconditionally go back to state 000.

The K-maps for the next-state equations and the next-state equations for Q_{2next}, Q_{1next}, and Q_{0next} are shown in Figure 7.32(c). For the next-state equations, $Q_{next} = D$ because we are using D flip-flops to implement the state memory.

The control words and output table, having the six control signals, are shown in Figure 7.32(d). State 000 performs both inputs of X and Y. The two multiplexer select lines *In_X* and *In_Y* must be asserted so that the data comes from the two primary inputs. The two numbers are loaded into the two corresponding registers by asserting the *XLoad* and *YLoad* lines. State 001 is for testing the two conditions, so no operations are performed. The no-op is accomplished by not loading the two registers and not outputting a value. For states 010 and 011, the *XY* multiplexer select line is used to select which of the two subtraction operations is to be performed. Asserting *XY* performs the operation $X - Y$; whereas, de-asserting *XY* performs the operation $Y - X$. The corresponding *In_X* or *In_Y* line is de-asserted to route the result from the subtractor back to the input of the register. The corresponding *XLoad* or *YLoad* line is asserted to store the result of the subtraction into the correct register. State 100 outputs the result from X by asserting the *Out* line.

The output equations, as derived from the output table, are shown in Figure 7.32(e). There is one equation for each of the six control signals, and each one is dependent only on the current state (i.e., the current values in Q_2, Q_1, and Q_0). We have assumed that the control signals have don't-care values in all of the unused states.

The complete control unit circuit is shown in Figure 7.32(f). The state memory consists of three D flip-flops. The inputs to the flip-flops are the next-state circuits derived from the three next-state equations. The output circuits for the six control signals are derived from the six output equations. The two status signals $(X = Y)$ and $(X > Y)$ come from the comparator in the datapath.

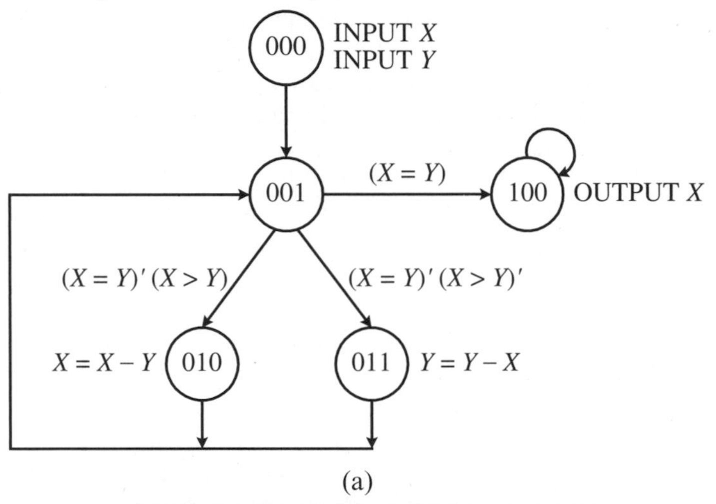

(a)

Current State $Q_2Q_1Q_0$	Next State (Implementation) $Q_{2next}Q_{1next}Q_{0next}$ $(D_2D_1D_0)$			
	$(X = Y), (X > Y)$			
	00	**01**	**10**	**11**
000	001	001	001	001
001	011	010	100	100
010	001	001	001	001
011	001	001	001	001
100	100	100	100	100
101 Unused	000	000	000	000
110 Unused	000	000	000	000
111 Unused	000	000	000	000

(b)

FIGURE 7.32 **Control unit for solving the GCD problem: (a) state diagram; (b) next-state (implementation) table; (c) K-maps and next-state equations; (d) control words and output table; (e) output equations; (f) circuit.** *(continued on next page)*

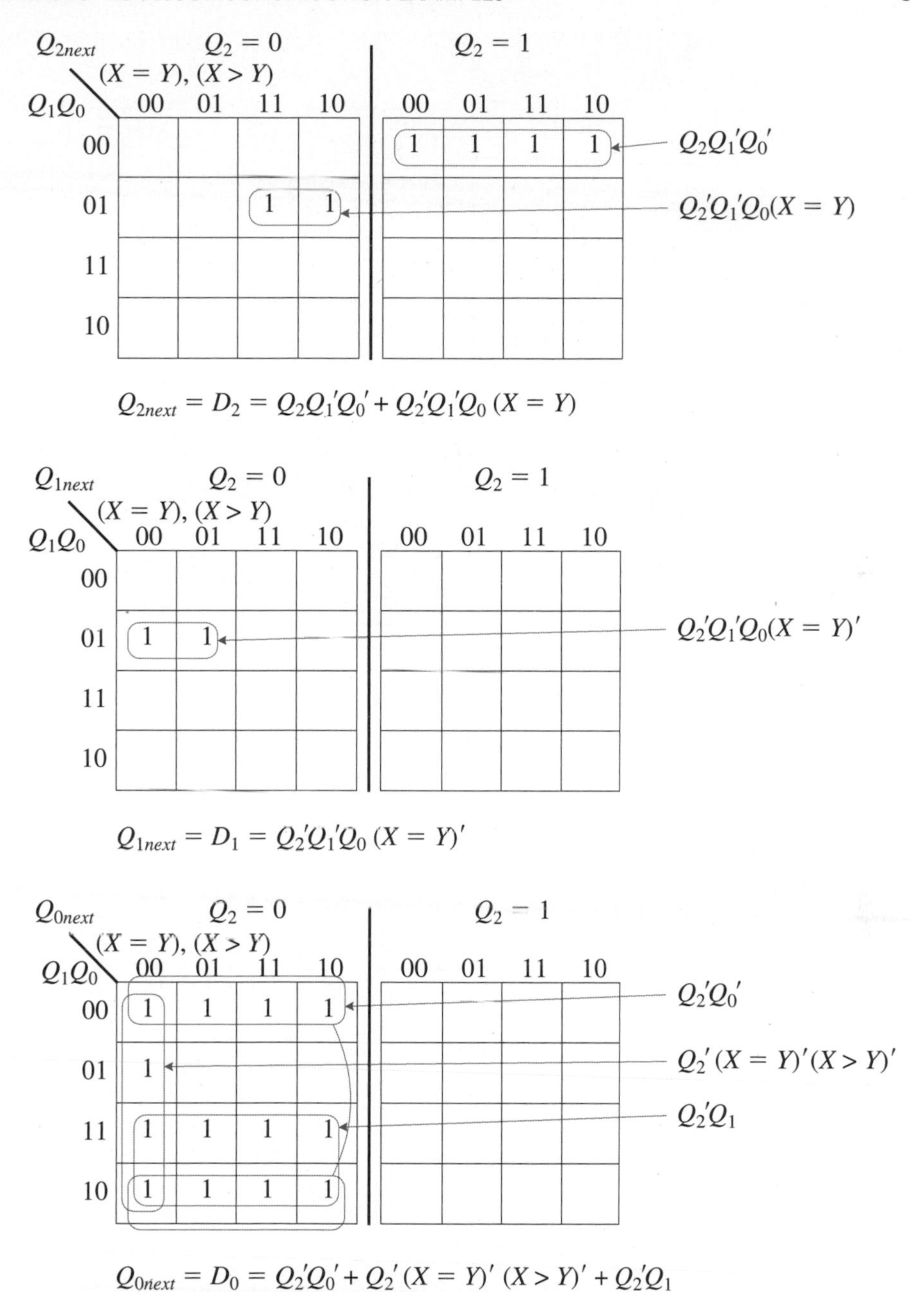

FIGURE 7.32 Control unit for solving the GCD problem: (a) state diagram; (b) next-state (implementation) table; (c) K-maps and next-state equations; (d) control words and output table; (e) output equations; (f) circuit. *(continued on next page)*

Control Word	State $Q_2Q_1Q_0$	Instruction	*In_X*	*In_Y*	*XLoad*	*YLoad*	*XY*	*Out*
0	000	INPUT *X*, INPUT *Y*	1	1	1	1	×	0
1	001	No operation	×	×	0	0	×	0
2	010	$X = X - Y$	0	×	1	0	1	0
3	011	$Y = Y - X$	×	0	0	1	0	0
4	100	OUTPUT *X*	×	×	0	0	×	1

(d)

$In\_X = Q_1'$ $\quad XLoad = Q_2'Q_0'$

$In\_Y = Q_0'$ $\quad YLoad = Q_2'Q_1'Q_0' + Q_2'Q_1Q_0$

$XY = Q_0'$ $\quad Out = Q_2$

(e)

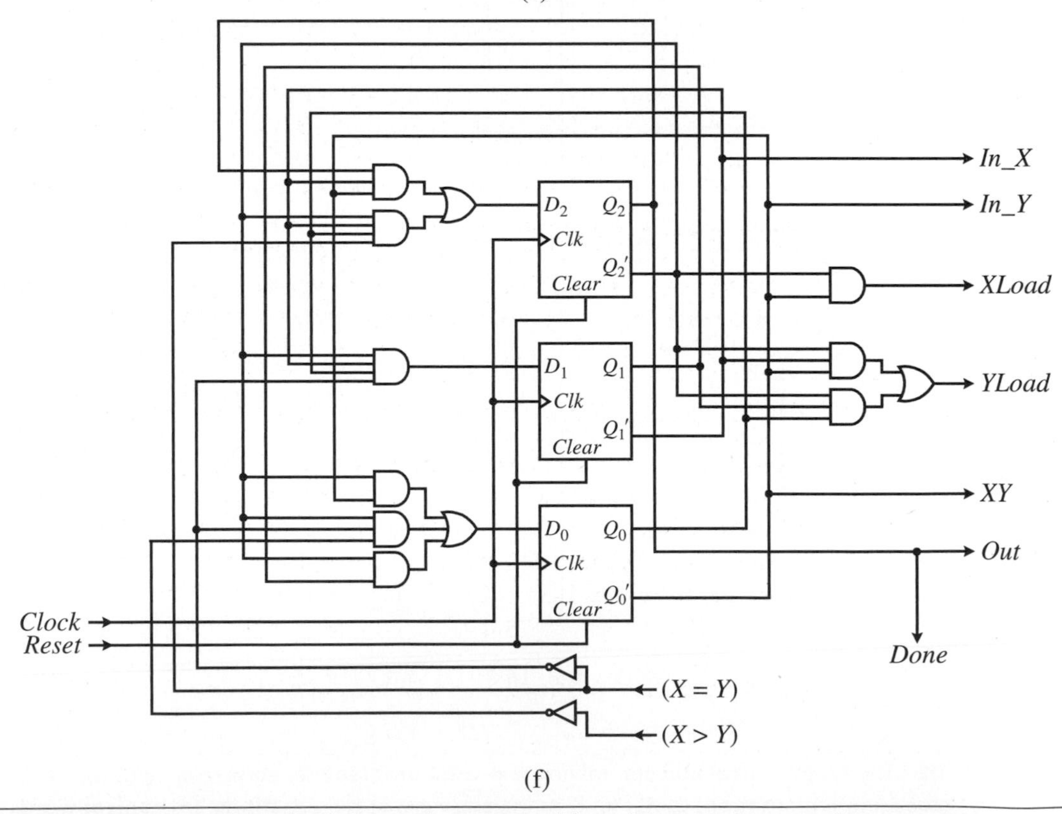

(f)

FIGURE 7.32 Control unit for solving the GCD problem: (a) state diagram; (b) next-state (implementation) table; (c) K-maps and next-state equations; (d) control words and output table; (e) output equations; (f) circuit.

The final microprocessor can now be formed easily by connecting the control unit and the datapath together using the designated control and status signals, as shown in Figure 7.33. To implement and test this circuit on an FPGA development board, you will need to map the two 8-bit inputs for *X* and *Y* to two sets of 8 switches, the output to 8 LEDs, a push button to the *Reset*, and a clock source to the *Clock* signal. A simulation trace of the microprocessor calculating the GCD of the two numbers 12 and 4 is shown in Figure 7.34.

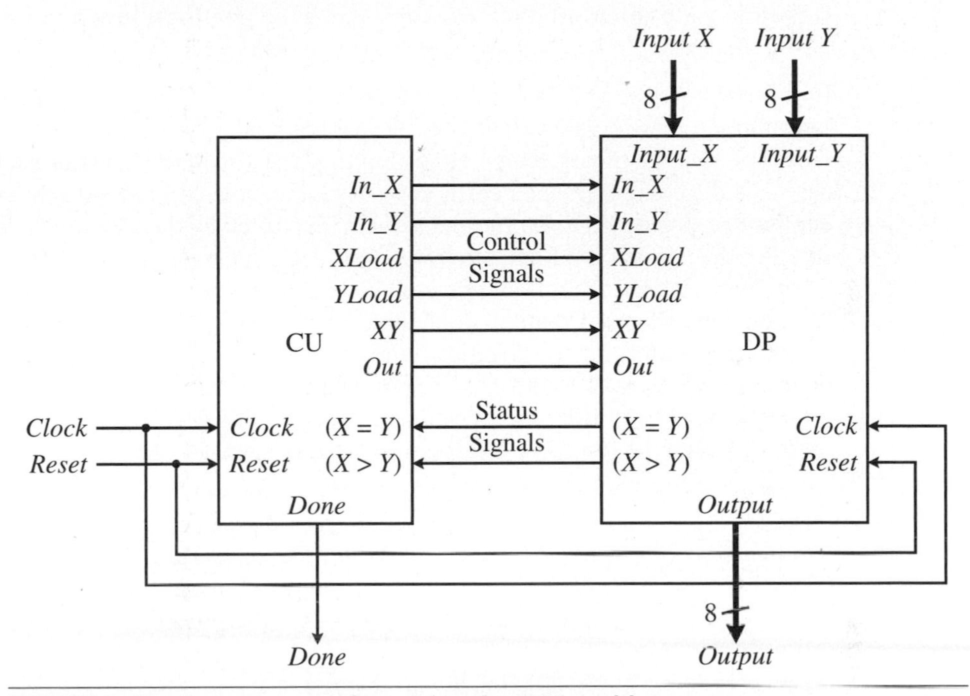

FIGURE 7.33 Microprocessor for solving the GCD problem.

FIGURE 7.34 Sample simulation for the GCD problem for the two input numbers 4 and 12. The GCD of these two numbers is 4.

7.5.2 High-Low Number Guessing Game

Example 7.15 shows the manual construction of a dedicated microprocessor to play the high-low number guessing game.

EXAMPLE 7.15

Designing a dedicated microprocessor to play the high-low number guessing game

In this example, we will design a complete dedicated microprocessor to play the high-low number guessing game. The user picks a number between 0 and 99, and the computer will use the binary search algorithm to guess the number. After each guess, the user tells the computer whether the guess is high or low compared to the picked number. Two push buttons *hi_button* and *lo_button* are used by the user to tell the computer whether the guess is too high, too low, or correct. The *hi_button* is pressed if the guess is too high, and the *lo_button* is pressed if the guess is too low. If the guess is correct, both buttons are pressed at the same time.

The algorithm for this high-low guessing game is shown in Figure 7.35. The two boundary variables *Low* and *High* are initialized to 0 and 100, respectively. The loop between lines 3 to 11 will keep repeating until both buttons, *hi_button* and *lo_button*, are pressed. Inside the loop, line 4 calculates the next guess by finding the middle number

```
1      Low = 0                      // initialize Low
2      High = 100                   // initialize High
       // repeat until both buttons are pressed
3      REPEAT {
          // calculate guess using binary search
4         Guess = (Low + High) / 2
5         OUTPUT Guess
6         IF (lo_button = '1' AND hi_button = '0') THEN
             // low button pressed
7            Low = Guess
8         ELSE IF (lo_button = '0' AND hi_button = '1') THEN
             // hi button pressed
9            High = Guess
10        END IF
11      } UNTIL (lo_button = '1' AND hi_button = '1')
12      WHILE (lo_button = '0' AND hi_button = '0')
        // blink correct guess
13         OUTPUT Guess
14         turn off display
15      END WHILE
```

FIGURE 7.35 Algorithm for the high-low number guessing game.

between the lower and upper boundaries and assigns it to the variable *Guess*. Line 5 outputs this new *Guess*. Lines 6 to 10 checks which button is pressed. If the *lo_button* is pressed, that means the guess is too low, so line 7 changes the *Low* boundary to the current *Guess*. Otherwise, if the *hi_button* is pressed, that means the guess is too high, and line 9 changes the *High* boundary to the current *Guess*. The loop is then repeated with the calculation of the new *Guess* in line 4. When both buttons are pressed, the condition in line 11 is true, and the loop is exited. Lines 12 to 15 simply cause the display to blink the correct guess by turning it on and off until either one of the buttons is pressed again.

The algorithm shown in Figure 7.35 has eight data manipulation operations in lines 1, 2, 4, 5, 7, 9, 13, and 14. The dedicated datapath for realizing this algorithm is shown in Figure 7.36. It requires three 8-bit registers (*Low*, *High*, and *Guess*) to store the low and high range boundary values and the guess, respectively. Two 2-to-1 multiplexers are used for the inputs to the *Low* and *High* registers to select between the initialization values for lines 1 and 2, and the new *Guess* values for lines 7 and 9.

The only arithmetic operations needed are the addition and division-by-2 in line 4. Hence, the outputs from the two registers *Low* and *High* go to the inputs of an adder for the addition, and the output of the adder goes to a shifter. The division-by-2 is performed by doing a right shift of 1 bit. The result from the shifter is stored in the register *Guess*. Depending on the condition in line 6, the value in the *Guess* register is loaded into either the *Low* or the *High* register by asserting the corresponding load signal for that register.

Eight 2-input AND gates are used to control the output of the *Guess* number. One input from each of the eight AND gates are connected to the 8-bit output of the *Guess* register. The other input from each of the eight AND gates are connected together in common to the output enable *Out* signal. By asserting *Out*, the data from the *Guess* register is passed to the output port. To blink the output display in lines 13 and 14, we just toggle the *Out* line.

The datapath shown in Figure 7.36 requires five control signals, *Init*, *LowLoad*, *HighLoad*, *GuessLoad*, and *Out*. Together, these five control signals form the control word for this datapath. The *Init* signal controls the two multiplexers to determine whether to load in the initialization values or the new guess. The three load signals *LowLoad*, *HighLoad*, and *GuessLoad* control the writing of the three respective registers. Finally, *Out* controls the output of the guess value.

Note that for this datapath, no comparator is used and no status signal is generated by the datapath. The conditional tests for the loop at line 11, and the high/low guess at lines 6 and 8, are provided by the two external push buttons, *hi_button* and *lo_button*. The input signals provided by these two push buttons will control the sequencing of the FSM.

The state diagram for this algorithm requires six states, as shown in Figure 7.37(a). State 000 is the starting initialization state. State 001 executes lines 4 and 5 by calculating the new guess and outputting it. State 001 also waits for the user keypress. If only the *lo_button* is pressed, then the FSM goes to state 010 to assign the guess as the new low value. If only the *hi_button* is pressed, then the FSM goes to state 011 to assign the guess as the new high value. If both buttons are pressed, then the FSM goes to state 100 to output the guess. From state 100, the FSM turns on and off the output by cycling between states 100 and 101 until a button is pressed. When a button

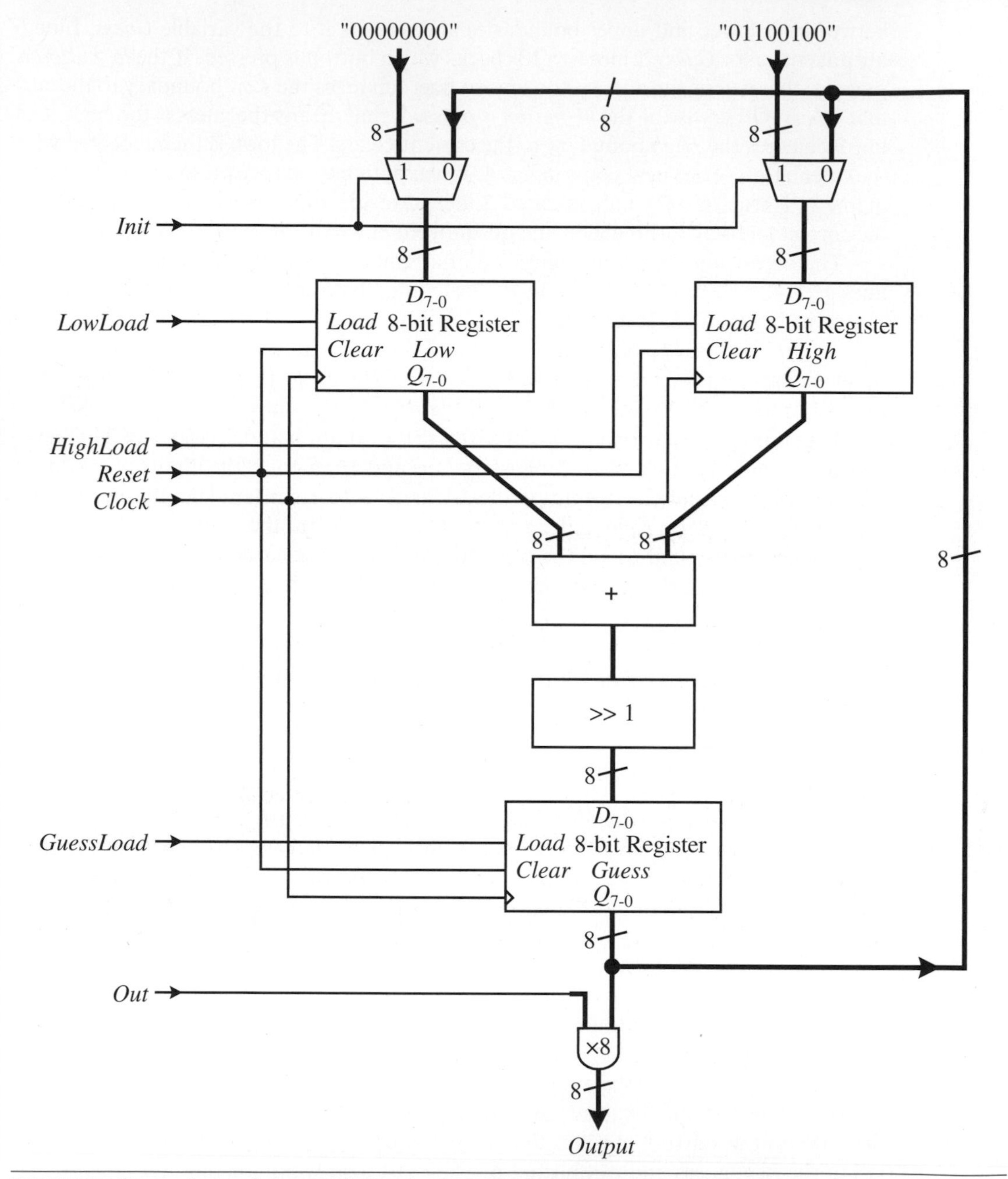

FIGURE 7.36 Dedicated datapath for the high-low number guessing game.

is pressed from either state 100 or 101, the FSM goes back to the initialization state for a new game.

The output table showing the five output signals, *Init*, *LowLoad*, *HighLoad*, *GuessLoad*, and *Out*, to be generated in each state is shown in Figure 7.37(d).

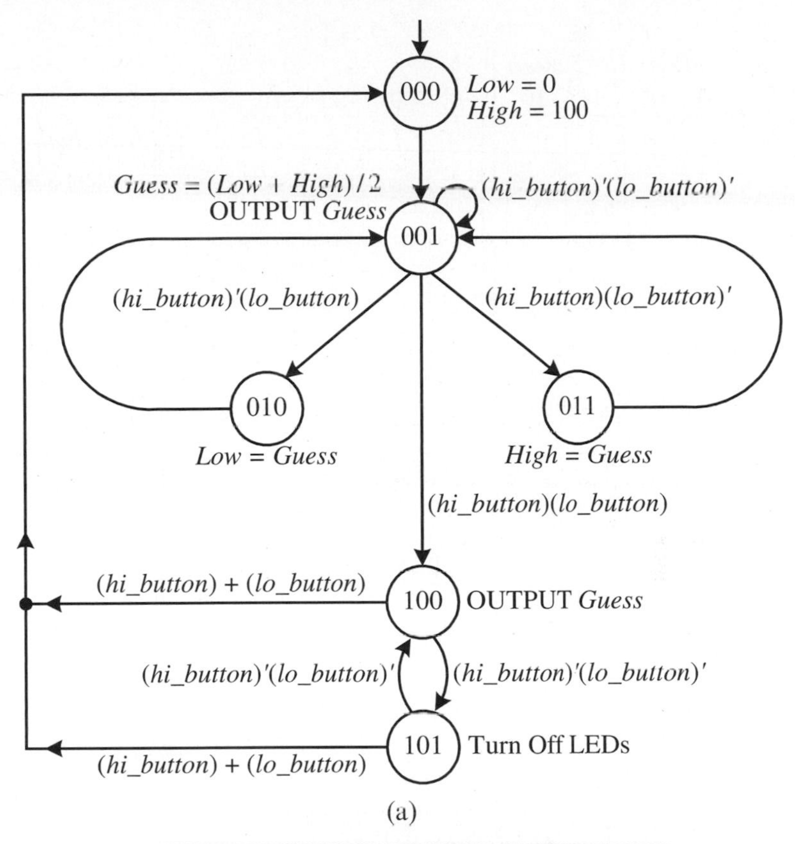

(a)

Current State $Q_2Q_1Q_0$	Next State (Implementation) $Q_{2next}Q_{1next}Q_{0next}$ $(D_2D_1D_0)$			
	hi_button, lo_button			
	00	01	10	11
000	001	001	001	001
001	001	010	011	100
010	001	001	001	001
011	001	001	001	001
100	101	000	000	000
101	100	000	000	000
110 Unused	000	000	000	000
111 Unused	000	000	000	000

(b)

FIGURE 7.37 Control unit for the high-low number guessing game: (a) state diagram; (b) next-state (implementation) table; (c) K-maps and next-state equations; (d) control words and output table; (e) output equations; (f) circuit. *(continued on next page)*

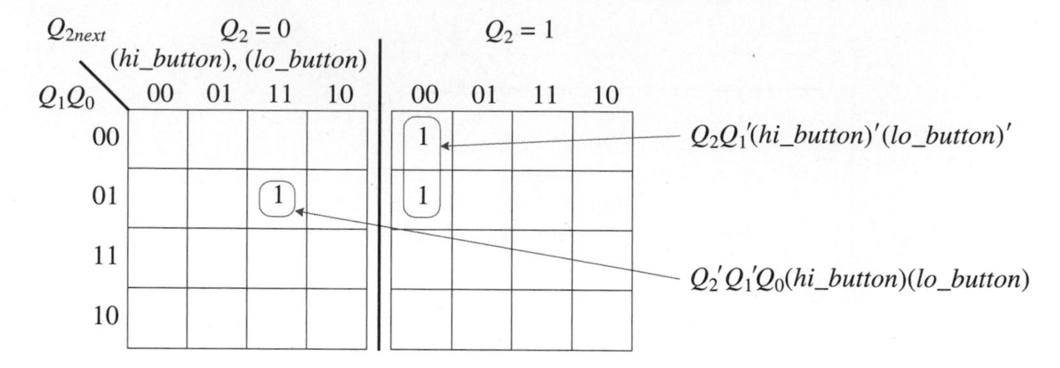

$Q_{2next} = D_2 = Q_2Q_1'(hi\_button)'(lo\_button)' + Q_2'Q_1'Q_0(hi\_button)(lo\_button)$

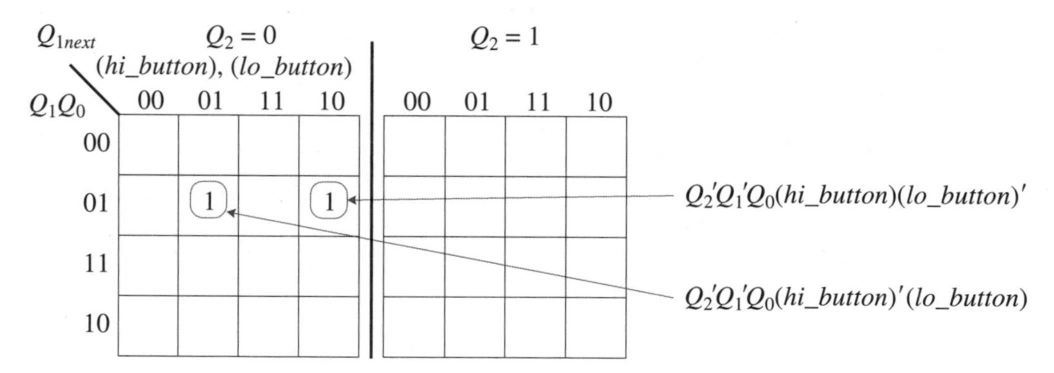

$Q_{1next} = D_1 = Q_2'Q_1'Q_0(hi\_button)(lo\_button)' + Q_2'Q_1'Q_0(hi\_button)'(lo\_button)$

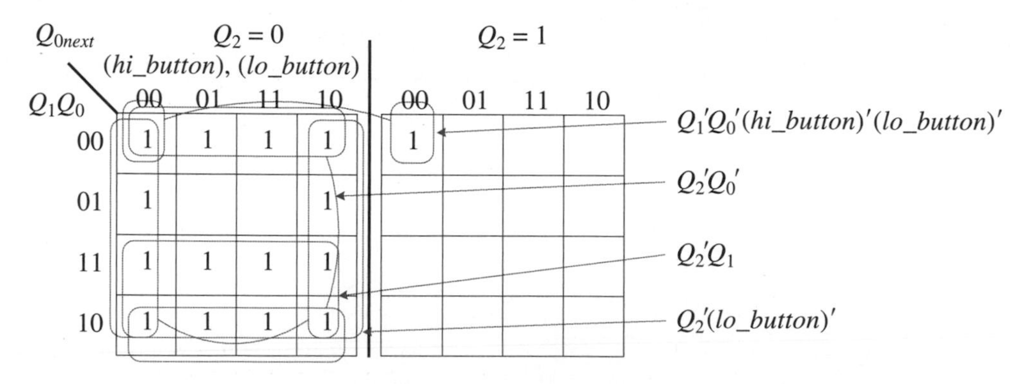

$Q_{0next} = D_0 = Q_1'Q_0'(hi\_button)'(lo\_button)' + Q_2'Q_0' + Q_2'Q_1 + Q_2'(lo\_button)'$

(c)

FIGURE 7.37 Control unit for the high-low number guessing game: (a) state diagram; (b) next-state (implementation) table; (c) K-maps and next-state equations; (d) control words and output table; (e) output equations; (f) circuit. *(continued on next page)*

Control Word	State $Q_2Q_1Q_0$	Instruction	*Init*	*HighLoad*	*LowLoad*	*GuessLoad*	*Out*
0	000	*Low* = 0, *High* = 100	1	1	1	0	1
1	001	*Guess* = (*Low* + *High*) / 2	0	0	0	1	1
2	010	*Low* = *Guess*	0	0	1	0	1
3	011	*High* = *Guess*	0	1	0	0	1
4	100	OUTPUT *Guess*	0	0	0	0	1
5	101	Turn off LEDs	0	0	0	0	0

(d)

$$Init = Q_2'Q_1'Q_0'$$
$$HighLoad = Q_2'Q_1'Q_0' + Q_2'Q_1Q_0$$
$$LowLoad = Q_2'Q_1'Q_0' + Q_2'Q_1Q_0'$$
$$GuessLoad = Q_2'Q_1'Q_0$$
$$Out = Q_2' + Q_0'$$

(e)

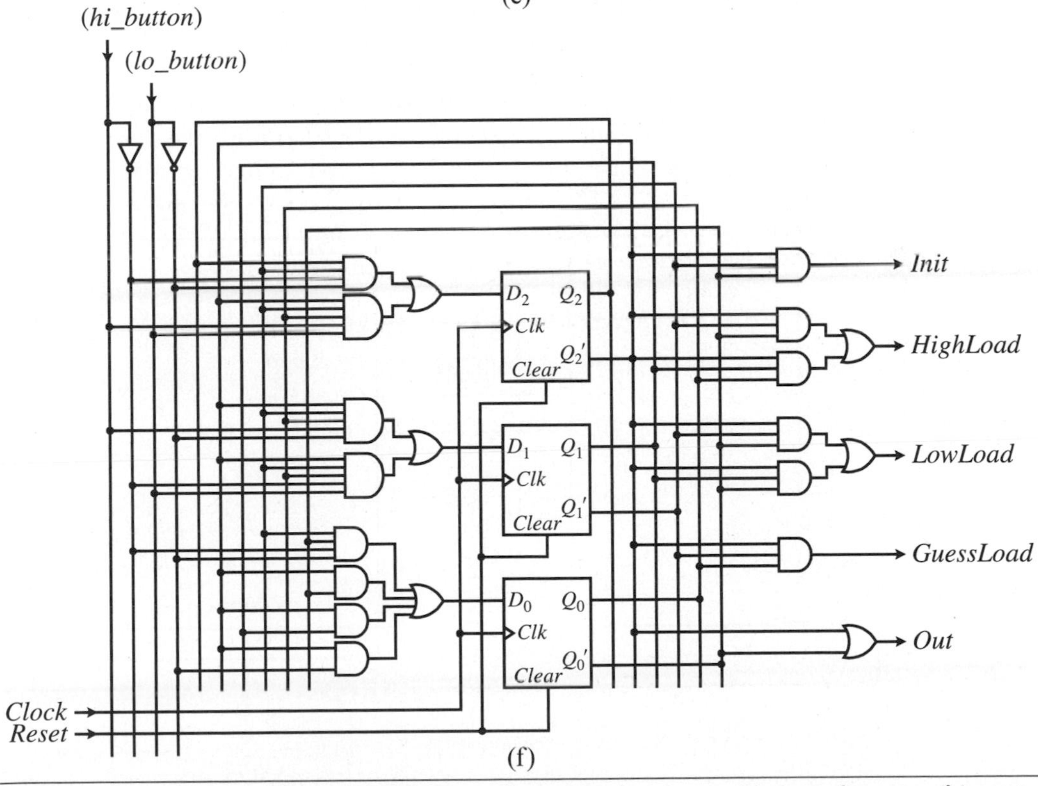

(f)

FIGURE 7.37 Control unit for the high-low number guessing game: (a) state diagram; (b) next-state (implementation) table; (c) K-maps and next-state equations; (d) control words and output table; (e) output equations; (f) circuit.

The corresponding output equations derived from the output table are shown in Figure 7.37(e).

Three D flip-flops are used to implement the state memory with only six of the encodings used for the six states. The next-state table is shown in Figure 7.37(b). The three K-maps and next-state equations are shown in Figure 7.37(c).

Using the three next-state equations to derive the next-state logic circuit, the three D flip-flops for the state memory, and the five output equations to derive the output logic circuit, we get the complete control unit circuit for the high-low number guessing game, as shown in Figure 7.37(f).

Connecting the datapath circuit shown in Figure 7.36 and the control unit circuit shown in Figure 7.37(f) together using the respective control signals produces the final microprocessor shown in Figure 7.38. Note that there are no status signals needed for this microprocessor, instead there are the two control input signals, *hi_button* and *lo_button*.

Figure 7.39 shows the interface needed to test this high-low number guessing game dedicated microprocessor on an FPGA development board. In order for the *hi_button* and *lo_button* to work correctly, they must be de-bounced with a one-shot circuit. Recall that the one-shot circuit outputs a one-cycle clock pulse when given an input of arbitrary time length. (See Section 6.6.3 for a detail description of the one-shot circuit.) If we do not use the one-shot circuit, then when the user presses, say, the *hi_button*, the FSM will cycle through the two states 001 and 011 many times before stopping in state 001 when the user finally releases the button. This is because the clock speed is very fast, so even though the user presses and releases the button immediately, the FSM would have gone through many clock cycles. Furthermore, the clock speed must be slow enough (around 4 Hz) for the FSM to be able to distinguish the difference

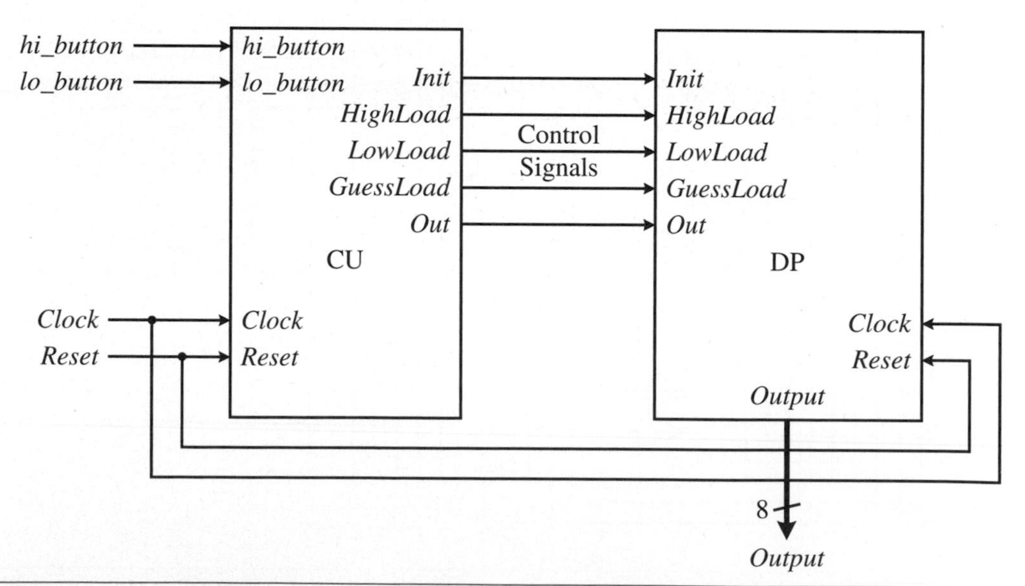

FIGURE 7.38 Microprocessor for the high-low number guessing game.

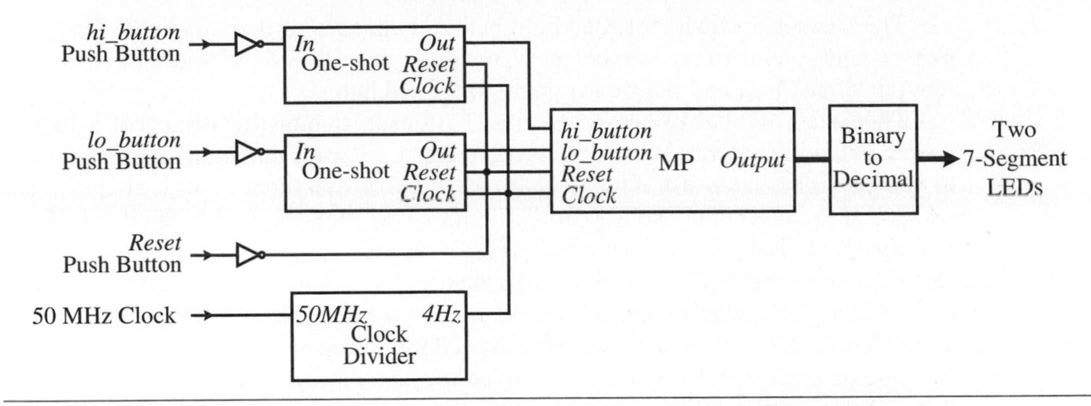

FIGURE 7.39 Interface to the high-low number guessing game microprocessor.

between a two-button press as opposed to just one button press after another. If the clock speed is too fast, then the FSM will never see the two-button press. The push buttons are assumed to be active-low, so an inverter is added to each of them to change the signals to active-high. To see the output *Guess* as a decimal number, a Binary-to-Decimal number converter is used, and the output from the converter is connected to two 7-segment LED displays.

7.5.3 Traffic Light Controller

Example 7.16 shows the manual construction of a dedicated traffic light microcontroller.

EXAMPLE 7.16

Designing a dedicated traffic light microcontroller

In this example, we will design the complete dedicated microprocessor to control a traffic light system at an intersection. For our traffic light system, we will have two sets of lights with each consisting of the red, yellow, and green lights. The sequence for the lights to turn on in both sets is shown next.

Set 1	Set 2	Time Delay	State
$Yellow_1$	Red_2	1 second	s_0
Red_1	$Green_2$	16 seconds	s_1
Red_1	$Yellow_2$	1 second	s_2
$Green_1$	Red_2	16 seconds	s_3
$Yellow_1$	Red_2	1 second	s_0
etc.	etc.	etc.	etc.

There is a short delay for each light before it changes to the next light. The green light remains on for 16 seconds before switching to the yellow light, and the yellow light remains on for 1 second before switching to the red light.

The system also has two crosswalk push buttons (matching the two sets of lights) for people to cross the street. When a button is pressed, we want that matching set of lights to turn red immediately if it is not already red, that is, if that set of lights is at green, then we want it to immediately go to yellow without completing the 16 seconds delay. If it is already at red, then it should remain at red. It will remain at red (while the other set is at green) for 16 seconds for the people to cross the street, after which the normal cycle repeats.

No data manipulation operations are needed for this microcontroller, so no datapath is required. However, a 4-bit counter is used for timing purposes, as discussed below.

The state diagram for our traffic light controller is shown in Figure 7.40(a). Notice that the state diagram has no reference to any timing except that the FSM will transition to the next state on every active clock edge. If the clock frequency is very high, then the FSM will cycle through the states so fast that you wouldn't see the lights stepping through the sequence. In the description, there are two situations where we want the lights to pause for 1 second, and this can be accomplished by using a 1 Hz clock frequency, because after 1 second, the FSM will move on to the next state and so the lights will change. However, when we want the green lights to pause for 16 seconds, we can either designate 16 states for keeping the same lights on for 16 seconds, or have only one state but use a counter to count 16 times for the FSM to stay in that one state before moving on to the next state. In our design, we have used the latter solution. A 4-bit binary counter is used to count from 0 to 15, after which it will cycle back to 0 and signify the overflow by asserting the *CountOverflow* bit.

On reset, the FSM will start in state s_0 with $Yellow_1$ and Red_2 turned on, and the 4-bit counter is reset to zero by setting *ClearCount* to a 1. After 1 second at the next clock cycle, the FSM transitions to state s_1 turning on Red_1 and $Green_2$. At this point, the counter will increment at a rate of 1 second. The FSM will remain in state s_1 until either the counter overflows by asserting the *CountOverflow* bit, which means that the time is up for the 16 seconds, or a pedestrian has pushed PB_2 and so we want the light to change immediately. These situations are covered by the three combinations of the three bits $PB_2PB_1CountOverflow = 001, 100$, and 101, and is labeled on the edge going from state s_1 to s_2. If PB_1 is pressed when the FSM is in state s_1, the FSM will loop back to s_1 (disregarding the other two control signals PB_2 and *CountOverflow*) and clears the counter. This essentially gives PB_1 priority over the other two signals, and it is an implementation decision that the designer makes, which depends on how you want the traffic light to operate.

States s_2 and s_3 are mirror images of states s_0 and s_1 for the other set of lights, and they work in a similar fashion.

The next-state table as derived from the state diagram is shown in Figure 7.40(b), and the two next-state equations as derived from the next-state table are shown in Figure 7.40(c).

The output table is shown in Figure 7.40(d). There are two sets of *Red*, *Yellow*, and *Green* signals to turn on and off the corresponding color lights, and a *ClearCount* signal to zero the 4-bit counter. Notice that for the *ClearCount* signal, it is set to a 1 on

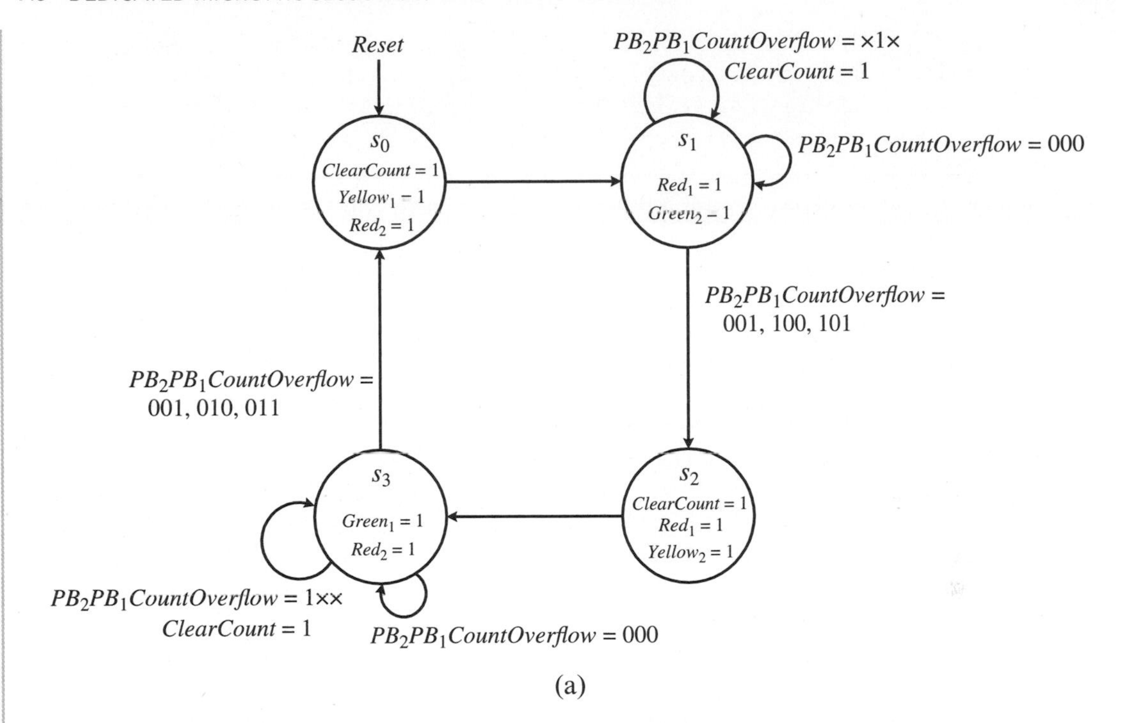

(a)

Current State Q_1Q_0	Next State (Implementation) $Q_{1next}Q_{0next}$ (D_1D_0)							
	PB_2, PB_1, CountOverflow							
	000	001	010	011	100	101	110	111
s_0 00	01	01	01	01	01	01	01	01
s_1 01	01	10	01	01	10	10	01	01
s_2 10	11	11	11	11	11	11	11	11
s_3 11	11	00	00	00	11	11	11	11

(b)

$$Q_{1next} = D_1 = Q_1Q_0' + Q_1PB_2 + Q_1PB_1'CountOverflow' + Q_0PB_2PB_1' + Q_1'Q_0PB_1'CountOverflow$$

$$Q_{0next} = D_0 = Q_0' + Q_1'PB_1 + Q_1PB_2 + Q_0PB_2'PB_1'CountOverflow'$$

(c)

FIGURE 7.40 Control unit for the traffic light controller: (a) state diagram; (b) next-state (implementation) table; (c) next-state equations; (d) output table; (e) output equations; (f) circuit. *(continued on next page)*

Current State Q_1Q_0	Output Signals						
	Red_1	$Yellow_1$	$Green_1$	Red_2	$Yellow_2$	$Green_2$	$ClearCount$
s_0 00	0	1	0	1	0	0	1
s_1 01	1	0	0	0	0	1	1 if (PB_1)
s_2 10	1	0	0	0	1	0	1
s_3 11	0	0	1	1	0	0	1 if (PB_2)

(d)

$$Red_1 = Q_1 \oplus Q_0$$
$$Yellow_1 = Q_1'Q_0'$$
$$Green_1 = Q_1Q_0$$
$$Red_2 = Q_1 \odot Q_0$$
$$Yellow_2 = Q_1Q_0'$$
$$Green_2 = Q_1'Q_0$$
$$ClearCount = Q_0' + Q_1'Q_0PB_1 + Q_1Q_0PB_2$$

(e)

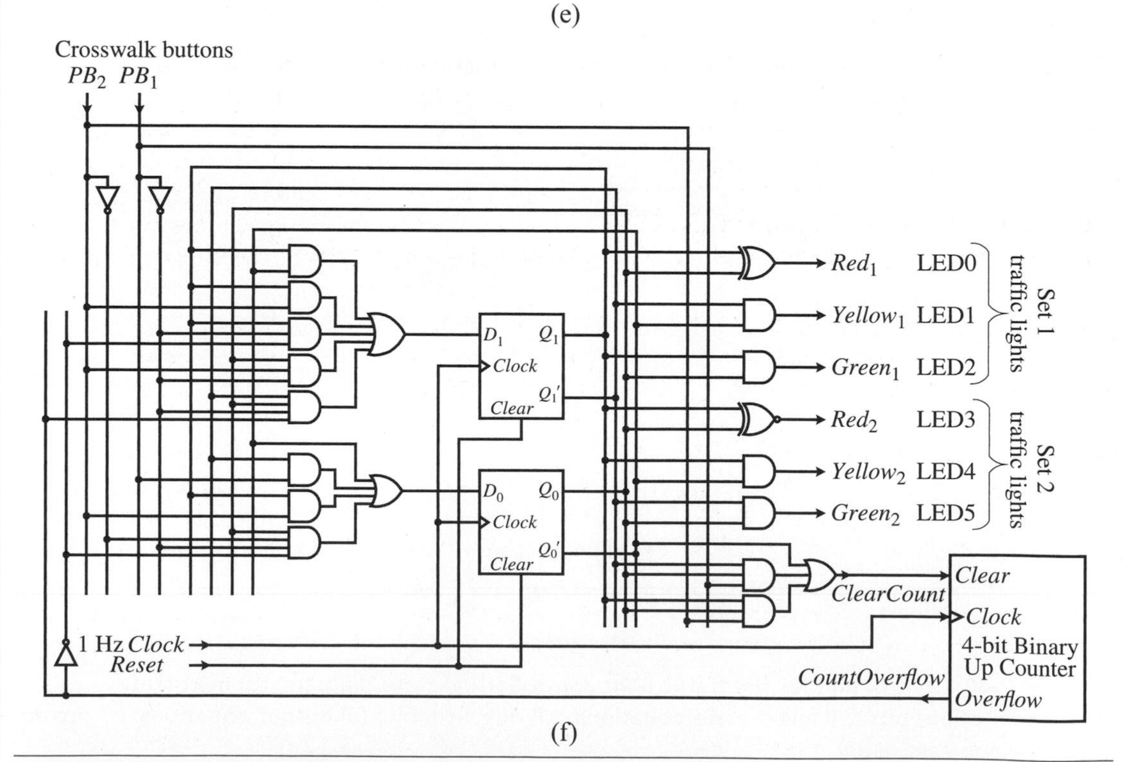

(f)

FIGURE 7.40 Control unit for the traffic light controller: (a) state diagram; (b) next-state (implementation) table; (c) next-state equations; (d) output table; (e) output equations; (f) circuit.

an edge instead of in a state, that is, it is dependent on both the current state and an input signal, making it a Mealy machine. For example, in state s_1, *ClearCount* is set to a 1 only when PB_1 also is a 1. The seven output equations as derived from the output table are shown in Figure 7.40(e).

As mentioned previously, a 4-bit binary up counter is needed to support the count delay. The *ClearCount* signal will assert the *Clear* input signal to zero the counter, and the *CountOverflow* output signal is connected to the counter's *Overflow* output bit. This counter is the only component needed for the datapath, so instead of drawing it as a separate circuit, it is combined with the controller circuit. The complete traffic light microcontroller circuit is shown in Figure 7.40(f).

7.6 Verilog and VHDL Code for Dedicated Microprocessors

Having manually designed the circuit for a dedicated microprocessor, we can implement it with a hardware development tool by either drawing the schematic circuit or writing the HDL code using the FSM+D (FSM *plus* Datapath) model. However, in so doing, we will not be using the full power of the HDL synthesizer. To use the automatic synthesis capability of a hardware development tool, we can design a microprocessor by writing HDL code at a higher level either using the FSMD (FSM *with* Datapath) model or the algorithmic model. Given a high-level description of a microprocessor, the synthesis tool can automatically synthesize the circuit for the microprocessor.

Designing a microprocessor using the FSM+D model involves separately writing behavioral HDL code to describe the FSM, and structural HDL code to construct the datapath. You start with the state diagram for the FSM, and in each state the FSM will generate the appropriate control signals to control the datapath. The FSM and the datapath are then connected together via the control and status signals using structural HDL code.

Designing a microprocessor using the FSMD model involves writing behavioral HDL code to describe both the FSM and the datapath together as one unit. You start with the state diagram for the FSM and the data operations that are to be executed in each state. The state diagram is translated into HDL statements, and the data operations are performed with the built-in HDL operators. During the synthesis process, the synthesizer will generate a separate FSM unit and a datapath unit automatically, and then connects these two units together as a complete microprocessor. The advantage of this model is that you do not have to design the FSM or the datapath manually, but you still have full control as to what datapath operation is executed in what state or in what clock cycle. In other words, you have full control over the timing of the FSM circuit.

A microprocessor also can be described algorithmically at the behavioral level using HDL. Using this model, the operation of the complete microprocessor is described algorithmically, and there is no need to know about the control unit or the datapath. The HDL synthesizer will synthesize the complete microprocessor automatically, with its control unit and datapath together. The advantage of designing microprocessors this way is that you do not need to know how to manually design a microprocessor. In other words, you do not need to know most of the materials presented in this book. Instead, you only need to know how to write HDL codes and the operations to be implemented in the microprocessor. The disadvantage is that you do not have control over the timing of the circuit. You can no longer specify what datapath operation is executed in what clock cycle.

The following subsections list the complete Verilog and VHDL code for the dedicated microprocessor for the summation problem to generate and sum the numbers from 1 to 10. This summation problem was first introduced in Example 7.3. The algorithm, state diagram, and datapath for this summation problem are repeated here in Figure 7.41 for convenience.

7.6.1 **FSM+D Model**

We will now create the dedicated microprocessor to generate and sum the numbers from 1 to 10 using the FSM+D model. For the FSM+D model, we first create a separate datapath and control unit. The microprocessor is then formed by combining these two units together using the respective control signals and status signals. The datapath circuit needs first to be designed manually as shown in Figure 7.41(c). Based on the circuit, the datapath is then constructed by connecting the components together using structural HDL coding. All of the components used in the datapath must, of course, already have been defined in separate files. The components needed in this example are Mux21, Register, Add, and TriState_Buffer.

```
1    sum = 0
2    i = 1
3    DO {
4       sum = sum + i
5       i = i + 1
6    } WHILE (i ≠ 11)
7    OUTPUT sum
```

(a)

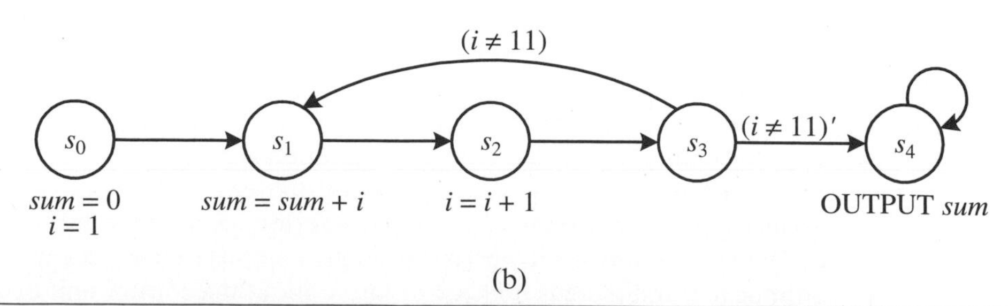

(b)

FIGURE 7.41 Summation problem to generate and sum the numbers from 1 to 10: (a) algorithm; (b) state diagram; (c) datapath. *(continued on next page)*

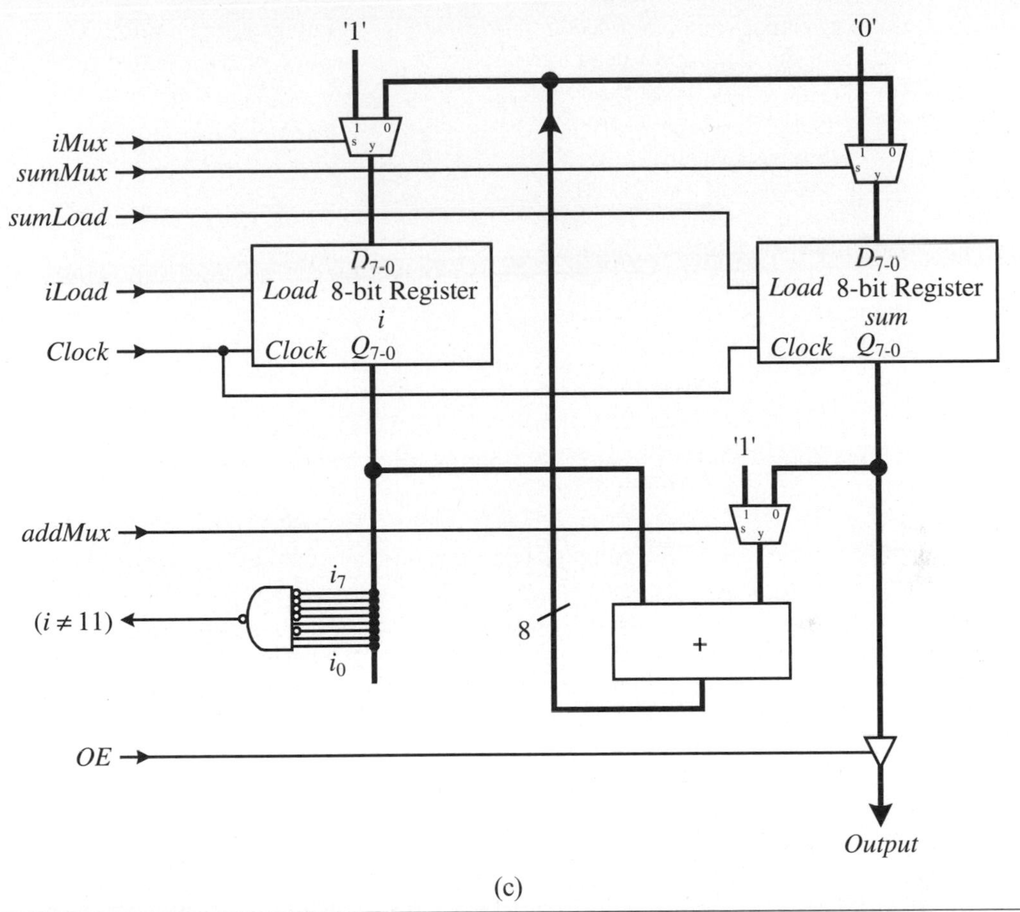

FIGURE 7.41 Summation problem to generate and sum the numbers from 1 to 10: (a) algorithm; (b) state diagram; (c) datapath.

The control unit is based on the state diagram derived for the FSM. For Verilog there are two `always` blocks, and for VHDL there are two PROCESS blocks, one for the next-state logic and one for the output logic. The main code in both logic blocks consists of a `case` statement to select the current state that the FSM is in. For each current state in the next-state logic block, a new state value is assigned to the state variable based on the transitions in the state diagram. In the output logic block, all of the datapath control signals are generated for every state based on the output table.

Finally, again using structural HDL coding, the control unit and the datapath are combined together to form the complete microprocessor. The Verilog codes for these three components, datapath, control unit, and microprocessor, are shown in Figures 7.42, 7.43, and 7.44, respectively. The VHDL codes for these three components are shown in Figures 7.45, 7.46, and 7.47, respectively.

```
module DP (
  input Clock, Reset,
  input iMux, sumMux, sumLoad, iLoad, addMux, OE,
  output ine11,
  output [n-1:0] Output,
  output [3:0] debug
);
  parameter n = 8;
  wire [n-1:0] dp_iMux, dp_sumMux, dp_i, dp_sum, dp_addMux,
     dp_add;

  assign debug = dp_i[3:0];

  Mux21 U0_iMux(.S(iMux), .D1(8'd1), .D0(dp_add), .Y(dp_iMux));
  Mux21 U1_sumMux(.S(sumMux), .D1(8'd0), .D0(dp_add),
    .Y(dp_sumMux));
  Register #(8) U2_iReg(.Clock(Clock), .Clear(Reset),
    .Write(iLoad), .D(dp_iMux), .Q(dp_i));
  Register #(8) U3_sumReg(.Clock(Clock), .Clear(Reset),
    .Write(sumLoad), .D(dp_sumMux), .Q(dp_sum));
  Mux21 U4_addMux(.S(addMux), .D1(8'd1), .D0(dp_sum),
    .Y(dp_addMux));
  Add U5_Add(.A(dp_i), .B(dp_addMux), .F(dp_add));
  TriState_Buffer #(8) U6_Tri(.E(OE), .D(dp_sum), .Y(Output));

  nand(ine11,~dp_i[7],~dp_i[6],~dp_i[5],~dp_i[4],dp_i[3],
    ~dp_i[2],dp_i[1],dp_i[0]);

endmodule
```

FIGURE 7.42 Verilog code for the datapath for the FSM+D model of the summation problem.

```
module CU (
  input Clock, Reset,
  output reg iMux, sumMux, sumLoad, iLoad, addMux, OE,
  input ine11,
  output [2:0] debug
);

// Declare state encodings
parameter s0=0, s1=1, s2=2, s3=3, s4=4;
  reg [2:0] state;

  assign debug = state;
```

FIGURE 7.43 Verilog code for the control unit for the FSM+D model of the summation problem. *(continued on next page)*

```
// next-state logic
always @ (posedge Clock or posedge Reset) begin
   if (Reset == 1) begin
      state <= s0;
      end
   else begin
      case (state)
      s0: begin
         state <= s1;
         end
      s1: begin
         state <= s2;
         end
      s2: begin
         state <= s3;
         end
      s3: begin
         if (ine11 = 1)
            state <= s1;
         else
            state <= s4;
         end
      s4: begin
         state <= s4;
         end
      default: begin
         state <= s0;
         end
      endcase
   end
end  // always

// output logic
always @ (state) begin
   case (state)
   s0: begin
      iMux = 1;
      sumMux = 1;
      sumLoad = 1;
      iLoad = 1;
      addMux = 0;
      OE = 1;
      end
```

FIGURE 7.43 Verilog code for the control unit for the FSM+D model of the summation problem. *(continued on next page)*

```
            s1: begin
               iMux = 0;
               sumMux = 0;
               sumLoad = 1;
               iLoad = 0;
               addMux = 0;
               OE = 1;
               end
            s2: begin
               iMux = 0;
               sumMux = 0;
               sumLoad = 0;
               iLoad = 1;
               addMux = 1;
               OE = 1;
               end
            s3: begin
               iMux = 0;
               sumMux = 0;
               sumLoad = 0;
               iLoad = 0;
               addMux = 0;
               OE = 1;
               end
            s4: begin
               iMux = 0;
               sumMux = 0;
               sumLoad = 0;
               iLoad = 0;
               addMux = 0;
               OE = 1;
               end
            default: begin
               iMux = 0;
               sumMux = 0;
               sumLoad = 0;
               iLoad = 0;
               addMux = 0;
               OE = 1;
               end
            endcase
         end // always
      endmodule
```

FIGURE 7.43 Verilog code for the control unit for the FSM+D model of the summation problem.

```
module MP (
  input Clock, Reset,
  output [n-1:0] Output,
  output [2:0] debug_cu,
  output [3:0] debug_dp
);
  parameter n = 8;
  wire mp_iMux, mp_sumMux, mp_sumLoad, mp_iLoad, mp_addMux,
    mp_ine11, mp_OE;

  DP U0(Clock, Reset, mp_iMux, mp_sumMux, mp_sumLoad, mp_iLoad,
    mp_addMux, mp_OE, mp_ine11, Output, debug_dp);
  CU U1(Clock, Reset, mp_iMux, mp_sumMux, mp_sumLoad, mp_iLoad,
    mp_addMux, mp_OE, mp_ine11, debug_cu);

endmodule
```

FIGURE 7.44 Verilog code for the microprocessor for the FSM+D model of the summation problem.

```
LIBRARY IEEE;
USE IEEE.STD_LOGIC_1164.ALL;

ENTITY DP IS
GENERIC (n: INTEGER := 8);
PORT(
  Clock, Reset: IN STD_LOGIC;
  iMux, sumMux, sumLoad, iLoad, addMux, OE: IN STD_LOGIC;
  ine11: OUT STD_LOGIC;
  Output: OUT STD_LOGIC_VECTOR(n-1 DOWNTO 0);
  debug: OUT STD_LOGIC_VECTOR(3 DOWNTO 0));
END DP;

ARCHITECTURE Structural OF DP IS
  COMPONENT Mux21 IS PORT(
       S: IN STD_LOGIC;
       D1, D0: IN STD_LOGIC_VECTOR(n-1 DOWNTO 0);
       Y: OUT STD_LOGIC_VECTOR(n-1 DOWNTO 0));
  END COMPONENT;

  COMPONENT Reg IS PORT (
       Clock, Clear, Load: IN STD_LOGIC;
       D: IN STD_LOGIC_VECTOR(n-1 DOWNTO 0);
       Q: OUT STD_LOGIC_VECTOR(n-1 DOWNTO 0));
  END COMPONENT;
```

FIGURE 7.45 VHDL code for the datapath for the FSM+D model of the summation problem. *(continued on next page)*

```
    COMPONENT Add IS PORT(
         A, B: IN STD_LOGIC_VECTOR(n-1 DOWNTO 0);
         F: OUT STD_LOGIC_VECTOR(n-1 DOWNTO 0);
         Unsigned_Overflow: OUT STD_LOGIC);
    END COMPONENT;

    COMPONENT TriState_Buffer IS
         GENERIC (n: INTEGER := 8);
         PORT (
                E: IN STD_LOGIC;
                D: IN STD_LOGIC_VECTOR(n-1 DOWNTO 0);
                Y: OUT STD_LOGIC_VECTOR(n-1 DOWNTO 0));
    END COMPONENT;

    SIGNAL dp_iMux, dp_sumMux, dp_i, dp_sum, dp_addMux, dp_add:
      STD_LOGIC_VECTOR(n-1 DOWNTO 0);
  BEGIN
    debug <= dp_i(3 DOWNTO 0);

    U0: Mux21 PORT MAP(S=>iMux, D1=>"00000001", D0=>dp_add,
      Y=>dp_iMux);
    U1: Mux21 PORT MAP(S=>sumMux, D1=>"00000000", D0=>dp_add,
      Y=>dp_sumMux);
    U2: Reg GENERIC MAP(8) PORT MAP(Clock=>Clock, Clear=>Reset,
      Load=>iLoad, D=>dp_iMux, Q=>dp_i);
    U3: Reg GENERIC MAP(8) PORT MAP(Clock=>Clock, Clear=>Reset,
      Load=>sumLoad, D=>dp_sumMux, Q=>dp_sum);
    U4: Mux21 PORT MAP(S=>addMux, D1=>"00000001", D0=>dp_sum,
      Y=>dp_addMux);
    U5: Add PORT MAP(A=>dp_i, B=>dp_addMux, F=>dp_add);
    U6: TriState_Buffer GENERIC MAP(8) PORT MAP(E=>OE, D=>dp_sum,
      Y=>Output);

    ine11 <= NOT ((NOT dp_i(7)) AND (NOT dp_i(6)) AND
      (NOT dp_i(5)) AND (NOT dp_i(4)) AND (dp_i(3)) AND
      (NOT dp_i(2)) AND (dp_i(1)) AND (dp_i(0)));
  END Structural;
```

FIGURE 7.45 VHDL code for the datapath for the FSM+D model of the summation problem.

```
  LIBRARY IEEE;
  USE IEEE.STD_LOGIC_1164.ALL;

  ENTITY CU IS PORT(
    Clock, Reset: IN STD_LOGIC;
    iMux, sumMux, sumLoad, iLoad, addMux, OE: OUT STD_LOGIC;
    ine11: IN STD_LOGIC;
    debug: OUT STD_LOGIC_VECTOR(2 DOWNTO 0));
  END CU;
```

FIGURE 7.46 VHDL code for the control unit for the FSM+D model of the summation problem. *(continued on next page)*

```
ARCHITECTURE Behavioral OF CU IS
  TYPE state_type IS (s0, s1, s2, s3, s4);
  SIGNAL state: state_type;
BEGIN
  -- next-state logic
  PROCESS (Clock, Reset)
  BEGIN
       IF (Reset = '1') THEN
          state <= s0;
       ELSIF (Clock'EVENT AND Clock = '1') THEN
             CASE state IS
             WHEN s0 =>
                debug <= "000";
                state <= s1;
             WHEN s1 =>
                debug <= "001";
                state <= s2;
             WHEN s2 =>
                debug <= "010";
                state <= s3;
             WHEN s3 =>
                debug <= "011";
                IF (ine11 = '1') THEN
                   state <= s1;
                ELSE
                   state <= s4;
                END IF;
             WHEN s4 =>
                debug <= "100";
                state <= s4;
             WHEN OTHERS =>
                debug <= "111";
                state <= s0;
             END CASE;
       END IF;
  END PROCESS;

  -- output logic
  PROCESS (state)
  BEGIN
       CASE state IS
       WHEN s0 =>
          iMux <= '1';
          sumMux <= '1';
          sumLoad <= '1';
```

FIGURE 7.46 VHDL code for the control unit for the FSM+D model of the summation problem. *(continued on next page)*

```
                    iLoad <= '1';
                    addMux <= '0';
                    OE <= '1';
                WHEN s1 =>
                    iMux <= '0';
                    sumMux <= '0';
                    sumLoad <= '1';
                    iLoad <= '0';
                    addMux <= '0';
                    OE <= '1';
                WHEN s2 =>
                    iMux <= '0';
                    sumMux <= '0';
                    sumLoad <= '0';
                    iLoad <= '1';
                    addMux <= '1';
                    OE <= '1';
                WHEN s3 =>
                    iMux <= '0';
                    sumMux <= '0';
                    sumLoad <= '0';
                    iLoad <= '0';
                    addMux <= '0';
                    OE <= '1';
                WHEN s4 =>
                    iMux <= '0';
                    sumMux <= '0';
                    sumLoad <= '0';
                    iLoad <= '0';
                    addMux <= '0';
                    OE <= '1';
                WHEN OTHERS =>
                    iMux <= '0';
                    sumMux <= '0';
                    sumLoad <= '0';
                    iLoad <= '0';
                    addMux <= '0';
                    OE <= '1';
                END CASE;
        END PROCESS;
    END Behavioral;
```

FIGURE 7.46 VHDL code for the control unit for the FSM+D model of the summation problem.

```
LIBRARY IEEE;
USE IEEE.STD_LOGIC_1164.ALL;

ENTITY MP IS
GENERIC (n: INTEGER := 8);
PORT(
  Clock, Reset: IN STD_LOGIC;
  Output: OUT STD_LOGIC_VECTOR(n-1 DOWNTO 0);
  debug_cu: OUT STD_LOGIC_VECTOR(2 DOWNTO 0);
  debug_dp: OUT STD_LOGIC_VECTOR(3 DOWNTO 0));
END MP;

ARCHITECTURE Structural OF MP IS
  COMPONENT DP IS PORT(
       Clock, Reset: IN STD_LOGIC;
       iMux, sumMux, sumLoad, iLoad, addMux, OE: IN STD_LOGIC;
       ine11: OUT STD_LOGIC;
       Output: OUT STD_LOGIC_VECTOR(n-1 DOWNTO 0);
       debug: OUT STD_LOGIC_VECTOR(3 DOWNTO 0));
  END COMPONENT;

  COMPONENT CU IS PORT(
       Clock, Reset: IN STD_LOGIC;
       iMux, sumMux, sumLoad, iLoad, addMux, OE: OUT STD_LOGIC;
       ine11: IN STD_LOGIC;
       debug: OUT STD_LOGIC_VECTOR(2 DOWNTO 0));
  END COMPONENT;

  SIGNAL mp_iMux, mp_sumMux, mp_sumLoad, mp_iLoad, mp_addMux,
    mp_OE, mp_ine11: STD_LOGIC;
BEGIN
  U0: DP PORT MAP(Clock, Reset, mp_iMux, mp_sumMux, mp_sumLoad,
    mp_iLoad, mp_addMux, mp_OE, mp_ine11, Output, debug_dp);
  U1: CU PORT MAP(Clock, Reset, mp_iMux, mp_sumMux, mp_sumLoad,
    mp_iLoad, mp_addMux, mp_OE, mp_ine11, debug_cu);
END Structural;
```

FIGURE 7.47 VHDL code for the microprocessor for the FSM+D model of the summation problem.

7.6.2 **FSMD Model**

The construction of a microprocessor using the FSMD model does not require us to construct the circuit for the control unit and the datapath manually. We only need to derive the state diagram for the FSM and know what data operations are performed in each state. We can then write the high-level HDL code based on the state diagram and the data operations. The translation process from the state diagram to HDL code is quite straightforward. The FSM state diagram is translated into a HDL `case` statement with one case for each state. The data operations are embedded within the FSM code using built-in HDL operators. As a result, no control signals or status signals needs to be connected between the control unit and the datapath. From the HDL code, the HDL synthesizer automatically generates the circuit for the complete microprocessor.

Figures 7.48 and 7.49 show the Verilog and VHDL codes, respectively, for the complete dedicated microprocessor using the FSMD model for the summation problem. The format of this code follows very closely to that of the FSM code as discussed in Chapter 6. Here, we have one always/PROCESS block, which contains not only the state transition assignment statements, but also the data manipulation statements. Because the control unit and the datapath are combined into one module, the control signals and status signals are no longer needed to join them together.

```
module Summation (
  input Clock, Reset,
  output reg [7:0] Output
);
  // Declare state encodings
  parameter s0=0, s1=1, s2=2, s3=3, s4=4;
  reg [2:0]state;
  reg [7:0]sum;
  reg [7:0]i;
  always @ (posedge Clock or posedge Reset) begin
     if (Reset == 1) begin
         Output <= 0;
         state <= s0;
         end
     else begin
         case (state)
             s0: begin
                sum <= 0;
                i <= 1;
                Output <= 0;
                state <= s1;
                end
             s1: begin
                sum <= sum + i;
                Output <= 0;
                state <= s2;
                end
             s2: begin
                // outputs 55=1+2+3+4+5+6+7+8+9+10
                i <= i + 1;
                Output <= 0;
                state <= s3;
                end
             s3: begin
                //outputs 66=1+2+3+4+5+6+7+8+9+10+11
                //i <= i + 1;
```

FIGURE 7.48 Verilog code for the FSMD model of the summation problem.
(continued on next page)

```
                //outputs 55=1+2+3+4+5+6+7+8+9+10
                //i = i + 1;
                Output <= 0;
                if (i != 11)
                   state <= s1;
                else
                   state <= s4;
                end
             s4: begin
                Output <= sum;
                state <= s4;
                end
             default: begin
                Output <= 0;
                state <= s0;
                end
             endcase
         end
    end
endmodule
```

FIGURE 7.48 Verilog code for the FSMD model of the summation problem.

```
LIBRARY IEEE;
USE IEEE.STD_LOGIC_1164.ALL;
USE IEEE.STD_LOGIC_UNSIGNED.ALL;

ENTITY Summation IS PORT (
  Clock: IN STD_LOGIC;
  Reset: IN STD_LOGIC;
  Output: OUT STD_LOGIC_VECTOR(7 DOWNTO 0));
END Summation;

ARCHITECTURE FSMD OF Summation IS
  TYPE state_type IS (s0, s1, s2, s3, s4);
  SIGNAL state: state_type;
  SIGNAL sum: STD_LOGIC_VECTOR(7 DOWNTO 0);
  SIGNAL i: STD_LOGIC_VECTOR(7 DOWNTO 0);
BEGIN
  PROCESS (Clock, Reset)
  BEGIN
       IF (Reset = '1') THEN
          state <= s0;
          Output <= (OTHERS => '0');
       ELSIF (Clock'EVENT AND Clock = '1') THEN
          CASE state IS
          WHEN s0 =>
              sum <= (OTHERS => '0');
```

FIGURE 7.49 VHDL code for the FSMD model of the summation problem.
(continued on next page)

```
                i <= "00000001";
                Output <= (OTHERS => '0');
                state <= s1;
            WHEN s1 =>
                sum <= sum + i;
                Output <= (OTHERS => '0');
                state <= s2;
            WHEN s2 =>
                i <= i + 1;
                Output <= (OTHERS => '0');
                state <= s3;
            WHEN s3 =>
                Output <= (OTHERS => '0');
                IF (i /= 11) THEN
                    state <= s1;
                ELSE
                    state <= s4;
                END IF;
            WHEN s4 =>
                Output <= sum;
                state <= s4;
            WHEN OTHERS =>
                Output <= (OTHERS => '0');
                state <= s0;
            END CASE;
        END IF;
    END PROCESS;
END FSMD;
```

FIGURE 7.49 VHDL code for the FSMD model of the summation problem.

7.6.3 **Algorithmic Model**

The complete microprocessor also can be designed by writing HDL code in a truly algorithmic behavioral style that has no reference to either the control unit or the datapath. Using the algorithmic model to design a circuit is similar to writing computer programs using a high-level language. It offers all of the basic language constructs that are available in most high-level computer programming languages, such as variable assignments, loops, and conditional tests. The HDL synthesizer, like the compiler, will translate the HDL algorithmic description of the circuit automatically to a netlist, which then can be programmed directly onto an FPGA chip. Using this model to design a microprocessor is very simple and powerful, but it has its limitations.

One limitation is with the use of loops. The synthesizer can synthesize loops only when the number of times to repeat the loop is fixed at compile time. In other words, you cannot have a loop where a variable is used in the testing of the ending condition and the value of that variable is not known at compile time. For example, we would not be able to write algorithmic HDL code to implement a microprocessor to sum the numbers from 1 to n, where n is a user input number, and therefore is not known before the run.

Another problem is that because the synthesizer will construct both the FSM and the datapath from the algorithmic code automatically, you do not need to specify the states for the FSM. Furthermore, you have no control over what components are used in the datapath, and what control words are executed in what state of the FSM. Not being able to decide what components are used in the datapath is not too big of a problem, because the synthesizer does do a good job in deciding that for you. The issue is with not being able to specify the states and what control words are executed in what state of the FSM. This is purely a timing issue. In some timing-critical applications (such as communication protocols and real-time controls) we need to control exactly in what clock cycle a certain register-transfer operation is performed. In other words, we need to be able to assign a control word to a specific state of the FSM.

Figures 7.50 and 7.51 show the Verilog and VHDL code, respectively, for the complete dedicated microprocessor using the algorithmic model for the summation problem.

```
module Summation (
  input Reset,
  output reg [7:0] Output
);

  reg [7:0]sum;
  reg [3:0]i;

  always @ (Reset) begin
     // Must use blocking assignment =
     sum = 0;
     for (i=1; i<=10; i=i+1) begin
       sum = sum + i;
     end
     Output = sum;
  end
endmodule
```

FIGURE 7.50 Verilog code for the algorithmic model of the summation problem.

```
LIBRARY IEEE;
USE IEEE.STD_LOGIC_1164.ALL;
USE IEEE.STD_LOGIC_UNSIGNED.ALL;

ENTITY Summation IS PORT (
  Reset: IN STD_LOGIC;
  Output: OUT STD_LOGIC_VECTOR(7 DOWNTO 0));
END Summation;
```

FIGURE 7.51 VHDL code for the algorithmic model of the summation problem.
(continued on next page)

```
ARCHITECTURE Algorithmic OF Summation IS
BEGIN
  PROCESS (Reset)
     VARIABLE sum: STD_LOGIC_VECTOR(7 DOWNTO 0);
     VARIABLE i: STD_LOGIC_VECTOR(7 DOWNTO 0);
  BEGIN
     sum := "00000000";
     FOR i IN 1 TO 10 LOOP
       sum := sum + i;
     END LOOP;
     Output <= sum;
  END PROCESS;
END Algorithmic;
```

FIGURE 7.51 VHDL code for the algorithmic model of the summation problem.

7.7 PROBLEMS

7.1. Derive the truth table for the datapath circuit shown next. The truth table should have columns only for the control signal inputs A_0, E, *Subtract*, and *OutE*, and the output signal *Output*. The data inputs, D_0 to D_3, are written in the table entries.

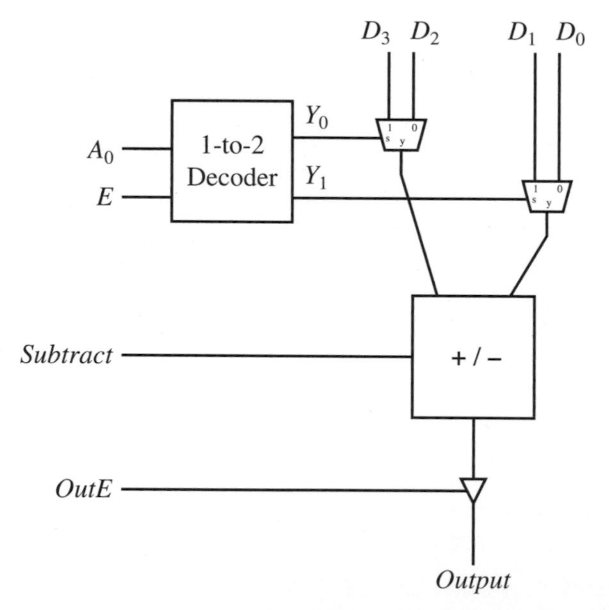

7.2. Implement on an FPGA the dedicated microprocessor for the GCD program from Section 7.5.1.

7.3. Implement on an FPGA the dedicated microprocessor for the high-low number guessing game from Section 7.5.2.

7.4. Implement on an FPGA the dedicated microprocessor for the traffic light controller from Section 7.5.3.

7.5. This question refers to the FSMD code listed in Section 7.6.2 for solving the summation of the numbers from 1 to 10 problem.

a) What is the total number of clock cycles required for this microprocessor to complete execution?

b) Modify the FSMD code to optimize this microprocessor so that it will require the least number of clock cycles to generate and sum the numbers from 1 to 10. You can change anything you like as long as it can generate and sum the numbers correctly. How many clock cycles does it require?

7.6. Manually design and implement on an FPGA a dedicated microprocessor to count from 1 to 10.

7.7. Manually design and implement on an FPGA a dedicated microprocessor using the following datapath to enter an 8-bit number n, and then output

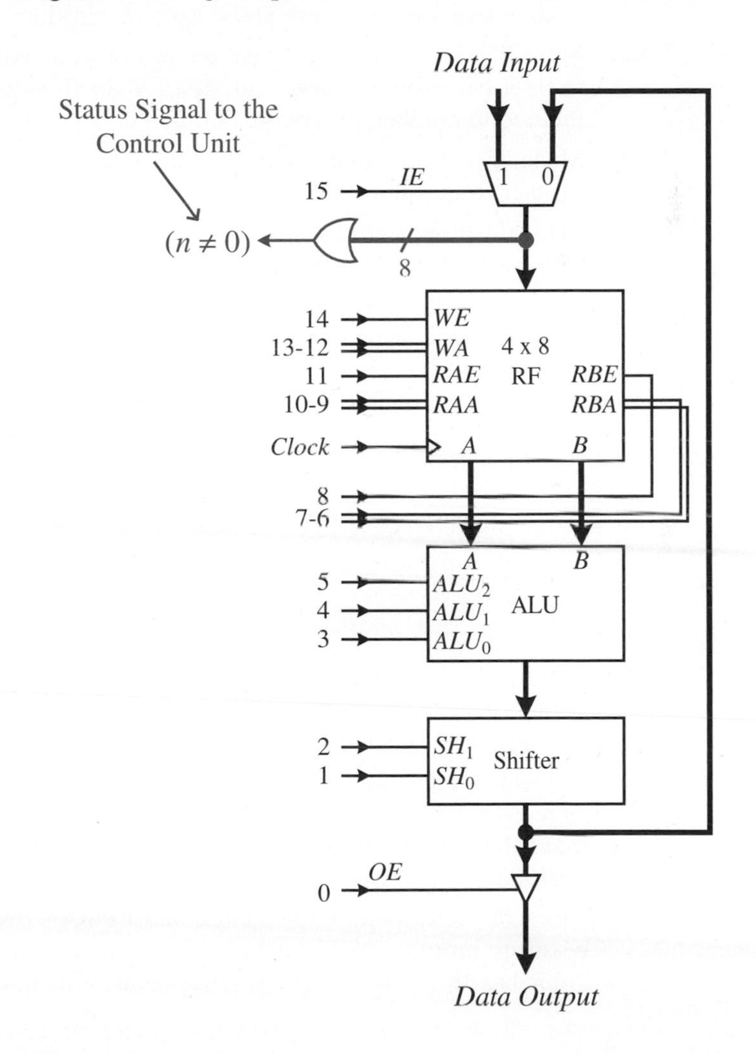

the sum of all the numbers from *n* down to 1. The operations of the ALU and the shifter are defined in Chapter 4. The operation of the register file is defined in Chapter 5.

7.8. Manually design and implement on an FPGA a dedicated microprocessor to enter two 8-bit numbers, and output the larger of the two numbers. The two numbers are entered through one input port.

7.9. Manually design and implement on an FPGA a dedicated microprocessor to enter one 8-bit number. Output a 1 if the number has five 1 bits; otherwise, output a 0.

7.10. Manually design and implement on an FPGA a dedicated microprocessor to enter two 8-bit numbers. Output a 1 if the two numbers together have five 1 bits; otherwise, output a 0.

7.11. Manually design and implement on an FPGA a dedicated microprocessor to enter two 8-bit numbers, and output the product of these two numbers.

7.12. Manually design and implement on an FPGA a dedicated microprocessor to enter two 8-bit numbers. Output a 1 if the first number is divisible by the second number; otherwise, output a 0.

7.13. Manually design and implement on an FPGA a dedicated microprocessor to enter three numbers, and output the larger of the three numbers.

7.14. Manually design and implement on an FPGA a dedicated microprocessor to enter three numbers. Output the three numbers in ascending order.

7.15. Manually design and implement on an FPGA a dedicated microprocessor to evaluate the factorial of *n*. The algorithm is shown next.

```
product = 1
INPUT n
WHILE (n > 1){
   product = product * n
   n = n - 1
   }
OUTPUT product
```

7.16. Manually design and implement on an FPGA a dedicated microprocessor to enter several numbers until a 0 is entered. Output the largest and second largest of the numbers entered.

7.17. Manually design and implement on an FPGA a dedicated microprocessor to read from eight input DIP switches. Output on the 7-segment the decimal number that represents the number of DIP switches that are in the on position.

7.18. Manually design and implement on an FPGA a dedicated microprocessor to input one 8-bit value, and then determine whether the input value has an equal number of 0 and 1 bits. The microprocessor outputs a 1 if the input value has the same number of 0's and 1's; otherwise, it outputs a 0. For example, the number 10111011 will produce a 0 output; whereas, the number 00110011 will produce a 1 output. The algorithm is shown next.

```
1 Count = 0                    // for counting the number of 1 bits
2 INPUT N
3 WHILE (N ≠ 0){
4    IF (N(0) = 1) THEN        // least significant bit of N
5       Count = Count + 1
6    END IF
7    N = N >> 1                // shift N right one bit
8 }
  // output 1 if the test (Count = 4) is true
9 OUTPUT (Count = 4)
```

7.19. Assume that the control unit and datapath circuits shown next are used to construct a dedicated microprocessor. Determine the instructions being executed in each state of the FSM. Write out the complete algorithm that the resulting microprocessor will execute. In other words, write out the pseudocode for the algorithm. Briefly describe what the algorithm does.

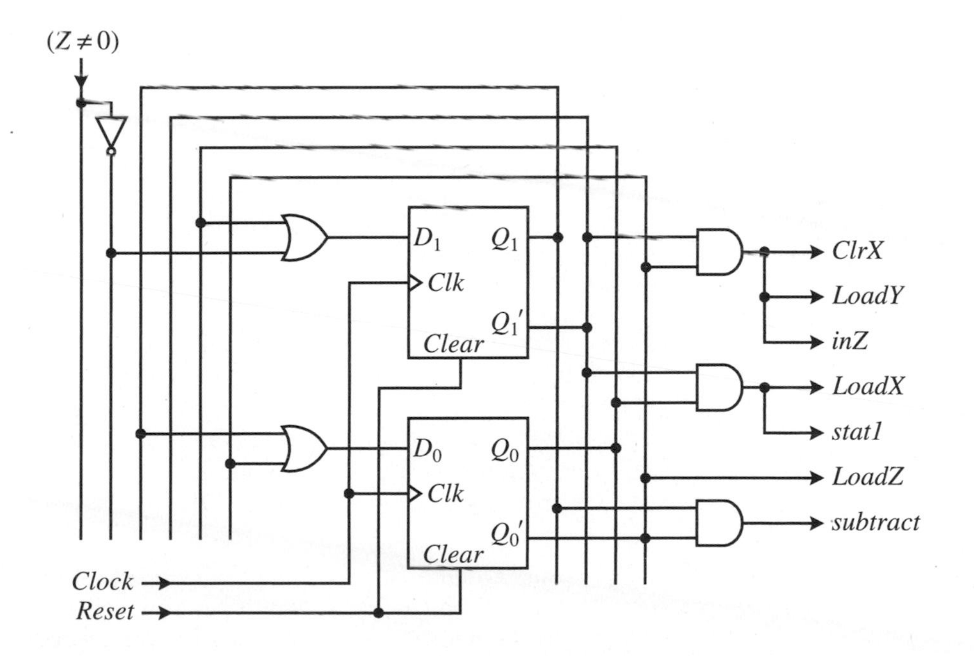

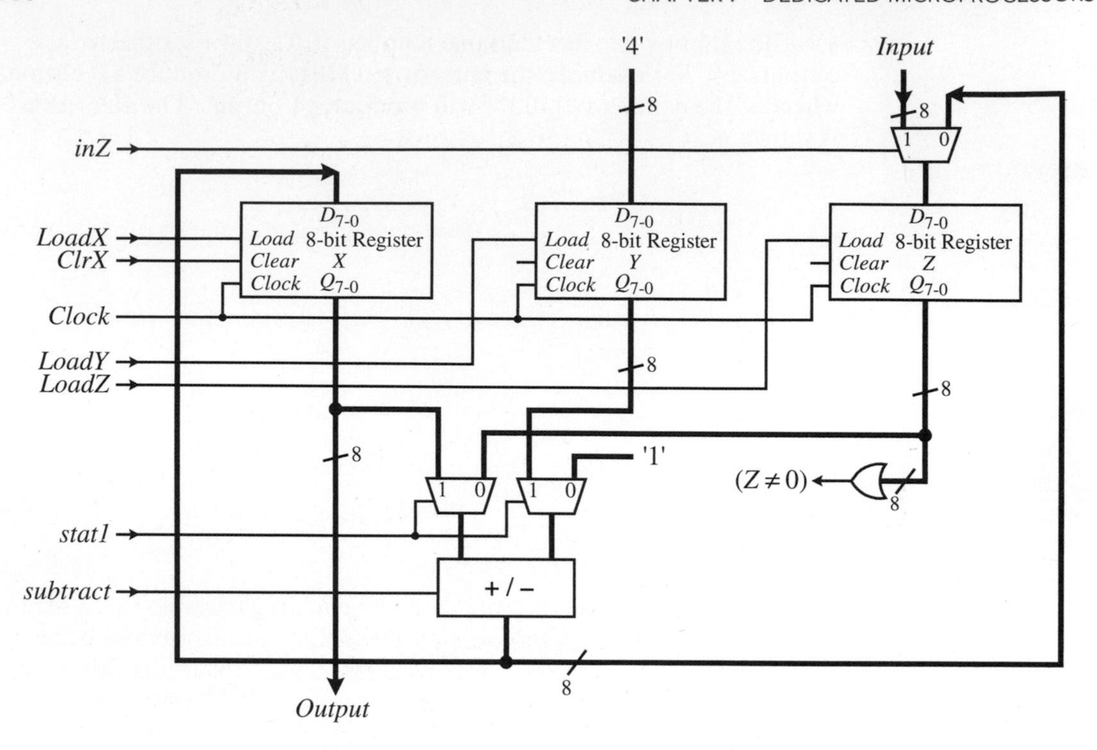

7.20. Manually design and implement on an FPGA a dedicated microprocessor for executing the algorithm shown next. Use only one adder-subtractor unit for all of the addition and subtraction operations. Label clearly all of the control and status signals.

```
w = 0
x = 0
y = 0
INPUT z
WHILE (z ≠ 0) {
  w = w - 2
  IF (z is an odd number) THEN
     x = x + 2
  ELSE
     y = y + 1
  END IF
  z = z - 1
  }
```

7.21. Manually design and implement on an FPGA a dedicated microprocessor for executing the algorithm shown next. Use only one adder (i.e., no adder-subtractor and no ALU) for all of the arithmetic operations. The datapath is 4 bits wide.

```
s1 = 0;
s2 = 0;
FOR(i=0; i ≠ 10; i++){
    INPUT j;
    IF (j is even) THEN
        s1++;
    ELSE
        s2++;
    END IF
    }
OUTPUT s1;
OUTPUT s2;
```

7.22. Manually design and implement on an FPGA a dedicated microprocessor for implementing a stack of size 10. When the *Push* signal is asserted, the input value is pushed onto the stack. When the *Pop* signal is asserted, the value at the top of the stack is output.

7.23. Manually design and implement on an FPGA a dedicated microprocessor to execute the Bubble Sort algorithm shown next.

```
int A[10];      // A is an integer array of size 10
                // you need to store A in 10 memory locations

// assume that A is initialized with random numbers

// this is the Bubble Sort routine to sort A to ascending order
repeat {
   swapped = false;
   for (j=1; j<10; j++) {
         if (A[j-1] > A[j]) {
                  temp = A[j];
                  A[j] = A[j-1];
                  A[j-1] = temp;
                  swapped = true;
         }
   }
until (!swapped);
}
```

7.24. Write the Verilog code for the above problems using the FSM+D model.

7.25. Write the Verilog code for the above problems using the FSMD model.

7.26. Write the VHDL code for the above problems using the FSM+D model.

7.27. Write the VHDL code for the above problems using the FSMD model.

CHAPTER 8

General-Purpose Microprocessors

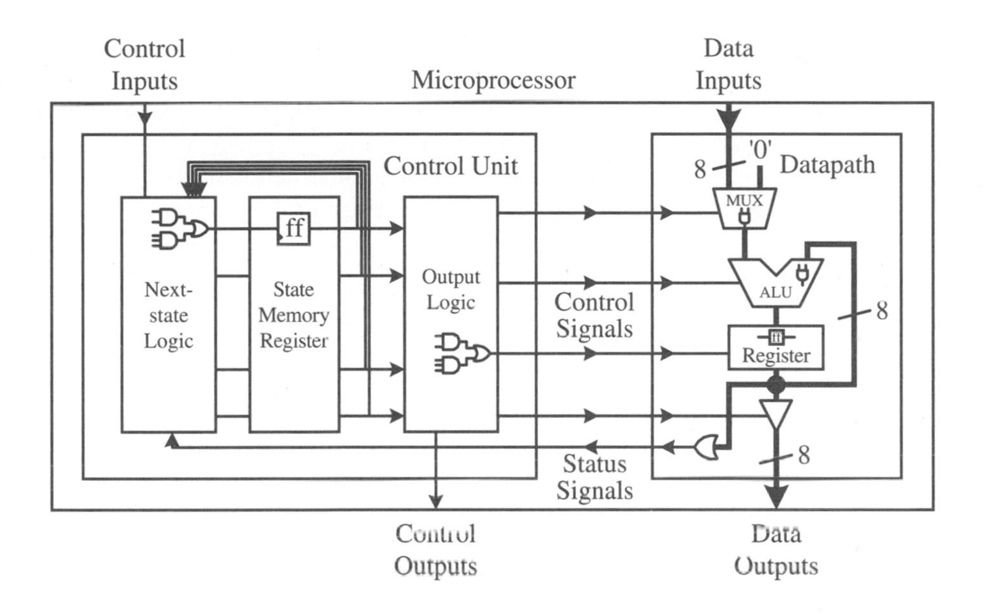

Unlike a dedicated or custom microprocessor that is capable of performing only one function, a general-purpose microprocessor, such as the Intel Core™i7 CPU, is capable of performing many different functions under the direction of program instructions. Given a different instruction set or program, the general-purpose microprocessor will perform a different function. However, a general-purpose microprocessor also can be viewed as a dedicated microprocessor, because it is made to perform only one function, and that is to execute the program instructions. In this sense, we can design and construct a general-purpose microprocessor in the same way that we constructed the dedicated microprocessors as discussed in Chapter 7.

8.1 Overview of the CPU Design

A general-purpose microprocessor is often referred to as the **central processing unit** (CPU). The CPU is simply a dedicated microprocessor that executes only software instructions. Figure 8.1 shows an overview of a general-purpose microprocessor. The following discussion references this diagram.

In designing a CPU, we must first define its instruction set and how the instructions are encoded and executed. We need to answer questions such as:

- How many instructions do we want?
- What are the instructions?
- What binary encoding (normally referred to as the **operation code** or **opcode**) do we assign to each of the instructions?

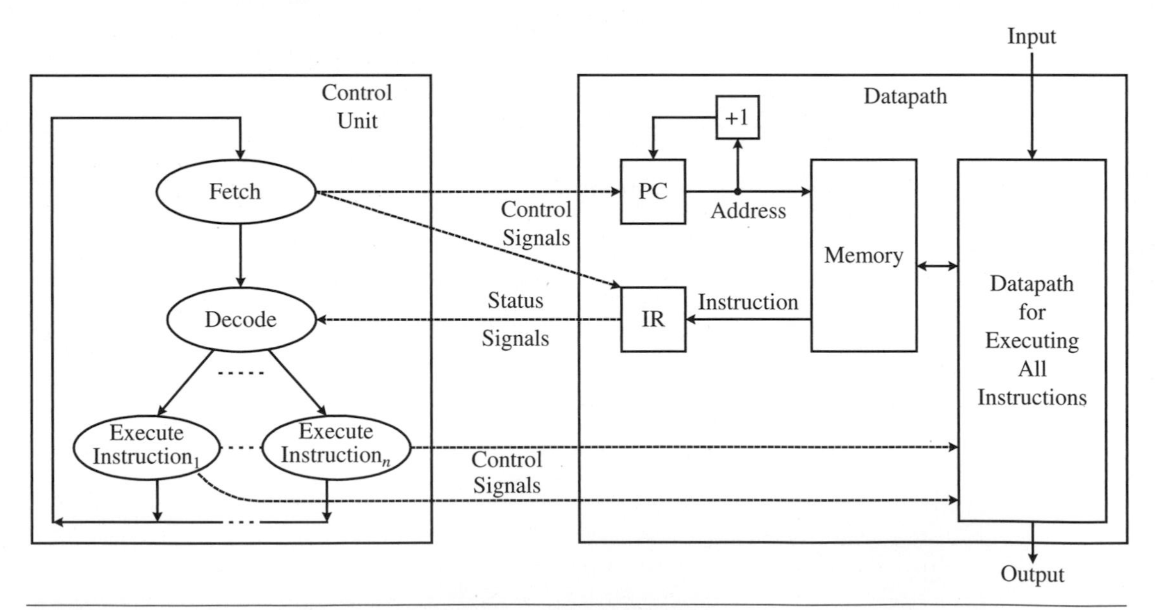

FIGURE 8.1 Overview of a general-purpose microprocessor.

- What are the operands in the instruction?
- How are the operands encoded?
- How many bits do we use to encode an instruction?

After we have decided on the instruction set, we can proceed to design a datapath that can execute all of the instructions in the instruction set. In this step, we are creating a custom datapath, so we need to answer questions such as:

- What functional units do we need?
- How many registers do we need?
- Do we use a single register file or separate registers?
- How are the different units connected together?

Creating the datapath for a general-purpose microprocessor is exactly the same as creating the datapath for a dedicated microprocessor. However, in addition to being able to perform all of the instructions in the instruction set, other data operations and registers must be included in the datapath for the general-purpose microprocessor. These data operations and registers deal with how the general-purpose microprocessor fetches the instructions from memory and executes them. In particular, the **program counter** (*PC*) register contains the memory location of where the next instruction is stored, and the **instruction register** (*IR*) stores the instruction being fetched from the memory. Usually after an instruction is fetched from the memory location pointed to by the *PC*, the *PC* is incremented to the next memory location, ready for the next instruction fetch. However, if the instruction is a jump instruction, the *PC* is loaded with the new memory address for the jump.

The control unit for a general-purpose microprocessor basically cycles through three main steps, usually referred to as the **instruction cycle**:

Step 1 fetches an instruction.
Step 2 decodes the instruction.
Step 3 executes the instruction.

Steps 1 and 2 are each executed in one state of the finite-state machine (FSM). For Step 3, most instructions will execute in one clock cycle, although some memory access instructions might require two or more clock cycles to complete. Therefore, they might require several states for correct timing.

For fetching the instruction in Step 1, the control unit simply reads the memory location specified by the *PC* and copies the content of that memory location into the *IR*. The *PC* is then incremented by 1 (assuming that each instruction occupies one memory location). For decoding the instruction in Step 2, the control unit extracts the opcode bits (which uniquely identifies the instruction) from the instruction register and determines what the current instruction is by jumping to the state that is assigned for executing that instruction. Once in that particular state, the control unit performs Step 3 simply by asserting the appropriate control signals to control the datapath to execute that instruction.

Instructions for the program usually are stored in external memory, so in addition to the CPU, external memory is connected to the CPU via an address bus and a data bus. Therefore, Step 1 (fetch an instruction) usually involves the control unit setting up a memory address on the address bus and telling the external memory to output the instruction from that memory location onto the data bus. The control unit then reads the instruction from the data bus. To keep our design simple, instead of having an external memory, we will include the memory as part of the datapath so that we do not have to worry about the handshaking and timing issues involved for accessing external memory.

8.2 The EC-1 General-Purpose Microprocessor

This first version of the EC[1] computer is extremely small and very limited as to what it can do, and therefore, its general-purpose microprocessor is very "EC"[2] to design manually. In order to keep the manual design of the microprocessor manageable, we have to keep the number of variables small. Because these variables determine the number of states and input signals for the finite-state machine, these factors have to be kept to the bare minimum. Nevertheless, the manual building of this computer demonstrates how a general-purpose microprocessor is designed and how the different components are put together. After this exercise, you will appreciate the power of designing with HDL at a higher abstraction level and the use of an automatic HDL synthesizer.

We first will manually design the general-purpose microprocessor for our EC-1 computer. Then we will interface this microprocessor to external I/Os, and implement the complete computer in an FPGA (field-programmable gate array) chip on a development board to make it into a real-working general-purpose microprocessor. Using the few instructions available in its instruction set, we then will write a program in machine language to execute on the EC-1 and see that it actually works.

8.2.1 Instruction Set

The instructions that our EC-1 general-purpose microprocessor can execute and the corresponding encodings for them are defined in Figure 8.2. The *Instruction* column shows the syntax and mnemonic to use for the instruction when writing a program in assembly language. The *Encoding* column shows the binary encoding defined for the instruction, and the *Operation* column shows the operation of the instruction.

As we can see from Figure 8.2, our EC-1's instruction set has only five instructions. To encode five instructions, the **operation code** (or **opcode**) will require three bits—giving us eight different combinations. As shown in the *Encoding* column, the first three most significant bits is the opcode given to the instruction. For example, the opcode for

[1] "EC" is the acronym for Enoch's Computer

[2] "EC" sounds like "easy"

Instruction	Encoding	Operation	Comment
IN A	011 ×××××	$A \leftarrow Input$	Input to A
OUT A	100 ×××××	$Output \leftarrow A$	Output from A
DEC A	101 ×××××	$A \leftarrow A - 1$	Decrement A
JNZ address	110 ×aaaa	IF (A ! = 0) THEN PC = aaaa	Jump to address if A is not zero
HALT	111 ×××××	Halt	Halt execution

Notations:

A = accumulator

PC = program counter

aaaa = four bits for specifying a memory address

× = don't-cares

FIGURE 8.2 Instruction set for the EC-1.

the IN A instruction is 011, the opcode for OUT A is 100, and so on. The three encodings, 000, 001, and 010, are not defined and so can be used as a "no-operation" (NOP) instruction. Because the width of each instruction is fixed at 8 bits, the last 5 bits are not used by all of the instructions, except for the JNZ (Jump Not Zero) instruction. Normally, for a more extensive instruction set, these extra bits are used as operand bits to specify what registers or other resources to use. In our case, only the JNZ instruction uses the last 4 bits, designated as aaaa, to specify an address in the memory to jump to.

The IN A instruction inputs an 8-bit value from the data input port, *Input*, and stores it into the **accumulator** (A). The accumulator is an 8-bit register for performing data operations. The OUT A instruction enables a tri-state buffer to output the content of the accumulator to the output port, *Output*. The DEC A instruction decrements the content of A by 1 and stores the result back into A. The JNZ instruction tests to see if the value in A is equal to 0 or not. If A is equal to 0, then nothing is done, but if A is not equal to 0, then the last 4 bits (aaaa) of the instruction are loaded into the PC. When this value is loaded into the PC, we essentially are performing a jump to this new memory address, because the value stored in the PC is the memory location for the next fetch operation. Finally, the HALT instruction halts the CPU by having the control unit stay in the *Halt* state indefinitely until reset.

8.2.2 Datapath

Having defined the instruction set for the EC-1 general-purpose microprocessor, we now are ready to design the custom datapath that will execute all of the operations as defined by all of the instructions. The custom datapath for the EC-1 is shown in Figure 8.3.

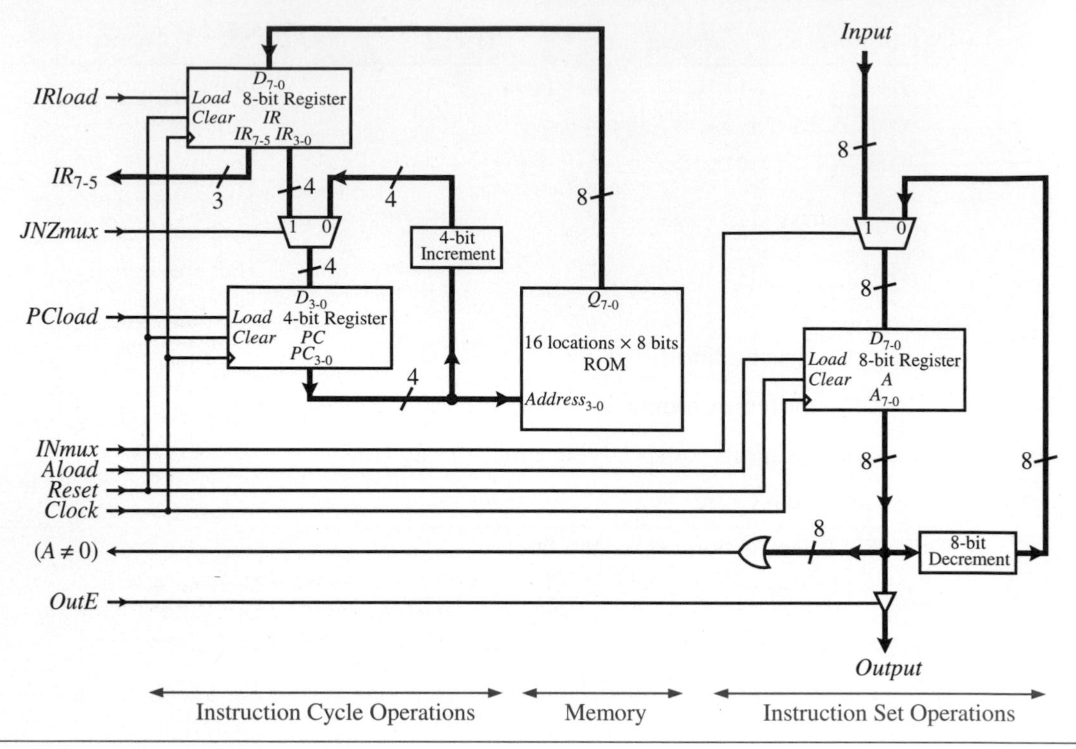

FIGURE 8.3 Datapath for the EC-1.

The datapath can be viewed as having three separate parts: (1) the portion that performs the instruction cycle operations of fetching an instruction and either incrementing or loading the *PC*, (2) the memory, and (3) the portion that performs the data operations for all of the instructions in the instruction set.

The portion of the datapath that performs the instruction cycle operations basically contains the instruction register (*IR*) and the program counter (*PC*). The bit width of the instructions determines the size of the *IR*; whereas, the number of addressable memory locations determines the size of the *PC*. For this datapath, we want a memory with 16 locations, each being 8 bits wide, so we need a 4-bit ($2^4 = 16$) address. Therefore, the *PC* is 4 bits wide, and the *IR* is 8 bits wide. A 4-bit increment unit is used to increment the *PC* by 1. The *PC* needs to be loaded with either the result of the increment unit for the next instruction in memory or the 4-bit address from the JNZ instruction; therefore, a 2-to-1 multiplexer is needed for this purpose. One input of the multiplexer is from the increment unit, and the other input is from the four least significant bits of the *IR*, IR_{3-0}.

Instead of having external memory, we have included the memory as part of the datapath in order to keep our first design simple. In this design, the memory is a 16-location × 8-bit wide read-only memory (ROM). We use a ROM here instead of a RAM because the instruction set does not have an instruction that writes to memory. The output of the *PC* is connected directly to the 4-bit memory address lines, because the memory location always is determined by the content of the *PC*. The 8-bit memory output, Q_{7-0}, is connected to the input of the *IR* to execute the instruction fetch operation (Step 1 of the instruction cycle). The construction of this memory was discussed in Section 5.14.1.

The portion of the datapath that performs the instruction set operations includes the 8-bit register for the accumulator *A*, and an 8-bit decrement unit. A 2-to-1 multiplexer is used to select the input to the accumulator. For the IN A instruction, the input to the accumulator is from the data input port, *Input*; whereas for the DEC A instruction, the input is from the output of the decrement unit, which performs the decrement of *A*. The output of the accumulator is connected via a tri-state buffer to the data output port, *Output*. The JNZ instruction requires an 8-input OR gate connected to the output of the accumulator to test for the condition $(A \neq 0)$. The actual operation required by the JNZ instruction is to load the *PC* with the four least significant bits of the *IR*. The HALT instruction also does not require any specific datapath actions. It simply asserts a *Halt* signal to notify the external world that the program has halted.

The control word for this custom datapath has six control signals: *IRload*, *PCload*, *INmux*, *Aload*, *JNZmux*, and *OutE*. The datapath provides two status signals, IR_{7-5} and $(A \neq 0)$, to the control unit. The three *IR* bits, IR_{7-5}, which forms the opcode, are sent to the control unit for the instruction decode operation (Step 2 of the instruction cycle). The control words for executing the instruction cycle operations and the instruction set operations are discussed in the next section.

8.2.3 **Control Unit**

The state diagram for the control unit is shown in Figure 8.4(a), and the actions that are executed, specifically the control signals that are asserted in each state, are shown in Figure 8.4(d).

In the *Fetch* state, 000, the *IR* is loaded with the memory content from the location specified by the *PC* by asserting the *IRload* signal. Also in this state, the *PC* is incremented by 1, and the result is loaded back into the *PC* by asserting the *PCload* signal. In the output table shown in Figure 8.4(d), these two assertions are denoted by the two 1s under the *IRload* and *PCload* columns for control word 1. There is no timing conflict in asserting both the *IRload* and *PCload* signals together in the same clock cycle because the *PC* will not be updated with the incremented value until the next clock cycle, and at that time, the *IR* already will have been written with the newly fetched instruction. This is correct only if the memory can be accessed in one clock cycle.

After fetching, the FSM goes to the *Decode* state unconditionally. In the *Decode* state, the FSM tests the three most significant bits of the *IR*, IR_{7-5}, and goes to the corresponding state as encoded by the 3-bit opcode for executing the instruction. Testing of the opcode bits does not involve any datapath operations and so the control word in the output table for this state contains all 0s.

In the five instruction execute states corresponding to the five instructions, the appropriate control signals for the datapath are asserted to execute that instruction. For example, the IN A instruction requires setting the *INmux* signal to a 1 for the input multiplexer, and setting the *Aload* signal to a 1 to load the input value into *A*. In order

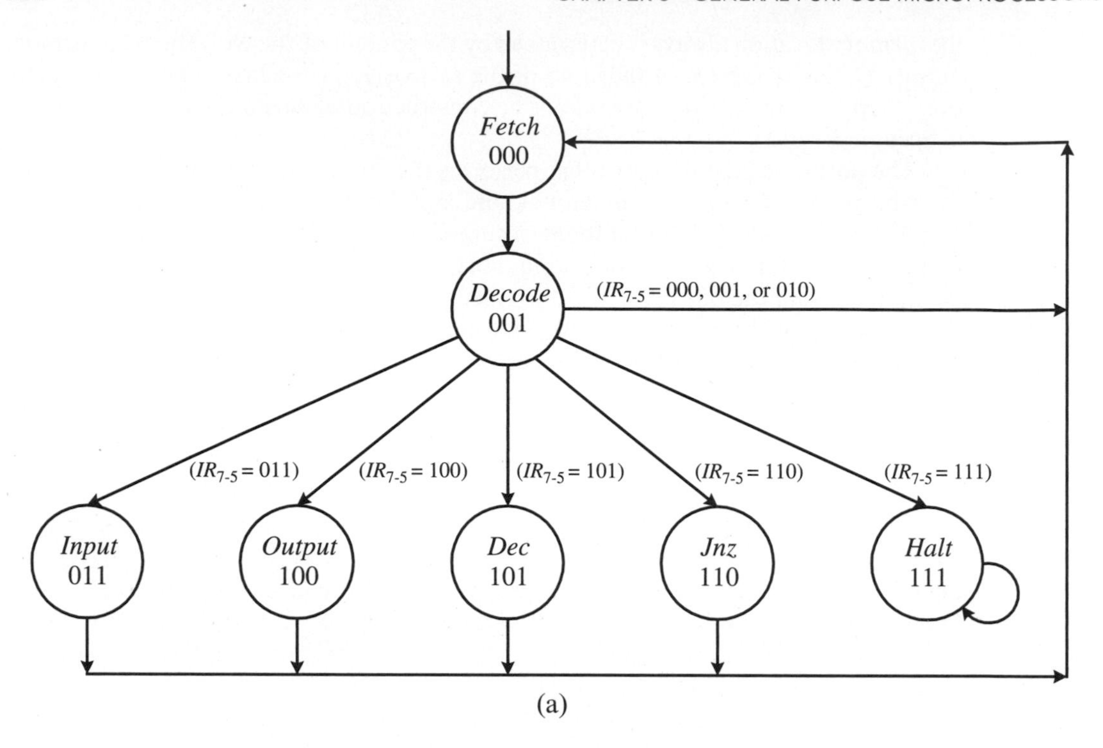

(a)

Current State $Q_2Q_1Q_0$	Next State (Implementation) $Q_{2next}Q_{1next}Q_{0next}$ ($D_2D_1D_0$)							
	IR_7, IR_6, IR_5							
	000 NOP	**001** NOP	**010** NOP	**011** INPUT	**100** OUTPUT	**101** DEC	**110** JNZ	**111** HALT
000 *Fetch*	001	001	001	001	001	001	001	001
001 *Decode*	000	000	000	011	100	101	110	111
011 *Input*	000	000	000	000	000	000	000	000
100 *Output*	000	000	000	000	000	000	000	000
101 *Dec*	000	000	000	000	000	000	000	000
110 *Jnz*	000	000	000	000	000	000	000	000
111 *Halt*	111	111	111	111	111	111	111	111

(b)

FIGURE 8.4 Control unit for the EC-1: (a) state diagram; (b) next-state (implementation) table; (c) next-state equations; (d) control words and output table; (e) output equations; (f) circuit.
(continued on next page)

$$Q_{2next} = D_2 = Q_2'Q_1'Q_0IR_7 + Q_2Q_1Q_0$$
$$Q_{1next} = D_1 = Q_2'Q_1'Q_0(IR_6IR_5 + IR_7IR_6) + Q_2Q_1Q_0$$
$$Q_{0next} = D_0 = Q_2'Q_1'Q_0' + Q_2'Q_1'Q_0(IR_6IR_5 + IR_7IR_5) + Q_2Q_1Q_0$$

(c)

Control Word	State $Q_2Q_1Q_0$	*IRload*	*PCload*	*INmux*	*Aload*	*JNZmux*	*OutE*	*Halt*
1	000 *Fetch*	1	1	0	0	0	0	0
2	001 *Decode*	0	0	0	0	0	0	0
3	011 *Input*	0	0	1	1	0	0	0
4	100 *Output*	0	0	0	0	0	1	0
5	101 *Dec*	0	0	0	1	0	0	0
6	110 *Jnz*	0	IF ($A \neq 0$) THEN 1 ELSE 0	0	0	1	0	0
7	111 *Halt*	0	0	0	0	0	0	1

(d)

$$IRload = Q_2'Q_1'Q_0'$$
$$PCload = Q_2'Q_1'Q_0' + Q_2Q_1Q_0'(A \neq 0)$$
$$INmux = Q_2'Q_1Q_0$$
$$Aload = Q_2'Q_1Q_0 + Q_2Q_1'Q_0$$
$$JNZmux = Q_2Q_1Q_0'$$
$$OutE = Q_2Q_1'Q_0'$$
$$Halt = Q_2Q_1Q_0$$

(e)

FIGURE 8.4 Control unit for the EC-1: (a) state diagram; (b) next-state (implementation) table; (c) next-state equations; (d) control words and output table; (e) output equations; (f) circuit. *(continued on next page)*

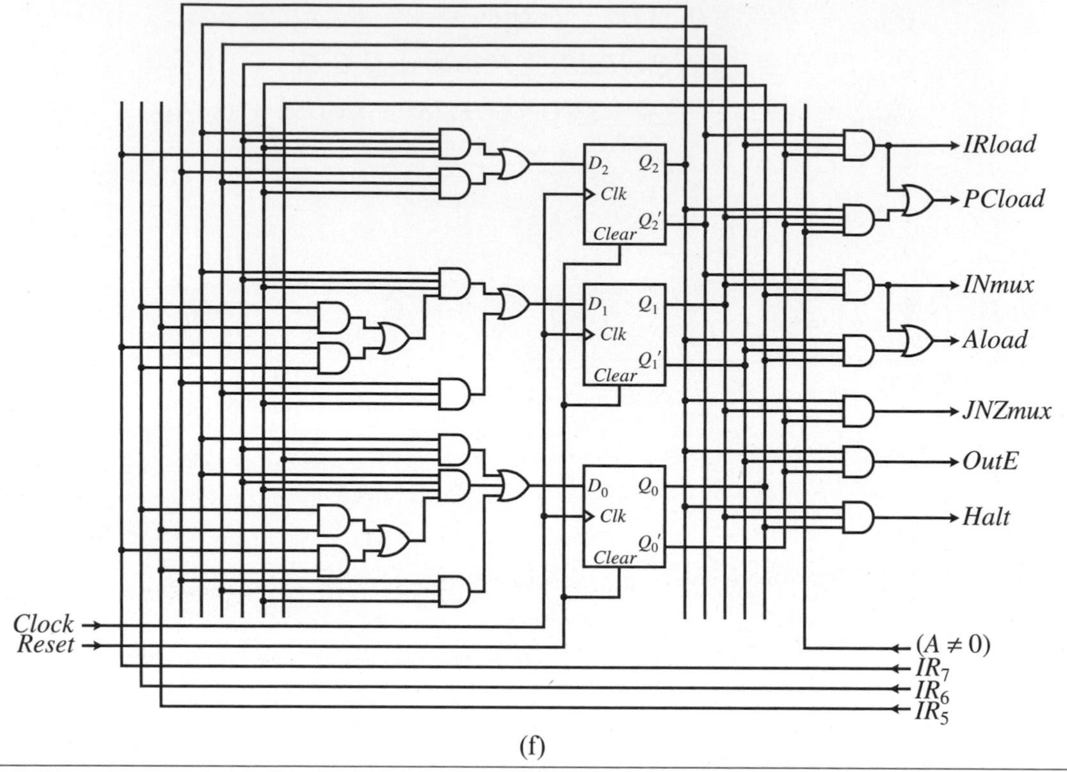

FIGURE 8.4 Control unit for the EC-1: (a) state diagram; (b) next-state (implementation) table; (c) next-state equations; (d) control words and output table; (e) output equations; (f) circuit.

for the input instruction to read in the correct value, the input value must be set up first before resetting the CPU. Furthermore, because the *Input* state does not wait for an Enter key signal, only one value can be read in, even if there are multiple input statements. The reason for these limitations is because the FSM clock speed is very fast and so there is not enough time for the user to change the input value between input statements. The OUT A instruction simply asserts the enable signal on the output tri-state buffer. The DEC A instruction requires setting *INmux* to 0 and *Aload* to 1, so the output from the decrement unit is routed back to the accumulator and gets loaded in. Control words 3 to 5 in the output table shown in Figure 8.4(d) show the assertions of these control signals.

The JNZ instruction asserts the *JNZmux* signal to route the four address bits from the *IR*, $IR_{3\text{-}0}$, to the *PC*. Whether the *PC* actually gets loaded with this new address depends on the condition of the $(A \neq 0)$ status signal. Therefore, the *PCload* control signal is asserted only if $(A \neq 0)$ is a 1. If we use the Moore FSM model, the FSM will require two states for the JNZ instruction: one state for asserting the *PCload* signal when $(A \neq 0)$ is true, and one state for de-asserting the *PCload* signal when $(A \neq 0)$ is false. However, if we use the Mealy FSM model by asserting the *PCload* signal conditionally based on the status signal $(A \neq 0)$, that is, asserting the *PCload* signal on an edge rather than in a state, then only one state is needed to execute the JNZ instruction. To do this, we see that in the output table shown in Figure 8.4(d), control word 6 for executing the JNZ instruction has a conditional value under the *PCload* control signal column showing that *PCload* gets a 1 only if the condition $(A \neq 0)$ is true, otherwise, it gets a 0.

Once the FSM enters the *Halt* state, it unconditionally loops back to itself, giving the impression that the CPU has halted. In this state, we also assert the *Halt* signal to notify the external world that execution has stopped.

With seven states, three flip-flops are needed for the state memory of the control unit. The next-state table shown in Figure 8.4(b) lists the current state with the three flip-flops Q_2, Q_1, and Q_0. The entries in the table are the next-state values Q_{2next}, Q_{1next}, and Q_{0next}. By using D flip-flops for the state memory, the next-state values (Q_{next}) will be the same as the implementation values (*D*) because the characteristic equation for the D flip-flop is $Q_{next} = D$. Notice that the next-state entries in the table are quite simple to follow because almost all of them are unconditional, that is, the same value across each row. Only the *Decode* state has conditional edges that depend on the opcode bits.

There is one next-state equation for each of the three D flip-flops used. The three next-state equations, as derived from the next-state table, are shown in Figure 8.4(c). The derivation of the next-state equations is fairly easy, because most of the entries in the next-state table are 0s. The output equations shown in Figure 8.4(e) are derived directly from the output table shown in Figure 8.4(d).

Finally, we can derive the circuit for the control unit based on the next-state equations and the output equations. The complete control unit circuit for the EC-1 general-purpose microprocessor is shown in Figure 8.4(f).

8.2.4 **Complete Circuit**

The complete circuit for the EC-1 general-purpose microprocessor shown in Figure 8.5 is constructed by connecting the datapath from Figure 8.3 and the control unit from Figure 8.4(f) together using the designated control and status signals. The complete schematic circuit and HDL code for the EC-1 microprocessor can be downloaded from the book website.

8.2.5 **Sample Program**

Dedicated microprocessors, as discussed in Chapter 7, have the algorithm built into the hardware circuit of the microprocessor. General-purpose microprocessors, on the other hand, do not have a built-in algorithm. They are designed only to execute program instructions fetched from the memory. Therefore, in order to test the EC-1 computer, we need to write a program using the instructions available in the instruction set, and have this program loaded into the memory.

Only five instructions are defined in the EC-1 instruction set, as shown in Figure 8.2. For our sample program, we will use these five instructions to write a countdown program to input a number and then count down from this input number to 0. The program is shown in Figure 8.6(a).

Because we do not have a compiler for the EC-1, we need to manually compile this program. The binary executable code for this program is shown in Figure 8.6(b).

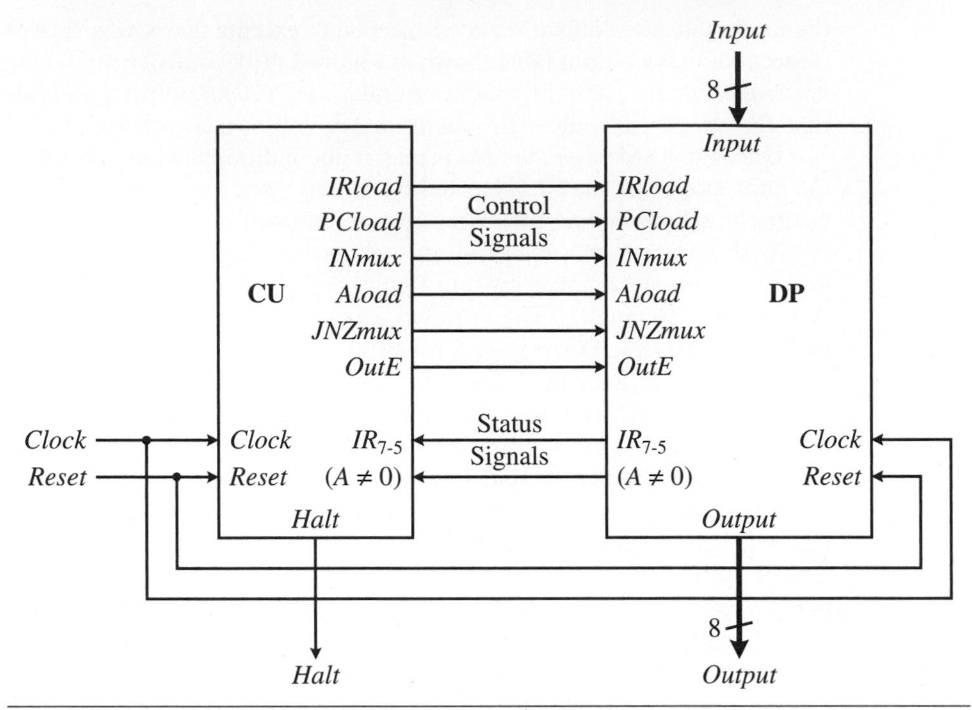

FIGURE 8.5 Complete circuit for the EC-1 general-purpose microprocessor.

```
        IN A          -- input a value into the A register
loop:   OUT A         -- output the value from the A register
        DEC A         -- decrement A by one
        JNZ loop      -- go back to loop if A is not zero
        HALT          -- halt
```

(a)

```
memory        instruction
address       encoding
0000          01100000;     -- IN A
0001          10000000;     -- OUT A
0010          10100000;     -- DEC A
0011          11000001;     -- JNZ 0001
0100          11111111;     -- HALT
```

(b)

FIGURE 8.6 Countdown program to run on the EC-1: (a) assembly code; (b) binary executable code.

The binary code is obtained by replacing each instruction with its corresponding 3-bit opcode, as listed in Figure 8.2, followed by five bits for the operand. None of the instructions, except for the JNZ instruction, uses these five operand bits, so either a 0 or a 1 can be used. From Figure 8.2, we find that the opcode for the IN A instruction is 011; therefore, the encoding for this first instruction is 01100000. Similarly, the opcode for the OUT A instruction is 100; therefore, the encoding used is 10000000.

For the JNZ instruction, the four least significant bits represent the memory address to jump to if the condition is true. In the example, we have put the first instruction, IN A, in memory location 0000. Because the JNZ instruction jumps to the second instruction, OUT A, which is stored in memory location 0001, therefore, the four address bits for the JNZ instruction are 0001. The opcode for the JNZ instruction is 110, therefore, the encoding for the complete JNZ instruction is 11000001.

Typically, with the memory being external to the CPU, the computer (with the help of the operating system) will provide means to independently load the program instructions into the memory. However, to keep our design simple, we have included the memory as part of the CPU inside the datapath. Furthermore, we do not have an operating system for loading the instructions into the memory separately. In our EC-1 design, we have chosen to use a simple ROM for the memory that already has been initialized with the countdown program instructions. Refer to Section 5.14.1 for a full discussion and HDL code of the ROM with the program instructions built in. Because the program is synthesized together with the ROM, each time you change the program, you will have to re-synthesize the whole system.

8.2.6 **Simulation**

Figure 8.7 shows a sample simulation of the EC-1, showing the countdown from the input 3 on the *Output* signal.

8.2.7 **Hardware Implementation**

A complete computer system includes not only the microprocessor but also the memory, input, and output devices. So far, we have constructed the general-purpose microprocessor with the built-in memory, as shown in Figure 8.5. To see our own EC-1 microprocessor work, we need to connect it to input and output devices.

Figure 8.8 shows the interface between the EC-1 microprocessor with the input and output devices on an FPGA development board. The input device consists of eight switches, and the output device is three 7-segment LED displays. Because the microprocessor outputs an 8-bit binary number, we need a converter to convert the

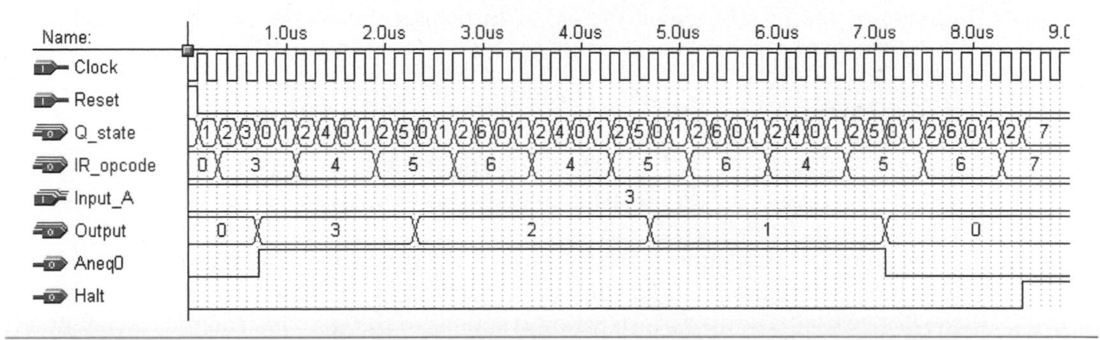

FIGURE 8.7 A sample simulation trace of the countdown program running on the EC-1 starting at the input 3.

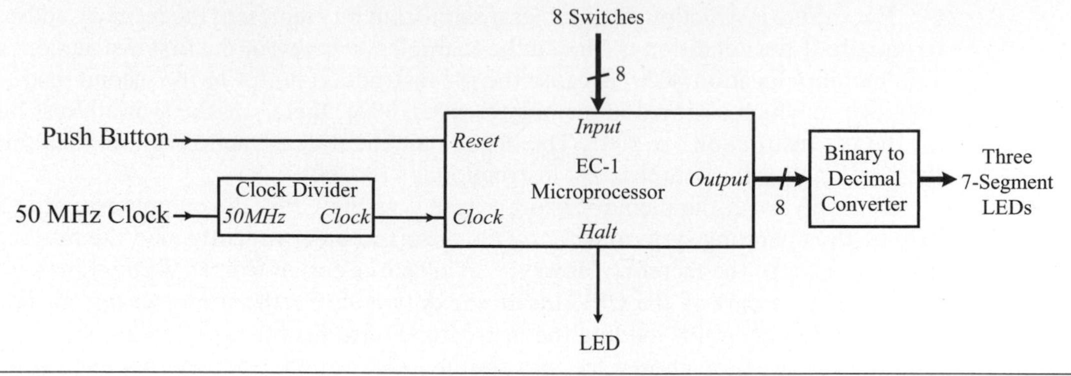

FIGURE 8.8 Hardware implementation of the EC-1 computer.

8-bit binary number to three BCD (binary coded decimal) digits. With this converter, we will be able to see the output as a 3-digit decimal number. See Problem 3.36 for the HDL code for this BCD converter. The HDL code for this converter is also available from the book website. A single LED is connected to the *Halt* signal to show when the microprocessor has finished running the program. A push button is connected to the *Reset* signal to reset the microprocessor. A clock divider circuit is used to slow down the clock frequency so that we can see the program being executed, that is, to actually see the numbers counting down. Otherwise, the countdown will finish so quickly that we will see only the last number, zero, on the display. To run the program, first set up the binary input number on the eight switches, and then press the *Reset* button. The countdown will begin with the input number. When the countdown reaches zero, the program stops and the *Halt* light turns on.

8.3 The EC-2 General-Purpose Microprocessor

For our next example, we will design the general-purpose microprocessor for a second version of the EC computer, the EC-2. In this second version, we have added a few more instructions, and we will be able to load and store data to the memory.

8.3.1 Instruction Set

The instruction set for the EC-2 general-purpose microprocessor has eight instructions, as shown in Figure 8.9. We keep this number at eight so that we can still use only three bits to encode them.

The LOAD instruction loads the content of the memory at the specified address into the accumulator *A*. The address is specified by the five least significant bits of the instruction. The STORE instruction is similar to the LOAD instruction, except that it stores the value in *A* to the memory at the specified address. The ADD and SUB instructions, respectively, add and subtract the content of *A* with the content in a memory

Instruction	Encoding	Operation	Comment
LOAD A, address	000 aaaaa	$A \leftarrow M[\text{aaaaa}]$	Load *A* with content of memory location aaaaa
STORE A, address	001 aaaaa	$M[\text{aaaaa}] \leftarrow A$	Store *A* into memory location aaaaa
ADD A, address	010 aaaaa	$A \leftarrow A + M[\text{aaaaa}]$	Add *A* with *M*[aaaaa] and store result back into *A*
SUB A, address	011 aaaaa	$A \leftarrow A - M[\text{aaaaa}]$	Subtract *A* with *M*[aaaaa] and store result back into *A*
IN A	100 ×××××	$A \leftarrow Input$	Input to *A*
JZ address	101 aaaaa	IF ($A = 0$) THEN PC = aaaaa	Jump to address if *A* is zero
JPOS address	110 aaaaa	IF ($A \geq 0$) THEN PC = aaaaa	Jump to address if *A* is zero or a positive number
HALT	111 ×××××	Halt	Halt execution

Notations:

A = accumulator

M = memory

PC = program counter

aaaaa = five bits for specifying a memory address

× = don't cares

FIGURE 8.9 Instruction set for the EC-2.

location and store the result back into *A*. The IN instruction inputs a value from the data input port, *Input*, and stores it into *A*. The JZ (Jump if Zero) instruction loads the *PC* with the specified address if *A* is zero. Loading the *PC* with a new address causes the CPU to jump to this new memory location. The JPOS (Jump if Positive) instruction loads the *PC* with the specified address if *A* is zero or a positive number. The value in *A* is taken as a two's complement signed number, so a positive number is one where the most significant bit of the number is a 0, which includes the number zero. Finally, the HALT instruction halts the CPU. The value in the accumulator *A* is continually sent to the output so no output instruction is necessary.

8.3.2 Datapath

The custom datapath for the EC-2 is shown in Figure 8.10. The portion of the datapath for performing the instruction cycle operations is similar to that of the EC-1, with the instruction register (*IR*), the program counter (*PC*), and the increment unit for incrementing the *PC*. The minor differences between the two are in the size of the *PC*

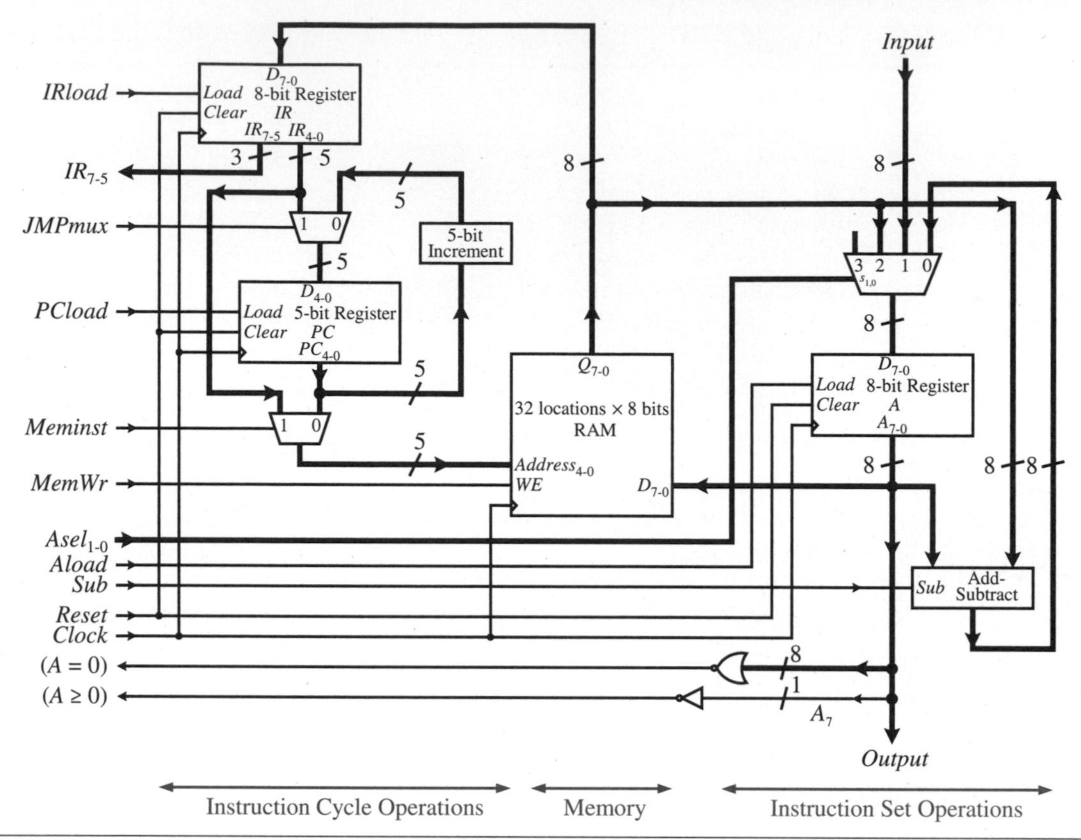

FIGURE 8.10 Datapath for the EC-2.

and the increment unit. For the EC-2, we want a memory with 32 locations, therefore, the memory address, the size of the *PC*, and the increment unit must all be 5 bits wide.

The main modification to this portion of the datapath is the addition of a second 2-to-1 multiplexer that is connected between the output of the *PC* and the memory address input. One input of this multiplexer comes from the *PC*, and the other input comes from the five least significant bits of the *IR*, IR_{4-0}. The reason for this is because there are now two different types of operations that can access the memory. The first is still for the fetch operation, where the memory address is given by the content of the *PC*. The second type is for the four instructions, LOAD, STORE, ADD, and SUB, where they use the memory as an operand. The memory address for these four instructions is from the five least significant bits of the *IR*, IR_{4-0}. The select signal for this multiplexer is *Meminst*.

The memory size for the EC-2 is increased to 32 locations, thus requiring five address bits. The memory is still included as part of the datapath rather than as an independent external unit to the CPU. The memory also has separate read and write data ports so that a bidirectional data bus is not required. In order to accommodate the STORE instruction for storing the value of *A* into the memory, we need to use a RAM instead of a ROM as in the EC-1. To realize this operation, the output of the accumulator A is connected to the memory data input, $D_{7\text{-}0}$. The signal *MemWr*, when asserted, causes the memory to write the value from register A into the location specified by the address in the instruction.

The output of the memory at $Q_{7\text{-}0}$ is connected to both the input of the IR and to the input of the accumulator, A, through a 4-to-1 multiplexer. The connection to the IR is for the fetch operation just like in the EC-1 design. The connection to the accumulator is to perform the LOAD instruction, where the content of the memory is loaded into A. Because the memory is only one source among two other sources that are loaded into A, the multiplexer is needed. The construction of this RAM was discussed in Section 5.14.2.

The portion of the datapath for performing the instruction set operations includes the 8-bit accumulator A, an 8-bit adder-subtractor combination unit, and a 4-to-1 multiplexer. The adder-subtractor unit performs the ADD and SUB instructions. The *Sub* signal, when asserted, selects the subtraction operation, and when de-asserted, it selects the addition operation. The 4-to-1 multiplexer allows the accumulator input to come from one of three sources. For the ADD and SUB instructions, the A input comes from the output of the adder-subtractor unit. For the IN instruction, the A input comes from the data input port, *Input*. For the LOAD instruction, the A input comes from the output of the memory, $Q_{7\text{-}0}$. The selection of this multiplexer is through the two signal lines, $Asel_{1\text{-}0}$. The fourth input of the multiplexer is not used.

The output of the accumulator is connected directly to the output port, so the value of the accumulator is always available at the output port. Therefore, no specific output instruction is necessary to output the value in A.

For the two conditional jump instructions JZ and JPOS, the datapath provides the two status signals $(A = 0)$ and $(A \geq 0)$, respectively, that are generated from two comparators. The $(A = 0)$ status signal outputs a 1 if the value in A is a 0, so an 8-input NOR gate is used. The $(A \geq 0)$ status signal outputs a 1 if the value in A, which is treated as a two's complement signed number, is a zero or positive number. Because for a two's complement signed number, a leading 0 means positive and a leading 1 means negative, the condition $(A \geq 0)$ is simply the negated value of bit A_7 (the most significant bit of A).

The control word for this custom datapath has eight control signals, *IRload*, *JMPmux*, *PCload*, *Meminst*, *MemWr*, $Asel_{1\text{-}0}$, *Aload*, and *Sub*, but requires nine bits because $Asel_{1\text{-}0}$ has two bits. The datapath provides three status signals $IR_{7\text{-}5}$, $(A = 0)$, and $(A \geq 0)$ to the control unit. The three IR bits, $IR_{7\text{-}5}$, which forms the opcode, are sent to the control unit for the instruction decode operation (Step 2 of the instruction cycle). The control words for executing the instruction cycle operations and the instruction set operations are discussed in the next section.

8.3.3 **Control Unit**

The state diagram for the control unit is shown in Figure 8.11(a), and the actions that are executed, specifically the control signals that are asserted in each state, are shown in Figure 8.11(d). States for executing the instructions are given the same name as the

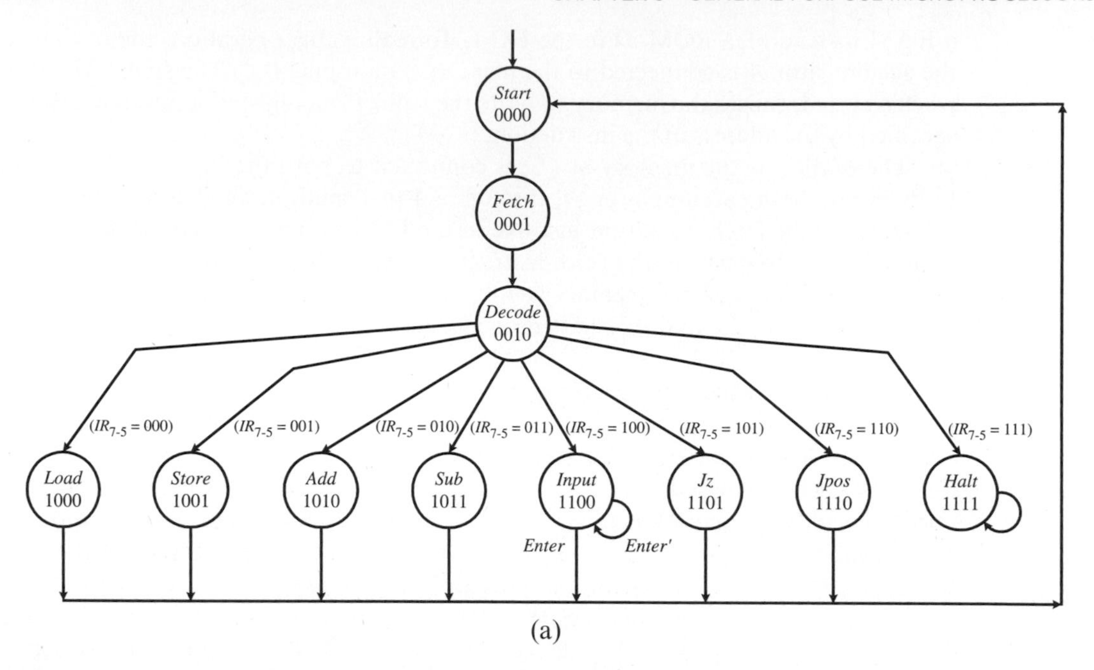

(a)

Current State $Q_3Q_2Q_1Q_0$	Next State (Implementation) $Q_{3next}Q_{2next}Q_{1next}Q_{0next}$ ($D_3D_2D_1D_0$)									
	IR_7, IR_6, IR_5								*Enter*	
	000 LOAD	001 STORE	010 ADD	011 SUB	100 INPUT	101 JZ	110 JPOS	111 HALT	0	1
0000 *Start*	0001	0001	0001	0001	0001	0001	0001	0001		
0001 *Fetch*	0010	0010	0010	0010	0010	0010	0010	0010		
0010 *Decode*	1000	1001	1010	1011	1100	1101	1110	1111		
1000 *Load*	0000	0000	0000	0000	0000	0000	0000	0000		
1001 *Store*	0000	0000	0000	0000	0000	0000	0000	0000		
1010 *Add*	0000	0000	0000	0000	0000	0000	0000	0000		
1011 *Sub*	0000	0000	0000	0000	0000	0000	0000	0000		
1100 *Input*									1100	0000
1101 *Jz*	0000	0000	0000	0000	0000	0000	0000	0000		
1110 *Jpos*	0000	0000	0000	0000	0000	0000	0000	0000		
1111 *Halt*	1111	1111	1111	1111	1111	1111	1111	1111		

(b)

FIGURE 8.11 Control unit for the EC-2: (a) state diagram; (b) next-state (implementation) table; (c) next-state equations; (d) control word and output table; (e) output equations; (f) circuit. *(continued on next page)*

$$Q_{3next} = D_3 = Q_3'Q_2'Q_1Q_0' + Q_3Q_2Q_1'Q_0'Enter' + Q_3Q_2Q_1Q_0$$
$$Q_{2next} = D_2 = Q_3'Q_2'Q_1Q_0'IR_7 + Q_3Q_2Q_1'Q_0'Enter' + Q_3Q_2Q_1Q_0$$
$$Q_{1next} = D_1 = Q_3'Q_2'Q_1'Q_0 + Q_3'Q_2'Q_1Q_0'IR_6 + Q_3Q_2Q_1Q_0$$
$$Q_{0next} = D_0 = Q_3'Q_2'Q_1'Q_0' + Q_3'Q_2'Q_1Q_0'IR_5 + Q_3Q_2Q_1Q_0$$

(c)

State $Q_3Q_2Q_1Q_0$	*IRload*	*JMPmux*	*PCload*	*Meminst*	*MemWr*	$Asel_{1\text{-}0}$	*Aload*	*Sub*	*Halt*
0000 *Start*	0	0	0	0	0	00	0	0	0
0001 *Fetch*	1	0	1	0	0	00	0	0	0
0010 *Decode*	0	0	0	1	0	00	0	0	0
1000 *Load*	0	0	0	1	0	10	1	0	0
1001 *Store*	0	0	0	1	1	00	0	0	0
1010 *Add*	0	0	0	1	0	00	1	0	0
1011 *Sub*	0	0	0	1	0	00	1	1	0
1100 *Input*	0	0	0	0	0	01	1	0	0
1101 *Jz*	0	1	$(A = 0)$	0	0	00	0	0	0
1110 *Jpos*	0	1	$(A \geq 0)$	0	0	00	0	0	0
1111 *Halt*	0	0	0	0	0	00	0	0	1

(d)

$$IRload = Q_3'Q_2'Q_1'Q_0$$
$$JMPmux = Q_3Q_2Q_1'Q_0 + Q_3Q_2Q_1Q_0'$$
$$PCload = Q_3'Q_2'Q_1'Q_0 + Q_3Q_2Q_1'Q_0(A = 0) + Q_3Q_2Q_1Q_0'(A \geq 0)$$
$$Meminst = Q_3'Q_2'Q_1Q_0' + Q_3Q_2'Q_1'Q_0' + Q_3Q_2'Q_1'Q_0 + Q_3Q_2'Q_1$$
$$MemWr = Q_3Q_2'Q_1'Q_0$$
$$Asel_1 = Q_3Q_2'Q_1'Q_0'$$
$$Asel_0 = Q_3Q_2Q_1'Q_0'$$
$$Aload = Q_3Q_1'Q_0' + Q_3Q_2'Q_1$$
$$Sub = Q_3Q_2'Q_1Q_0$$
$$Halt = Q_3Q_2Q_1Q_0$$

(e)

FIGURE 8.11 Control unit for the EC-2: (a) state diagram; (b) next-state (implementation) table; (c) next-state equations; (d) control word and output table; (e) output equations; (f) circuit.
(continued on next page)

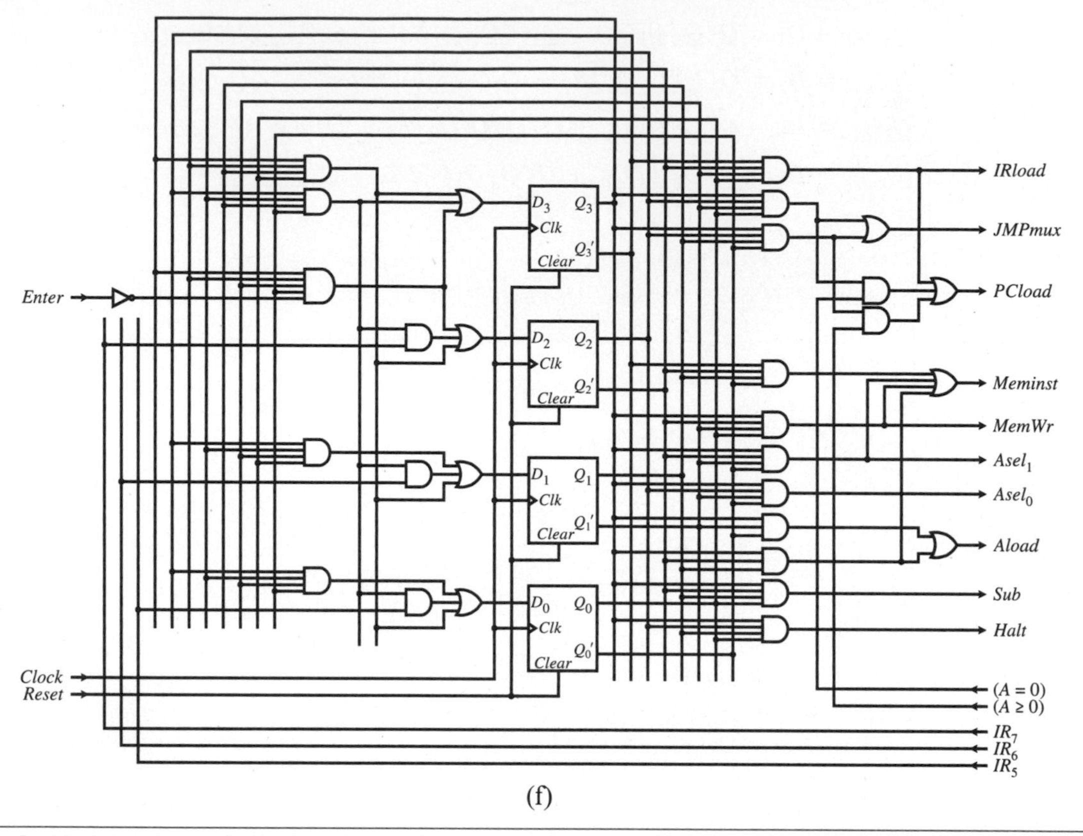

(f)

FIGURE 8.11 Control unit for the EC-2: (a) state diagram; (b) next-state (implementation) table; (c) next-state equations; (d) control word and output table; (e) output equations; (f) circuit.

instruction mnemonics. An extra *Start* state is added for timing purposes. The *Decode* state for this design needs to decode eight opcodes by branching to eight different states to execute the corresponding eight instructions. Like before, the decoding of the opcodes depends on the three most significant bits of the IR, IR_{7-5}.

An important timing issue for this control unit has to do with the memory accesses of the four instructions, LOAD, STORE, ADD, and SUB. The problem is that only after fetching these instructions will the address of the memory location for these instructions be available. Furthermore, only after decoding the instruction will the control unit know that the memory needs to be read. If we change the memory address during the *Execute* state, the memory will not have enough time to output the value for the instruction to operate on.

Usually, when instructions require a memory access for one of its operands, an extra memory read state will be inserted between the *Decode* state and the *Execute* state. This way, the memory will have one clock cycle to output the data for the instruction to operate on in the following clock cycle. This assumes that the memory requires only one clock cycle for a read operation. If the memory is slower, then more clock cycles must be inserted.

To minimize the number of states in our design, we use the *Decode* state to also perform the memory read. This way, when the control unit gets to the *Execute* state, the memory already will have the data ready. Whether the data from the memory actually is used will depend on the instruction being executed. If the instruction does not require the data from the memory, then it is simply ignored. On the other hand, if the instruction needs the data, then the data is there and ready to be used. This solution works in this design because it does not conflict with the operations for the rest of the instructions in our instruction set. The memory read operation performed in the *Decode* state is accomplished by asserting the *Meminst* signal from this state. Looking at the output table in Figure 8.11(d), this is reflected by the 1 under the *Meminst* column for the *Decode* state. By asserting *Meminst*, $IR_{4\text{-}0}$ will provide the address for the memory, and the value from that memory location will be read and made available for use in the next clock cycle.

The actual execution of each instruction is accomplished by asserting the correct control signals to control the operation of the datapath, as shown by the assignments made for the respective rows in the output table in Figure 8.11(d). At this point, you should be able to understand why each assignment is made by looking at the operation of the datapath. For example, to execute the LOAD instruction, *Meminst* is asserted in order to read from the memory location addressed by $IR_{4\text{-}0}$. The $Asel_1$ signal is asserted and the $Asel_0$ signal is de-asserted in order to select input 2 of the multiplexer so that the output from the memory can pass to the input of the accumulator A. The actual loading of A is done by asserting the *Aload* signal. To perform the STORE instruction, the memory address is taken from $IR_{4\text{-}0}$ by asserting *Meminst*. The writing into memory takes place when *MemWr* is asserted.

The *Input* state for this state diagram waits for the Enter key signal before looping back to the *Start* state. In so doing, we can read in several values by having multiple input statements in the program. After the *Enter* signal is asserted, there is no state that waits for the *Enter* signal to be de-asserted (i.e., for the Enter key to be released). Therefore, in order for this controller to work correctly, the user must release the Enter key before the execution of the next input instruction. This is almost impossible to do if the clock speed is very fast, so we need to slow down the clock to give the user enough time to release the Enter key. Ideally, there should be another state to wait for the release of the Enter key before continuing.

The next-state (implementation) table for the state diagram and the four next-state equations, as derived from the next-state table, are shown in Figures 8.11(b) and (c), respectively. To keep the table reasonably small, all of the possible combinations of the input signals are not listed. All of the states, except the *Input* state, depend only on the three *IR* bits, $IR_{7\text{-}5}$; whereas, the *Input* state depends only on the *Enter* signal. The blank entries in the table, therefore, can be viewed as having all 0s. With 11 states, four D flip-flops are used to implement the state memory for the control unit circuit. The output equations shown in Figure 8.11(e) are derived directly from the output table in Figure 8.11(d).

Finally, we can derive the circuit for the control unit based on the next-state equations and the output equations. The complete control unit circuit for the EC-2 general-purpose microprocessor is shown in Figure 8.11(f).

8.3.4 Complete Circuit

The complete circuit for the EC-2 general-purpose microprocessor is constructed by connecting the datapath from Figure 8.10 and the control unit from Figure 8.11(f) together using the designated control and status signals as shown in Figure 8.12. The complete schematic circuit and HDL code for the EC-2 microprocessor can be downloaded from the book website.

8.3.5 Sample Program

The EC-2 uses an internal RAM for its memory. Refer to Section 5.14.2 for a full discussion and HDL code of the RAM. The Verilog code for the RAM, including three EC-2 programs, is repeated in Figure 8.13. On reset, the RAM is initialized with the instructions for the program to be executed. The three programs included in the code are: (1) COUNT, which displays the count from input *n* down to 0; (2) SUM, which evaluates the sum of all of the numbers between an input number *n* and 1; and (3) GCD, which calculates the greatest common divisor of two input numbers. The last two programs have been commented out in the code. To try out any one of the three

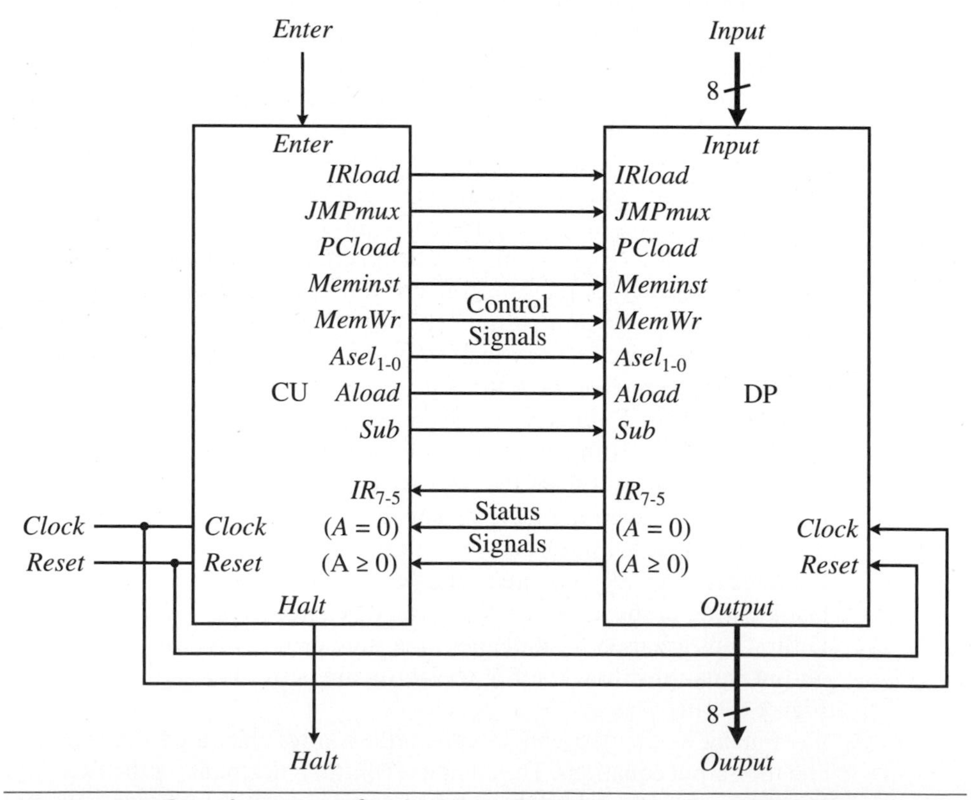

FIGURE 8.12 Complete circuit for the EC-2 general-purpose microprocessor.

```
module ram(
#(parameter size=5)
  input Clock,
  input Reset,
  input WE,
  input [size-1:0] Address,
  input [7:0] D,
  output reg [7:0] Q
);

  reg [7:0] ram1[2**size-1:0];

  always @(posedge Clock or posedge Reset) begin
    // this reset block and the Reset signal
    // is only needed to initialize the RAM locations
    if (Reset) begin
      // initialize RAM with EC-2 program

      ////////////////////////////////////////////////////////////
      // COUNT
      // Program to countdown from input n to 0
      ram1[0] = 8'b10000000; // IN A
      ram1[1] = 8'b01111111; // SUB A,11111
      ram1[2] = 8'b10100100; // JZ 00100
      ram1[3] = 8'b11000001; // JPOS 00001
      ram1[4] = 8'b11111111; // HALT
      ram1[31] = 8'b00000001; // storage for the constant 1

      ////////////////////////////////////////////////
      // SUM
      // Program to sum n downto 1 where n is an input number
//    ram1[0] = 8'b00011101; // LOAD A,one    // to zero sum
//    ram1[1] = 8'b01111101; // SUB A,one     // by doing 1 - 1
//    ram1[2] = 8'b00111110; // STORE A,sum
//
//    ram1[3] = 8'b10000000; // IN A
//    ram1[4] = 8'b00111111; // STORE A,n
//
//    ram1[5] = 8'b00011111; // loop: LOAD A,n    // n + sum
//    ram1[6] = 8'b01011110; // ADD A,sum
//    ram1[7] = 8'b00111110; // STORE A,sum
//    ram1[8] = 8'b00011111; // LOAD A,n          // decrement A
//    ram1[9] = 8'b01111101; // SUB A,one
//    ram1[10]= 8'b00111111; // STORE A,n
//
```

FIGURE 8.13 Verilog description of the RAM including three programs with which to initialize the RAM. *(continued on next page)*

```
//    ram1[11] = 8'b10101101;  // JZ out
//    ram1[12] = 8'b11000101;  // JPOS loop
//    ram1[13] = 8'b00011110;  // out: LOAD A,sum
//    ram1[14] = 8'b11111111;  // HALT
//
//    ram1[29] = 8'b00000001;  // storage for the constant 1
//    ram1[30] = 8'b00000000;  // storage for variable sum
//    ram1[31] = 8'b00000000;  // storage for variable n


      ///////////////////////////////////////////////////////
      // GCD
      // Program to calculate the GCD of two input
      // numbers, x and y
//    ram1[0] = 8'b10000000;  // IN A              // input x
//    ram1[1] = 8'b00111110;  // STORE A,x
//    ram1[2] = 8'b10000000;  // IN A              // input y
//    ram1[3] = 8'b00111111;  // STORE A,y
//
//    ram1[4] = 8'b00011110;  // loop: LOAD A,x    // x=y?
//    ram1[5] = 8'b01111111;  // SUB A,y
//    ram1[6] = 8'b10110000;  // JZ out            // x=y
//    ram1[7] = 8'b11001100;  // JPOS xgty         // x>y
//
//    ram1[8] = 8'b00011111;  // LOAD A,y          // y>x
//    ram1[9] = 8'b01111110;  // SUB A,x           // y-x
//    ram1[10] = 8'b00111111; // STORE A,y
//    ram1[11] = 8'b11000100; // JPOS loop
//
//    ram1[12] = 8'b00011110; // xgty: LOAD A,x    // x>y
//    ram1[13] = 8'b01111111; // SUB A,y           // x-y
//    ram1[14] = 8'b00111110; // STORE A,x
//    ram1[15] = 8'b11000100; // JPOS loop
//
//    ram1[16] = 8'b00011110; // out: LOAD A,x
//    ram1[17] = 8'b11111111; // HALT
//
//    ram1[30] = 8'b00000000; // storage for variable x
//    ram1[31] = 8'b00000000; // storage for variable y
```

FIGURE 8.13 Verilog description of the RAM including three programs with which to initialize the RAM. *(continued on next page)*

```
    end else begin
      if (WE)
        ram1[Address] = D;
    end
  end  // always

  always @ (Address) begin
    Q = ram1[Address];
  end
endmodule
```

FIGURE 8.13 Verilog description of the RAM including three programs with which to initialize the RAM.

programs, simply uncomment the program that you want to execute and re-synthesize the entire microprocessor circuit.

It is assumed that you already are familiar with writing computer programs using either machine language or assembly language, so we will not go into details here. The comments annotated throughout the three sample programs should be sufficient for you to understand what the programs are doing.

8.3.6 Hardware Implementation

Figure 8.14 shows the interface between the EC-2 microprocessor and the input and output devices on an FPGA development board. The input device consists of eight switches, and the output device is three 7-segment LED displays. Because the microprocessor outputs an 8-bit binary number, we need a converter to convert the 8-bit binary number to three BCD (binary coded decimal) digits. With this converter, we will be able to see the output as a 3-digit decimal number. See Problem 3.36 for the HDL code for this BCD converter. The HDL code for this converter is also available from the book website. A single LED is connected to the *Halt* signal to show when the

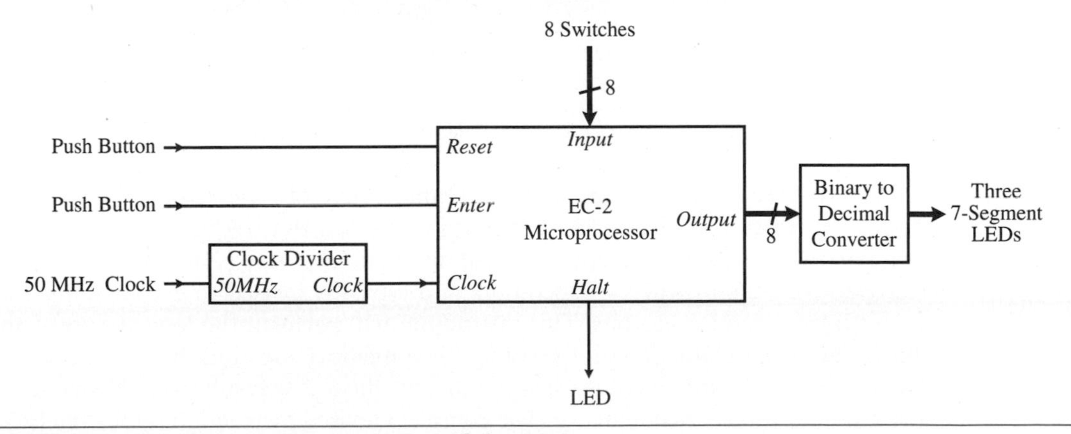

FIGURE 8.14 Hardware implementation of the EC-2.

microprocessor has finished running the program. A push button is connected to the *Reset* signal to reset the microprocessor, and a second push button is used as the Enter key. A clock divider circuit is used to slow down the clock frequency to approximately 1 Hz for the correct operation of the Enter key, and to see the program being executed. When executing the IN A instruction, the processor will stay in the *Input* state, allowing the user to set up the input number on the input switches. When the input number has been set up, the Enter push button needs to be pressed for the processor to read in the input number and then continue execution of the program.

8.4 Extending the EC-2 Instruction Set

The instruction set for the EC-2 is quite limited, so to make the EC-2 a bit more useful, we will add a few more instructions. To still keep our design simple, we will try to minimize the number of changes needed for the datapath and the control unit. In the original design, three bits are used to encode the opcode, giving only eight different combinations to encode the eight different instructions that we already have. So is it possible to add more instructions without having to add another bit for the opcode? The answer will depend on how many more instructions we want to add, what the instruction operations are that we want to add, and whether there are any unused bits in our current instruction encodings.

We can see that both the HALT and IN A instructions have unused bits, so we can use them to encode more instructions. We can combine the HALT and IN A instructions to use the same 3-bit opcode, and use a fourth bit to distinguish between them. Because there are still unused bits, we can continue with this encoding scheme to add another instruction, the OUT A instruction to output the value from the accumulator. This new encoding scheme will work because the execution of these three instructions does not require any additional operands. The revised opcode encodings for these three instructions are shown next.

Instruction	Encoding	Operation	Comment
IN A	111 00×××	$A \leftarrow Input$	Input to A
OUT A	111 01×××	$Output \leftarrow A$	Output from A
HALT	111 1××××	Halt	Halt execution

Now that we have freed up opcode 100 that originally was assigned to the IN A instruction, we can use it for another new instruction that requires the use of the remaining five bits for its operand. A useful instruction to have is the AND instruction, which performs the logical AND operation, and typically is used to extract individual bits out from several other bits. This instruction will perform the logical AND of the value in the accumulator A with the content in a memory location that is specified by the five least significant bits of the instruction encoding. The result from the operation is stored back into the accumulator. The complete revised instruction set for the EC-2 is shown in Figure 8.15.

Instruction	Encoding	Operation	Comment
LOAD A, address	000 aaaaa	$A \leftarrow M[\text{aaaaa}]$	Load A with content of memory location aaaaa
STORE A, address	001 aaaaa	$M[\text{aaaaa}] \leftarrow A$	Store A into memory location aaaaa
ADD A, address	010 aaaaa	$A \leftarrow A + M[\text{aaaaa}]$	Add A with M[aaaaa] and store result back into A
SUB A, address	011 aaaaa	$A \leftarrow A - M[\text{aaaaa}]$	Subtract A with M[aaaaa] and store result back into A
AND A, address	100 aaaaa	$A \leftarrow A$ AND M[aaaaa]	AND A with M[aaaaa] and store result back into A
JZ address	101 aaaaa	IF $(A = 0)$ THEN PC = aaaaa	Jump to address if A is zero
JPOS address	110 aaaaa	IF $(A \geq 0)$ THEN PC = aaaaa	Jump to address if A is zero or a positive number
IN A	111 00×××	$A \leftarrow Input$	Input to A
OUT A	111 01×××	$Output \leftarrow A$	Output from A
HALT	111 1××××	Halt	Halt execution

Notations:

A = accumulator

M = memory

PC = program counter

aaaaa = five bits for specifying a memory address

× = don't cares

FIGURE 8.15 Revised instruction set for the EC-2.

The modified datapath for the revised instruction set is shown in Figure 8.16. No changes in the datapath are necessary to perform the HALT and IN A instructions. A tri-state buffer is added between the accumulator and the output port for controlling the OUT A instruction. To realize the new AND instruction, we need to add a logical AND functional unit. The inputs to this functional unit are from the accumulator and the memory, and its output is connected to the currently unused input, input 3, of the 4-input multiplexer that goes to the accumulator. Finally, to decode the IN A, OUT A, and HALT instructions, we need to send five bits, $IR_{7\text{-}3}$, to the control unit instead of the original three opcode bits. These changes are highlighted in the datapath figure.

The modified state diagram is shown in Figure 8.17 with the changes highlighted. The final construction of the complete control unit circuit is left as an exercise for the reader. (See Problem 8.5.)

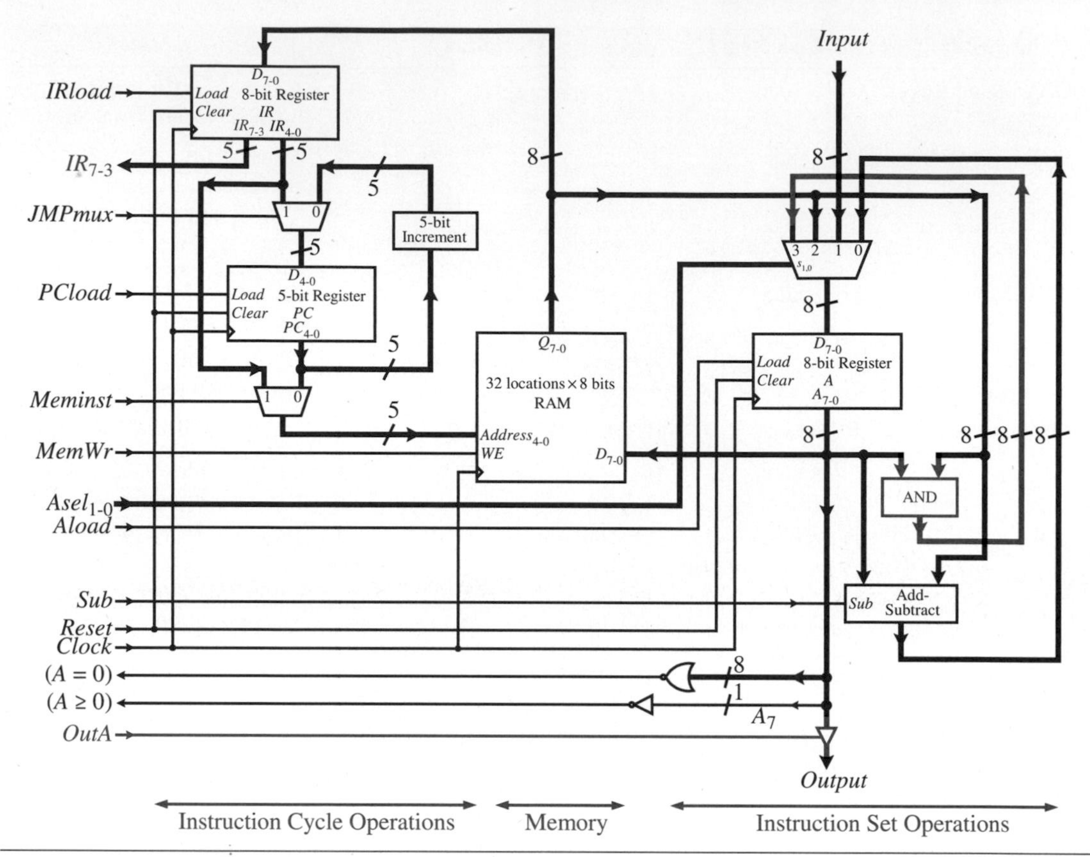

FIGURE 8.16 Modified datapath for the EC-2.

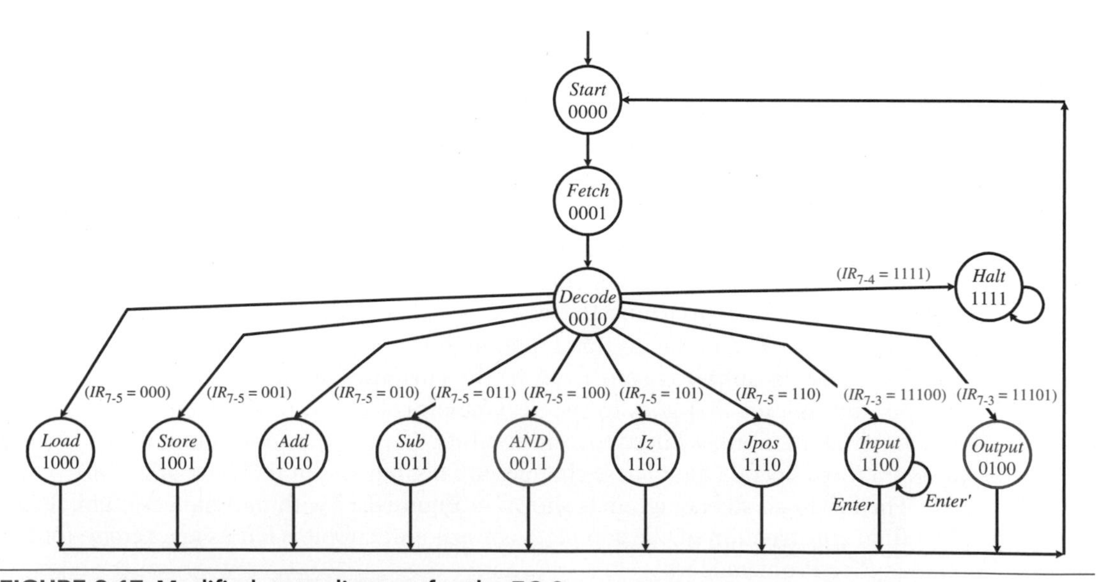

FIGURE 8.17 Modified state diagram for the EC-2.

8.5 Using and Interfacing the EC-2

We saw in Section 8.3.5 how we can use the EC-2 to run simple machine language programs. Three sample programs were given in that section: (1) COUNT, which displays the count from input n down to 0; (2) SUM, which evaluates the sum of all of the numbers between an input number n and 1; and (3) GCD, which calculates the greatest common divisor of two input numbers. In this section, we will look at how to interface the EC-2 to control external devices. Specifically, we will use the EC-2 to control the elevator problem introduced in Section 6.6.5.

Interfacing a general-purpose microprocessor to external devices requires a full understanding of the operations of the external device with which you want to interface, the resources available on your microprocessor, both hardware and the instruction set, and how the external device is connected to your microprocessor. There are usually many different ways of connecting the external device to the microprocessor, and the software program for controlling the device can be very different depending on how they are connected. The software program for controlling the device is usually referred to as the device driver for that device.

The elevator problem controls an elevator moving between two floors. A picture of the elevator setup and a summary of the I/O signals are shown next.

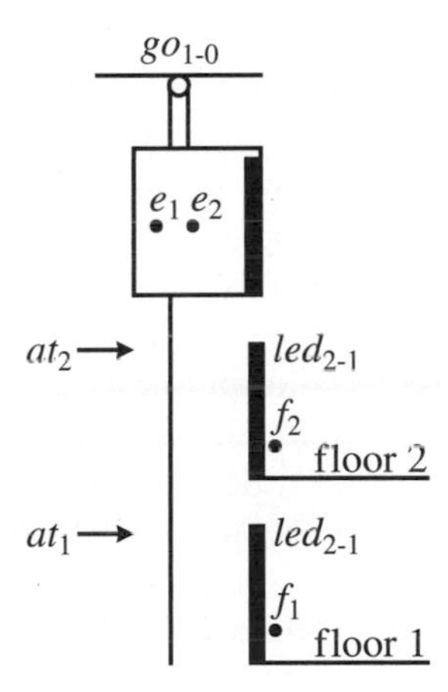

Inputs:

- f_1, f_2: Buttons at each floor to call the elevator. f_1 is on floor 1 and f_2 is on floor 2.
- e_1, e_2: Buttons inside the elevator to tell the elevator which floor to go to.
- at_1, at_2: Signals from the elevator mechanism to say which floor the elevator is on.

Outputs:

- go_1: 0 to turn off the elevator motor and 1 to turn on the motor.
- go_0: 0 to go to floor 1 and 1 to go to floor 2.
- led_1, led_2: LEDs on each of the two floors to show which floor the elevator is on.

We noted that the corresponding f_i and e_i input signals are basically the same and therefore can be combined into one signal, f_i.

The first thing we must do is to decide how to connect the I/O signals from the elevator to the EC-2 microprocessor. The EC-2 already has eight input lines and eight output lines via the *Input* and *Output* ports, respectively, so we can connect the elevator I/O signals to these two I/O ports on the microprocessor. Originally, these two I/O ports were connected to eight switches and eight LEDs for the user to enter numbers and to see the results from the accumulator. By changing these connections, we will not be able to enter numbers or see the results. If we still need this functionality, then we will have to modify the datapath and add another input port and output port to our EC-2 microprocessor. We also will need to modify the control unit and the instruction set so that we can access the second I/O port. For the current problem, this modification is not necessary. We will connect the four input signals from the elevator, f_2, f_1, at_2, and at_1, to the first four bits, $Input_{3\text{-}0}$, of the *Input* port of the EC-2 in this given order. So f_2 is connected to $Input_3$, f_1 is connected to $Input_2$, at_2 is connected to $Input_1$, and at_1 is connected to $Input_0$. Similarly, we will connect the four output signals from the elevator, go_1, go_0, led_2, and led_1, to the first four bits, $Output_{3\text{-}0}$, of the *Output* port in this given order.

Having decided on the connections between the microprocessor and the elevator, we are now ready to write the machine language program to control the elevator. We will use the extended instruction set for the modified EC-2 microprocessor as discussed in Section 8.4.

The operation of the elevator is based on the state diagram derived in Section 6.6.5 and repeated here in Figure 8.18 for convenience. The complete machine language program listing for controlling the elevator is shown in Figure 8.19. The logical progression of the program basically follows the state diagram. One difficulty in writing the program is to know how to send out the correct output signals, and how to test for the conditions of the input signals.

To send out the correct output signals, we have defined four constant values stored in memory locations 24 to 27 (in decimal). Only the four least significant bits of these constants are used because the four output signals from the elevator go_1, go_0, led_2, and led_1 are connected to $Output_{3\text{-}0}$. For example, when we are in state 00, we want to turn on led_1 and turn off the remaining three outputs, therefore, we need to output the binary value ××××0001. Because led_1 is connected to $Output_0$, sending a 1 to $Output_0$ will turn on led_1. The other outputs are turned off. This constant value, 00000001, is stored in memory location 24, and the two instructions

```
LOAD A, $24
OUT A
```

are executed in state 00 to turn on only led_1.

To test the conditions of the input signals, after executing the IN A instruction, we need to mask out the bit that we are interested in by performing the AND operation and then executing the conditional jump. The four constants for the four masks that we need are stored in memory locations 28 to 31 (in decimal). Only the four least significant bits of these constants are used because the four input signals from the elevator f_2, f_1, at_2,

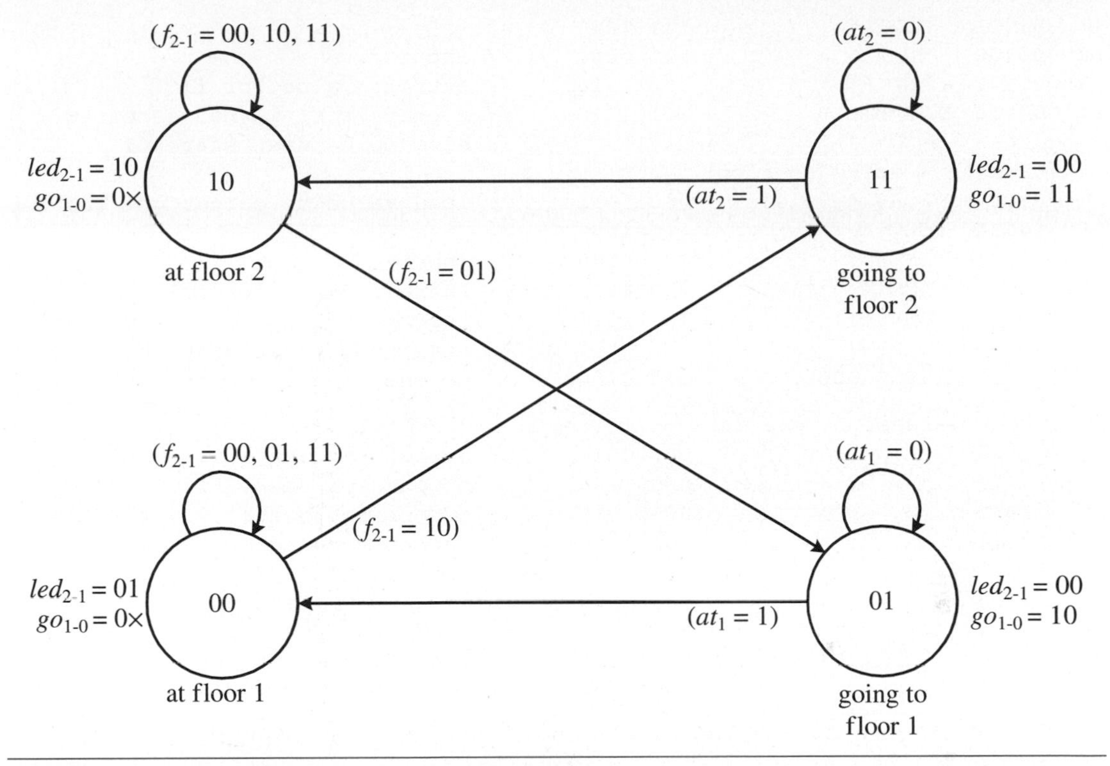

FIGURE 8.18 **State diagram for the elevator problem.**

```
memory          instruction        instruction  Comments
address                            encoding
// state 00 - at floor 1
0    00000      LOAD A, 11000      00011000     // load value to output led1
1    00001      OUT A              11101000
2    00010      IN A               11100000     // input
3    00011      AND A, 11100       10011100     // extract value for f2
4    00100      JZ 00000           10100000     // repeat if f2 is not asserted
                                                // else go to next state 11
// state 11 - going to floor 2
5    00101      LOAD A, 11001      00011001     // load value to output go1 and go0
6    00110      OUT A              11101000
7    00111      IN A               11100000     // input
8    01000      AND A, 11101       10011101     // extract value for at2
9    01001      JZ 00101           10100101     // repeat if at2 is not asserted
                                                // else go to next state 10
// state 10 - at floor 2
10   01010      LOAD A, 11010      00011010     // load value to output led2
```

FIGURE 8.19 **Program for controlling the elevator using the EC-2 general-purpose microprocessor.** *(continued on next page)*

```
11  01011   OUT A             11101000
12  01100   IN A              11100000    // input
13  01101   AND A, 11110      10011110    // extract value for f1
14  01110   JZ 01010          10101010    // repeat if f1 is not asserted
                                          // else go to next state 01
// state 01 - going to floor 1
15  01111   LOAD A, 11011     00011011    // load value to output go1 and go0
16  10000   OUT A             11101000
17  10001   IN A              11100000    // input
18  10010   AND A, 11111      10011111    // extract value for at1
19  10011   JZ 01111          10101111    // repeat if at1 is not asserted
                                          // else go to next state 00
20  10100   JPOS 00000        11000000    // unconditional jump to state 00

// output values, bits 3 to 0 = go1, go0, led2, led1
24  11000                     00000001    // output for state 00
25  11001                     00001100    // output for state 11
26  11010                     00000010    // output for state 10
27  11011                     00001000    // output for state 01
// input values, bits 3 to 0 = f2, f1, at2, at1
28  11100                     00001000    // input for state 00
29  11101                     00000010    // input for state 11
30  11110                     00000100    // input for state 10
31  11111                     00000001    // input for state 01
```

FIGURE 8.19 Program for controlling the elevator using the EC-2 general-purpose microprocessor.

and at_1 are connected to $Input_{3\text{-}0}$. So for example, when we are in state 00, we want to test whether f_2 is asserted, therefore, we need to AND the input value with the mask 00001000. This constant mask value, 00001000, is stored in memory location 11100, and the three instructions

IN A
AND A, 11100
JZ 00000

are executed in state 00 to test for the input signal f_2.

The result after the AND operation will extract bit 3 of the input, which is the f_2 input signal. If f_2 is a 0, then the entire result after the AND operation will be 0. On the other hand, if f_2 is a 1, then the result after the AND operation will not be a 0. Thus, after the AND operation, we can execute the JZ conditional jump instruction either to stay in the current state or go to the next state. If the result is 0, then we need to loop back to the current state with the instruction JZ 00000. Otherwise, the jump is not performed and the program continues with the next instruction, which is to go to the next state 11.

This elevator problem does not have any strict timing issues, so it is not necessary to consider the time that it takes to execute each instruction. In other problems where the timing for sending and receiving signals is critical, we will need to be more precise. For example, if we use the EC-2 to control an RS232 protocol communication, then we need to know exactly how long it takes to execute the instructions needed to output a bit because the RS232 protocol requires a certain predetermined speed or baud rate to send out each bit.

8.6 Pipelining

Today's general-purpose microprocessors, such as the Intel Core™i7 CPU, are designed to execute more instructions over a shorter period of time by using a technique known as **pipelining**. This is an implementation technique in which multiple instructions are overlapped in execution, similar to the production of devices in a factory assembly line. The computer pipeline is divided into stages, and each stage will perform a part of the instruction execution. Instructions pass through the stages as they get executed. As one instruction completes in one stage and moves down to the next stage, another instruction will move in to take its place. Because multiple instructions are in the pipeline at any one time, concurrent execution of instructions is achieved.

Pipelining does not decrease the time to execute one instruction—the time from when an instruction first enters the pipeline until it exits the pipeline. In fact, it takes a little more time to execute one instruction as compared with a nonpipelined processor because of the overhead needed to implement pipelining. Pipelining, however, increases the instruction throughput rate, which is measured by how fast instructions exit the pipeline. With an increase in instruction throughput, a program will run faster and will have a lower total execution time.

In order for pipelining to work efficiently, all of the stages must be balanced in terms of the time needed for each stage. If all of the stages are balanced, then they all can be synchronized and ready to proceed at the same time as instructions move from one stage to the next. If this is the case, then the time between instructions on the pipelined machine—assuming ideal conditions—is equal to

$$\text{Time between instructions}_{\text{pipelined}} = \frac{\text{Time between instructions}_{\text{nonpipelined}}}{\text{Number of pipeline stages}}$$

In other words, the speedup from pipelining under ideal conditions is equal to the number of pipeline stages; so a five-stage pipeline is five times faster. However, the actual time per instruction in the pipelined processor will be greater than the minimum possible, and the speedup will be less than the number of pipeline stages because the stages usually are not balanced and overheads are involved with the addition of pipeline stages.

8.6.1 Basic Pipelined Processor

Based on the MIPS architecture model proposed by Hennessy and Patterson,[3] the general steps for executing an instruction can be divided into five pipeline stages: (1) instruction fetch (IF); (2) instruction decode (ID); (3) execute or effective address

[3] Computer Organization & Design: The Hardware/Software Interface, D. Patterson and J. Hennessy.

calculation (EX); (4) memory access (MEM); and (5) write back (WB) of result. Instructions are fetched from memory in the IF stage. In the ID stage, besides decoding of the instruction, the register file is accessed to read in the instruction operands. The reading of registers and decoding occur simultaneously. In the EX stage, the instruction is either executed, or if it is a memory access instruction, then the address of the memory location is calculated. If an instruction requires memory access, then after calculating the memory address in the EX stage, the data from that memory location is accessed in the MEM stage. Finally, in the WB stage, the result from executing the instruction is written back into a register.

Instructions entering the pipeline start in the IF stage. At the end of the first clock cycle, as the instruction moves from the IF to the ID stage, a new instruction enters the pipeline and fills the IF stage. After five clock cycles, all of the stages will have a different instruction, and the first instruction will now be in the final WB stage. This basic pipeline execution of instructions is shown graphically in Figure 8.20. The clock cycles are listed across the top of the table. The sequence of instructions entering the pipeline is listed down the rows of the table. The entries in the table show the stage that an instruction is in during a particular clock cycle. For example, in clock cycle 5, instruction $i+2$ is in the EX stage. The table clearly shows that each instruction requires five clock cycles to execute (because of the five stages), but because of the overlapping of instruction executions, every clock cycle starting from clock cycle 5 will have a new instruction completing. So the throughput is one instruction per cycle under ideal situations.

Because of the overlapping operations, the datapath for the pipelined machine needs to have enough resources to handle the parallel execution of instructions. Several issues need to be considered. First, note that starting from clock cycle 4 in Figure 8.20, every clock cycle will have both an instruction fetch (IF) and a memory access (MEM). If there is only a single memory, a conflict will occur between the instruction fetch and data memory access because both of these operations require the use of the memory. Thus, to resolve this problem, the datapath needs to have a separate instruction and data memory.

A second issue, similar to the first, is that in every clock cycle starting from clock cycle 5 onward, the register file is used in two stages; first in the ID stage where a register is read, and second in the WB stage where a register is written. We need to make sure that the register file can handle both reading and writing within the same clock cycle.

Clock Cycle	1	2	3	4	5	6	7	8	9
Instruction i	IF	ID	EX	MEM	WB				
Instruction $i+1$		IF	ID	EX	MEM	WB			
Instruction $i+2$			IF	ID	EX	MEM	WB		
Instruction $i+3$				IF	ID	EX	MEM	WB	
Instruction $i+4$					IF	ID	EX	MEM	WB

FIGURE 8.20 Execution of instructions in a pipelined machine.

A more serious problem is that a new instruction is fetched in every clock cycle. In order for this to be possible, the program counter (PC) needs to be incremented and written back in every clock cycle during the IF stage in preparation for the next instruction. The problem arises when we consider the effect of branches, which also might change the PC. The PC changes when a conditional branch is taken, but whether the condition is true or not is not known until the EX stage, and the new branch address is not known until the MEM stage. As a result, the PC will not have the correct value in the next clock cycle.

Finally, pipelining the datapath requires that values passed from one pipe stage to the next be placed in registers. These registers, known as pipeline registers, carry both data and control signals from one pipeline stage to the next. Any instruction is active in exactly one stage of the pipeline at a time; therefore, any action taken on behalf of an instruction occurs between a pair of pipeline registers.

8.6.2 **Pipeline Hazards**

In a pipelined machine, situations called **hazards** prevent the next instruction in the instruction stream from being executed in the following clock cycle. There are three types of hazards: (1) structural hazards; (2) data hazards; and (3) control hazards.

When a machine is pipelined, the overlapped execution of instructions requires pipelining of functional units and duplication of resources to allow for all possible combinations of instructions in the pipeline. **Structural hazards** occur when instructions in the pipeline cannot execute because the hardware cannot support the combination of instructions that we want to execute in the same clock cycle. An example would be if two instructions in the pipeline need to use an adder to add numbers but the datapath has only one adder available. In order to resolve a structural hazard, either more hardware is added to the datapath, or stalls are inserted into the pipeline to delay the execution of the conflicting instruction by one clock cycle. A structural hazard also occurs if there is only one memory for storing both data and instructions. This is resolved, as discussed previously, by having two separate memories.

Data hazards occur when an instruction depends on the result of a previous instruction that is still being executed in the pipeline. An example would be if an ADD instruction is followed immediately by a SUBTRACT instruction, and the SUBTRACT instruction requires, as one of its operands, the result from the ADD instruction. Because the ADD instruction is still being executed in the pipeline, its result is not yet available for the SUBTRACT instruction to use as shown in Figure 8.21(a).

The result of the ADD instruction is not written into the destination register R1 until the WB stage in clock cycle 5. However, the value in register R1 is needed as an operand by the SUBTRACT instruction at the beginning of the EX stage in clock cycle 4. A naive solution would be to simply stall the pipeline for two clock cycles so that the SUBTRACT instruction does not start until clock cycle 4, as shown in Figure 8.21(b). By doing this, the EX stage for the SUBTRACT instruction will occur in clock cycle 6, which is after the WB stage for the ADD instruction in clock cycle 5.

A better solution to resolve data hazards is a technique known as **forwarding**. This technique involves forwarding the result from the first functional unit (in this case the

Clock Cycle	1	2	3	4	5	6
ADD R1 ← R2 + R3	IF	ID	EX	MEM	WB	
SUB R4 ← R1 − R5		IF	ID	EX	MEM	WB

(a)

Clock Cycle	1	2	3	4	5	6	7	8
ADD R1 ← R2 + R3	IF	ID	EX	MEM	WB			
SUB R4 ← R1 − R5		stall	stall	IF	ID	EX	MEM	WB

(b)

Clock Cycle	1	2	3	4	5	6
ADD R1 ← R2 + R3	IF	ID	EX ↘	MEM	WB	
SUB R4 ← R1 − R5		IF	ID	EX	MEM	WB

(c)

FIGURE 8.21 Data hazard: (a) source register R1 for the subtraction does not have the correct value ready in clock cycle 4; (b) delaying the subtraction instruction by inserting two stalls; (c) the connection shows the forwarding of the result directly from the EX stage of the addition to the EX stage of the subtraction.

adder) directly to the input of the second functional unit (in this case the subtractor) without going through any registers. With forwarding, no stalls are required to remove the data hazard as shown by the arrow in Figure 8.21(c). The result from the adder is available at the end of the ADD instruction's EX stage (cycle 3), and this value is needed by the subtractor at the beginning of the SUBTRACT instruction's EX stage (cycle 4). Therefore, the output from the adder can be routed directly to the input of the subtractor, and the subtraction can continue without any stalls.

Another technique known as pipeline scheduling can also be used to remove data hazards. In this case, the compiler can rearrange the instructions so that two dependent instructions are not scheduled one after the other. This way, no data hazards will occur.

Typically, the PC is incremented by one (assuming that each instruction occupies one memory location) after each instruction fetch. However, this might not be the case for a branch instruction. When executing a branch instruction, if the condition is true and the branch is taken, a new address for fetching the next instruction must be calculated and stored into the PC; but this new address is not known until later in the EX stage. In the meantime, several new instructions immediately after the branch instruction already will have entered the pipeline. When the branch is taken, these instructions (which should not have been executed) already have entered the pipeline and are being partially executed. Thus, they must be removed from the pipeline and the operations that they already have performed must be undone.

This problem is a **control hazard**, and its effect creates the greatest performance loss for a pipelined machine. Even if we add extra hardware so that we can calculate the branch address and update the PC during the second ID stage, we still need to add one stall in the pipeline. The cost for stalls from branch instructions is too high, and so various branch prediction techniques have been suggested to improve on the performance of branch instructions. Using branch prediction techniques, if the prediction of a branch is correct, then the pipeline can proceed normally at full speed. The pipeline is stalled and some operations need to be undone only if the prediction is wrong.

8.7 Verilog and VHDL Code for General-Purpose Microprocessors

This section presents the Verilog and VHDL codes for the EC-2 general-purpose microprocessor using both the structural FSM+D model and the behavioral FSMD model. When writing HDL code using the FSM+D model, the microprocessor with its separate control unit and datapath circuits first must be designed manually, as we did in Section 8.3. These two separate circuits are translated to HDL code, and then combined together using structural level coding to form the microprocessor. In practice, we seldom want to create a microprocessor using the FSM+D model. This exercise is just to show how the complete HDL code is written. Section 8.7.1 lists the Verilog code using the FSM+D model of the EC-2 microprocessor.

Section 8.7.2 lists both the Verilog and VHDL behavioral level code that defines the EC-2 microprocessor using the FSMD model. Using this model, we need to manually derive only the state diagram. From the state diagram, we easily can write the HDL code for the control unit at the behavioral level so that the control unit circuit can be generated by the synthesizer automatically. The datapath operations are embedded as HDL instructions within the FSM code, therefore, the name FSMD. Comparing the FSMD code with the FSM+D code, we can see how easier it is to construct a microprocessor circuit using the behavioral FSMD model and quickly appreciate the power of the HDL synthesizer.

8.7.1 FSM+D Model

Figures 8.22, 8.23, and 8.24 show the Verilog code for the EC-2 microprocessor, datapath, and control unit, respectively, using the FSM+D model. Figure 8.22 shows the structural Verilog code for connecting the datapath and the control unit to form the microprocessor. Figure 8.23 shows the structural Verilog code for the data-path. This code is based on the datapath circuit derived in Section 8.3.2 and shown in Figure 8.10. The code for the individual components used in the datapath can be found in Chapters 4 and 5. Figure 8.24 shows the behavioral Verilog code for the control unit as derived in Section 8.3.3. This control unit behavioral code follows the template for writing an FSM sequential circuit, as discussed in Section 6.7. The next-state logic portion of this code is based on the state diagram shown in Figure 8.11(a), and the output logic portion of this code is based on the output table shown in Figure 8.11(d). The complete FSM+D code for the EC-2 microprocessor for both Verilog and VHDL can be downloaded from the book website.

```
module mp (
   input Clock, Reset, Enter,
   input [7:0] Input,
   output [7:0] Output,
   output Halt,
   output [3:0] debug
   );

   wire [2:0] IR75;
   wire PCload, Meminst, MemWr, Aload;
   wire IRload, JMPmux, Sub, Aeq0, Apos;
   wire [1:0] Asel;

   dp U1(.Clock(Clock), .Reset(Reset), .IRload(IRload),
      .JMPmux(JMPmux), .PCload(PCload), .Meminst(Meminst),
      .MemWr(MemWr), .Asel(Asel), .Aload(Aload), .Sub(Sub),
      .IR75(IR75), .Apos(Apos), .Aeq0(Aeq0), .Input(Input),
      .Output(Output));
   cu U2(.Clock(Clock), .Reset(Reset), .Enter(Enter),
      .IRload(IRload), .JMPmux(JMPmux), .PCload(PCload),
      .Meminst(Meminst), .MemWr(MemWr), .Asel(Asel),
      .Aload(Aload), .Sub(Sub), .IR75(IR75), .Apos(Apos),
      .Aeq0(Aeq0), .Halt(Halt), .debug(debug));
endmodule
```

FIGURE 8.22 Structural Verilog code for the EC-2 microprocessor.

```
module dp (
   input   Clock, Reset,
   input   IRload, JMPmux, PCload, Meminst, MemWr,
   input   [1:0] Asel,
   input   Aload, Sub,
   output  [2:0] IR75,
   output  Apos, Aeq0,
   input   [7:0] Input,
   output  [7:0] Output
   );

   wire [7:0] dp_IR, dp_RAMQ;
   wire [4:0] dp_JMPmux, dp_PC, dp_increment, dp_meminst;
   wire [7:0] dp_Amux, dp_addsub, dp_A;

   register #(8) U0_IR(.Clock(Clock), .Clear(Reset),
      .Load(IRload), .D(dp_RAMQ), .Q(dp_IR));
   mux2 #(5) U1_JMPmux(.S(JMPmux), .D1(dp_IR[4:0]),
      .D0(dp_increment), .Y(dp_JMPmux));
   register #(5) U2_PC(.Clock(Clock), .Clear(Reset),
      .Load(PCload), .D(dp_JMPmux), .Q(dp_PC));
   mux2 #(5) U3_meminst(.S(Meminst), .D1(dp_IR[4:0]),
      .D0(dp_PC), .Y(dp_meminst));
```

FIGURE 8.23 Structural Verilog code for the EC-2 datapath. *(continued on next page)*

```
   increment U4_inc(.A(dp_PC), .F(dp_increment));
   ram U5_ram(.Clock(Clock), .Reset(Reset), .WE(MemWr),
      .Address(dp_meminst), .D(dp_A), .Q(dp_RAMQ));
   mux4 #(8) U6_Amux(.S(Asel), .D3(8'b00000000), .D2(dp_RAMQ),
      .D1(Input), .D0(dp_addsub), .Y(dp_Amux));
   register #(8) U7_A(.Clock(Clock), .Clear(Reset),
      .Load(Aload), .D(dp_Amux), .Q(dp_A));
   addsub #(8) U8_addsub(.S(Sub), .A(dp_A), .B(dp_RAMQ),
      .F(dp addsub));

   assign IR75 = dp_IR[7:5];
   assign Aeq0 = (dp_A == 0)? 1:0;
   assign Apos = ~dp_A[7];
   assign Output = dp_A;
endmodule
```

FIGURE 8.23 Structural Verilog code for the EC-2 datapath.

```
module cu (
   input Clock, Reset, Enter,
   output reg IRload, JMPmux, PCload, Meminst, MemWr,
   output reg [1:0] Asel,
   output reg Aload, Sub,
   input [2:0] IR75,
   input Apos, Aeq0,
   output reg Halt,
   output [3:0] debug
   );

   reg [3:0] state;

   // next state logic
   always @ (posedge Clock, posedge Reset) begin
     if (Reset) begin
       state<=4'b0000;
       end
     else
       case (state)
       4'b0000: begin // Start
         state<=4'b0001;
         end
       4'b0001: begin // Fetch
         state<=4'b0010;
         end
       4'b0010: begin // Decode
         case (IR75)
           3'b000: state<=4'b1000; // Load
           3'b001: state<=4'b1001; // Store
```

FIGURE 8.24 Behavioral Verilog code for the EC-2 control unit. *(continued on next page)*

```
          3'b010: state<=4'b1010; // Add
          3'b011: state<=4'b1011; // Sub
          3'b100: state<=4'b1100; // Input
          3'b101: state<=4'b1101; // Jz
          3'b110: state<=4'b1110; // Jpos
          3'b111: state<=4'b1111; // Halt
          default:state<=4'b0000; // Start
        endcase
        end
      4'b1000: begin // Load
        state<=4'b0000;
        end
      4'b1001: begin // Store
        state<=4'b0000;
        end
      4'b1010: begin // Add
        state<=4'b0000;
        end
      4'b1011: begin // Sub
        state<=4'b0000;
        end
      4'b1100: begin // Input
        if (Enter) begin
          state<=4'b0000;
          end
        else begin
          state<=4'b1100;
          end
        end
      4'b1101: begin // Jz
        state<=4'b0000;
        end
      4'b1110: begin // Jpos
        state<=4'b0000;
        end
      4'b1111: begin // Halt
        state<=4'b1111;
        end
      default: begin
        state<=4'b0000;
        end
      endcase
  end // always

  // output logic
  always @ (state) begin
    case (state)
    4'b0001: begin // Fetch
```

FIGURE 8.24 Behavioral Verilog code for the EC-2 control unit. *(continued on next page)*

```
         IRload<=1;
         JMPmux<=0;
         PCload<=1;
         Meminst<=0;
         MemWr<=0;
         Asel<=2'b00;
         Aload<=0;
         Sub<=0;
         Halt<=0;
         end
       4'b0010: begin // Decode
         IRload<=0;
         JMPmux<=0;
         PCload<=0;
         Meminst<=1;
         MemWr<=0;
         Asel<=2'b00;
         Aload<=0;
         Sub<=0;
         Halt<=0;
         end
       4'b1000: begin // Load
         IRload<=0;
         JMPmux<=0;
         PCload<=0;
         Meminst<=1;
         MemWr<=0;
         Asel<=2'b10;
         Aload<=1;
         Sub<=0;
         Halt<=0;
         end
       4'b1001: begin // Store
         IRload<=0;
         JMPmux<=0;
         PCload<=0;
         Meminst<=1;
         MemWr<=1;
         Asel<=2'b00;
         Aload<=0;
         Sub<=0;
         Halt<=0;
         end
       4'b1010: begin // Add
         IRload<=0;
         JMPmux<=0;
         PCload<=0;
```

FIGURE 8.24 Behavioral Verilog code for the EC-2 control unit. *(continued on next page)*

```
        Meminst<=1;
        MemWr<=0;
        Asel<=2'b00;
        Aload<=1;
        Sub<=0;
        Halt<=0;
        end
      4'b1011: begin // Sub
        IRload<=0;
        JMPmux<=0;
        PCload<=0;
        Meminst<=1;
        MemWr<=0;
        Asel<=2'b00;
        Aload<=1;
        Sub<=1;
        Halt<=0;
        end
      4'b1100: begin // Input
        IRload<=0;
        JMPmux<=0;
        PCload<=0;
        Meminst<=0;
        MemWr<=0;
        Asel<=2'b01;
        Aload<=1;
        Sub<=0;
        Halt<=0;
        end
      4'b1101: begin // Jz
        IRload<=0;
        JMPmux<=1;
        PCload<=Aeq0; // load PC if condition is true
        Meminst<=0;
        MemWr<=0;
        Asel<=2'b00;
        Aload<=0;
        Sub<=0;
        Halt<=0;
        end
      4'b1110: begin // Jpos
        IRload<=0;
        JMPmux<=1;
        PCload<=Apos; // load PC if condition is true
        Meminst<=0;
        MemWr<=0;
        Asel<=2'b00;
```

FIGURE 8.24 Behavioral Verilog code for the EC-2 control unit. *(continued on next page)*

```
        Aload<=0;
        Sub<=0;
        Halt<=0;
        end
      4'b1111: begin // Halt
        IRload<=0;
        JMPmux<=0;
        PCload<=0;
        Meminst<=0;
        MemWr<=0;
        Asel<=2'b00;
        Aload<=0;
        Sub<=0;
        Halt<=1;
        end
      default: begin
        IRload<=0;
        JMPmux<=0;
        PCload<=0;
        Meminst<=0;
        MemWr<=0;
        Asel<=2'b00;
        Aload<=0;
        Sub<=0;
        Halt<=0;
        end
      endcase
    end

    assign debug = state;

endmodule
```

FIGURE 8.24 Behavioral Verilog code for the EC-2 control unit.

8.7.2 FSMD Model

Figures 8.25 and 8.26 show the complete behavioral FSMD Verilog and VHDL code, respectively, for the EC-2 general-purpose microprocessor. The RAM used is the same component as shown in Figure 8.13. The three registers *IR*, *PC*, and *A* are declared as `reg`/SIGNAL. The `always`/PROCESS block is structured just like a regular FSM following the state diagram from Figure 8.11(a). The *Decode* state uses the `case` statement to check the opcode, which is the first three bits of the *IR*. From there, the FSM jumps to the state for executing the corresponding instruction. The actual execution of an instruction simply uses built-in operators such as `A <= A + memory_data` to perform the addition command. The complete FSMD code for the EC-2 general-purpose microprocessor written in both Verilog and VHDL can be downloaded from the book website.

```
module mp (
  input Clock, Reset, Enter,
  input [7:0] Input,
  output [7:0] Output,
  output reg Halt,
  output [3:0] debug
  );

  reg [3:0] state;
  reg [7:0] IR;
  reg [4:0] PC;
  reg [7:0] A;
  reg [4:0] memory_address;
  wire [7:0] memory_data;
  reg MemWr;

  ram U5_ram(.Clock(Clock), .Reset(Reset), .WE(MemWr),
      .Address(memory_address), .D(A), .Q(memory_data));

  always @ (posedge Clock, posedge Reset) begin
   if (Reset) begin
    PC <= 5'b00000;
    IR <= 8'b00000000;
    A  <= 8'b00000000;
    MemWr <= 1'b0;
    Halt <= 1'b0;
    state<=4'b0000;
    end
   else
    case (state)
    4'b0000: begin // Start
      memory_address <= PC;
      MemWr <= 1'b0;
      state <= 4'b0001;
      end
    4'b0001: begin // Fetch
      IR <= memory_data;
      PC <= PC + 1;
      state<=4'b0010;
      end
    4'b0010: begin // Decode
      // memory access using last 5 bits of IR
      memory_address <= IR[4:0];
      case (IR[7:5])
        3'b000: state<=4'b1000; // Load
        3'b001: state<=4'b1001; // Store
        3'b010: state<=4'b1010; // Add
        3'b011: state<=4'b1011; // Sub
        3'b100: state<=4'b1100; // Input
```

FIGURE 8.25 Behavioral FSMD Verilog code for the EC-2 general-purpose microprocessor. *(continued on next page)*

```
      3'b101: state<=4'b1101; // Jz
      3'b110: state<=4'b1110; // Jpos
      3'b111: state<=4'b1111; // Halt
      default:state<=4'b0000; // Start
     endcase
     end
4'b1000: begin // Load
  A <= memory_data;
  state<=4'b0000;
  end
4'b1001: begin // Store
  MemWr <= 1'b1;
  state<=4'b0000;
  end
4'b1010: begin // Add
  A <= A + memory_data;
  state<=4'b0000;
  end
4'b1011: begin // Sub
  A <= A - memory_data;
  state<=4'b0000;
  end
4'b1100: begin // Input
  A <= Input;
  if (Enter) begin
    state<=4'b0000;
    end
  else begin
    state<=4'b1100;
    end
  end
 4'b1101: begin // Jz
  if (A == 0)
    PC <= IR[4:0];
  state<=4'b0000;
  end
  4'b1110: begin // Jpos
   if (A[7] == 1'b0)
     PC <= IR[4:0];
  state<=4'b0000;
  end
  4'b1111: begin // Halt
   Halt <= 1'b1;
   state<=4'b1111;
   end
```

FIGURE 8.25 Behavioral FSMD Verilog code for the EC-2 general-purpose microprocessor. *(continued on next page)*

```
      default: begin
        state<=4'b0000;
        end
      endcase
  end // always

  assign Output = A; // send value of Accumulator to the output
  assign debug = state;

endmodule
```

FIGURE 8.25 Behavioral FSMD Verilog code for the EC-2 general-purpose microprocessor.

```
LIBRARY IEEE;
USE IEEE.STD_LOGIC_1164.ALL;
USE IEEE.STD_LOGIC_ARITH.ALL;
USE IEEE.STD_LOGIC_UNSIGNED.ALL;

ENTITY mp IS PORT (
  Clock, Reset: IN STD_LOGIC;
  Enter: IN STD_LOGIC;
  Input: IN STD_LOGIC_VECTOR(7 DOWNTO 0);
  Output: OUT STD_LOGIC_VECTOR(7 DOWNTO 0);
  Halt: OUT STD_LOGIC;
  debug: OUT STD_LOGIC_VECTOR(3 DOWNTO 0));
END mp;

ARCHITECTURE FSMD OF mp IS
  SIGNAL state: STD_LOGIC_VECTOR(3 DOWNTO 0);
  SIGNAL IR: STD_LOGIC_VECTOR(7 DOWNTO 0);
  SIGNAL PC: STD_LOGIC_VECTOR(4 DOWNTO 0);
  SIGNAL A: STD_LOGIC_VECTOR(7 DOWNTO 0);
  SIGNAL memory_address: STD_LOGIC_VECTOR(4 DOWNTO 0);
  SIGNAL memory_data: STD_LOGIC_VECTOR(7 DOWNTO 0);
  SIGNAL MemWr: STD_LOGIC;

  COMPONENT ram
  GENERIC (size: INTEGER := 5);
  PORT (
    Clock: IN STD_LOGIC;
    Reset: IN STD_LOGIC;
    WE: IN STD_LOGIC;
    Address: IN STD_LOGIC_VECTOR(size-1 DOWNTO 0);
    D: IN STD_LOGIC_VECTOR(7 DOWNTO 0);
```

FIGURE 8.26 Behavioral FSMD VHDL code for the EC-2 general-purpose microprocessor. *(continued on next page)*

```
      Q: OUT STD_LOGIC_VECTOR(7 DOWNTO 0));
    END COMPONENT;
  BEGIN
    U5_RAM: ram
      GENERIC MAP(5)
      PORT MAP (
        Clock   => Clock,
        Reset   => Reset,
        WE      => MemWr,
        Address => memory_address,
        D       => A,
        Q       => memory_data
        );

      PROCESS(Clock, Reset)
      BEGIN
      IF(Reset = '1') THEN
        PC <= "00000";
        IR <= "00000000";
        A <= "00000000";
        MemWr <= '0';
        Halt <= '0';
        state <= "0000";
      ELSIF(Clock'EVENT AND Clock = '1') THEN
        CASE state IS
        WHEN "0000" => -- reset, start
          memory_address <= PC;
          MemWr <= '0';
          state <= "0001";
        WHEN  "0001" => -- fetch
          IR <= memory_data;
          PC <= PC + 1;
          state <= "0010";
        WHEN "0010" => -- decode
          -- using last 5 bits of IR
          memory_address <= IR(4 DOWNTO 0);
          CASE IR(7 DOWNTO 5) IS
            WHEN "000" => state <= "1000"; -- s_load;
            WHEN "001" => state <= "1001"; -- s_store;
            WHEN "010" => state <= "1010"; -- s_add;
            WHEN "011" => state <= "1011"; -- s_sub;
            WHEN "100" => state <= "1100"; -- s_in;
            WHEN "101" => state <= "1101"; -- s_jz;
            WHEN "110" => state <= "1110"; -- s_jpos;
            WHEN "111" => state <= "1111"; -- s_halt;
```

FIGURE 8.26 Behavioral FSMD VHDL code for the EC-2 general-purpose microprocessor. *(continued on next page)*

```
          WHEN OTHERS => state <= "0000";  -- s_start;
        END CASE;
      WHEN "1000" => -- load A from memory
        A <= memory_data;
        state <= "0000";
      WHEN "1001" => -- store A to memory
        MemWr <= '1';
        state <= "0000";
      WHEN "1010" => -- add
        A <= A + memory_data;
        state <= "0000";
      WHEN "1011" => -- subtract
        A <= A - memory_data;
        state <= "0000";
      WHEN "1100" =>
        A <= input;
        IF (Enter = '0') THEN -- wait for Enter key
          state <= "1100";
        ELSE
          state <= "0000";
        END IF;
      WHEN "1101" =>
        IF (A = 0) THEN -- jump if A is 0
          PC <= IR(4 DOWNTO 0);
        END IF;
        state <= "0000";
      WHEN "1110" =>
        IF (A(7) = '0') THEN -- jump if MSB(A) is 0
          PC <= IR(4 DOWNTO 0);
        END IF;
        state <= "0000";
      WHEN "1111" =>
        halt <= '1';
        state <= "1111";
      WHEN OTHERS =>
        state <= "0000";
      END CASE;
    END IF;
  END PROCESS;

  Output <= A; -- send value of Accumulator to the output
  debug <= state;
END FSMD;
```

FIGURE 8.26 Behavioral FSMD VHDL code for the EC-2 general-purpose microprocessor.

8.8 PROBLEMS

8.1. Manually redesign the EC-1 microprocessor to accommodate each of the following changes. The changes are to be done separately.

a) Replace the INPUT instruction with a LOAD constant instruction. The encoding for this instruction is 011ccccc. The operation for this instruction is A ← ccccc, where ccccc is the five least significant bits from the instruction encoding. These five bits are zero extended to eight bits and then loaded into the accumulator.

b) Modify the INPUT instruction so that it will wait for an external Enter key signal before continuing to the next instruction.

c) Add an extra INC instruction, using the opcode 000, to the EC-1 instruction set. The INC instruction increments the accumulator.

d) Add an extra LOAD instruction, using the opcode 001, to the EC-1 instruction set. The LOAD instruction loads the accumulator with the content of memory location aaaa, where aaaa are the four least significant bits of the instruction encoding.

8.2. Write the behavioral Verilog/VHDL code for the EC-1 microprocessor.

8.3. Rewrite the behavioral Verilog/VHDL code for the EC-1 microprocessor with each of the changes from Problem 8.1.

8.4. Write and run the following programs on the EC-2 microprocessor:

a) Input two numbers, and output the sum of these two numbers.

b) Input two numbers, and output the larger of the two numbers.

c) Input two numbers, and output the product of these two numbers.

d) Keep inputting numbers until a 0. Output the total number of numbers entered.

e) Keep inputting numbers until a 0. Output the sum of these numbers.

f) Keep inputting numbers until a 0. Output the largest of these numbers.

g) Keep inputting numbers until a 0. Output the largest and second largest of these numbers.

h) Input three numbers, and output these numbers in ascending order.

8.5. Finish the manual construction of the modified EC-2 microprocessor circuit as discussed in Section 8.4.

8.6. Manually redesign the original EC-2 microprocessor to accommodate each of the following changes. The changes are to be done separately.

a) Replace the SUB instruction in the EC-2 instruction set with a LSHIFT instruction. The LSHIFT instruction shifts the content of the accumulator left by one bit and the rightmost bit is filled with a 0. The result of the shift operation is written back into the accumulator.

b) Replace the SUB instruction in the EC-2 instruction set with an OUTPUT instruction. The OUTPUT instruction outputs the content of the accumulator to the output port. The output port should not show anything when the OUTPUT instruction is not being executed.

c) Add an extra LSHIFT instruction to the EC-2 instruction set. The LSHIFT instruction is defined in part (a) above. You need to use four bits for the opcode.

d) Add two instructions, LSHIFT and RSHIFT, to the EC-2 instruction set. The LSHIFT instruction is defined in part (a) above. The RSHIFT instruction is similar but shifts to the right instead. You need to use four bits for the opcode.

e) Add an extra NOT A instruction that performs the logical NOT operation on the accumulator A. The result is written back into A.

f) The IN A instruction has five bits that are not used. Therefore, we can use bit 4 to differentiate between two different instructions. If bit 4 is a 0, then it is the original IN A instruction, otherwise it is a LOAD constant instruction where the first four bits, bit 0 to bit 3, are first signed extended to eight bits and then loaded into A. The encoding for this new LOAD instruction is 100 1cccc, and the operation is A ← ssssсссс, where ssss is the sign extension of the constant cccc.

g) Use separate instruction and data memories. The program instructions are hardwired and stored in a ROM. All data memory accesses (i.e., all load, store, add, and sub instructions) are from a RAM.

h) Use an external RAM with a single bidirectional data bus instead of the internal RAM.

8.7. Rewrite the behavioral FSMD Verilog/VHDL code for the EC-2 microprocessor with each of the changes from Problem 8.6.

8.8. Given the instruction set as defined below, manually design and implement a general-purpose microprocessor that can execute this instruction set.

Instruction	Encoding	Operation	Comment
Data Movement Instructions			
LDA A,rrr	0001 0rrr	$A \leftarrow R[\text{rrr}]$	Load accumulator from register rrr
STA rrr,A	0010 0rrr	$R[\text{rrr}] \leftarrow A$	Load register rrr from accumulator
LDM A,aaaaaa	0011 0000 00aaaaaa	$A \leftarrow M[\text{aaaaaa}]$	Load accumulator from memory location aaaaaa
STM aaaaaa, A	0100 0000 00 aaaaaa	$M[\text{aaaaaa}] \leftarrow A$	Load memory location aaaaaa from accumulator
LDI A,iiiiiii	0101 0000 iiiiiii	$A \leftarrow \text{iiiiiii}$	Load accumulator with immediate value (iiiiiii is a signed number)
Jump Instructions			
JMP absolute	0110 0000 00 aaaaaa	$PC = \text{aaaaaa}$	Absolute unconditional jump to address aaaaaa
JMPR relative	0110 mmmm	$PC = PC + \text{mmmm}$	Relative unconditional jump (mmmm is two's complement offset from PC)
JZ absolute	0111 0000 00 aaaaaa	IF $(A = 0)$ THEN $PC = \text{aaaaaa}$	Absolute jump to address aaaaaa if A is zero
JZR relative	0111 mmmm	IF $(A = 0)$ THEN $PC = PC + \text{mmmm}$	Relative jump if A is zero (mmmm is two's complement offset from PC)
JNZ absolute	1000 0000 00 aaaaaa	IF $(A \mathrel{!=} 0)$ THEN $PC = \text{aaaaaa}$	Absolute jump to address aaaaaa if A is not zero
JNZR relative	1000 mmmm	IF $(A \mathrel{!=} 0)$ THEN $PC = PC + \text{mmmm}$	Relative jump if A is not zero (mmmm is two's complement offset from PC)
JP absolute	1001 0000 00 aaaaaa	IF $(A = \text{positive})$ THEN $PC = \text{aaaaaa}$	Absolute jump to address aaaaaa if A is positive
JPR relative	1001 mmmm	IF $(A = \text{positive})$ THEN $\text{PC} = \text{PC} + \text{mmmm}$	Relative jump if A is positive (mmmm is two's complement offset from PC)
Arithmetic and Logical Instructions			
AND A,rrr	1010 0rrr	$A \leftarrow A \text{ AND } R[\text{rrr}]$	Accumulator AND register
OR A,rrr	1011 0rrr	$A \leftarrow A \text{ OR } R[\text{rrr}]$	Accumulator OR register
ADD A,rrr	1100 0rrr	$A \leftarrow A + R[\text{rrr}]$	Accumulator + register
SUB A,rrr	1101 0rrr	$A \leftarrow A - R[\text{rrr}]$	Accumulator − register
NOT A	1110 0000	$A \leftarrow \text{NOT } A$	Invert accumulator
INC A	1110 0001	$A \leftarrow A + 1$	Increment accumulator
DEC A	1110 0010	$A \leftarrow A - 1$	Decrement accumulator
SHFL A	1110 0011	$A \leftarrow A \ll 1$	Shift accumulator left, pad with 0
SHFR A	1110 0100	$A \leftarrow A \gg 1$	Shift accumulator right, pad with 0
ROTR A	1110 0101	$A \leftarrow \text{Rotate\_right}(A)$	Rotate accumulator right

(*Continued*)

Instruction	Encoding	Operation	Comment
Input/Output and Miscellaneous			
In A	1111 0000	$A \leftarrow$ Input	Input to accumulator
Out A	1111 0001	Output $\leftarrow A$	Output from accumulator
HALT	1111 0010	Halt	Halt execution
NOP	0000 0000	No operation	No operation

Notations:

A = accumulator

R = general register

M = memory

PC = program counter

rrr = three bits for specifying the general register number (0 – 7)

aaaaaa = six bits for specifying the absolute memory address

iiiiiiii = an 8-bit signed number

mmmm = four bits for specifying the relative jump offset in two's complement format. The offset is relative to the current PC location.

8.9. Write the behavioral FSMD Verilog/VHDL code for a microprocessor that can execute the instructions in the instruction set defined in Problem 8.8.

CHAPTER 9

Interfacing Microprocessors

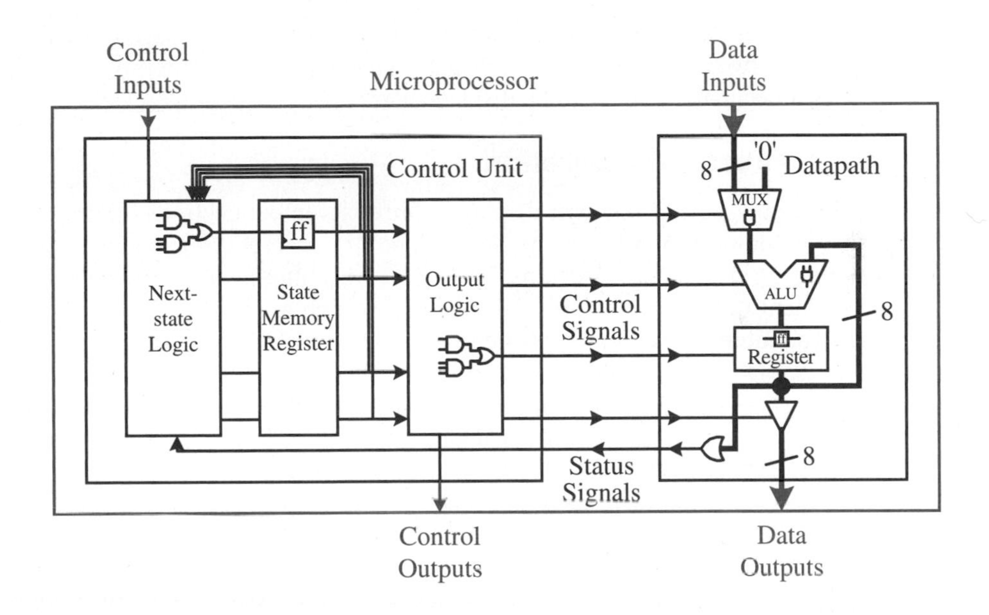

For microprocessors to be useful, there must be an interface between the microprocessor and the human; between the electronic world and the real world. Measurements made in the real world usually are analog and in a continuous range of values, whereas information stored inside the microprocessor is in discrete digital values. Furthermore, for microprocessors to communicate with other electronic components there must be a standard communication protocol between them. In this chapter, we will look at some common human I/O interfaces and standard communication protocols, and see how microprocessors interface with them.

9.1 Multiplexing 7-segment LED Display

So far, we have worked with discrete LEDs and individual 7-segment LED displays. Typically, two or more 7-segment displays are combined together in a package so that a larger decimal number can be displayed, as shown in Figure 9.1.

9.1.1 Theory of Operation

Individually, each 7-segment display requires seven connections for the seven LEDs (not including ground and the decimal point), so eight 7-segment displays would require 56 (8×7) connections. To reduce the number of connections, the eight 7-segment displays are connected together in such a way that fewer connections are needed. Internally, all of the same segments for each display are connected together in common. In other words, segment *a* for all of the displays are connected together, segment *b* for all of the displays are connected together, and so on. As a result, no matter how many digits are in a package, there will be only seven connections for the seven segments. Furthermore, either the negative or the positive ends for all of the LED segments in a 7-segment display are connected together, resulting in either a common cathode or common anode display, respectively.

Figure 9.2 shows the internal connections of a three-digit 7-segment display with a common cathode. For a common cathode display, when a digit connection is set to a 1, then that entire corresponding digit is turned off. A particular LED is turned on by setting the digit connection to a 0 and the segment connection to a 1. Because all of the same segments for all of the digits are connected together, if you connect, for example, segment *a* to a 1 and the three digit connections to a 0, then segment *a* for all three digits will be turned on. The problem is how to turn on a particular segment on one digit and turn off the same segment on another digit. For example, to display the number 45, we need to turn off segment *a* of digit 1 for the number 4, and turn on

FIGURE 9.1 An eight-digit 7-segment display showing the number 3.1415.

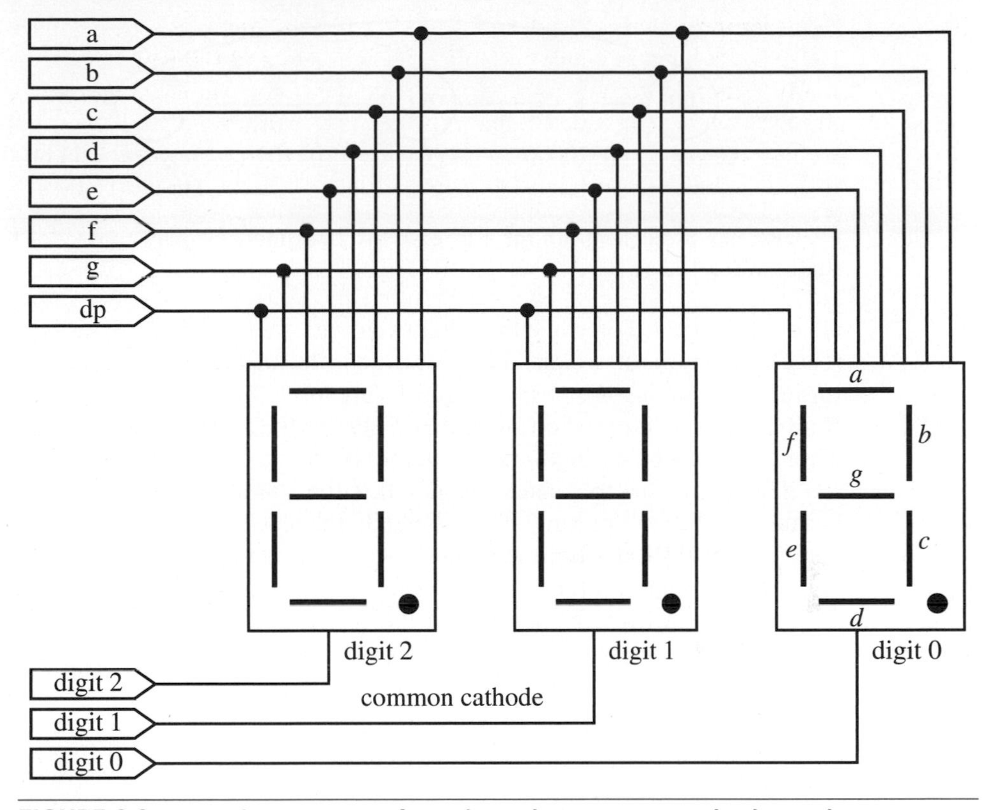

FIGURE 9.2 Internal connections for a three-digit 7-segment display with common cathode.

segment *a* of digit 0 for the number 5. But in order to turn on both digits, the two digit connections must be connected to 0. But if they are both connected to 0 then segment *a* on both digits will be either turned on or off together.

The solution to this problem is known as time multiplexing. Time multiplexing turns on only one digit at a time for only a short period of time. This turning on of one digit at a time will cycle continuously through all of the digits in the display. Although the digits are being turned on one at a time, we will get the impression that they are all turned on at the same time because they are cycling through so quickly. The tradeoff is that the digits will be slightly dimmer than if they are on continuously.

9.1.2 Controller Design

The state diagram, shown in Figure 9.3, for a three-digit 7-segment LED display controller to display the decimal number 123 is extremely simple. It continuously cycles through the three states turning on and off the appropriate *Digit* and *Segment* connections. The *Digit* connection has 3 bits where the left-most bit is connected to digit 2. The *Segment* connection has 7 bits where the left-most bit is connected to segment *a*.

State 00 displays the decimal number 1 on digit 2, state 01 displays the decimal number 2 on digit 1, and state 10 displays the decimal number 3 on digit 0. In state

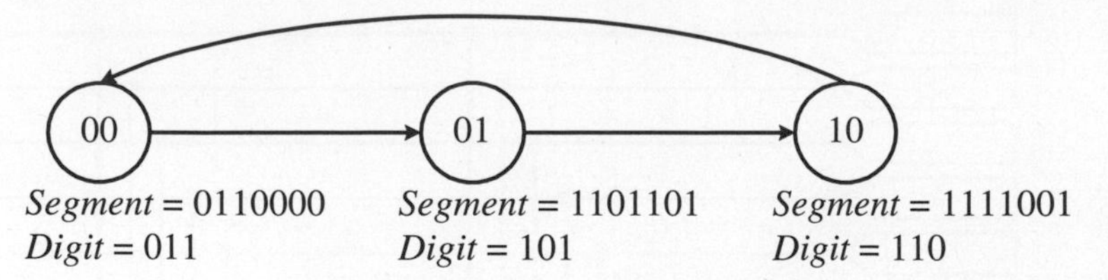

FIGURE 9.3 State diagram for a three-digit 7-segment display controller to display the number 123.

00, digit 2 is turned on while the other two digits are turned off by setting *Digit* = 011 (because a 0 turns on a digit). For showing the decimal number 1, segments *b* and *c* are turned on by setting *Segment* = 0110000 (because a 1 turns on a segment). Similarly in state 01, digit 1 is turned on by setting *Digit* = 101, and the segments for the number 2 are turned on by setting *Segment* = 1101101.

The main issue to consider is how fast the FSM cycles through the three states because of the multiplexing of the digits. If the clock speed is too slow, then you will see only one digit at a time. As you slowly increase the clock speed, you will start to see all three digits, but they will have a flicker. Increasing the clock speed a little more will cause the flicker to go away, and you will see all of the digits on. If you continue to increase the clock speed, then the digits will begin to get dimmer.

Figures 9.4 and 9.5 show the Verilog and VHDL code, respectively, of this simple three-digit 7-segment LED display controller to display the decimal number 123.

```
module fsm (
  input Clock, Reset,
  output reg [2:0] Digit,
  output reg [0:6] Segment
    );

  reg [1:0] state;

  always @ (posedge Clock, posedge Reset) begin
    if (Reset) begin
      Digit <= 3'b111;        // off
      Segment <= 7'b0000000; // off
      state <= 2'b00;
      end
    else
      case (state)
      2'b00: begin
        // display the number 1 on digit 2
        Segment <= 7'b0110000; // 1
        Digit <= 3'b011;        // D2
        state <= 2'b01;
        end
      2'b01: begin
        // display the number 2 on digit 1
```

FIGURE 9.4 Behavioral Verilog code for a three-digit 7-segment display controller.
(continued on next page)

```
          Segment <= 7'b1101101; // 2
          Digit <= 3'b101;       // D1
          state <= 2'b10;
          end
        2'b10: begin
          // display the number 3 on digit 0
          Segment <= 7'b1111001; // 3
          Digit <= 3'b110;       // D0
          state <= 2'b00;
          end
        default: begin
          state <= 2'b00;
          end
        endcase
  end // always
endmodule
```

FIGURE 9.4 **Behavioral Verilog code for a three-digit 7-segment display controller.**

```
LIBRARY IEEE;
USE IEEE.STD_LOGIC_1164.ALL;

ENTITY fsm IS PORT (
  Clock, Reset: IN STD_LOGIC;
  Digit: OUT    STD_LOGIC_VECTOR (2 DOWNTO 0);
  Segment: OUT  STD_LOGIC_VECTOR (0 TO 6));
END fsm;

ARCHITECTURE Behavioral OF fsm IS
  SIGNAL state: STD_LOGIC_VECTOR(1 DOWNTO 0);
BEGIN
    PROCESS(Clock, Reset)
    BEGIN
    IF(Reset = '1') THEN
      Digit <= "111";
      Segment <= "0000000";
      state <= "00";
    ELSIF(Clock'EVENT AND Clock = '1') THEN
      CASE state IS
      WHEN "00" =>
      -- display the number 1 on digit 2
      Segment <= "0110000";  -- 1
      Digit <= "011";        -- D2
      state <= "01";
    WHEN "01" =>
      -- display the number 2 on digit 1
      Segment <= "1101101";  -- 2
      Digit <= "101";        -- D1
      state <= "10";
```

FIGURE 9.5 **Behavioral VHDL code for a three-digit 7-segment display controller.**
(continued on next page)

```
        WHEN "10" =>
          -- display the number 3 on digit 0
          Segment <= "1111001"; -- 3
          Digit <= "110";       -- D0
          state <= "00";
        WHEN OTHERS =>
          state <= "00";
        END CASE;
      END IF;
    END PROCESS;
  END Behavioral;
```

FIGURE 9.5 Behavioral VHDL code for a three-digit 7-segment display controller.

9.2 Issues with Interfacing Switches

Switches are common, simple input devices, and should be very easy to interface with a microprocessor. If the timing in the control unit is not done correctly, however, these simple input devices might not function as expected. In this section, we will look at some problems that might occur when interfacing with switches.

To illustrate the problems, we will design a dedicated microprocessor for inputting many 8-bit unsigned numbers through one input port and then output the sum of these numbers. The algorithm continues to input numbers as long as the number entered is not a 0. Each number entered also is displayed on the output. When the number entered is a 0, the algorithm stops and outputs the sum of all of the numbers entered. The algorithm for solving this problem is shown in Figure 9.6.

The algorithm shown in Figure 9.6 has five data manipulation statements in lines 1, 3, 4, 8, and 10. There is one conditional test in line 5. The algorithm requires an adder and two 8-bit registers: one for variable *X* and one for variable *sum*. The dedicated datapath is shown in Figure 9.7.

Line 1 in the algorithm is performed by asserting the *Reset* signal to initialize both registers to zero so no control word is needed. Line 3 is performed by asserting the *XLoad* signal. The input operands to the adder are from the two registers, and line 4 is performed by asserting the *sumLoad* signal. The *OutSelect* signal selects the 2-to-1 multiplexer for one of the two sources, register *X* or register *sum*, to output. The *Out*

```
1    sum = 0
2    BEGIN LOOP
3      INPUT X
4      sum = sum + X
5      IF (X = 0) THEN
6        EXIT LOOP
7     END IF
8     OUTPUT X
9    END LOOP
10   OUTPUT sum
```

FIGURE 9.6 Algorithm for solving the summing input numbers problem.

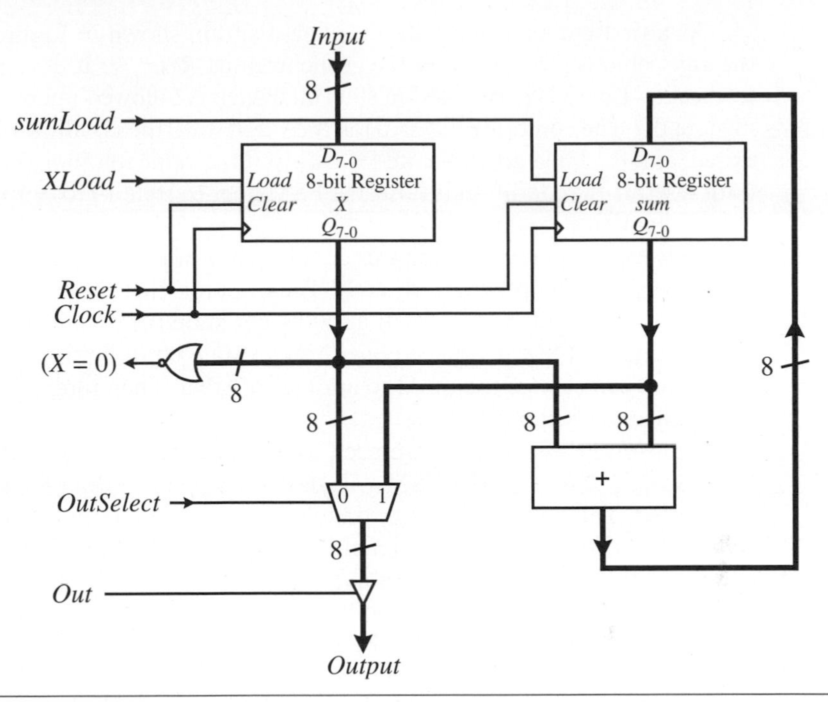

FIGURE 9.7 Dedicated datapath for solving the summing input numbers problem.

control signal is used to enable the tri-state buffer to output the value from the selected source. The conditional test $(X = 0)$ is generated by the 8-input NOR gate that is connected to the output of the X register.

This dedicated datapath for solving the summing input numbers problem requires four control signals, *sumLoad*, *XLoad*, *OutSelect*, and *Out*, and generates one status signal, $(X = 0)$. The control words are shown in Figure 9.8.

At first glance, this algorithm seems very simple and straightforward. Because of the requirements of this problem, however, the actual hardware implementation of this microprocessor is a bit tricky. Specifically, the requirement that many different numbers be input through one input port requires careful timing considerations and an understanding about how mechanical switches behave. Also, the timing for outputting the value in register X needs some careful thought.

Control Word	Instruction	*XLoad*	*sumLoad*	*OutSelect*	*Out*
0	INPUT X	1	0	×	0
1	$sum = sum + X$	0	1	×	0
2	OUTPUT X	0	0	0	1
3	OUTPUT *sum*	0	0	1	1

FIGURE 9.8 Control words for solving the summing input numbers problem.

As a first try, we begin with the state diagram shown in Figure 9.9(a). Line 1 of the algorithm is performed by the asynchronous *Reset*, so it does not require a state to execute. Line 3 is performed in state 00, which is followed unconditionally by line 4 in state 01. The condition $(X = 0)$ is then tested. If the condition is true, the loop is exited, and the FSM goes to state 11 to output the value for *Sum* and stays in that state until reset. If the condition is false, the FSM goes to state 10 to output *X*, and the loop repeats back to state 00.

However, if you implement this circuit in hardware using, for example, a 50 MHz clock speed, it will not work correctly. The reason is that the FSM cycles through the three loop states (00, 01, and 10) at 20 ns per state (i.e., 1/50000000 seconds/cycle). As a result, the FSM will have gone through state 00 to input a number many times before you can change the input to another number. Therefore, the same number will be summed many times.

To resolve this problem, we need to add another input signal that acts like the Enter switch. This way, the FSM will stay in state 00, waiting for the *Enter* signal to be asserted. This will give the user time to set up the input number before pressing the Enter switch. When the *Enter* signal is asserted, the FSM will exit state 00 with the new number to be processed. This modified state diagram is shown in Figure 9.9(b).

This modified state diagram still has a timing problem because of the fast clock speed. Starting from state 00, the FSM waits for the Enter switch to be pressed. After entering a nonzero number and pressing the Enter switch, the FSM goes to state 01, but before you have time to release it, the FSM will again have cycled through the complete loop and is back at state 00 in 60 ns. Since you have not yet released the switch, the FSM will continue on another loop with the same input number. We need to break the loop by waiting for the Enter switch to be released. This is shown in the state diagram in Figure 9.10(a). State 10 will wait for the Enter switch to be released before continuing on and looping back to state 00.

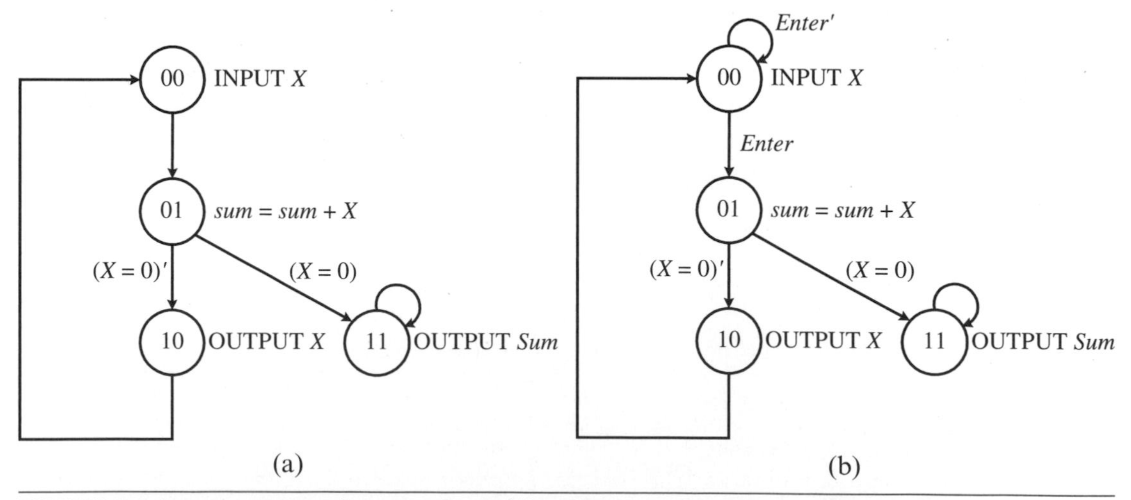

FIGURE 9.9 Incorrect state diagrams for solving the summing input numbers problem.

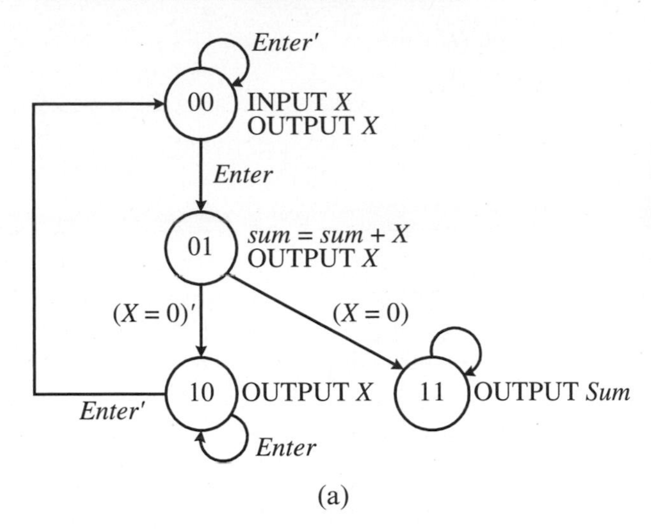

(a)

Current State Q_1Q_0	Next State (Implementation) $Q_{1next}Q_{0next}$ (D_1D_0)			
	Enter, (X = 0)			
	00	01	10	11
00	00	00	01	01
01	10	11	10	11
10	00	00	10	10
11	11	11	11	11

(b)

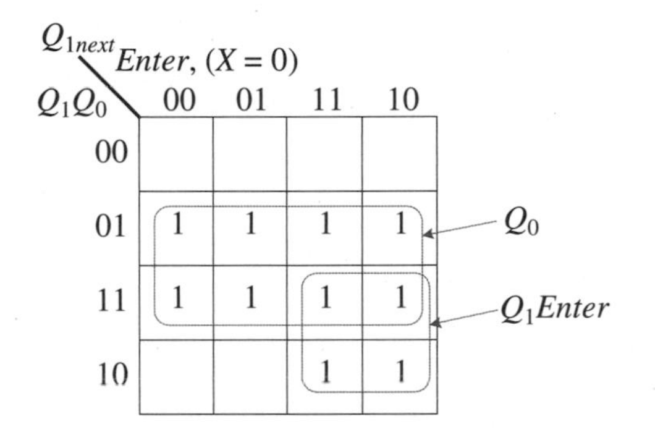

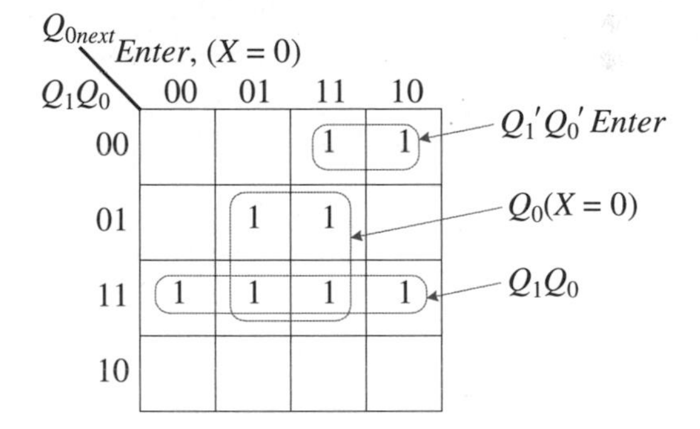

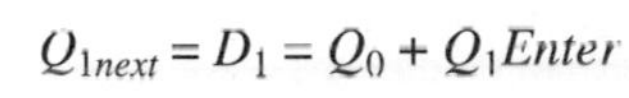

$$Q_{0next} = D_0 = Q_1Q_0 + Q_0(X = 0) + Q_1'Q_0'\,Enter$$

(c)

Control Word	State Q_1Q_0	Instruction	*XLoad*	*sumLoad*	*OutSelect*
0	00	INPUT *X*, OUTPUT *X*	1	0	0
1	01	*sum* = *sum* + *X* OUTPUT *X*	0	1	0
2	10	OUTPUT *X*	0	0	0
3	11	OUTPUT *sum*	0	0	1

(d)

FIGURE 9.10 Control unit for solving the summing input numbers problem: (a) state diagram; (b) next-state (implementation) table; (c) K-maps and next-state equations; (d) control words and output table; (e) output equations; (f) circuit. *(continued on next page)*

$$XLoad = Q_1'Q_0'$$
$$sumLoad = Q_1'Q_0$$
$$OutSelect = Q_1Q_0$$

(e)

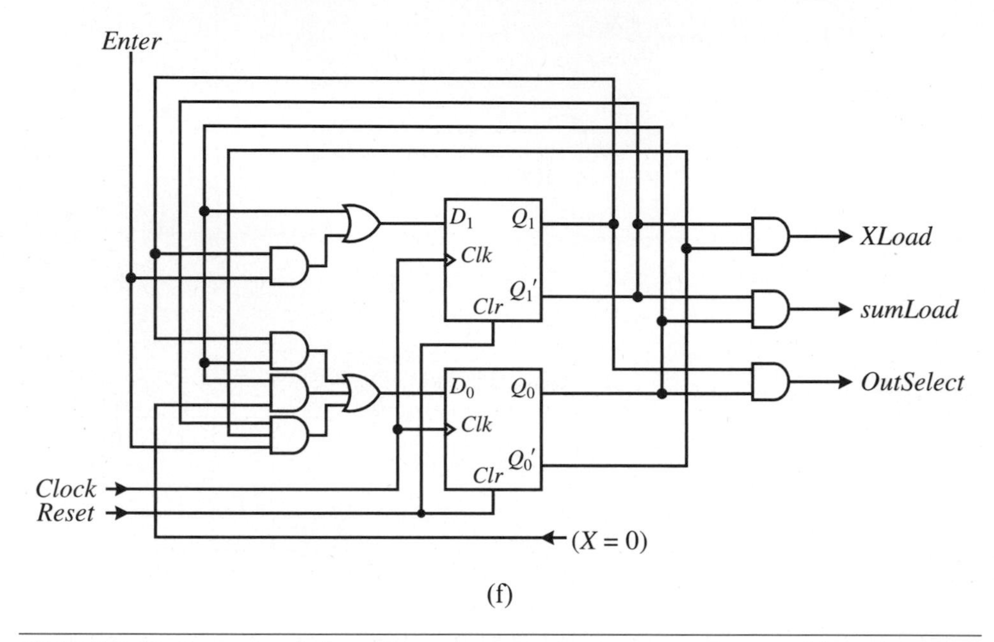

(f)

FIGURE 9.10 Control unit for solving the summing input numbers problem: (a) state diagram; (b) next-state (implementation) table; (c) K-maps and next-state equations; (d) control words and output table; (e) output equations; (f) circuit.

Theoretically, this last state diagram is correct. In practice, however, there still might be a problem with the operation of the mechanical switch used for the *Enter* signal. When a mechanical switch is pressed, it usually goes on and off several times before settling down in the on position. This is referred to as the "debounce" problem. When the switch is mechanically fluctuating between the on and the off positions after it is pressed, the FSM again can go through the loop many times. We need to debounce the switch. This, however, is not done in the FSM circuit itself but in the interface circuit between the FSM and the switch. We will address this problem when we build the interface circuit.

Another timing issue has to do with outputting X in state 10. The FSM stays in state 10 only for as long as the Enter key is still pressed. As soon as the user releases the Enter key, which normally will occur in less than a second, the FSM will exit state 10. Therefore, the X value will be displayed only for a fraction of a second, which is too short to be seen. A better solution is to output X in all of the three states inside the loop, and to output *Sum* when the loop exits as shown in Figure 9.10(a) with the four OUTPUT operations. In making this change, we no longer need the tri-state buffer along with the *Out* control signal in the datapath because we are always outputting a value.

We now will construct the control unit circuit based on the state diagram shown in Figure 9.10(a). Four states are used for the five data manipulation statements. All of the states except for 11 will output X. State 00 inputs X and waits for the *Enter* signal. This allows the user to set up the input number and then press the Enter switch. When the Enter switch is pressed, the FSM goes to state 01, to sum X, and tests for the condition $(X = 0)$. If the condition is true, the FSM terminates in state 11 and outputs *sum*; otherwise, it goes to state 10 to wait for the *Enter* signal to be de-asserted when the user releases the Enter switch. After exiting state 10, the FSM continues on to repeat the loop in state 00.

The next-state table, as derived from the state diagram, is shown in Figure 9.10(b). The table requires four variables: two to encode the four states, Q_1 and Q_0, and two for the status signals, *Enter* and $(X = 0)$.

Using D flip-flops to implement the state memory, the implementation table is the same as the next-state table, except that the values in the table entries are the inputs to the flip-flops, D_1 and D_0, instead of the flip-flops outputs, Q_{1next} and Q_{0next}. The K-maps and the next-state equations for Q_{1next} and Q_{0next} are shown in Figure 9.10(c).

The modified control words and output table for the three control signals are shown in Figure 9.10(d). State 00 performs line 3 of the algorithm in Figure 9.6 by asserting *XLoad*, and line 8 by de-asserting *OutSelect*. When *OutSelect* is de-asserted, X is passed to the output. State 01 performs line 4 and line 8. Line 4 is executed by asserting *sumLoad*, and line 8 is executed by de-asserting *OutSelect*. State 10 again performs line 8 by de-asserting *OutSelect*. Finally, state 11 performs line 10 by asserting *OutSelect*.

The output equations, as derived from the output table, are shown in Figure 9.10(e). There is one equation for each of the three control signals. Each equation is dependent only on the current state (i.e., the current values in Q_1 and Q_0).

The complete control unit circuit is shown in Figure 9.10(f). The state memory consists of two D flip-flops. The inputs to the flip-flops are the next-state circuits derived from the two next-state equations. The output circuits for the three control signals are derived from the three output equations. The status signal $(X = 0)$ comes from the comparator in the datapath.

The final microprocessor is formed by connecting the control unit and the datapath together using the designated control and status signals, as shown in Figure 9.11.

In order to implement the circuit onto the FPGA development board, we need to connect the microprocessor's I/Os to the switches, LEDs, and clock source. The most important interface circuit for this problem is to debounce the Enter switch. A simple circuit to debounce a switch is to use a D flip-flop, as shown in Figure 9.12. The clock frequency for the D flip-flop clock input must be slow enough for the switch bounce to settle, so that the flip-flop will latch in a single value. The exact clock frequency is not too critical. The clock divider circuit used in the example slows down a 50 MHz clock to approximately 4 Hz. Figure 9.13 shows a simulation trace of the microprocessor executing the summing input numbers problem with the two inputs 3 and 5.

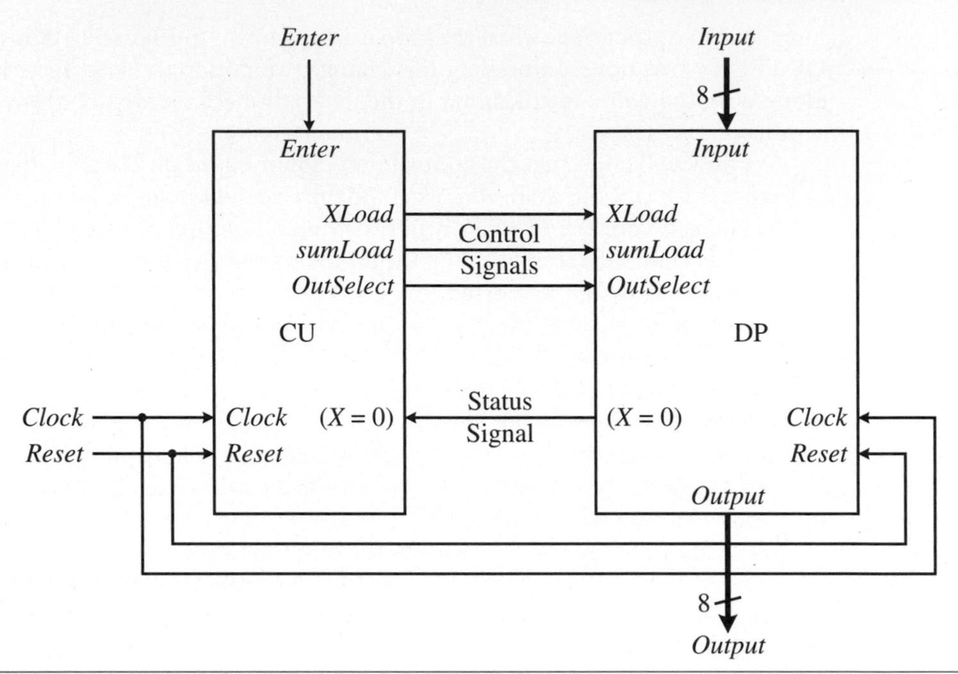

FIGURE 9.11 Microprocessor for solving the summing input numbers problem.

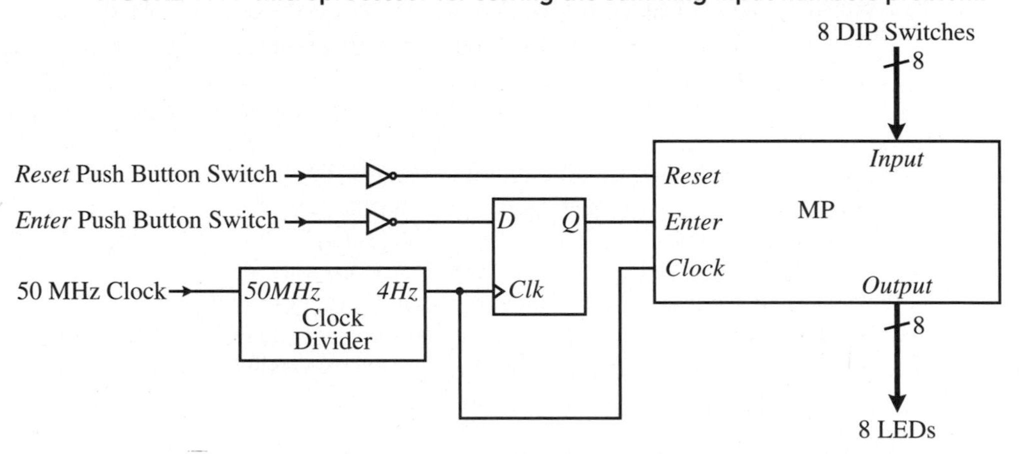

FIGURE 9.12 External I/O interface to the summing input numbers microprocessor.

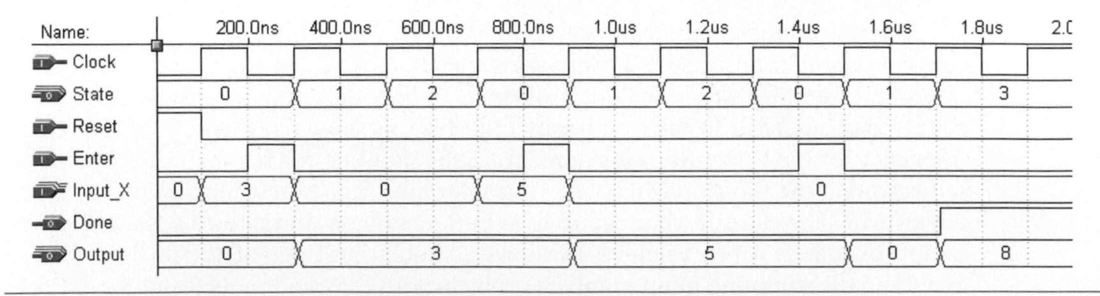

FIGURE 9.13 Sample simulation for the summing input numbers problem.

9.3 3 × 4 Keypad Controller

Keypads are just like push buttons, but with several packaged together into one unit. Figure 9.14(a) shows a picture of a typical 12-key keypad arranged in a 3 × 4 grid. Normally-opened push buttons usually are used for the keys.

9.3.1 Theory of Operation

Twelve discreet push buttons usually will have 24 connections because each push button requires two connections. In order to reduce the number of connections, the keys in a keypad are connected such that those in the same row will have one connection point connected in common, and those in the same column will have the other connection point connected in common. The internal connections of a 3 × 4 keypad is shown in Figure 9.14(b). This connection configuration has only seven connections for the 12 keys: three connections for the three columns and four connections for the four rows.

Using normally-opened push buttons, the intersections for each column and row are normally disconnected when not pressed. When a particular key is pressed, it will connect between the column and the row where that key is located. For example, when key 1 is pressed, it will connect row 0 with column 2; when key 8 is pressed, it will connect row 2 with column 1.

We will look at how we can connect a keypad so that we can distinguish which key has been pressed. Again we need to keep in mind that regardless of whether a key

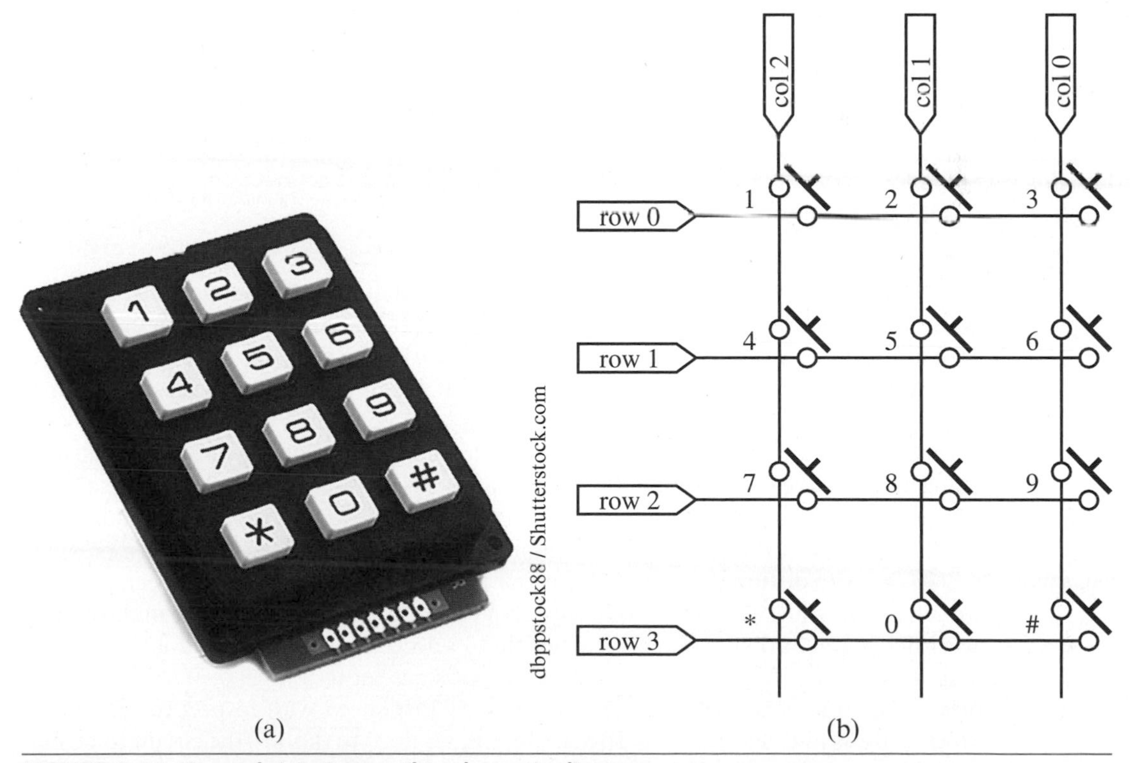

FIGURE 9.14 Keypad: (a) picture; (b) schematic diagram.

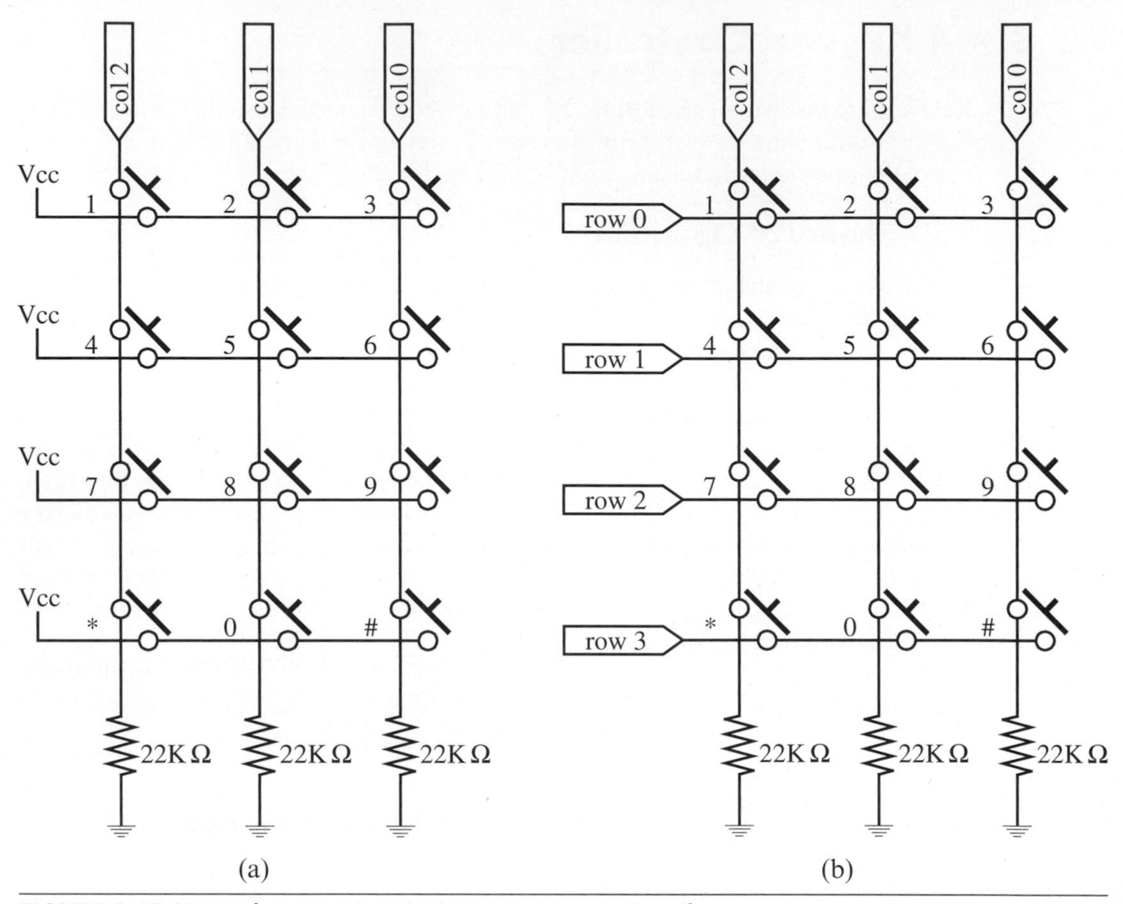

FIGURE 9.15 Keypad connection: (a) incorrect connection; (b) correct connection.

is pressed, we cannot have undefined logic values, that is, a high-impedance Z value. As a first attempt, we might be tempted to connect the circuit shown in Figure 9.15(a) with all of the row inputs connected to Vcc, and all of the columns are pulled down to ground via a 22K Ω resistor. We first note that at the three column connection points, no undefined logic value is possible. For example, at column 2, when none of the keys in that column is pressed, then the signal is a logic 0 because of the pull-down resistor, and when any one of the keys in that column is pressed, then column 2 will have a logic 1 because of the row connection to Vcc. The problem, however, is that we would not be able to distinguish which of the four keys in that column (1, 4, 7, or *) has been pressed.

In order to distinguish which of the four keys in the same column has been pressed, we cannot have all of the rows connected to Vcc at the same time. Just like multiplexing several 7-segment digits, we need to be able to selectively set one row at a time to the logic 1 value (i.e., Vcc) while the rest of the rows are set to a logic 0. With any multiplexing scheme, this need to cycle at a relatively fast speed with respect to the time it takes to press and release a key. To achieve this, we need to modify the circuit to be like that shown in Figure 9.15(b). Replacing the Vcc connections for the four rows with just an input connection, these four inputs can now either be set to a logic 0 or a logic 1 by the microprocessor. So, for example, when key 0 is pressed, a logic 1 will be seen at the column 1 connection point only if the row 3 connection is a logic 1. At this point, you might wonder how is this different from the original circuit in which all of the rows are connected to a logic 1? For both cases, you see a logic 1 at the column connection point when any one of the four column keys is pressed. The important difference is that if we set row 3 to a logic 0, column 1 will have a logic 0 regardless of whether key 0 is pressed or not. In other words, because we know to which row we are sending a logic 1, if we get a logic 1 at column 1, then we will know which intersection point in that column got connected by a key press.

For this multiplexing scheme to work, we need to be able to send either a logic 1 or a logic 0 to each of the rows at a fast speed and then read the column connection values to see if it is a 1 or a 0. A controller is used to repeatedly cycle through each of the four rows by sending a logic 1 value to one row while sending a logic 0 to the rest of the rows. After sending a logic 1 to a row, it will read the three column signals to see if any of them has a logic 1 value. If there is a logic 1 value for a particular column, then the key at that intersection of the grid has been pressed.

9.3.2 Controller Design

The state diagram for a 3 × 4 keypad controller is shown in Figure 9.16. It continuously cycles through the eight states, which are grouped in pairs. The first state in the pair will send out the *Row* value and the second state in the pair will test for the three *Column* values. For example, state 000 sets row 0 to a 1, and then state 001 tests which of the three columns has a 1. If column 0 has a 1 then the value 3 in binary is assigned to the output *Key*. Figure 9.17 shows the behavioral Verilog code for a 3 × 4 keypad controller.

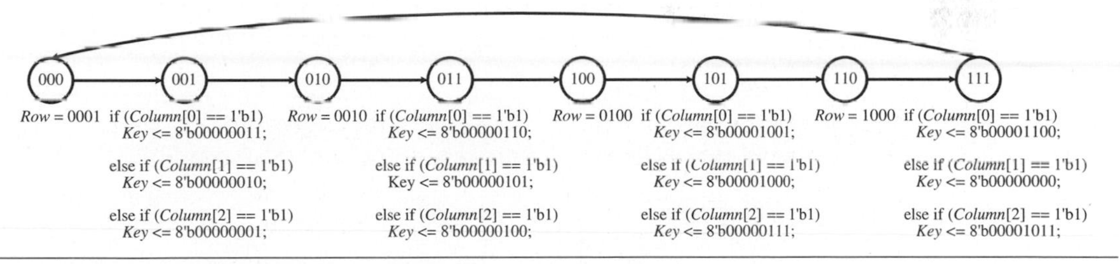

FIGURE 9.16 State diagram for a 3 × 4 keypad controller.

```
module fsm (
  input Clock, Reset,
  input [2:0] Column,
  output reg [3:0] Row,
  output reg [7:0] Key
    );
```

FIGURE 9.17 Behavioral Verilog code for a 3 × 4 keypad controller. *(continued on next page)*

```
reg [2:0] state;

always @ (posedge Clock, posedge Reset) begin
  if (Reset) begin
    state <= 3'b000;
    end
  else
    case (state)
    3'b000: begin
      // set row0 to 1
      Row <= 4'b0001;
      state <= 3'b001;
      end
    3'b001: begin
      // check key press
      if (Column[0] == 1'b1) Key <= 8'b00000011;
      else if (Column[1] == 1'b1) Key <= 8'b00000010;
      else if (Column[2] == 1'b1) Key <= 8'b00000001;
      state <= 3'b010;
      end
    3'b010: begin
      // set row1 to 1
      Row <= 4'b0010;
      state <= 3'b011;
      end
    3'b011: begin
      // check key press
      if (Column[0] == 1'b1) Key <= 8'b00000110;
      else if (Column[1] == 1'b1) Key <= 8'b00000101;
      else if (Column[2] == 1'b1) Key <= 8'b00000100;
      state <= 3'b100;
      end
    3'b100: begin
      // set row2 to 1
      Row <= 4'b0100;
      state <= 3'b101;
      end
    3'b101: begin
      // check key press
      if (Column[0] == 1'b1) Key <= 8'b00001001;
      else if (Column[1] == 1'b1) Key <= 8'b00001000;
      else if (Column[2] == 1'b1) Key <= 8'b00000111;
      state <= 3'b110;
      end
    3'b110: begin
      // set row3 to 1
      Row <= 4'b1000;
      state <= 3'b111;
      end
```

FIGURE 9.17 Behavioral Verilog code for a 3 × 4 keypad controller. *(continued on next page)*

```
        3'b111: begin
          // check key press
          if (Column[0] == 1'b1) Key <= 8'b00001100;
          else if (Column[1] == 1'b1) Key <= 8'b00000000;
          else if (Column[2] == 1'b1) Key <= 8'b00001011;
          state <= 3'b000;
          end
        default: begin
          state <= 3'b000;
          end
        endcase
  end // always
endmodule
```

FIGURE 9.17 Behavioral Verilog code for a 3 × 4 keypad controller.

9.4 PS2 Keyboard and Mouse

Older computer keyboards and mice use the PS2 protocol and connector to communicate with the computer. Although this is an old technology, it is a good example of how a serial communication protocol works. Although the keyboard and the mouse both use the PS2 protocol, the controller for the mouse is slightly more difficult because it requires an initialization sequence and a bidirectional communication channel, whereas the keyboard does not. In this section, we will construct controllers for both.

9.4.1 Theory of Operation—PS2 Keyboard

The communication between the PS2 keyboard and the controller (which will be implemented on an FPGA chip) uses two signals, *KeyboardClock* and *KeyboardData*. When there is no activity, that is, when no key is pressed on the keyboard, both *KeyboardClock* and *KeyboardData* are at a 1. When a key is pressed (or released), the keyboard sends a unique code for that key to the controller serially over the *KeyboardData* line. The serial data on the *KeyboardData* line is synchronized between the keyboard and the controller by clock pulses that the keyboard sends over the *KeyboardClock* line.

The data for each key that is sent over the *KeyboardData* line consists of eleven bits. These eleven bits are: a 0 for the start bit, eight data bits for the key code starting with the least significant bit to the most significant bit, an odd parity bit, and lastly, a 1 for the stop bit. Figure 9.18 lists some of the key codes generated by the keyboard when the corresponding key is pressed. When a key is released, a different code is generated. The odd parity bit is set such that the total number of 1 bits in the eight data bits plus the parity bit is an odd number.

Figure 9.19 shows a sample timing diagram for the data transmission of the key code 4E (01001110 in binary) for the hyphen key. Starting from the inactive state, where both the *KeyboardData* and *KeyboardClock* lines are at a 1, the transmission begins by setting the *KeyboardData* line to a 0 for the start bit. The keyboard then sends out the data and parity bit on the *KeyboardData* line at a rate of one bit per clock cycle on the *KeyboardClock* line. The clock pulses on the *KeyboardClock* line are generated by

Key	Key Code	Key	Key Code	Key	Key Code	Key	Key Code
1	16	A	1C	K	42	U	3C
2	1E	B	32	L	4B	V	2A
3	26	C	21	M	3A	W	1D
4	25	D	23	N	31	X	22
5	2E	E	24	O	44	Y	35
6	36	F	2B	P	4D	Z	1A
7	3D	G	34	Q	15	Esc	76
8	3E	H	33	R	2B	BS	66
9	46	I	43	S	1B	CR	5A
0	45	J	3B	T	2C	Ctrl	14

FIGURE 9.18 A partial list of key codes generated by the keyboard.

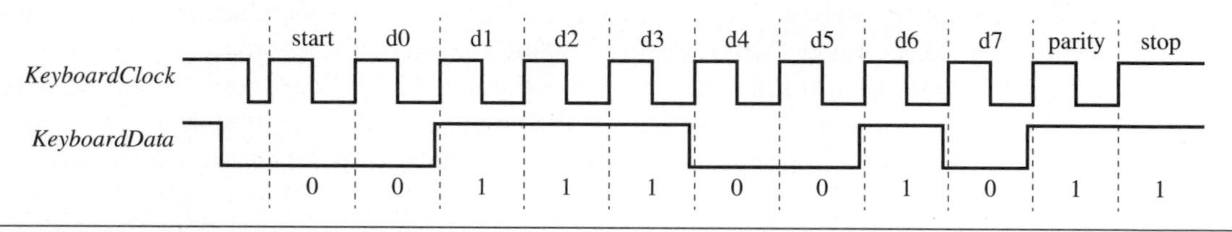

FIGURE 9.19 Sample timing diagram for the data transmission of the key code 4E.

the keyboard. The parity bit for the key code 4E is a 1, because the eight data bits have four (an even number) 1 bits, therefore, to make the parity odd, the parity bit must be a 1. Finally, a 1 stop bit is sent.

9.4.2 Controller Design—PS2 Keyboard

The state diagram for our PS2 keyboard controller shown in Figure 9.20(a) is derived by following the timing diagram shown in Figure 9.19. In each of the eight data states, d0, d1, . . . , d7, we will get one corresponding data bit from the *KeyboardData* input line. For example, suppose we use an 8-bit register named *Keycode* to store the eight data bits. Then in state d0, we will assign *KeyboardData* to $Keycode_0$ (i.e., the 0th bit of *Keycode*), and in state d1, we will assign *KeyboardData* to $Keycode_1$, and so on for all eight data bits. This is possible because the transition of the FSM from one state to the next is synchronized by the keyboard clock signal *KeyboardClock*. For simplicity, we will not check for the start bit, parity bit, nor the stop bit, but will just skip over them.

Note that in the state diagram there is not a wait state for waiting for the initial start bit. This might seem incorrect at first because without an initial wait state to wait for the start bit, the FSM is seen to be continuously going through all of the states even when no key is pressed. However, this is not the case because the clock for driving the FSM is not running continuously. We use the keyboard clock signal *KeyboardClock* to drive our FSM clock (see Figure 9.20(f) for the FSM clock connection), and this signal is at a constant 1 when there is no activity, so our FSM will remain at the start state waiting for a key to be pressed. When a key is pressed, the *KeyboardClock* will begin to toggle, thus activating the FSM and clocking in the data bits.

The next-state table using four D flip-flops to encode the eleven states is shown in Figure 9.20(b). The K-maps and next-state equations derived from the next-state table are shown in Figure 9.20(c).

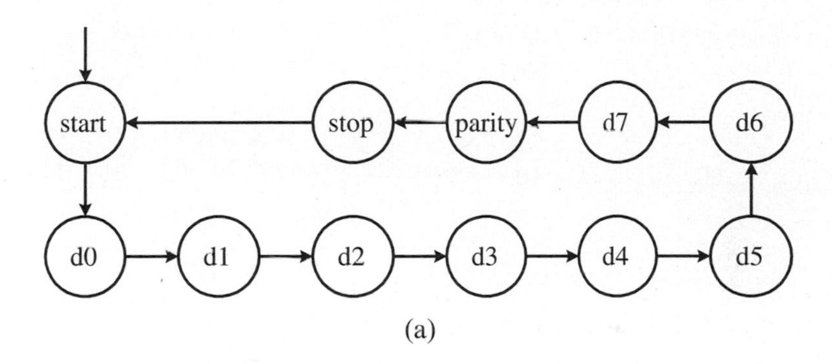

(a)

Current State $Q_3Q_2Q_1Q_0$		Next State (Implementation) $Q_{3next}Q_{2next}Q_{1next}Q_{0next}$ $(D_3D_2D_1D_0)$
0000	start	0001
0001	d0	0010
0010	d1	0011
0011	d2	0100
0100	d3	0101
0101	d4	0110
0110	d5	0111
0111	d6	1000
1000	d7	1001
1001	parity	1010
1010	stop	0000

(b)

FIGURE 9.20 **Controller for PS2 keyboard: (a) state diagram; (b) next-state (implementation) table; (c) K-maps and next-state equations; (d) output table; (e) output equations; (f) controller circuit; (g) interface circuit.** *(continued on next page)*

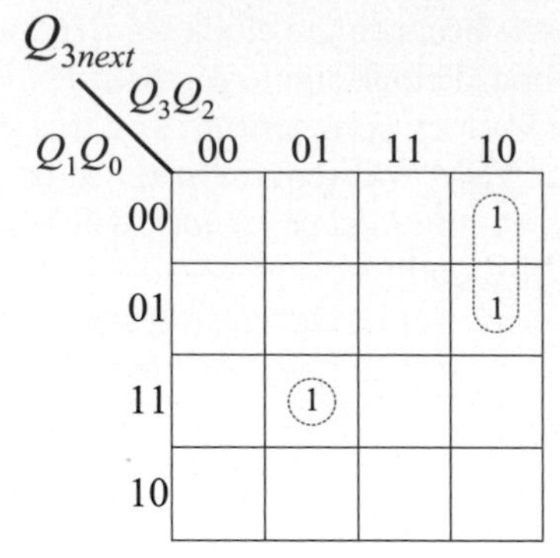

$Q_{3next} = D_3 = Q_3Q_2'Q_1' + Q_3'Q_2Q_1Q_0$

$Q_{2next} = D_2 = Q_3'Q_2Q_1' + Q_3'Q_2Q_0' + Q_3'Q_2'Q_1Q_0$

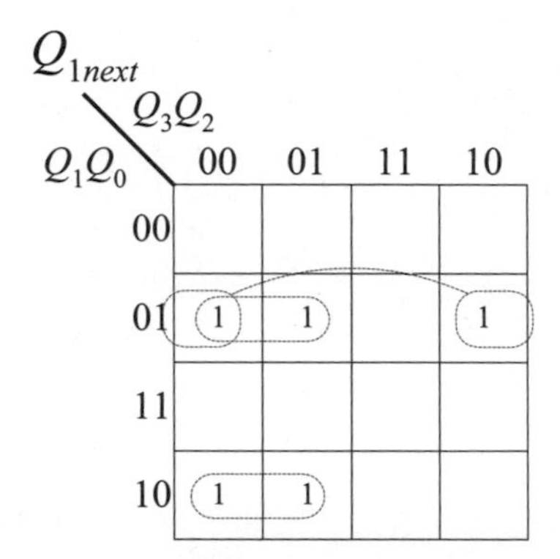

$Q_{1next} = D_1 = Q_3'Q_1'Q_0 + Q_2'Q_1'Q_0 + Q_3'Q_1Q_0'$

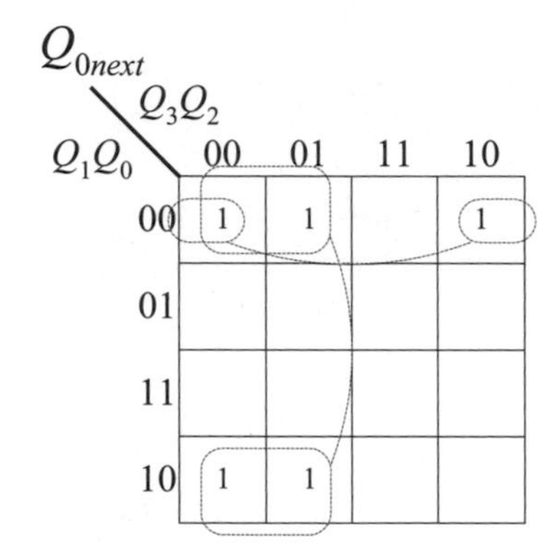

$Q_{0next} = D_0 = Q_3'Q_0' + Q_2'Q_1'Q_0'$

(c)

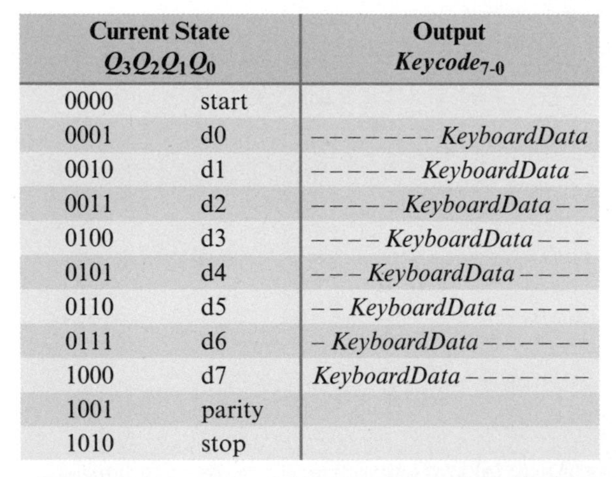

Current State $Q_3Q_2Q_1Q_0$		Output $Keycode_{7\text{-}0}$
0000	start	
0001	d0	– – – – – – – *KeyboardData*
0010	d1	– – – – – – *KeyboardData* –
0011	d2	– – – – – *KeyboardData* – –
0100	d3	– – – – *KeyboardData* – – –
0101	d4	– – – *KeyboardData* – – – –
0110	d5	– – *KeyboardData* – – – – –
0111	d6	– *KeyboardData* – – – – – –
1000	d7	*KeyboardData* – – – – – – –
1001	parity	
1010	stop	

A dash (–) means no change to that *Keycode* bit.

(d)

$$Keycode_7 = Q_3Q_2'Q_1'Q_0'\ KeyboardData$$
$$Keycode_6 = Q_3'Q_2Q_1Q_0\ KeyboardData$$
$$Keycode_5 = Q_3'Q_2Q_1Q_0'\ KeyboardData$$
$$Keycode_4 = Q_3'Q_2Q_1'Q_0\ KeyboardData$$
$$Keycode_3 = Q_3'Q_2Q_1'Q_0'\ KeyboardData$$
$$Keycode_2 = Q_3'Q_2'Q_1Q_0\ KeyboardData$$
$$Keycode_1 = Q_3'Q_2'Q_1Q_0'\ KeyboardData$$
$$Keycode_0 = Q_3'Q_2'Q_1'Q_0\ KeyboardData$$

(e)

FIGURE 9.20 Controller for PS2 keyboard: (a) state diagram; (b) next-state (implementation) table; (c) K-maps and next-state equations; (d) output table; (e) output equations; (f) controller circuit; (g) interface circuit. *(continued on next page)*

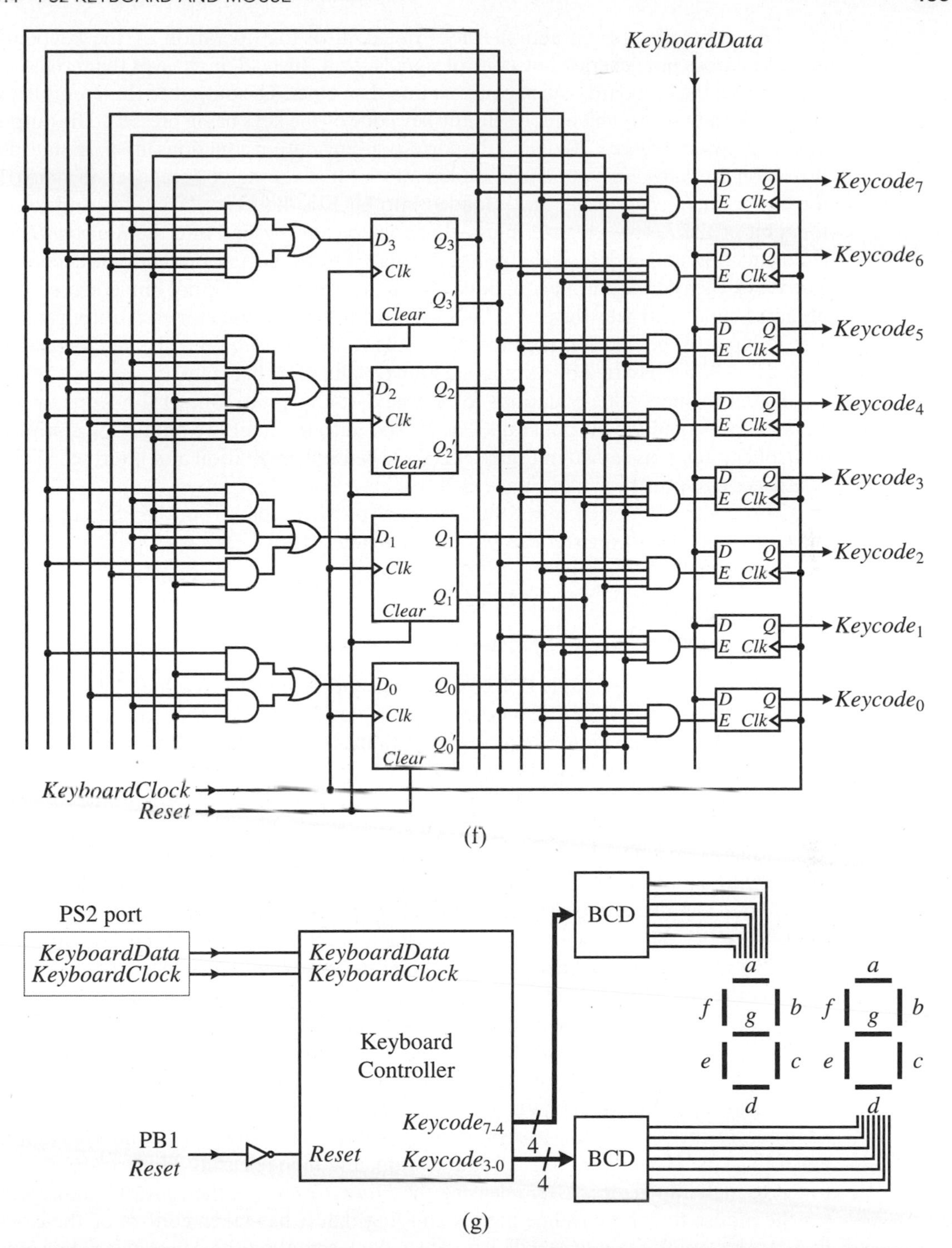

FIGURE 9.20 **Controller for PS2 keyboard: (a) state diagram; (b) next-state (implementation) table; (c) K-maps and next-state equations; (d) output table; (e) output equations; (f) controller circuit; (g) interface circuit.**

This controller circuit actually does not control the operation of the keyboard because it does not generate any control signals for it. Instead, it receives the serial data signals from the keyboard, and packages it into data bytes. The output of this controller is simply the data bytes, which represent the key code of the keys being pressed. The output table is shown in Figure 9.20(d) and the corresponding output equations in (e). In state d0, the bit on the *KeyboardData* line is loaded into bit 0 of the *Keycode* register; in state d1, the bit on the *KeyboardData* line is loaded into bit 1 of the *Keycode* register; and so on. Each bit of the *Keycode* register must, therefore, be able to load into the *KeyboardData* independently, and each load enable line is asserted by the corresponding state encoding. Therefore, each of the output equations shown in Figure 9.20(e) is not implemented simply as a 5-input AND gate. Instead, a D flip-flop with enable is used for each output signal. The D input to the flip-flop is connected to the *KeyboardData* line. The load enable, E, on the flip-flop is asserted by ANDing the four Q values for that state.

Using the next-state equations for the next-state circuit, four D flip-flops for the state memory, and the output equations for the output circuit, we obtain the complete controller circuit, as shown in Figure 9.20(f). The implementation and interface of this controller circuit is shown in Figure 9.20(g).

This controller design is not the only way to receive data from the PS2 keyboard. In fact, a simple shift register can be used instead.

9.4.3 **Theory of Operation—PS2 Mouse**

The PS2 mouse uses the same PS2 connections and protocol as for the PS2 keyboard. However, the design of the PS2 mouse controller is a little bit more interesting than the keyboard because the mouse requires an initialization command from the controller, and the signals on both the data and clock lines are bidirectional.

On power up, the keyboard will immediately send data bytes to the controller when a key is pressed. The mouse, on the other hand, needs an initialization command to tell it what to do before it will send data bytes to the controller. After initialization, the mouse will send data serially to the controller over the *MouseData* line and clocked by the *MouseClock* line just like the keyboard. Here, we use different names for the data and clock lines, but they are actually the same signal lines on the PS2 connector. Both the controller and the mouse use the same *MouseData* and *MouseClock* lines to send signals to each other, therefore, these two lines must be bidirectional, making the controller more complicated to design. Furthermore, because the controller has to control the *MouseClock* line, the FSM cannot be clocked by the *MouseClock* line, but needs a separate clock.

When the controller wants to send data to the mouse, which is the case for the initialization command on power up, the controller first must put the *MouseClock* and *MouseData* lines in a request-to-send state by pulling both *MouseClock* and *MouseData* low for at least 100 μs to inhibit communication, and then releasing *MouseClock* by setting it to high impedance. After releasing the *MouseClock* line, the controller must wait for the mouse to bring this line high, signifying that it has taken control of the clock line. At this point, the mouse will generate a clock signal on the *MouseClock* line, and the controller will send out the initialization command F4 on the *MouseData* line, one bit per clock pulse, starting with the least significant bit. Just like with the keyboard, each byte sent must be followed by an odd parity bit and finally a 1 stop bit. For the command F4, the odd parity bit is a 0, and the stop bit is always a 1.

After sending the stop bit, the mouse controller will release the *MouseData* line by setting it to high impedance. The mouse then will take control of the *MouseData* line and will send the acknowledge byte FA to the controller. Again, the acknowledge byte is sent out with least significant bit first, then followed by an odd parity bit and a stop bit. This concludes the initialization sequence for the mouse.

After the mouse is initialized, it will continuously send out data to the controller on the *MouseData* line clocked by the *MouseClock* line. The data bits are grouped into frames of three packets each, and each packet consists of eleven bits (a 0 start bit, eight data bits, an odd parity bit, and a 1 stop bit). The three bytes of data bits in each frame are interpreted as follows:

Bit	**7**	**6**	**5**	**4**	**3**	**2**	**1**	**0**
Byte 1	Yo	Xo	Ys	Xs	1	M	R	L
Byte 2	X_7	X_6	X_5	X_4	X_3	X_2	X_1	X_0
Byte 3	Y_7	Y_6	Y_5	Y_4	Y_3	Y_2	Y_1	Y_0

where:

- M, R, and L are the status of the Middle, Right, and Left mouse buttons, respectively (1 = pressed; 0 = released),
- Yo and Xo are the overflow bits in the Y and X directions,
- Ys is the sign bit in the Y direction (1 = moving down; 0 = moving up)
- Xs is the sign bit in the X direction (1 = moving left; 0 = moving up)
- $X_7 - X_0$ is the horizontal moving distance of X in two's complement (negative = moving left; positive = moving right)
- $Y_7 - Y_0$ is the vertical moving distance of Y in two's complement (negative = moving down; positive = moving up)

The following table shows the three bytes received for each operation of the mouse.

Operation	**Byte 3**	**Byte 2**	**Byte 1**
Middle mouse button click	00	00	0C
Right mouse button click	00	00	0A
Left mouse button click	00	00	09
Move left	00	FF	18
Move right	00	01	08
Move down	FF	00	28
Move up	01	00	08

The steps for the controller to take, as described above, are summarized next. The step numbers are also annotated as comments in the code listing shown in Figure 9.22.

1. Bring the clock and data line low for at least 100 μs.
2. Release the clock line.
3. Wait for the clock line to be released (high).
4. Send initialization command byte F4 hex, least significant bit first.
5. Wait for the mouse to bring clock low.
6. Wait for the mouse to bring clock high.
7. Repeat steps 5–7 for the remaining seven data bits, the odd parity bit, and the stop bit.
8. Release the data line.
9. Receive acknowledge byte FA.
10. Receive 3-packet data of 11 bits each.
11. Repeat steps 10 and 11.

9.4.4 **Controller Design—PS2 Mouse**

The dedicated microcontroller for controlling the PS2 mouse does not require a datapath. The FSM, written in behavioral Verilog and VHDL code are shown in Figures 9.21 and 9.22, respectively. Note that the *MouseClock* and *MouseData* signals are declared as `inout`/INOUT for the bidirectional signal. The mouse controller FSM state transitions follow exactly the steps as described above. In all of the states where we check for the *MouseClock* signal, instead of checking it directly, we check the *FilteredMouseClock* signal, which follows the original *MouseClock* signal by a slight time delay. A second process in the code assigns the *MouseClock* signal to the *FilteredMouseClock* signal at the next rising clock edge, thus adding a slight time delay. This is needed to correct slight timing differences.

```
module MouseController (
  input Clock, Reset,
  inout reg MouseClock, MouseData,
  output LeftButton, RightButton, MidButton,
  output [7:0] Byte1, Byte2, Byte3,
  output reg [7:0] Debug
);

  reg [43:0] packet;
  reg [5:0] count;
  reg [9:0] command;
  reg FilteredMouseClock;

  reg [3:0]state;
  parameter s_init1=0, s_init2=1, s_init3=2,
```

FIGURE 9.21 Behavioral Verilog code for a PS2 mouse controller. *(continued on next page)*

```
        s_cmd1=3, s_cmd2=4,
        s_cmd20=5, s_cmd21=6,
        s_ack1=7, s_ack2=8,
        s_recv1=9, s_recv2=10,
        s_end=11;

  assign LeftButton = packet[1];
  assign RightButton = packet[2];
  assign MidButton = packet[3];
  assign Byte1 = packet[8:1];
  assign Byte2 = packet[19:12];
  assign Byte3 = packet[30:23];

  // Next-state logic
  always @ (posedge Clock or posedge Reset) begin
    if (Reset) begin
      MouseClock <= 1'b0; // 1)
      MouseData <= 1'b0;
      packet <= {44{1'b0}}; // 44 0's
      command <= 10'b1011110100;  // stop, odd parity, F4
      count <= 0;
      state <= s_init1;
      Debug <= 8'b00000000;
    end else
      case (state)
      s_init1: begin // 2)
        MouseClock <= 1'b0;
        MouseData <= 1'b0;
        state <= s_init2;
        end
      s_init2: begin // 3)
        MouseClock <= 1'bz;
        MouseData <= 1'b0;
        state <= s_init3;
        end
      s_init3: begin // 4) wait for clock to go hi
        if (FilteredMouseClock == 1'b1) begin
          count <= 0;
          state <= s_cmd1;
          end
        end
      s_cmd1: begin // 5) send out command F4, LSB first
        MouseData <= command[count];
        if (FilteredMouseClock == 1'b0) begin // 6)
          if (count < 10) begin
            count <= count + 1;
            state <= s_cmd2;
```

FIGURE 9.21 Behavioral Verilog code for a PS2 mouse controller. *(continued on next page)*

```
        end
      else
        state <= s_cmd21;
      end
    end
  s_cmd2: begin // 7)
    if (FilteredMouseClock == 1'b1) begin
      state <= s_cmd1;
      end
    end
  s_cmd21: begin // 9)
    Debug <= 8'b00000001;
    MouseData <= 1'bZ; // release the data line
    if (FilteredMouseClock == 1'b1) begin
      count <= 0;
      state <= s_ack1;
      end
    end
  // at this point, if you have a mouse with an IR instead
  // of a mechanical ball, the LED on the bottom should be on

  // get acknowledge message FA
  s_ack1: begin
    Debug <= 8'b00000010;
    packet[count+22] <= MouseData;
    if (FilteredMouseClock == 1'b0) begin
      if (count < 11) begin
        count <= count + 1;
        state <= s_ack2;
        end
      else begin
        count <= 0;
        state <= s_recv1;
        end
      end
    end
  s_ack2: begin
    Debug <= 8'b00000011;
    if (FilteredMouseClock == 1'b1) begin
      state <= s_ack1;
      end
    end
  // Receive data from mouse
  s_recv1: begin
    Debug <= 8'b00000100;
    packet[count] <= MouseData;
```

FIGURE 9.21 Behavioral Verilog code for a PS2 mouse controller. *(continued on next page)*

```
            if (FilteredMouseClock == 1'b0) begin
              if (count < 33) begin
                count <= count + 1;
                state <= s_recv2;
                end
              else begin
                count <= 0;
                state <= s_recv1;
                end
              end
            end
          s_recv2: begin
            Debug <= 8'b00000101;
            if (FilteredMouseClock == 1'b1) begin
              state <= s_recv1;
              end
            end
          default: begin
            Debug <= 8'b11111111;
            end
          endcase
    end // always

  // slight delay for the mouse clock
    always @ (posedge Clock) begin
      FilteredMouseClock <= MouseClock;
      end

  endmodule
```

FIGURE 9.21 Behavioral Verilog code for a PS2 mouse controller.

```
LIBRARY IEEE;
USE IEEE.STD_LOGIC_1164.ALL;

ENTITY MouseController IS PORT (
  Reset: IN STD_LOGIC;
  Clock: IN STD_LOGIC;
  MouseClock: INOUT STD_LOGIC;
  MouseData: INOUT STD_LOGIC;
  LeftButton,RightButton,MidButton: OUT STD_LOGIC;
  Byte1: OUT STD_LOGIC_VECTOR(7 DOWNTO 0);
  Byte2: OUT STD_LOGIC_VECTOR(7 DOWNTO 0);
  Byte3: OUT STD_LOGIC_VECTOR(7 DOWNTO 0);
  Debug: OUT STD_LOGIC_VECTOR(7 DOWNTO 0)
  );
END MouseController;
```

FIGURE 9.22 Behavioral VHDL code for a PS2 mouse controller. *(continued on next page)*

```
ARCHITECTURE Behavioral OF MouseController IS
  TYPE state_type IS (
      s_init1,s_init2,s_init3,
      s_cmd1,s_cmd2,
      s_cmd20,s_cmd21,
      s_ack1,s_ack2,
      s_recv1,s_recv2,
      s_end);
  SIGNAL state: state_type;

  SIGNAL packet: STD_LOGIC_VECTOR(43 DOWNTO 0);
  SIGNAL count: INTEGER RANGE 0 TO 64;
  SIGNAL command: STD_LOGIC_VECTOR(9 DOWNTO 0);
  SIGNAL FilteredMouseClock: STD_LOGIC;

BEGIN

  LeftButton <= packet(1);
  RightButton <= packet(2);
  MidButton <= packet(3);
  Byte1 <= packet(8 DOWNTO 1);
  Byte2 <= packet(19 DOWNTO 12);
  Byte3 <= packet(30 DOWNTO 23);

  FSM: PROCESS(Clock, Reset)
  BEGIN
    IF (Reset = '1') THEN
      MouseClock <= '0'; -- 1)
      MouseData <= '0';
      packet <= (OTHERS => '0');
      command <= "1011110100";  -- stop, odd parity, F4
      count <= 0;
      state <= s_init1;
      Debug <= "00000000";
    -- this FSM is driven by the DE2 clock signal
    ELSIF (Clock'EVENT AND Clock = '1') THEN
      CASE state is
      WHEN s_init1 => -- 2)
        MouseClock <= '0';
        MouseData <= '0';
        state <= s_init2;
      WHEN s_init2 => -- 3)
        MouseClock <= 'Z';
        MouseData <= '0';
        state <= s_init3;
      WHEN s_init3 => -- 4) wait for clock to go hi
        IF (FilteredMouseClock = '1') THEN
          count <= 0;
          state <= s_cmd1;
        END IF;
```

FIGURE 9.22 Behavioral VHDL code for a PS2 mouse controller. *(continued on next page)*

```
WHEN s_cmd1 => -- 5) send out command F4, LSB first
  MouseData <= command(count);
  IF (FilteredMouseClock = '0') THEN -- 6)
    IF (count < 10) THEN
      count <= count + 1;
      state <= s_cmd2;
    ELSE
      state <= s_cmd21;
    END IF;
  END IF;
WHEN s_cmd2 => -- 7)
  IF (FilteredMouseClock = '1') THEN
    state <= s_cmd1;
  END IF;

WHEN s_cmd21 => -- 9)
  Debug <= "00000001";
  MouseData <= 'Z'; -- release the data line
  IF (FilteredMouseClock = '1') THEN
    count <= 0;
    state <= s_ack1;
  END IF;
-- at this point, if you have a mouse with an IR instead
-- of a mechanical ball, the LED on the bottom should be on

-- get acknowledge message FA
WHEN s_ack1 =>
  Debug <= "00000010";
  packet(count+22) <= MouseData;
  IF (FilteredMouseClock = '0') THEN
    IF (count < 11) THEN
      count <= count + 1;
      state <= s_ack2;
    ELSE
      count <= 0;
      state <= s_recv1;
    END IF;
  END IF;
WHEN s_ack2 =>
  Debug <= "00000011";
  IF (FilteredMouseClock = '1') THEN
    state <= s_ack1;
  END IF;
-- Receive data from mouse
WHEN s_recv1 =>
  Debug <= "00000100";
  packet(count) <= MouseData;
  IF (FilteredMouseClock = '0') THEN
```

FIGURE 9.22 Behavioral VHDL code for a PS2 mouse controller. *(continued on next page)*

```
                IF (count < 33) THEN
                  count <= count + 1;
                  state <= s_recv2;
                ELSE
                  count <= 0;
                  state <= s_recv1;
                END IF;
              END IF;
            WHEN s_recv2 =>
              Debug <= "00000101";
              IF (FilteredMouseClock = '1') THEN
                state <= s_recv1;
              END IF;

            WHEN OTHERS =>
              Debug <= "11111111";
            END CASE;
          END IF;
        END PROCESS;

      -- slight delay for the mouse clock
        PROCESS
        BEGIN
          WAIT UNTIL Clock'event and Clock = '1';
          FilteredMouseClock <= MouseClock;
        END PROCESS;

      END Behavioral;
```

FIGURE 9.22 Behavioral VHDL code for a PS2 mouse controller.

9.5 RS-232 Controller for Bluetooth Communication

A Bluetooth connection often is used for wireless communication between two devices within a short range. For example, mobile phones often use a Bluetooth connection to wireless speakers and headphones. We will use the HC-06 Slave Bluetooth module to communicate wirelessly between a dedicated microprocessor controller that we will design, and a master Bluetooth host on a computer or a mobile phone. The Bluetooth module works like a RS-232 serial modem using a transmit and a receive lines, so in order to work with it, we need to understand how the RS-232 serial protocol works.

The RS-232 serial port was once a standard feature on personal computers to connect peripheral devices, such as modem, mouse, and printers to the computer. In modern personal computers, the USB connection has replaced the RS-232. Nevertheless, because of its simplicity over the USB protocol, many RS-232 devices are still used, especially in industrial machines, scientific instruments, telecommunication equipments, and, in our case, the Bluetooth module.

9.5.1 Theory of Operation—RS-232

The RS-232 protocol defines the communication between two devices. The main device, such as a computer, is referred to as the DTE (data terminal equipment), and the secondary peripheral device, such as a scientific instrument, is referred to as the DCE (data circuit-terminating equipment or originally defined as data communication equipment). The physical connector for the RS-232 has either 9 (DB-9) or 25 (DB-25) pins, but typically only three pins (pins 2, 3, and 5) are used to connect between the two communicating devices. Pins 2 and 3 are the data transmit and receive signals, and pin 5 is the ground. The designation of whether pin 2 is the transmit or receive signal depends on whether it is at the DTE or at the DCE end. On the DTE end, pin 2 is the receive (RxD) signal and pin 3 is the transmit (TxD) signal. On the DCE end, they are reversed. A typical connection between a computer or controller with a peripheral device is to connect pin 2 on one end with pin 3 on the other end, and vice versa. Pin 5 on both ends are connected together.

Unlike the PS2 connection, the RS-232 has no clock line to synchronize between the two devices. Instead, a fixed clock speed (referred to as the baud rate) must be predetermined and agreed upon between the two devices. Typical baud rates are: 9600, 38400, and 115200. The baud rate is often thought of as the number of bits per second, but this is not always the case. Furthermore, the parity bit and the number of stop bits also must be predetermined and agreed upon between the two devices.

The parity bit is used for error checking. The RS-232 protocol allows for either even parity, odd parity, or no parity. For even and odd parity, the parity bit is set or reset depending on the total number of 1 bits in the 8-bit data. For an even parity, the total number of 1 bits in the 8-bit data plus the parity bit must be an even number. While for the odd parity, the total number of 1 bits must be an odd number. For example, if the eight data bits are 01001011 (which has four 1s), then the parity bit must be reset for even parity (to make the total number of 1 bits an even number), but set for odd parity (to make the total number of 1 bits an odd number). By checking the parity bit, the receiver can detect when some bits in the data transmission have been flipped. The careful reader might see that some errors are not detected. No parity simply means that a parity bit is not included in the transmission.

The RS-232 data transmission always begins with a start bit, and then is followed by eight data bits (with the least significant bit sent out first). After the last data bit (which is the most significant bit) is sent out, there will optionally be a parity bit, and one or more stop bits. Whether there is a parity bit or how many stop bits to use is predetermined on and set up between the two communicating devices. The normal configuration used for RS-232 communication is reflected in the frequently used acronym "N81," which refers to "no parity bit, 8 data bits, and 1 stop bit." With this setup configuration, 10 bits (one start bit, eight data bits, and one stop bit) are used per byte of data sent. The waveform for the eight data bits, 01001011 using N81, is shown in Figure 9.23.

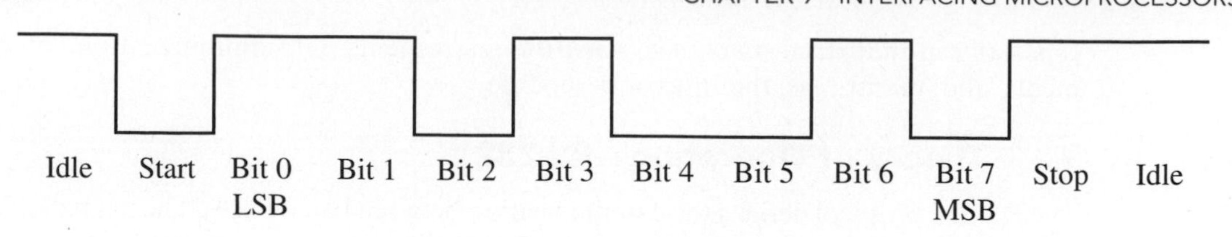

FIGURE 9.23 **RS-232 waveform for the eight data bits 01001011 using the N81 setup.**

9.5.2 Controller Design—RS-232

The RS-232 sender controller code written in Verilog is shown in Figure 9.24. The sender is basically a parallel-to-serial converter. When there is data to be sent, the 8-bit data is loaded into the *data* register and the *start* signal is asserted. The FSM, seeing the *start* signal asserted, will go to the *s_start* state to send out the start signal on the transmit *TxD* line. It then goes to the *s_send* state to send out the eight data bits from the 8-bit *data* register at one bit per clock cycle starting with the least significant bit. For this to work correctly, the clock speed of the FSM must be running at the correct baud rate.

```
// RS232 Sender
// Using N81
// The input clock must be set to the correct baud rate
module RS232Send (
   input clk,
   input reset,
   input [7:0] data,     // 8-bit input data
   input start,          // start transmission
   output reg TxD,       // serial transmit output signal
   output reg done
);

   parameter s_idle=0, s_start=1, s_send=2, s_stop=3;
   reg [1:0] state;
   reg [3:0] count;

   always @(posedge clk, posedge reset) begin
    if (reset) begin
      TxD = 1;
      done = 0;
      state = s_idle;
    end else
      case (state)
      s_idle: begin     // idle
```

FIGURE 9.24 **RS-232 sender controller code written in Verilog.** *(continued on next page)*

```
          TxD = 1;
          done = 0;
          count = 0;
          if (start) state = s_start;
          end
        s_start: begin   // Start bit
          TxD = 0;
          done = 0;
          count = 0;
          state = s_send;
          end
        s_send: begin
          // 8 data bits (least significant bit first)
          TxD = data[count];
          done = 0;
          // blocking assignment, so ordering matters
          count = count + 1;
          // test is AFTER the count increment
          if (count > 7) state = s_stop;
          end
        default: begin   // Stop bit
          TxD = 1;
          done = 1;
          state = s_idle;
          end
      endcase
    end
  endmodule
```

FIGURE 9.24 RS-232 sender controller code written in Verilog.

The RS-232 receiver controller code written in VHDL is shown in Figure 9.25. Unlike the sender, the receiver is a simple serial-to-parallel converter. Initially, the FSM waits in state *s_init* for the start bit on the *RxD* line. The start bit is a transition from a 1 to a 0. After receiving the start bit, the FSM enters the data receiving state *s_receive* to receive the eight data bits. The incoming data is a serial bit stream coming in at the predetermined baud rate. If the FSM clock is set at the correct baud rate, then all that the FSM has to do is to simply shift the incoming bits into the output register *data* at one bit per clock cycle. One important implementation note is that because the sender is sending out the bits on the rising edge of the clock, and if the receiver receives the bits also on the rising edge of the clock, the receiver might get the previous bit value from before the rising edge because the two clocks might be slightly off. To safeguard this problem, it is best to receive the bit in the middle of the clock cycle instead of at the start of the clock cycle. One easy way to implement this is to receive the bits at the falling edge of the clock, which occurs in the middle of the clock cycle. After receiving the last data bit and the stop bit, a *done* signal is asserted to inform the controller that it has received a byte of data.

```
-- RS232 Receiver
-- Using N81
-- The input clock must be set to the correct baud rate
LIBRARY IEEE;
USE IEEE.STD_LOGIC_1164.ALL;

Entity RS232Receive IS PORT (
  clk: IN STD_LOGIC;
  reset: IN STD_LOGIC;
  RxD: IN STD_LOGIC;  -- serial receive input signal
  data: OUT STD_LOGIC_VECTOR(7 DOWNTO 0);  -- 8-bit output data
  done: OUT STD_LOGIC);
END RS232Receive;

ARCHITECTURE Behavioral OF RS232Receive IS
  TYPE state_type IS (s_idle, s_receive, s_stop);
  SIGNAL state: state_type;
  SIGNAL count: INTEGER RANGE 0 TO 7;
BEGIN
  PROCESS(clk, reset)
  BEGIN
    IF (reset = '1') THEN
      data <= "00000000";
      done <= '0';
      state <= s_idle;
    -- receive bits at the falling clock edge
    ELSIF (clk'EVENT AND clk = '0') THEN
      CASE state IS
        WHEN s_idle =>
          IF (RxD = '0') THEN  -- wait for start bit
            data <= "00000000";
            done <= '0';
            count <= 0;
            state <= s_receive;
          END IF;
        WHEN s_receive =>
          data(count) <= RxD;
          done <= '0';
          -- non-blocking assignment,
          -- so ordering doesn't matter
          count <= count + 1;
          -- test is BEFORE the count increment
          IF (count > 6) THEN
            state <= s_stop;
          END IF;
        WHEN OTHERS =>  -- Stop bit
```

FIGURE 9.25 RS-232 receiver controller code written in VHDL. *(continued on next page)*

```
                done <= '1';
                state <= s_idle;
            END CASE;
        END IF;
    END PROCESS;
END Behavioral;
```

FIGURE 9.25 RS-232 receiver controller code written in VHDL.

9.5.3 **Implementation**

The HC-06 Bluetooth slave module can be paired with any Bluetooth master, such as a personal computer or a mobile phone. The maximum distance between the master and the slave is about 50 feet. The detailed operation of the Bluetooth communication is incorporated inside the module so there is no need for a user to understand how the actual Bluetooth communication works. The user needs to know only how to interface with the module using the RS-232 protocol because the module operates like a RS-232 DCE device. The connection setup between a PC acting as the Bluetooth master, the HC-06 Bluetooth slave module, and the RS-232 controller is shown in Figure 9.26.

The default settings for the HC-06 Bluetooth module are as follows:

Baud:	9600
Protocol:	N81 (no parity, eight data bits, and one stop bit)
Pairing name:	linvor
Password:	1234

The module can be paired easily with a Windows PC by going to the Devices and Printers option in the Control Panel. Select Add a device. After a moment, your Bluetooth device (with the default name "linvor") should show up. Select your device and continue. Enter the default password 1234 when it asks for it. Your Bluetooth device should now be in the list of installed devices. Bring up the Properties window for the device and note the COM port that is assigned to it. Before a connection is made, a LED on the Bluetooth module blinks, and after a connection is made, the LED stops blinking and stays on.

On the PC, a terminal emulator such as PuTTY[1] can be used to communicate with the Bluetooth module. In the PuTTY configuration screen, select the Serial connection type and enter the correct COM port and baud speed. Characters sent from PuTTY should be seen on the 8-bit data out lines from the RS-232 receiver controller, while data sent from the RS-232 sender controller should be seen on the PuTTY terminal screen.

[1] PuTTY is a free SSH and telnet client for Windows that you can download from the Web.

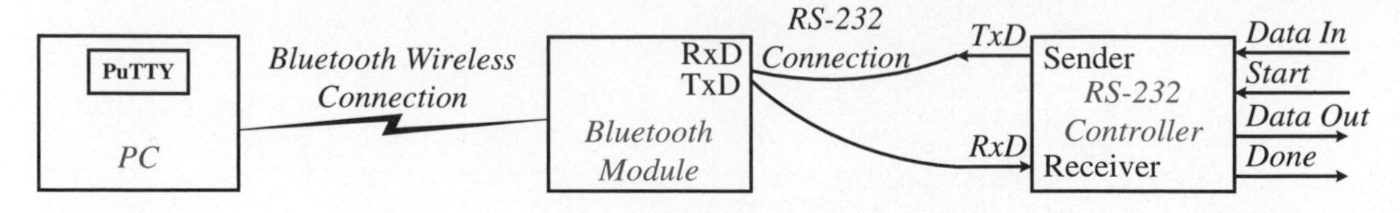

FIGURE 9.26 Connection setup between a PC acting as the Bluetooth master, the HC-06 Bluetooth slave module, and the RS-232 controller.

9.6 Liquid-Crystal Display Controller

The 16 × 2 liquid-crystal display (LCD) shown in Figure 9.27 is a common display module that can be interfaced easily to a microcontroller. In this section, we will describe its basic operation, and design a dedicated microprocessor to control and display some sample characters on it. This discussion does not cover the full operation of the module. Refer to the manufacturer's datasheet for the complete and detail description and operation of this LCD module.

9.6.1 Theory of Operation

The 16 × 2 LCD can display 16 characters on 2 lines, with each character being made up of a 5 × 7 dot matrix format. An internal IC controls its operations, such as how to turn on and off the individual dots in each character matrix to form the various characters, where to display the character, whether to show a cursor, whether to shift the characters, and other functions. Interfacing the LCD to a microprocessor is a matter of understanding the commands needed to communicate correctly with this internal IC.

Audrius Merfeldas / Shutterstock.com

FIGURE 9.27 A 16 × 2 liquid-crystal display.

Besides the power and ground connections, there are three control signals, RS, R/W, and Enable, and either eight or four data lines. If only four data lines are used, then only the four upper bits (D7 to D4) are connected. That means each byte will be sent as two separate hex digits, one after the other, with the most significant hex digit sent first. The RS signal determines whether the subsequent data bits are to be interpreted as a command or data. If RS = 0, then the bits are a command, and if RS = 1, then the bits are data. The R/W signal is for read or write. It is a read operation if R/W = 1, and a write operation if R/W = 0. The RS, R/W, and Data signals are latched into internal registers at the falling edge of the Enable signal.

The physical connections between the LCD and our dedicated microprocessor are shown in Figure 9.28. The connections are straightforward, with the three control signals and the eight data lines. If the 4-bit data option is used, then the lower four bits (D3 to D0) are not connected. Now it is just a matter of designing our dedicated microprocessor to communicate correctly with the LCD and to send to it the necessary commands.

On power up, the LCD needs to be initialized with a sequence of commands as shown in Figure 9.29. This initialization procedure consists of a sequence of hex codes to tell the LCD whether the 4-bit or 8-bit data will be used, the size format of the LCD, whether to turn the display, cursor, increment mode, shifting and blinking on or off, and then to clear the display. The initial hex code of 38 is sent two or more times to ensure that the LCD module enters the 4-bit or 8-bit data mode successfully. This is followed by 0E hex to set up the display, 06 hex to set up the entry mode, and finally 01 hex to clear the display.

After the initialization sequence, another command is sent to the LCD to position the cursor, and then the characters to be displayed are sent. In the HDL code shown in the next section, the words "HELLO WORLD" are displayed.

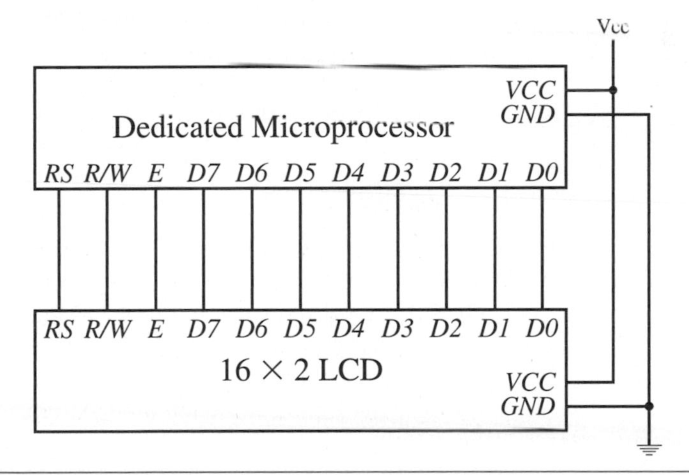

FIGURE 9.28 Connections between the LCD and our dedicated microprocessor.

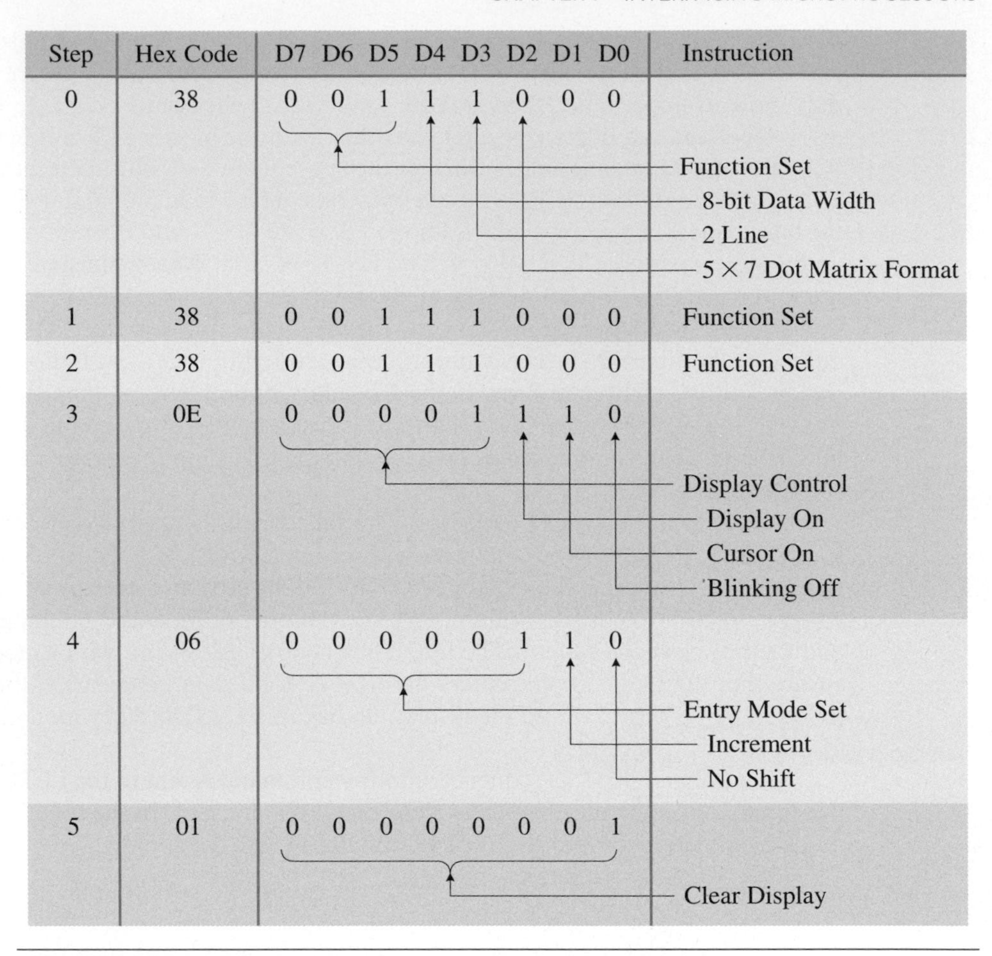

Step	Hex Code	D7	D6	D5	D4	D3	D2	D1	D0	Instruction
0	38	0	0	1	1	1	0	0	0	Function Set 8-bit Data Width 2 Line 5 × 7 Dot Matrix Format
1	38	0	0	1	1	1	0	0	0	Function Set
2	38	0	0	1	1	1	0	0	0	Function Set
3	0E	0	0	0	0	1	1	1	0	Display Control Display On Cursor On Blinking Off
4	06	0	0	0	0	0	1	1	0	Entry Mode Set Increment No Shift
5	01	0	0	0	0	0	0	0	1	Clear Display

FIGURE 9.29 Sequence of initialization commands for the LCD.

9.6.2 Controller Design

The dedicated microcontroller for controlling the LCD does not require a datapath. The FSM, written in behavioral Verilog and VHDL code are described in Figures 9.30 and 9.31, respectively. The initial sequence of the FSM states follows exactly the initialization steps shown in Figure 9.29. Two states are needed for each hex code sent because the RS, R/W, and Data signals are latched on the falling edge of the Enable signal, so we need to toggle the Enable line each time. After the initialization sequence, the characters for the two words, "HELLO WORLD" are sent.

```
module lcd_controller (
  input Clock, Reset,
  output reg [7:0] LCD_DATA,
  output reg LCD_RW, LCD_EN, LCD_RS,
  output LCD_ON, LCD_BLON
    );

  reg [3:0] state;
  reg [7:0] initcode[0:5];
  reg [7:0] line1[0:15];
  integer count;

  initial begin
    // LCD initialization sequence codes
    initcode[0] = 8'h38;  // Init three times
    initcode[1] = 8'h38;
    initcode[2] = 8'h38;
    //Display control: Display ON; Cursor ON; Blink OFF
    initcode[3] = 8'h0E;
    // Entry mode set: Increment One; No Shift
    initcode[4] = 8'h06;
    initcode[5] = 8'h01;  // Display clear
    // HELLO WORLD
    line1[0] = 8'h20;
    line1[1] = 8'h20;
    line1[2] = 8'h48;
    line1[3] = 8'h45;
    line1[4] = 8'h4C;
    line1[5] = 8'h4C;
    line1[6] = 8'h4F;
    line1[7] = 8'h20;
    line1[8] = 8'h20;
    line1[9] = 8'h57;
    line1[10] = 8'h4F;
    line1[11] = 8'h52;
    line1[12] = 8'h4C;
    line1[13] = 8'h44;
    line1[14] = 8'h20;
    line1[15] = 8'h20;
  end

  assign LCD_ON = 1'b1;
  assign LCD_BLON = 1'b0;

  always @ (posedge Clock, posedge Reset) begin
    if (Reset) begin
      count <= 0;
      state <= 4'b0000;
      end
```

FIGURE 9.30 Behavioral Verilog code for a 16 × 2 LCD controller. *(continued on next page)*

```
else
  case (state)
  // LCD initialization sequence
  // The LCD_DATA is written to the LCD
  // at the falling edge of the E line
  // therefore we need to toggle the E
  // line for each data write
  4'b0000: begin
    LCD_DATA <= initcode[count];
    LCD_EN <= 1'b1;  // EN=1;
    LCD_RS <= 1'b0;  // RS=0; an instruction
    LCD_RW <= 1'b0;  // R/W'=0; write
    state <= 4'b0001;
    end
  4'b0001: begin
    LCD_EN <= 1'b0;  // set EN=0;
    count <= count + 1;
    if (count+1 < 6)
      state <= 4'b0000;
    else
      state <= 4'b0010;
    end
  // move cursor to first line of display
  4'b0010: begin
    // x80 is address of 1st position on first line
    LCD_DATA <= 8'h80;
  // xBF is address of 1st position on second line
  //LCD_DATA <= 8'hBF;
    LCD_EN <= 1'b1;  // EN=1;
    LCD_RS <= 1'b0;  // RS=0; an instruction
    LCD_RW <= 1'b0;  // R/W'=0; write
    state <= 4'b0011;
    end
  4'b0011: begin
    LCD_EN <= 1'b0;  // EN=0; toggle EN
    count <= 0;
    state <= 4'b0100;
    end
  // write 1st line text
  4'b0100: begin
    LCD_DATA <= line1[count];
    LCD_EN <= 1'b1;  // EN=1;
    LCD_RS <= 1'b1;  // RS=1; data
    LCD_RW <= 1'b0;  // R/W'=0; write
    state <= 4'b0101;
    end
```

FIGURE 9.30 Behavioral Verilog code for a 16 × 2 LCD controller. *(continued on next page)*

```
      4'b0101: begin
        LCD_EN <= 1'b0;  // EN=0; toggle EN
        count <= count + 1;
        if (count+1 < 16)
          state <= 4'b0100;
        else
          state <= 4'b1010;
        end
      4'b1010: begin
        state <= 4'b1010;
        end
      default: begin
        state <= 4'b1010;
        end
      endcase
  end // always
endmodule
```

FIGURE 9.30 Behavioral Verilog code for a 16 × 2 LCD controller.

```
LIBRARY IEEE;
USE IEEE.STD_LOGIC_1164.ALL;

ENTITY lcd_controller IS
PORT (
  Clock, Reset: IN STD_LOGIC;
  LCD_DATA: OUT STD_LOGIC_VECTOR(7 DOWNTO 0);
  LCD_RW, LCD_EN, LCD_RS: OUT STD_LOGIC;
  LCD_ON, LCD_BLON: OUT STD_LOGIC);
END lcd_controller;

ARCHITECTURE FSMD OF lcd_controller IS
  TYPE state_type IS (s1,s2,s3,s4,s10,s11,s12,s13,s20,s21,s22,
    s23,s24);
  SIGNAL state: state_type;
  SUBTYPE ascii IS STD_LOGIC_VECTOR(7 DOWNTO 0);
  TYPE charArray IS array(0 to 15) OF ascii;
  TYPE initArray IS array(0 to 5) OF ascii;
  CONSTANT initcode: initArray := (x"38",x"38",x"38",x"0E",
    x"06",x"01");
  -- HELLO WORLD
  CONSTANT line1: charArray := (x"20",x"20",x"48",x"45",
    x"4C",x"4C",x"4F",x"20",x"20",x"57",x"4F",x"52",x"4C",
    x"44",x"20",x"20");
  SIGNAL count: INTEGER;

BEGIN

  LCD_ON <= '1';
```

FIGURE 9.31 Behavioral VHDL code for a 16 × 2 LCD controller. *(continued on next page)*

```
LCD_BLON <= '0';

lcd_control: PROCESS(Clock, Reset)
BEGIN
  IF (Reset = '1') THEN
    count <= 0;
    state <= s1;
  ELSIF (Clock'EVENT AND Clock = '1') THEN
    CASE state IS
    -- LCD initialization sequence
    -- The LCD_DATA is written to the LCD
    -- at the falling edge of the E line
    -- therefore we need to toggle the E
    -- line for each data write
    WHEN s1 =>
      LCD_DATA <= initcode(count);
      LCD_EN <= '1';  -- EN=1;
      LCD_RS <= '0';  -- RS=0; an instruction
      LCD_RW <= '0';  -- R/W'=0; write
      state <= s2;
    WHEN s2 =>
      LCD_EN <= '0';  -- set EN=0;
      count <= count + 1;
      IF (count + 1 < 6) THEN
        state <= s1;
      ELSE
        state <= s10;
      END IF;

    -- move cursor to first line of display
    WHEN s10 =>
      LCD_DATA <= x"80";  -- x80 is address of 1st position
      -- on first line
    --LCD_DATA <= x"BF";  -- xBF is address of 1st position
    --on second line
      LCD_EN <= '1';  -- EN=1;
      LCD_RS <= '0';  -- RS=0; an instruction
      LCD_RW <= '0';  -- R/W'=0; write
      state <= s11;
    WHEN s11 =>
      LCD_EN <= '0';  -- EN=0; toggle EN
      count <= 0;
      state <= s12;

    -- write 1st line text
    WHEN s12 =>
      LCD_DATA <= line1(count);
      LCD_EN <= '1';  -- EN=1;
      LCD_RS <= '1';  -- RS=1; data
```

FIGURE 9.31 Behavioral VHDL code for a 16 × 2 LCD controller. *(continued on next page)*

```
          LCD_RW <= '0';    -- R/W'=0; write
          state <= s13;
        WHEN s13 =>
          LCD_EN <= '0';    -- EN=0; toggle EN
          count <= count + 1;
          IF (count + 1 < 16) THEN
            state <= s12;
          ELSE
            state <= s20;
          END IF;
        WHEN s20 =>
          state <= s20;
        WHEN OTHERS =>
          state <= s20;
        END CASE;
      END IF;
    END PROCESS;
  END FSMD;
```

FIGURE 9.31 Behavioral VHDL code for a 16 × 2 LCD controller.

9.7 VGA Monitor Controller

In this section, we will design and implement a microcontroller to control a VGA monitor. This controller will allow simple graphics to be displayed on the VGA monitor.

9.7.1 Theory of Operation

The monitor screen for a low-resolution standard VGA format contains 640 columns × 480 rows of picture elements called pixels, as shown in Figure 9.32. VGA monitors with higher resolutions have more columns and rows, but the idea is the same. An image is displayed on the screen by turning on or off individual pixels. Turning on just one pixel doesn't represent much, but when many pixels are turned on at the same time, the combined pixels portray an image. The monitor continuously scans through the entire screen turning on or off one pixel at a time very quickly. Although only one pixel is turned on at any one time, you get the impression that all of the pixels are on at the same time because the monitor is scanning through the screen very fast.

Figure 9.32 shows that the scanning starts from row 0, column 0 at the top left corner and moves to the right until it reaches the last column in the row. When the scan reaches the end of a row, it retraces to the beginning of the next row. This repeats until the scan reaches the bottom row. When the scan reaches the last pixel at the bottom-right corner of the screen, it retraces back to the top-left corner of the screen and repeats the scanning process again. In order to reduce flicker on the screen, the entire screen must be scanned 60 times per second or higher. During the horizontal and the vertical retraces, all of the pixels are turned off.

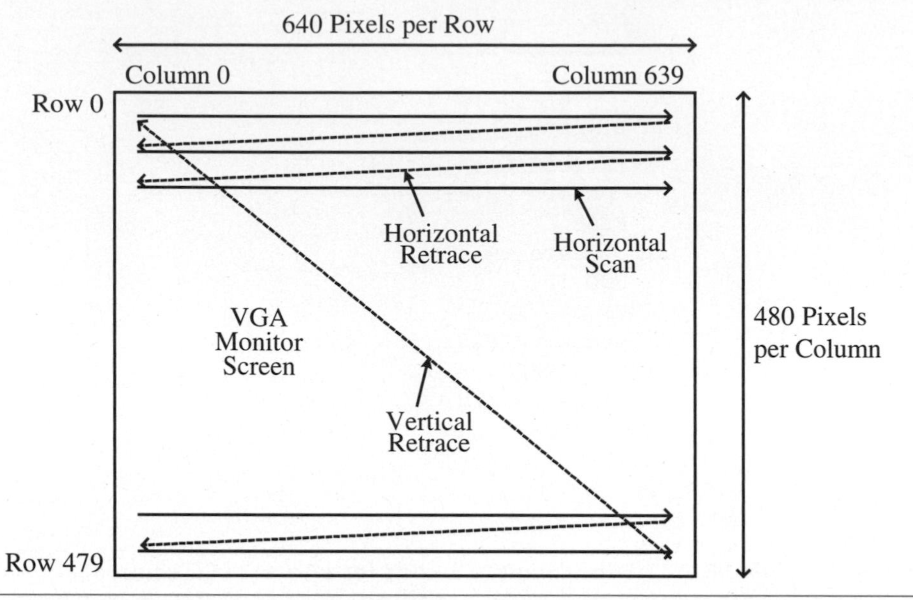

FIGURE 9.32 VGA monitor with 640 columns × 480 rows. The scan starts from row 0, column 0 and moves to the right and down until row 479, column 639.

The VGA monitor is controlled by five signals: red, green, blue, horizontal synchronization, and vertical synchronization. The three color signals, referred to collectively as the RGB signal, are used to control the color of a pixel at a given location on the screen. These three color signals can be turned on or off individually, therefore each pixel can display only one of eight colors. Newer VGA monitors use more bits per color signal, thus resulting in many more color combinations. The horizontal and vertical synchronization signals are used to control the timing of the scan rate. The horizontal synchronization signal determines the time to scan a row, while the vertical synchronization signal determines the time to scan the entire screen. By manipulating these five signals, images are formed on the monitor screen.

The horizontal and vertical synchronization signals timing diagram is shown in Figure 9.33. When inactive, both synchronization signals are at a 1. The start of a row scan begins with the horizontal synchronization signal going low for 3.77 μs, as shown in region *B* in Figure 9.33. This is followed by a 1.79 μs high on the signal, as shown in region *C*. Next, the data for the three color signals are sent (one pixel at a time) for the 640 columns while the horizontal signal remains high, as shown in region *D* for 25.42 μs. Finally, after the last column pixel, there is another 0.79 μs of inactivity on the RGB signal lines while the horizontal signal continues at a high, as shown in region *E*, before the horizontal synchronization signal goes low again for the next row scan. The total time to complete one row scan is 31.77 μs.

The timing for the vertical synchronization signal is analogous to the horizontal synchronization signal. The 64 μs active-low vertical synchronization signal resets the scan to the top-left corner of the screen, as shown in region *P*, followed by a 1020 μs

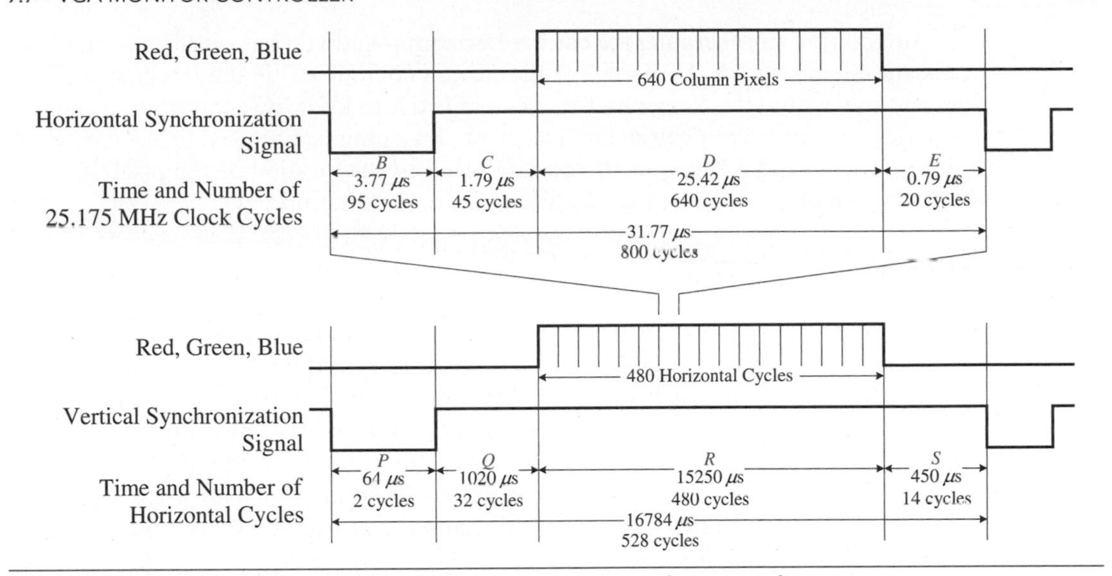

FIGURE 9.33 Horizontal and vertical synchronization signals timing diagram.

high on the signal as shown in region *Q*. Next, there are the 480 row scans of 31.77 μs each, giving a total of 15250 μs, as shown in region *R*. Finally, after the last row scan, there is another 450 μs, as shown in region *S*, before the vertical synchronization signal goes low again to start another complete screen scan starting at the top-left corner. The total time to complete one complete scan of the screen is 16784 μs.

In order to get the monitor to operate properly, we simply have to get the horizontal and vertical synchronization signals timing correct, and then send out the RGB data for each pixel at the given column and row position. It turns out that it is fairly simple to get the correct timing for the two synchronization signals if we use the correct clock frequency of 25.175 MHz. (A faster clock frequency needs to be used for higher resolution monitors.) For the 25.175 MHz clock frequency, the clock period is $1 / 25.175 \times 10^6$, which is about 0.0397 μs per clock cycle. For region *B* in the horizontal synchronization signal, we need 3.77 μs, which is about $3.77 / 0.0397 = 95$ clock cycles. For region *C*, we need 1.79 μs, which is about 45 clock cycles. Similarly, we need 640 clock cycles for region *D* for the 640 columns of pixels, and 20 clock cycles for region *E*. The total number of clock cycles needed for each row scan is, therefore, 800 clock cycles. Notice that with a 25.175 MHz clock, region *D* requires exactly 640 cycles, giving us the 640 columns per row that we want. Therefore, for higher resolution monitors, a faster clock speed will be needed because the total time for region *D* remains fixed at 25.42 μs.

The vertical timings are multiples of the horizontal cycles. For example, region *P* is 64 μs, which is about two horizontal cycles (2×31.77), and region *Q* is 32 ($1020 / 31.77$) cycles. The calculation for region *R* is 15250 μs / 31.77 μs = 480. Of course, it has to be exactly 480 times, because we need to have the 480 rows per screen. Finally, region *S* requires 14 cycles.

In addition to generating the correct horizontal and vertical synchronization signals, the circuit needs to keep track of the current column within the *D* region, and the current row within the *R* region of the scan in order to know when to turn on or off a specific pixel. To make a particular pixel green, for example, you need to test the values of the column and row counts. If they are equal to the location of the pixel that you want to turn on, then you assert the green signal, and that pixel will be green.

9.7.2 **Controller Design**

To get the horizontal and vertical synchronization timing correct, we can design a FSM with 800 states running at a clock speed of 25.175 MHz. For the first 95 states, we will output a 0 for the horizontal synchronization signal *H_Sync*. For the next $45 + 640 + 20 = 705$ states, we will output a 1 for *H_Sync*. The problem with this, however, is that it is difficult to manually derive the circuit for an 800-state FSM. A simple solution around this difficulty is to use just two states: one for when *H_Sync* is a 0 in region *B* and one for when it is a 1 in regions *C*, *D*, and *E*. We then use a counter that runs at the same clock speed as the FSM to keep count of how many times we have been in a state. For the first state, we will stay there for 95 counts before going to the next state, and for the second state, we will stay there for 705 counts before going back to the first state. In the first state, we will output a 0 for *H_Sync*, and in the second state, we will output a 1 for *H_Sync*.

To help keep track of when the three color signals can be enabled, we will generate an additional *H_Data_on* signal the same way we generated the *H_Sync* signal. The *H_Data_on* signal is asserted in region *D*, and de-asserted in regions *B*, *C*, and *E*. Thus, for 640 counts of repeating in one state for region *D*, we will set *H_Data_on* to a 1 and to a 0 in a second state for the remaining 160 counts for regions *B*, *C*, and *E*.

Combining the two states for *H_Sync* and two states for *H_Data_on* together results in the final state diagram for the horizontal synchronization timing, as shown in Figure 9.34(a). This FSM has four states corresponding to the four regions *B*, *C*, *D*, and *E*. The counter initially is set to 0 and increments by 1 at every clock cycle. In state *H_B* for region *B*, the FSM outputs a 0 for both *H_Data_on* and *H_Sync*. The FSM will stay in state *H_B* for 95 counts. The condition $(H\_cnt = B)$ checks to see whether the counter is equal to *B*, where *B* is equal to 95. When the count is equal to 95, the FSM goes to state *H_C*, which corresponds to region *C*. In state *H_C*, the FSM outputs a 0 for *H_Data_on* and a 1 for *H_Sync* for 45 counts (i.e., until *H_cnt* is $B + C = 95 + 45 = 140$). When *H_cnt* reaches 140, the FSM goes to state *H_D*, and outputs a 1 for both *H_Data_on* and *H_Sync*. When *H_cnt* reaches $B + C + D = 95 + 45 + 640 = 780$, the FSM goes to state *H_E*, and outputs a 0 for *H_Data_on* and a 1 for *H_Sync*. The FSM stays in state *H_E* for 20 more counts until $H\_cnt = B + C + D + E = 95 + 45 + 640 + 20 = 800$, and then it goes back to state *H_B*. When the FSM goes back to state *H_B*, *H_cnt* is reset back to 0, and the process starts over again for the next row scan.

The vertical synchronization timing is analogous to the horizontal synchronization timing, so we can do the same thing using a second counter and a second FSM. This second vertical FSM is identical to the horizontal FSM. The only difference is in the timing of the clock speed. Looking at the times for each region in the vertical

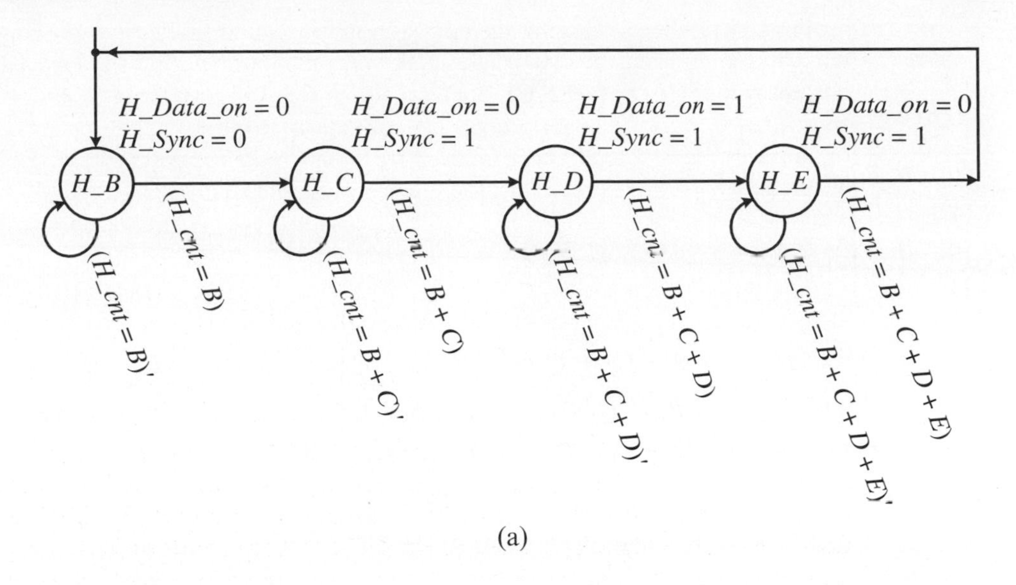

(a)

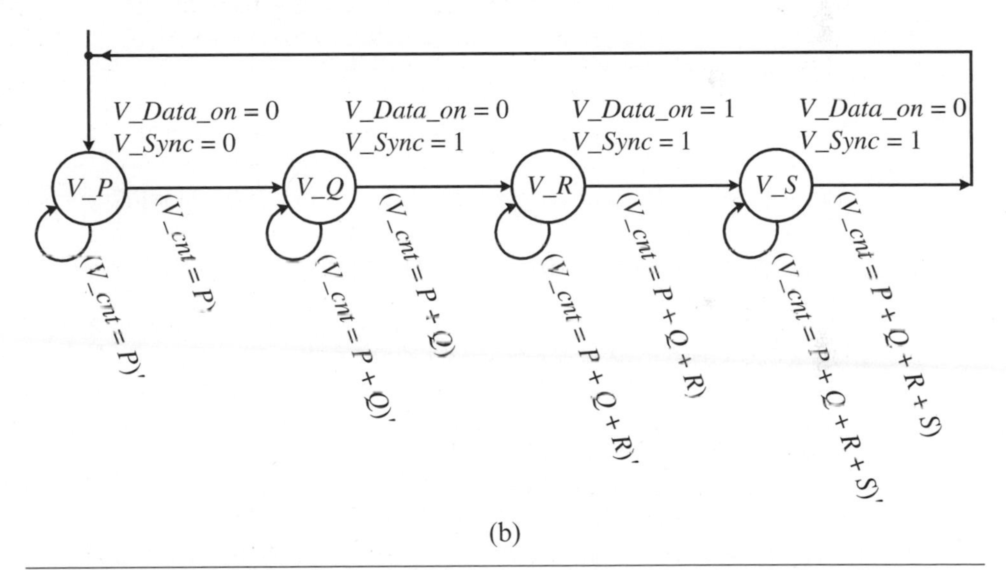

(b)

FIGURE 9.34 Controller for the VGA monitor: (a) state diagram for horizontal synchronization; (b) state diagram for vertical synchronization; (c) next-state table for horizontal synchronization; (d) next-state table for vertical synchronization; (e) output table; (f) FSM circuit for both the horizontal and vertical synchronization; (g) horizontal synchronization counter; (h) vertical synchronization counter; (i) complete circuit for the VGA controller. *(continued on next page)*

Current State Q_1Q_0		Next State (Implementation) $Q_{1next}Q_{0next}$ (D_1D_0)							
		$(H\_cnt = B)$		$(H\_cnt = B + C)$		$(H\_cnt = B + C + D)$		$(H\_cnt = B + C + D + E)$	
		0	1	0	1	0	1	0	1
00	H_B	00	01	—	—	—	—	—	—
01	H_C	—	—	01	10	—	—	—	—
10	H_D	—	—	—	—	10	11	—	—
11	H_E	—	—	—	—	—	—	11	00

$$Q_{1next} = Q_1'Q_0\,(H\_cnt = B + C) + Q_1Q_0' + Q_1Q_0\,(H\_cnt = B + C + D + E)'$$

$$Q_{0next} = Q_1'Q_0'\,(H\_cnt = B) + Q_1'Q_0\,(H\_cnt = B + C)' + Q_1Q_0'\,(H\_cnt = B + C + D) + Q_1Q_0\,(H\_cnt = B + C + D + E)'$$

(c)

Current State Q_1Q_0		Next State (Implementation) $Q_{1next}Q_{0next}$ (D_1D_0)							
		$(V\_cnt = P)$		$(V\_cnt = P + Q)$		$(V\_cnt = P + Q + R)$		$(V\_cnt = P + Q + R + S)$	
		0	1	0	1	0	1	0	1
00	V_P	00	01	—	—	—	—	—	—
01	V_Q	—	—	01	10	—	—	—	—
10	V_R	—	—	—	—	10	11	—	—
11	V_S	—	—	—	—	—	—	11	00

$$Q_{1next} = Q_1'Q_0\,(V\_cnt = P + Q) + Q_1Q_0' + Q_1Q_0\,(V\_cnt = P + Q + R + S)'$$

$$Q_{0next} = Q_1'Q_0'\,(V\_cnt = P) + Q_1'Q_0\,(V\_cnt = P + Q)' + Q_1Q_0'\,(V\_cnt = P + Q + R) + Q_1Q_0\,(V\_cnt = P + Q + R + S)'$$

(d)

Current State Q_1Q_0	*H_Data_on* or *V_Data_on*	*H_Sync* or V_Sync
00	0	0
01	0	1
10	1	1
11	0	1

(e)

FIGURE 9.34 Controller for the VGA monitor: (a) state diagram for horizontal synchronization; (b) state diagram for vertical synchronization; (c) next-state table for horizontal synchronization; (d) next-state table for vertical synchronization; (e) output table; (f) FSM circuit for both the horizontal and vertical synchronization; (g) horizontal synchronization counter; (h) vertical synchronization counter; (i) complete circuit for the VGA controller. *(continued on next page)*

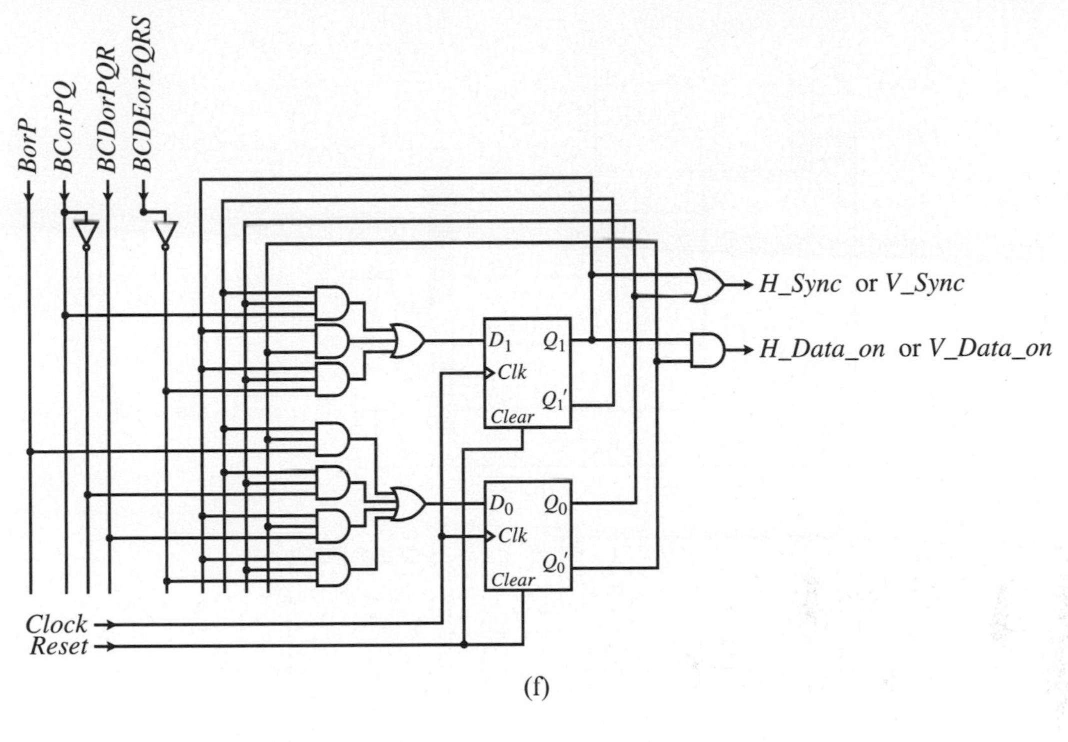

(f)

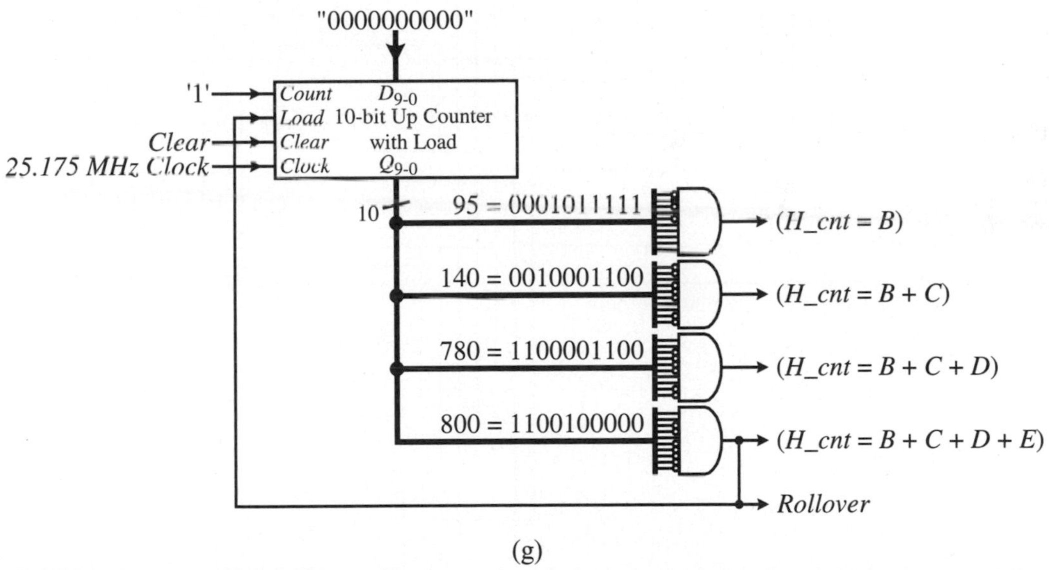

(g)

FIGURE 9.34 **Controller for the VGA monitor: (a) state diagram for horizontal synchronization; (b) state diagram for vertical synchronization; (c) next-state table for horizontal synchronization; (d) next-state table for vertical synchronization; (e) output table; (f) FSM circuit for both the horizontal and vertical synchronization; (g) horizontal synchronization counter; (h) vertical synchronization counter; (i) complete circuit for the VGA controller.** *(continued on next page)*

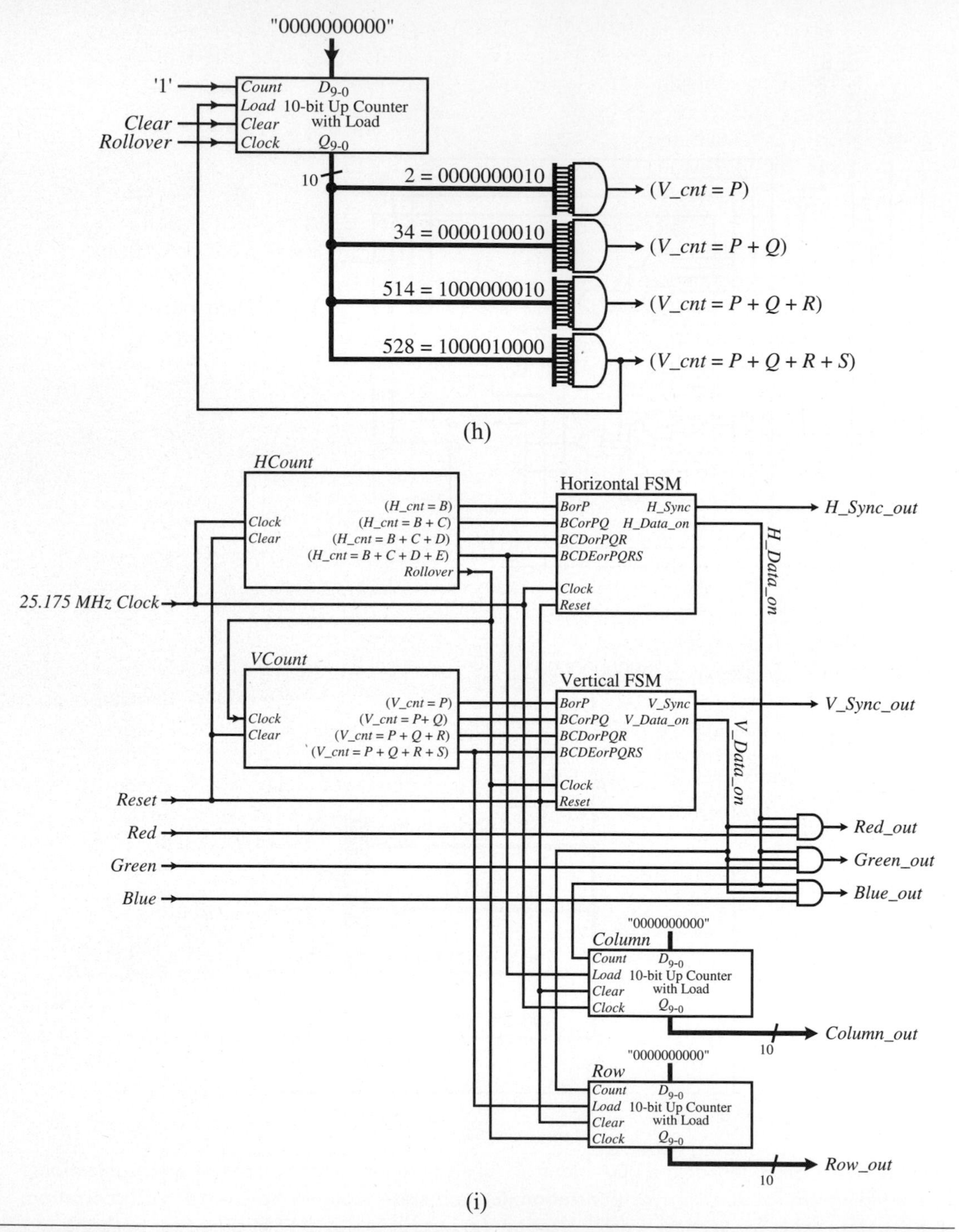

FIGURE 9.34 Controller for the VGA monitor: (a) state diagram for horizontal synchronization; (b) state diagram for vertical synchronization; (c) next-state table for horizontal synchronization; (d) next-state table for vertical synchronization; (e) output table; (f) FSM circuit for both the horizontal and vertical synchronization; (g) horizontal synchronization counter; (h) vertical synchronization counter; (i) complete circuit for the VGA controller.

synchronization signal in Figure 9.33, we see that the 64 μs for region P is approximately two times the total horizontal scan time of 31.77 μs each. The 1020 μs for region Q is approximately 32 horizontal scan time ($1020/31.77 \approx 32$). For region R, it is 480 horizontal cycles, and for region S, it is approximately 14 horizontal cycles. Therefore, the clock for both the vertical counter and the vertical FSM can be derived from the horizontal counter. The vertical clock ticks once for every 800 counts of the horizontal clock. The state diagram for the vertical timing is shown in Figure 9.34(b).

The next-state tables and next-state equations for the horizontal FSM and the vertical FSM are shown in Figures 9.34(c) and (d), respectively. Note that except for the four count conditions, the two tables are the same. Therefore, we can use the same circuit for both FSMs. The only difference is that their status signal inputs for the four count conditions come from different counter comparators. The output table is also the same for both FSMs and is shown in Figure 9.34(e). There are only two output signals to be generated: *H_Data_on* and *H_Sync* for the horizontal FSM, and *V_Data_on* and *V_Sync* for the vertical FSM. Finally, the FSM circuit is shown in Figure 9.34(f). We will need to use two instances of this FSM circuit: one for the horizontal FSM and one for the vertical FSM. The clock for the horizontal FSM is the 25.175 MHz clock, while the clock for the vertical FSM is derived from the *Rollover* signal from the horizontal FSM counter.

The four status signals for the FSM, *BorP*, *BCorPQ*, *BCDorPQR*, and *BCDEorPQRS*, are generated from two counters: a horizontal counter and a vertical counter. The horizontal counter, *HCount*, with the four comparators for ($H\_cnt = B$), ($H\_cnt = B + C$), ($H\_cnt = B + C + D$), and ($H\_cnt = B + C + D + E$), is shown in Figure 9.34(g). A 10-bit counter is needed to count from 0 up to 800. A 10-input AND gate is used for each of the four comparators. The inputs to each AND gate is set to the equivalent binary value for $B = 95$, $B + C = 95 + 45 = 140$, $B + C + D = 95 + 45 + 640 = 780$, and $B + C + D + E = 95 + 45 + 640 + 20 = 800$, respectively. The counter cycles back to 0 by asserting the *Load* line when the count reaches 800 and loading in the value 0. The clock signal for this horizontal counter is the 25.175 MHz clock. The output from the comparator ($H\_cnt = 800$) is the counter rollover signal *Rollover*, and is used as the vertical clock signal for the vertical counter, the vertical FSM, and the row counter.

The vertical counter, *VCount*, with the four comparators for ($V\_cnt = P$), ($V\_cnt = P + Q$), ($V\_cnt = P + Q + R$), and ($V\_cnt = P + Q + R + S$), is shown in Figure 9.34(h). The circuit for this counter is just like the horizontal counter, except that the comparator values are different, and we do not need to output a rollover clock signal. The clock for this counter is the *Rollover* signal from the horizontal counter.

The complete VGA monitor controller circuit is shown in Figure 9.34(i). To make sure that the three RGB signals to the monitor are valid, they have to be turned on (if needed) only in regions D and R. Therefore, the three color output signals *Red_out*, *Green_out*, and *Blue_out* are ANDed with *H_Data_on* and *V_Data_on*. For example, if the input *Red* signal is a 1, the output *Red_out* signal is a 1 only when the scan is within the regions D and R.

Finally, in order to turn on a specific pixel, the circuit needs to keep track of the current column within the D region and the current row within the R region of the scan. Two additional counters, *Column* and *Row*, are used for this purpose. Because they need to count only when the scan is in regions D and R, respectively, the *Count* input for the *Column* counter is asserted by the *H_Data_on* signal, while the *Count* input for the *Row* counter is asserted by the *V_Data_on* signal. Once the counts reach 640 and 480, respectively, they will have gone passed these two regions, and so the counters will not count. They will have to be reset to 0 anytime before the scan reaches the beginning of these two regions again. In the circuit, the two *Load* lines are asserted when the two respective counters roll back to 0. Finally, the *Column* counter clock is from the 25.175 MHz source, and the *Row* counter clock is from the *Rollover* clock signal derived from the horizontal counter.

9.7.3 Implementation

To display something on the screen, you simply have to check the current column and row values that the scan is at, and then assert the RGB signal if you want the pixel at that location to be turned on. For example, if you simply assert the *Red* signal continuously, all of the pixels will be red, and you will see the entire screen being red. On the other hand, if you just want the first row of pixels to be red, then you need to assert the *Red* signal only when the *Row* counter is 0. To get a red border around the screen, you would assert the *Red* signal when *Row* = 0, or *Row* = 639, or *Column* = 0, or *Column* = 479. Figure 9.35 shows the circuit to draw a red border around the entire screen using the VGA controller circuit from Figure 9.34(i).

Typically, a two-dimensional video memory stores the RGB color data for every pixel. The column and row counts then are used as the address to this video memory. The content of the memory location is the value to be set for the RGB signals.

To get more colors per pixel, more bits are used for each of the three color signals. For example, on the Altera DE1 FPGA development board, four bits are used for each

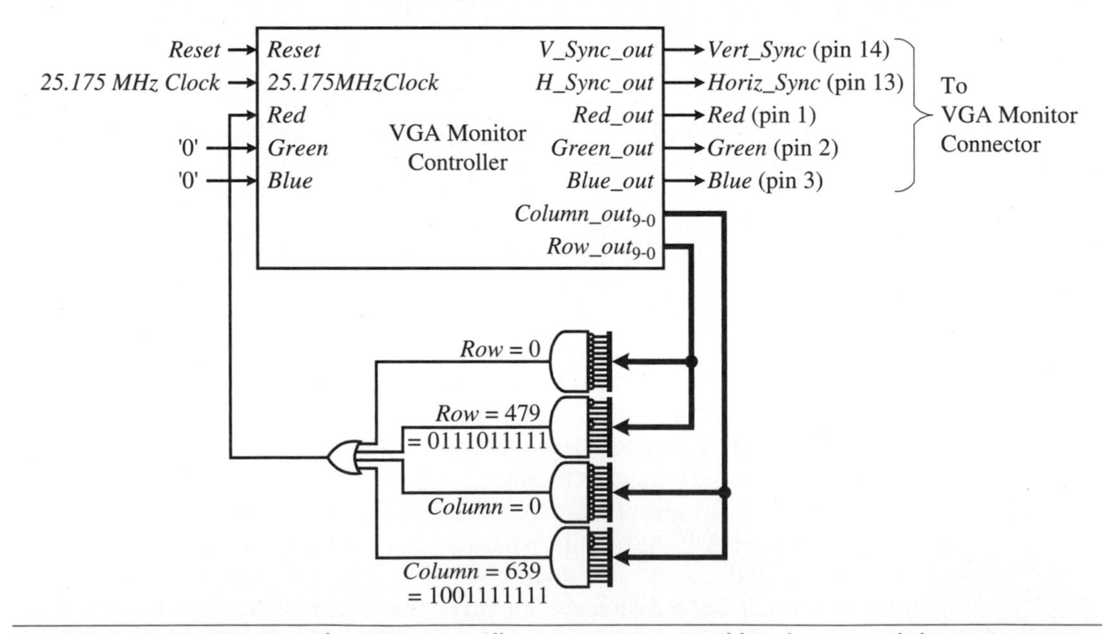

FIGURE 9.35 Circuit using the VGA controller to generate a red border around the entire screen.

of the three colors, while on the DE2 board, 10 bits are used. On the Xilinx Spartan 3E board only one bit is used for each of the three colors, while on the XUP Virtex-II Pro board, eight bits are used. Regardless of how many bits are used for each of the three colors, there are always only three lines connecting to the monitor because the three RGB color signals are actually analog signals. So a digital-to-analog converter is needed between the output of the controller to the VGA input.

For higher resolution monitors, a faster clock speed is needed. The following table shows the timing specifications for two other standard VGA resolutions.

Resolution H × V	Pixel clock (MHz)	*B* (μs/ clocks)	*C* (μs/ clocks)	*D* (μs/ clocks)	*E* (μs/ clocks)	*P* (ms/ lines)	*Q* (ms/ lines)	*R* (ms/ lines)	*S* (ms/ lines)
800×600	40	3.2/128	2.2/88	20/800	1/40	0.11/4	0.61/23	15.84/600	0.03/1
1024×768	65	2.09/136	2.46/160	15.75/1024	0.37/24	0.12/6	0.60/29	15.88/768	0.06/3

9.8 A/D Controller for Temperature Sensor

Many sensory devices such as the light sensor, temperature sensor, and pressure sensor output an analog signal rather than a digital signal. In order for a microprocessor to process analog signals, an analog-to-digital (A/D) converter must be used to convert the analog signal to digital signal. In this section, we will design a microcontroller to interface with an A/D converter to read values from an analog temperature sensor.

Different A/D converter chips have different functions, precisions, packaging, and interface. The A/D converter chip that we will use is the National Semiconductor ADC0832, which is an 8-bit A/D converter chip that interfaces with a microcontroller using a serial data link. Because of this, the A/D converter can be located at the analog signal source and through just a few wires can communicate with the microprocessor with a highly noise immune serial bit stream. This greatly minimizes circuitry to maintain accuracy of the analog signal, which otherwise is most susceptible to noise pickup.

9.8.1 Theory of Operation

The ADC0832 8-bit A/D converter chip interfaces with a microcontroller using a serial data link. Either three or four lines are necessary to connect and communicate between the microcontroller and the ADC chip. For our example, we will use four lines, of which three are inputs: clock (CLK), chip select (CSN), and data input (DI), and one is output: data out (DO).

The active-low CSN signal line enables the ADC chip. Commands from the controller are sent to the ADC via the DI signal line, and the converted digital data from the ADC is sent to the controller via the DO signal line. The clock synchronizes the data transmission on both the DI and DO lines. The interface between the microcontroller and the ADC chip is shown in Figure 9.36.

An analog to digital conversion is initiated by first pulling the chip select (CSN) line low, and remains low for the entire conversion process. The ADC chip then waits for the start bit and command. A clock is generated by the microcontroller (if not externally provided continuously as shown by the dotted line connection in Figure 9.36) on the clock (CLK) input line. A 1 start bit followed by a 2-bit MUX address command is sent to the ADC on the data in (DI) line on each rising edge of the clock. The first MUX address command bit selects between the single-ended or differential analog input options. A 1 bit selects the single-ended option, which uses a ground referenced input. After selecting the single-ended option, the second bit selects which of the chip's two input channels to use. A 0 selects channel 0 on the chip, and a 1 selects channel 1. The data out (DO) line now comes out of tri-state, and the converted digital data, starting with the most significant bit, is sent out on this line to the microcontroller. After this, the same converted digital data, this time starting with the least significant bit, is sent to the microcontroller on the DO line. At the end of the data transmission, the CSN line needs to be de-asserted before starting another conversion. The timing diagram for the communication between the microcontroller and the A/D converter chip is shown in Figure 9.37.

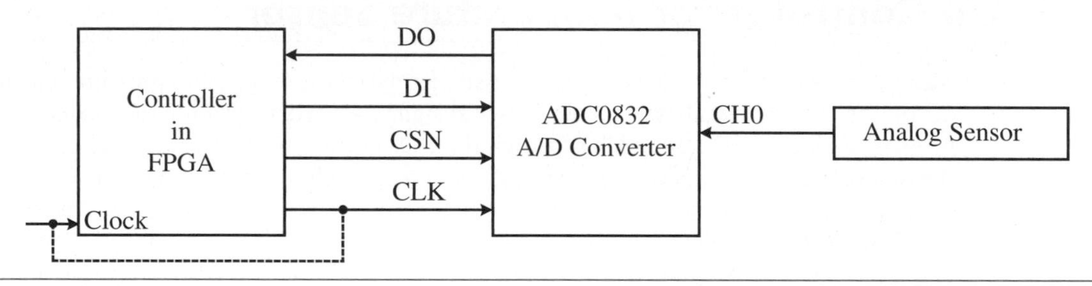

FIGURE 9.36 Serial data link between the microcontroller implemented in an FPGA and the A/D converter chip.

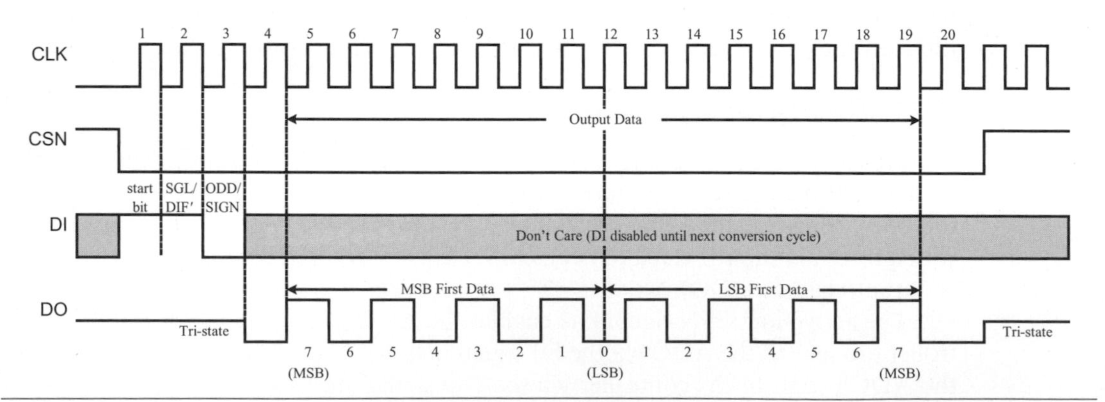

FIGURE 9.37 Timing diagram for communicating between the microcontroller and the A/D converter chip.

9.8.2 Controller Design

We will write HDL code using the FSMD model to describe the controller for the A/D converter chip. Figure 9.38 shows the Verilog code for this FSM controller. The state transitions and output signals generated in the FSM code follow very closely to the timing diagram shown in Figure 9.37.

The input clock speed for this FSM is not critical. The FSM generates the output clock (CLK) signal by toggling CLK at each state transition to synchronize the data transmission between it and the ADC chip.

Figure 9.39 shows the VHDL code for the A/D converter controller. It is quite similar to the Verilog version shown in Figure 9.38, except that instead of having the FSM generate a separate synchronizing CLK signal, it uses the same external clock signal to drive both the FSM and the ADC chip. Therefore, this FSM does not need to output a second CLK signal. The clock connection for this FSM is the dotted line shown in Figure 9.36. The sending of the command and the receiving of the converted data is done on the falling edge of the clock.

```
module ADC_FSM
(
  input Clock, Reset,
  input DO,
  output reg CLK, CSN, DI,
  output reg [7:0] DataOut,
  output reg DataReady
);
  reg [7:0] MSBF_Buffer, LSBF_Buffer;
  reg [2:0] index;
  reg [3:0] state;

  always @ (posedge Clock or posedge Reset) begin
    if (Reset) begin
      CSN <= 1;
      CLK <= 0;
      DI <= 1;        // don't care
      DataReady <= 0;
      state <= 4'b0000;
    end else begin
      case (state)
        4'b0000: begin
          CSN <= 0;
          CLK <= 0;
          DI <= 1; // start bit
          DataReady <= 0;
          state <= 4'b0001;
          end
```

FIGURE 9.38 Behavioral Verilog code for the A/D converter controller. *(continued on next page)*

```
4'b0001: begin
  CSN <= 0;
  CLK <= 1;
  DI <= 1;
  DataReady <= 0;
  state <= 4'b0010;
  end
4'b0010: begin
  CLK <= 0;
  DI <= 1; // set SGL/DIF'
  state <= 4'b0011;
  end
4'b0011: begin
  CLK <= 1;
  state <= 4'b0100;
  end
4'b0100: begin
  CLK <= 0;
  DI <= 0; // set ODD/SGN'; 0 = channel 0, 1 = channel 1
  state <= 4'b0101;
  end
4'b0101: begin
  CLK <= 1;
  state <= 4'b0110;
  end
4'b0110: begin
  CLK <= 0; // MUX settling time
  state <= 4'b0111;
  end
4'b0111: begin
  CLK <= 1;
  index <= 7;
  state <= 4'b1000;
  end
4'b1000: begin
  CLK <= 0;
  state <= 4'b1001;
  end
4'b1001: begin
  CLK <= 1;
  // receive 8 data bits, MSB first
  MSBF_Buffer[index] <= DO;
  if (index > 0) begin
```

FIGURE 9.38 Behavioral Verilog code for the A/D converter controller. *(continued on next page)*

```
                index <= index - 1;
                state <= 4'b1000;
              end else begin
                // receive 8 data bits, LSB first
                LSBF_Buffer[index] <= DO;
                index <= index + 1;
                state <= 4'b1010;
                end
              end
            4'b1010: begin
              CLK <= 0;
              state <= 4'b1011;
              end
            4'b1011: begin
              CLK <= 1;
              LSBF_Buffer[index] <= DO;
              if (index < 7) begin
                index <= index + 1;
                state <= 4'b1010;
              end else
                state <= 4'b1100;
              end
            4'b1100: begin
              CLK <= 1;
              CSN <= 1;
              if (MSBF_Buffer == LSBF_Buffer) begin
                DataOut <= MSBF_Buffer;
                DataReady <= 1;
              end else begin
                DataOut <= 8'b00000000;
                DataReady <= 0;
                end
              state <= 4'b0000;
              end
            default: begin
              state <= 4'b0000;
            end
          endcase
        end // else
      end // always
    endmodule
```

FIGURE 9.38 Behavioral Verilog code for the A/D converter controller.

```
LIBRARY IEEE;
USE IEEE.STD_LOGIC_1164.ALL;

ENTITY ADC_FSM IS PORT (
    Clock    : IN STD_LOGIC;
    Reset    : IN STD_LOGIC;
    DO       : IN STD_LOGIC;
    CSN      : OUT STD_LOGIC;
    DI       : OUT STD_LOGIC;
    DataOut  : OUT STD_LOGIC_VECTOR(7 DOWNTO 0);
    DataReady : OUT STD_LOGIC
  );
END ENTITY;

ARCHITECTURE FSM OF ADC_FSM IS
  TYPE state_type IS (s0, s1, s2, s3, s4, s5, s6, s7);
  SIGNAL state : state_type;

  SIGNAL MSBF_Buffer: STD_LOGIC_VECTOR(7 DOWNTO 0);
  SIGNAL LSBF_Buffer: STD_LOGIC_VECTOR(7 DOWNTO 0);
  SIGNAL index: INTEGER;

BEGIN
  PROCESS (Clock, Reset)
  BEGIN
    IF (Reset = '1') THEN
      CSN <= '1';
      DataReady <= '0';
      index <= 0;
      state <= s0;
    ELSIF (Clock'EVENT AND Clock = '0') THEN
      CASE state IS
        WHEN s0 =>
          CSN <= '0';
          DI <= '1'; -- start bit
          DataReady <= '0';
          state <= s1;
        WHEN s1 =>
          CSN <= '0';
          DI <= '1'; -- set SGL/DIF'
          DataReady <= '0';
          state <= s2;
        WHEN s2 =>
          CSN <= '0';
          DI <= '0'; -- set ODD/SIGN; 0 = channel 0, 1 = channel 1
          DataReady <= '0';
          state <= s3;
        WHEN s3 => -- wait for one clock for Mux setting
          CSN <= '0';
```

FIGURE 9.39 Behavioral VHDL code for the A/D converter controller. *(continued on next page)*

```
            DataReady <= '0';
            state <= s4;
          WHEN s4=> -- wait for one clock for Mux setting
            CSN <= '0';
            DataReady <= '0';
            index <= 7;
            state <= s5;
          WHEN s5 =>
            CSN <= '0';
            DataReady <= '0';
            IF (index > 0) THEN
              MSBF_Buffer(index) <= DO;
              index <= index - 1;
              state <= s5;
            ELSE
              MSBF_Buffer(index) <= DO;
              LSBF_Buffer(index) <= DO;
              index <= 1;
              state <= s6;
            END IF ;
          WHEN s6 =>
            CSN <= '0';
            DataReady <= '0';
            IF (index <= 7) THEN
              LSBF_Buffer(index) <= DO;
              index <= index + 1;
              state <= s6;
            ELSE
              state <= s7;
            END IF ;
          WHEN s7 =>
            CSN <= '1';
            DataReady <= '1';
            DataOut(0) <= MSBF_Buffer(0);
            DataOut(1) <= MSBF_Buffer(1);
            DataOut(2) <= MSBF_Buffer(2);
            DataOut(3) <= MSBF_Buffer(3);
            DataOut(4) <= MSBF_Buffer(4);
            DataOut(5) <= MSBF_Buffer(5);
            DataOut(6) <= MSBF_Buffer(6);
            DataOut(7) <= MSBF_Buffer(7);
            state <= s0;
          WHEN OTHERS =>
        END case;
      END IF ;
    END PROCESS;
  END FSM;
```

FIGURE 9.39 Behavioral VHDL code for the A/D converter controller.

9.8.3 Implementation

One way to test the FSM controller code simply is to connect an adjustable 10K Ω resistor to the channel 0 analog input of the ADC chip, as shown in Figure 9.40(a). One side of the potentiometer is connected to +5 V and the other side to ground. The center connection of the potentiometer connects to channel 0 of the ADC chip. Because the ADC0832 is an 8-bit A/D converter, the digital value output should range from 0 to 255. When the potentiometer is turned all the way to one side, it should give a reading of 0, and when turned all the way to the other side, it should give a reading of 255.

Different analog sensors can be used as inputs to the ADC chip. Figure 9.40(b) shows the connections of a thermistor temperature sensor, and Figure 9.40(c) shows

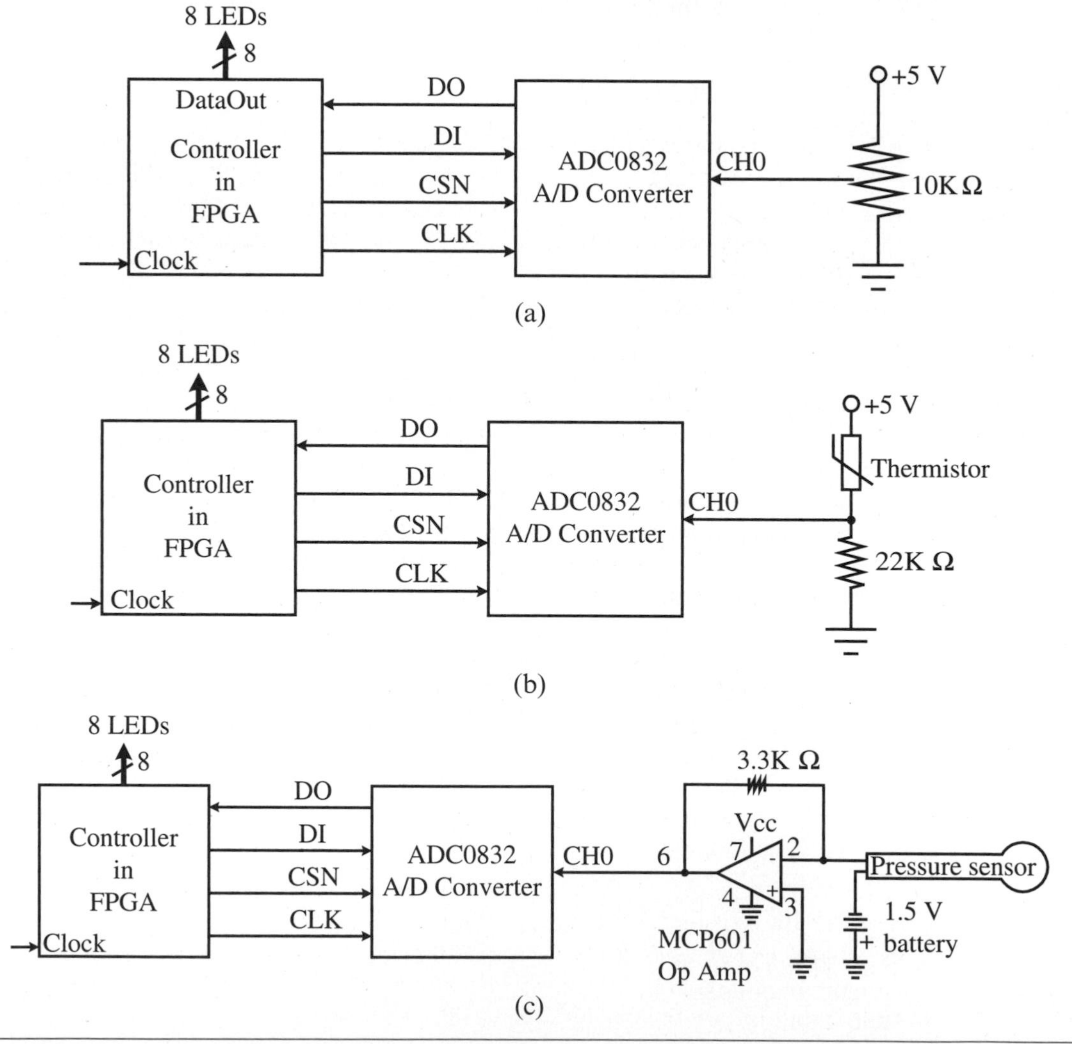

FIGURE 9.40 Analog sensor interface to the ADC chip: (a) potentiometer for testing; (b) thermistor temperature sensor; (c) a pressure sensor using a MCP601 Op Amp.

the connections for a pressure sensor. The pressure sensor requires an op amp to amplify the signal.

9.9 I^2C Bus Controller for Real-Time Clock

In this section, we will design and implement a microcontroller to control the DS3232 real-time clock chip. The real-time clock chip is connected to the microcontroller using the standard I^2C bus protocol. The microcontroller is connected on the I^2C bus as the master and the real-time clock chip is connected as a slave. Using the I^2C bus as the communication channel, the master controller will be able to send and receive data to and from the slave.

9.9.1 Theory of Operation

The I^2C (Inter IC) bus is a simple bidirectional serial bus that supports multiple masters and slaves. It consists of only two lines; a serial bidirectional data line (SDA) and a serial bidirectional clock line (SCL). Within the I^2C bus specifications, a standard mode with a maximum clock rate of 100 kHz and a fast mode with a maximum clock rate of 400 kHz are defined.

Each device connected to the I^2C bus is software addressable by a unique address, and a simple master/slave relationship exists at all times among the devices. The device that controls the sending and receiving of messages by controlling the bus access is the master. Devices that are controlled by the master are the slaves. Both the master and the slave can send and receive messages. A device that sends data onto the bus is referred to as the transmitter and a device receiving data is referred to as the receiver.

More than one master and more than one slave can coexist on the same I^2C bus. The bus, however, is always controlled by a single master at any one time, and is responsible for generating the serial clock (SCL) and controlling the bus access by initiating and terminating a message transfer.

Our system consists of only one master (the microcontroller that is implemented on an FPGA) and one slave (the Maxim DS3232 real-time clock chip) as shown in Figure 9.41. Both the SDA and SCL lines connecting between the master and the slave are open-drained, and must be pulled up to Vcc with a 5.6K Ω resistor. Regardless of how many devices are connected to the bus, only one pull-up resister is needed per line. This implementation of the master controller does not follow the full specifications of the I^2C protocol, so if you use this for other I^2C connections, the master controller might halt in the error state.

The I^2C bus is idle when both SCL and SDA are at a logic 1 level. The master initiates a data transfer by issuing a START condition, which is a high to low transition on the SDA line while the SCL line is high, as shown in Figure 9.42(a). The bus is considered to be busy after the START condition. After the START condition, a slave address is sent out on the bus by the master, as shown in Figure 9.43. This address is 7 bits long followed by an eighth bit that is a data direction bit ($R/\overline{W}$), where a 0 indicates a write from the master to the slave, and a 1 indicates a read from the slave to the

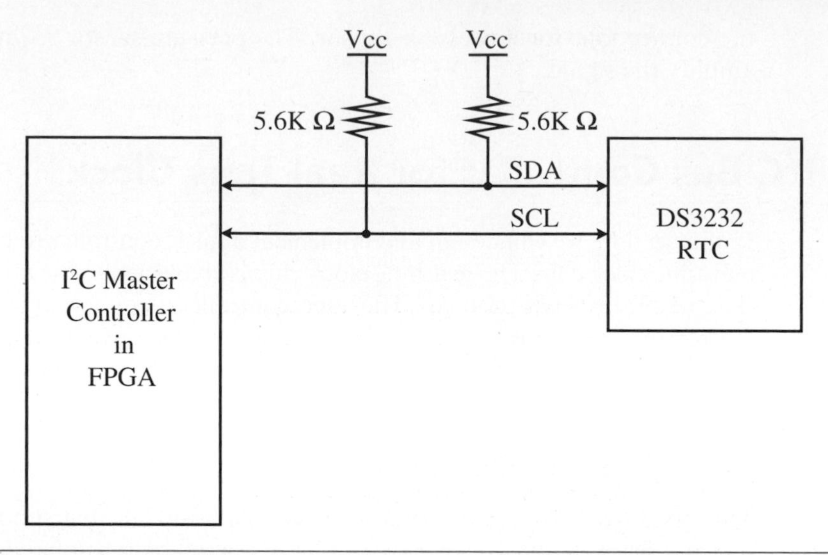

FIGURE 9.41 I^2C bus system with the I^2C master controller implemented in an FPGA and a real-time clock chip acting as the slave.

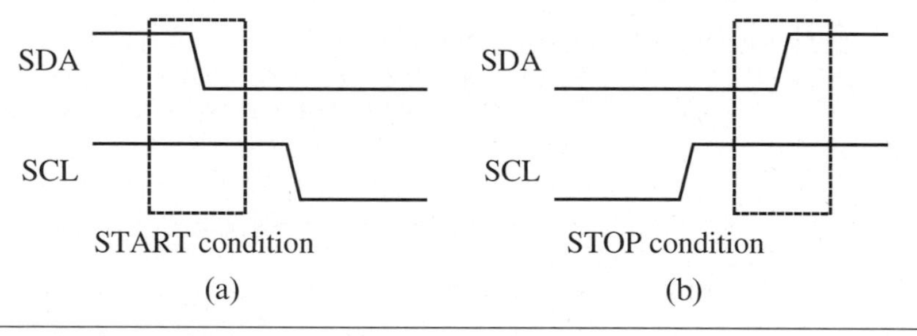

FIGURE 9.42 The START (a) and STOP (b) conditions on the I^2C bus.

master. The master, who is controlling the SCL line, will send out the bits on the SDA line, one bit per clock cycle of the SCL line, with the most significant bit sent out first. The value on the SDA line can be changed only when the SCL line is at a low.

The slave device whose address matches the address that is being sent out by the master will respond with an acknowledgment (ACK) bit on the SDA line by pulling the SDA line low during the ninth clock cycle of the SCL line. The direction bit ($R/\overline{W}$) determines whether the master or the slave will be the transmitter in the subsequent data transmission after the sending of the slave address.

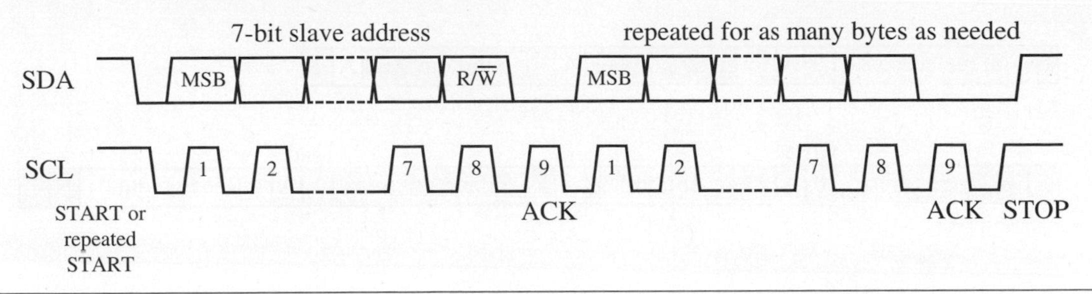

FIGURE 9.43 Timing diagram for communicating between the master and the slave on the I²C bus.

Every byte put on the SDA line for transmission must be 8 bits long with the most significant bit first. Except for the START and STOP conditions, the SDA line must not change when the SCL line is high. The number of bytes that can be transmitted is unrestricted. Each byte has to be followed by an acknowledge bit. The master generates the acknowledge-related clock pulse. The transmitter releases the SDA line (sets it to high impedance) during the acknowledge clock pulse, and the receiver must pull down the SDA line during the acknowledge clock pulse to acknowledge the receipt of the byte. The one exception is when a master-receiver is involved in a transfer. In this case the master-receiver must signal the end of data to the slave-transmitter by not generating an acknowledgment on the last byte clocked out of the slave.

To signal the end of data transfer, the master sends a STOP condition by pulling the SDA line from low to high, while the SCL line is at a high, as shown in Figure 9.42(b). Alternatively, instead of sending a STOP condition, the master can send a repeated START condition so that it can change the direction of the data transmission without having to release the bus.

Figure 9.44(a) shows the scenario in which the master writes 1 byte of data to the slave (i.e., the master is the transmitter and the slave is the receiver). The master initiates the data transfer by first issuing the START condition followed by the 7-bit slave address plus the write (0) bit. After receiving an acknowledgment from the slave, the master sends the register number to let the slave know which register the following data should be written into. The slave responds with an acknowledgment, and the master sends the data byte to the slave. After the slave acknowledges receipt of the data byte, the master sends the STOP condition.

Figure 9.44(b) shows the scenario in which the master reads 1 byte of data from the slave (i.e., the master is the receiver and the slave is the transmitter). The master initiates the data transfer by issuing the START condition followed by the 7-bit slave address plus the write (0) bit. Although the master wants to receive a byte, it first needs to send the register address byte to let the slave know which register the master wants to read from. After receiving an acknowledgment from the slave, the master sends the

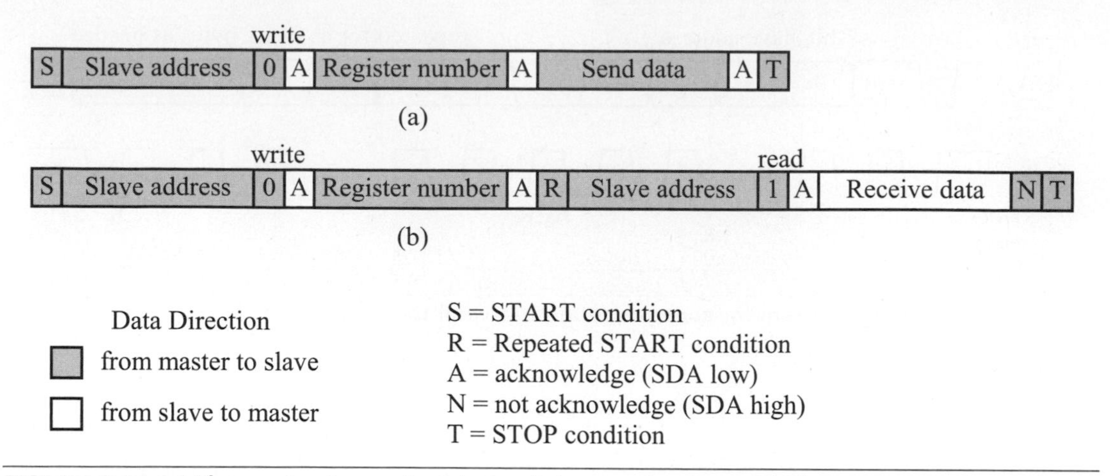

FIGURE 9.44 (a) The master transmits 1 byte of data to the slave. (b) The master receives 1 byte of data from the slave.

register number to the slave, and the slave responds again with an acknowledgment. This time the master has to do a repeated START condition because it needs to change the data direction from a write to a read. The repeated START is followed by the slave address again, but this time with the read (1) bit instead. The slave acknowledges and then sends the data byte from the addressed register to the master. This time, because the master is the receiver, the master has to acknowledge receipt of the data byte. If the master wants to receive more data bytes from the slave, it sends a 0 to acknowledge it. If the master doesn't want to receive any more bytes, it won't acknowledge by keeping SDA at a high. Finally, the master sends the STOP condition.

To summarize the master-transmitter and master-receiver scenarios, for both cases, the master first sends the 7-bit slave address and the write bit, followed by the register number to access. In the master-transmitter scenario, the master can immediately send out the data byte because the data direction is still a write. For the master-receiver scenario, the master has to do a repeated START and resend the slave address with the read bit in order to change the direction of the data transmission from a write to a read. After this, the master can receive a byte of data from the slave from the given register number.

9.9.2 **Controller Design**

Following the I^2C protocol specification as described above, we can derive the state diagram for our I^2C master controller. This state diagram will require many states, but many of them are identical and follow the same pattern. Figure 9.45(a) shows the initial portion of the state diagram. Starting from the idle state x00, the FSM goes to state x01 to send out the START condition, followed by the 7-bit slave address, and the write bit. The slave address and the write bit are previously stored in the 8-bit *SlaveAddress_Write* array, and is shifted out one bit at a time onto the *SDA* line in the two states x02 and x03. Two states are needed because the *SCL* line must toggle

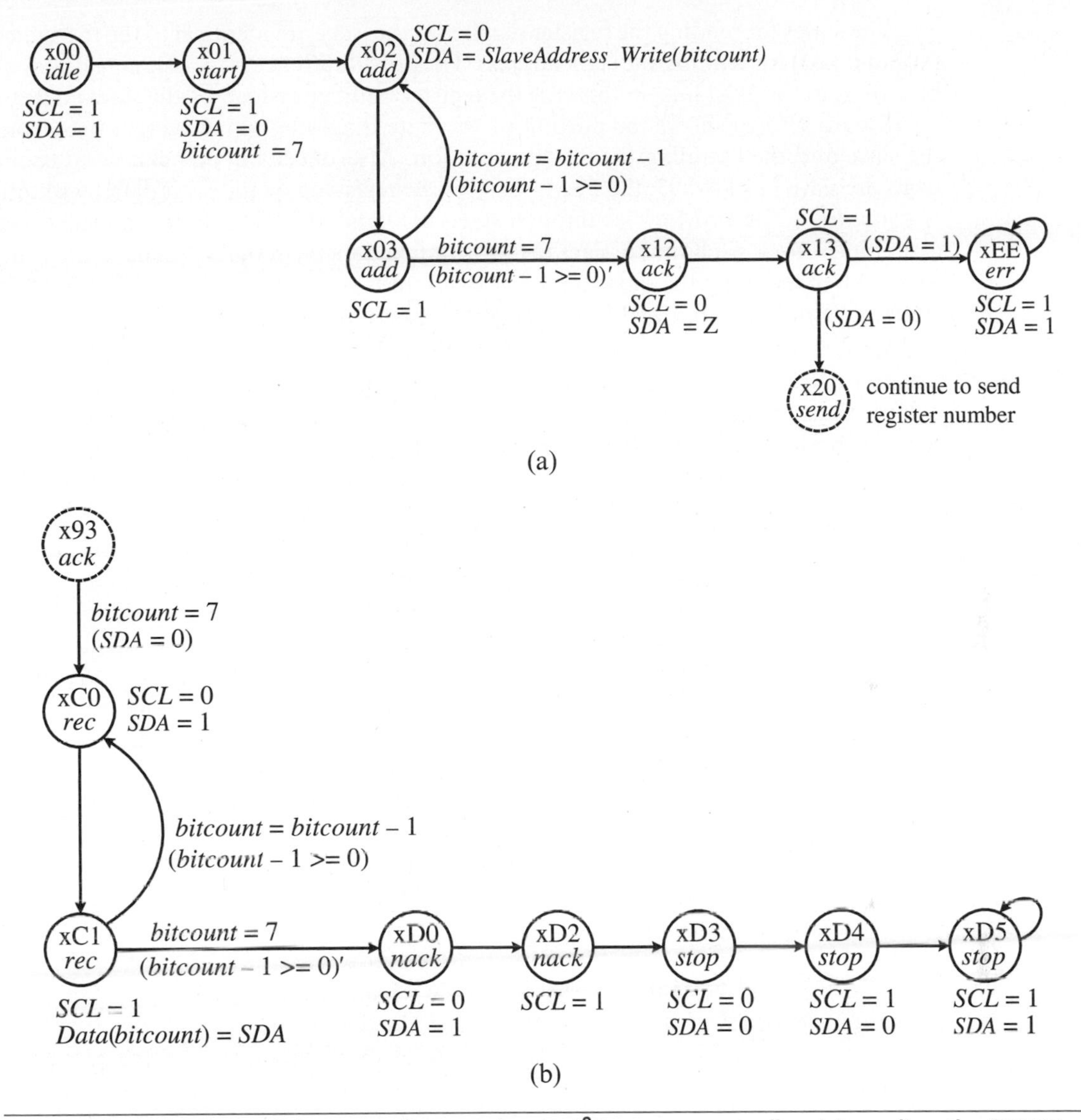

FIGURE 9.45 Portions of the state diagram for the I²C master controller: (a) sending the START condition and the slave address; (b) receiving a byte from the slave, and then sending the STOP condition.

between 0 and 1 for each bit. After sending out these eight bits, the FSM sets the *SDA* line to high-impedance Z in state x12 to receive the acknowledge signal from the slave in state x13. If the acknowledge signal is invalid (1), then the FSM goes to the error state xEE and terminates there, otherwise, the FSM continues to state x20 to send the register number to the slave.

The states for sending the register number to the slave are identical to the two states (x02 and x03) for sending the slave address. The only difference is in assigning different bit values to the *SDA* line; in this case the register number instead of the slave address.

Figure 9.45(b) shows the portion of the state diagram for receiving a byte from the slave and then sending the STOP condition. After receiving an acknowledgment from the slave in state x93, the FSM prepares the receiving of the 8-bit data by setting *bitcount* to 7. The FSM cycles through states xC0 and xC1 eight times to receive the eight bits of data, each time assigning the incoming data from the *SDA* line to the *Data* array. After receiving the data, the FSM then sends out a not acknowledge signal in states xD0 and xD2 followed by the STOP condition in the last three states.

We will write HDL code using the FSMD model to describe this I^2C master controller. Figures 9.46(a) and (b) show the code segment in VHDL corresponding to the two portions of the state diagram from Figure 9.45, respectively. The complete source code for both Verilog and VHDL can be downloaded from the book website.

```
output: PROCESS(CLK_200k_Hz, Go)
BEGIN
    IF(Go = '0') THEN
      -- when idle, both SDA and SCL = 1
      SCL <= '1';
      SDA01 <= '1';
      Ready <= '0';
      state <= x"00";
    ELSIF(CLK_200k_Hz'EVENT and CLK_200k_Hz = '1') THEN
      CASE state IS
      WHEN x"00" => -- Idle
        -- when idle, both SDA and SCL = 1
        SCL <= '1'; -- SCL = 1
        SDA01 <= '1'; -- SDA = 1
        Ready <= '0';
        state <= x"01";
      -- send start condition and slave address
      WHEN x"01" => -- Start
        SCL <= '1';  -- SCL stays at 1 while
        SDA01 <= '0'; -- SDA changes from 1 to 0
        Ready <= '0';
        bitcount <= 7; -- starting bit count
        state <= x"02";
      -- send 7-bit slave address followed by R/W' bit
      WHEN x"02" =>
        SCL <= '0';
        SDA01 <= SlaveAddress_Write(bitcount);
        state <= x"03";
      WHEN x"03" =>
```

FIGURE 9.46 Portions of the VHDL code corresponding to the two partial state diagrams shown in Figure 9.45: (a) sending the START condition and the slave address; (b) receiving a byte from the slave, and then sending the STOP condition. *(continued on next page)*

```
  SCL <= '1';
  IF (bitcount - 1) >= 0 THEN
    bitcount <= bitcount - 1;
    state <= x"02";
  ELSE
    bitcount <= 7;
    state <= x"12";
  END IF;
-- get acknowledgment' from slave
WHEN x"12" =>
  SCL <= '0';
  SDA01 <= '1';
  state <= x"13";
WHEN x"13" =>
  SCL <= '1';
  IF SDA = '1' THEN
    state <= x"EE"; -- acknowledge error
  ELSE
    state <= x"20"; -- send register address
  END IF;

-- send 8-bit register address to slave
WHEN x"20" =>
```

(a)

```
-- get acknowledge signal from slave
WHEN x"93" =>
  SCL <= '1';
  IF SDA = '1' THEN
    state <= x"EE"; -- acknowledge error
  ELSE
    bitcount <= 7;
    state <= x"C0"; -- go to receive data byte
  END IF;
-- receive byte from RTC slave
WHEN x"C0" =>
  SCL <= '0';
  SDA01 <= '1';
  state <= x"C1";
WHEN x"C1" =>
  SCL <= '1';
  Data(bitcount) <= SDA; -- MSB of data read in
  IF (bitcount - 1) >= 0 THEN
    bitcount <= bitcount - 1;
    state <= x"C0";
  ELSE
```

FIGURE 9.46 Portions of the VHDL code corresponding to the two partial state diagrams shown in Figure 9.45: (a) sending the START condition and the slave address; (b) receiving a byte from the slave, and then sending the STOP condition. *(continued on next page)*

```
        bitcount <= 7;
        state <= x"D0";
      END IF;
    WHEN x"D0" =>
      SCL <= '0';
      SDA01 <= '1'; -- send a not acknowledge' (1) signal
      state <= x"D2";
      DataOut <= Data;
    -- send a not acknowledge' (1) signal
    WHEN x"D2" =>
      SCL <= '1';
      state <= x"D3";
    -- send stop condition
    -- SDA goes from 0 to 1 while SCL is 1
    WHEN x"D3" =>
      SCL <= '0';
      -- SDA starts at 0 to prepare for the 0 to 1 transition
      SDA01 <= '0';
      state <= x"D4";
    WHEN x"D4" =>
      SCL <= '1';  -- SCL goes to 1
      SDA01 <= '0'; -- SDA starts from 0
      state <= x"D5";
    WHEN x"D5" =>
      SCL <= '1';  -- SCL stays at 1 while
      SDA01 <= '1'; -- SDA goes to 1
      Ready <= '1';
      state <= x"D5";
```

(b)

FIGURE 9.46 Portions of the VHDL code corresponding to the two partial state diagrams shown in Figure 9.45: (a) sending the START condition and the slave address; (b) receiving a byte from the slave, and then sending the STOP condition.

Assigning values to the *SDA* signal requires some special attention. You might have noticed in the code listing that for all of the assignments to the *SDA* signal, it instead uses the signal named *SDA01*. Recall that the SDA line in the I^{2}C bus is bidirectional so that it is capable of doing both input and output. Furthermore, the SDA line in the I^{2}C bus is open-drained and is pulled up by a 5.6K Ω resistor. So to output a logic 1 on this line, you need to set the line to a high impedance. To get a high impedance, you need to use a tri-state output and assign to it a *Z* value. The actual conversion from the internal *SDA01* signal to the *SDA* signal is done in the following assignment statement.

```
SDA <= 'Z' WHEN SDA01 = '1' ELSE '0';
```

This statement assigns a high-impedance value *Z* to the *SDA* signal when the internal signal *SDA01* has a logic 1 value; otherwise, the *SDA* signal gets the logic value 0. Therefore, to assign the logic value 0 or 1 to the *SDA* signal, the code instead assigns the value 0 or 1 to the internal signal *SDA01*, which in turn sets the *SDA* signal to 0 or *Z*, respectively.

The I^2C bus protocol specifies that the maximum clock rate for standard mode is 100 kHz. Therefore, we need to make sure that the *SCL* signal has this frequency. The *SCL* signal is generated in the FSM process by going back and forth between two states: the first state sets *SCL* to a 0 and the second state sets it to a 1. Therefore, for every *SCL* cycle, the FSM must go through two states or two cycles. Therefore, to get a 100 kHz speed for the *SCL* signal, the FSM clock speed must run two times faster or at 200 kHz. If the main clock speed is 50 MHz, then we will need a clock divider to count the 50 MHz clock ticks from 0 to 250 (i.e., 50,000,000/200,000). To get a 50% duty cycle for the 200 kHz clock, the clock signal CLK_200k_Hz toggles after every 125 counts (i.e., 250/2).

9.9.3 **Implementation**

Now that we have an I^2C master controller that can communicate with the real-time clock slave, we still need a user interface that will allow us to set the time on the real-time clock chip and to display the time on a 7-segment LED display. The top level interface is shown in Figure 9.47.

The user interface FSM has two functions: (1) to repeatedly send commands to the I^2C controller to read the date and time information from the real-time clock chip. The date and time information then is displayed on the six 7-segment LED display; (2) it waits for user inputs through the switches and push buttons. When a switch or push button is asserted, the FSM sends the appropriate command to the I^2C controller through the four control signal lines *Go*, *Read_WriteN*, *DataOut*[7..0], and *AddressOut*[7..0] to set the corresponding date and time registers in the real-time clock chip. The *DataIn* and *Ready* input signals are for reading the date and time information from the real-time clock chip. The four output signals *Hour_Month*, *Minute_Day*, *Second_Year*, and *Temperature_Day* will output the appropriate values to the BCD to 7-segment decoder for displaying the date and time information.

The user inputs consist of the *Date_TimeN* switch for selecting whether to display the date or the time, the *Hour12_24N* switch for selecting whether to display the time in 12- or 24-hour format, and the *SetHour_MonthN*, *SetMinute_DateN*,

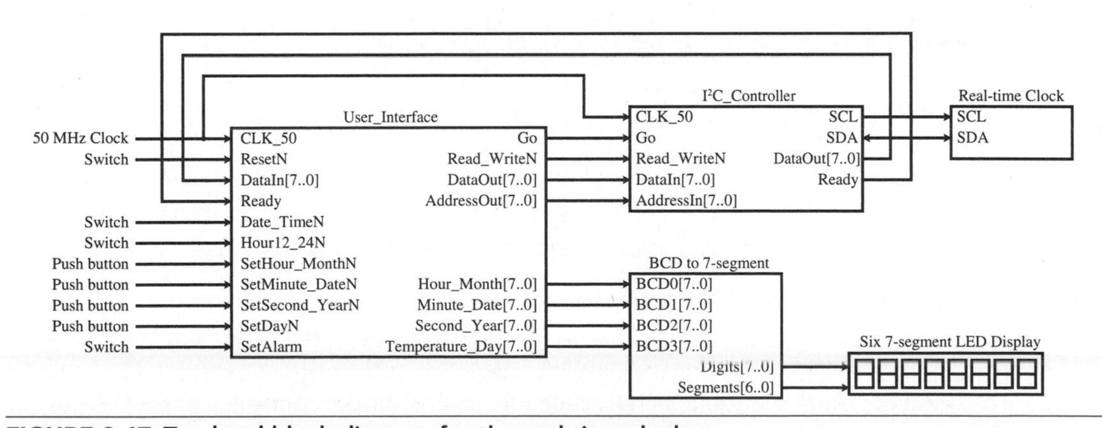

FIGURE 9.47 Top level block diagram for the real-time clock.

SetSecond_YearN, *SetDayN*, and *SetAlarm* push buttons and switches for setting the date and time information.

9.10 PROBLEMS

9.1. Experiment with different clock speeds for the three-digit 7-segment LED display controller and see its effects.

9.2. Write an HDL program to convert an 8-bit unsigned binary number to its decimal equivalent, and display the decimal number on the three-digit 7-segment LED display.

9.3. Implement the controller circuit for the incorrect state diagram for solving the summing input numbers problem shown in Figure 9.9(a) to verify that the Enter push button will not work.

9.4. Implement the controller circuit for the incorrect state diagram for solving the summing input numbers problem shown in Figure 9.9(b) to verify that the Enter push button will not work.

9.5. Implement the controller circuit for the correct state diagram for solving the summing input numbers problem shown in Figure 9.10 to verify that the *Enter* push button will now work.

9.6. Design and implement a microcontroller to combine the 3 × 4 keypad controller with the three-digit 7-segment LED display controller so that when a numeric key is pressed on the keypad the corresponding digit is displayed on the 7-segment LED display.

9.7. Design and implement a microcontroller to combine the 3 × 4 keypad controller with the three-digit 7-segment LED display controller so that any decimal number having at most three digits can be entered and displayed on the 7-segment LED display. Use the # key on the keypad as the enter key.

9.8. Design and implement a combination lock where a user has to enter the correct 3-digit code to open a lock. Use the 3 × 4 keypad for input. The correct code is 427. Use an LED to display the status of the lock. Turn on the LED when the lock is opened, and turn off when the lock is closed.

9.9. Design and implement a door entry alarm shut-off controller. An on-off switch is mounted on a door. When the door is opened, the switch sends out a 1 signal. The person now has 30 seconds to enter the correct three-digit code on a 3 × 4 keypad. If the correct code is entered within 30 seconds, the alarm is disabled, otherwise, the alarm will sound. Use an LED to show the status of the alarm, and another LED to show whether the alarm is enabled or disabled.

9.10. Implement the VHDL code for the PS2 mouse controller to see that it works.

9.11. Manually design the controller circuit for the PS2 mouse. Implement your circuit on an FPGA development board to make sure that it works correctly.

9.12. Drive the 16×2 LCD FSM controller with a 5 MHz clock speed and see what happens to the display. What is the fastest clock speed for the LCD controller in which the LCD will still display correctly?

9.13. Modify the behavioral HDL code for the 16×2 LCD FSM controller to use the 4-bit data format instead of the 8-bit data format.

9.14. Implement a microcontroller to input characters from a PS2 keyboard and display the characters on the LCD screen.

9.15. Implement the VGA monitor controller to see that it works.

9.16. Modify the VGA controller to drive a 1024 columns $\times$ 768 rows resolution VGA monitor.

9.17. Design and implement a microcontroller to input characters from a PS2 keyboard and display the characters on a VGA monitor.

9.18. Implement the A/D controller to see that it works.

9.19. Implement the I^2C controller for the real-time clock to see that it works.

APPENDIX A

Xilinx Development Tutorial

The Xilinx ISE Design Suite and a Xilinx FPGA (field-programmable gate array) development board provide all of the necessary tools for implementing and trying out all of the circuits presented in this book, including building the final general-purpose microprocessor. The ISE Design Suite software offers a completely integrated development tool and graphical-user interface for the design and synthesis of digital logic circuits. Together with the FPGA development board, these circuits actually can be implemented in hardware. The main component on the development board is an FPGA chip, which is capable of implementing very complex digital logic circuits. After synthesizing a circuit and downloading it onto the FPGA, you can see the operation of the circuit in hardware.

The WebPACK Edition of the ISE Design Suite software can be downloaded for free from the Xilinx website at http://www.xilinx.com/products/design-tools/ise-design-suite/ise-webpack.htm. You also will need to register and obtain a free license for the program from Xilinx's licensing support website.

This lab assumes that you are familiar with the Windows environment, and that the ISE Design Suite software already has been installed on your computer. The rest of this lab will provide a step-by-step instruction for the schematic and HDL entry of a 2-input AND gate circuit.

A.1 Starting ISE

After the successful installation of the ISE software, there should be a link to the program named **Project Navigator** under the Windows' **Start** button in the **Xilinx Design Tools > ISE Design Tools**' folder. Click on this link to start the program. You should see the main **ISE Project Navigator** window similar to Figure A.1.

A.1.1 Creating a New Project

Each circuit design in ISE is called a project. Each project should be placed in its own folder, because the program creates many associated working files for each project. Perform the following steps to create a new project and a new folder for storing the project files.

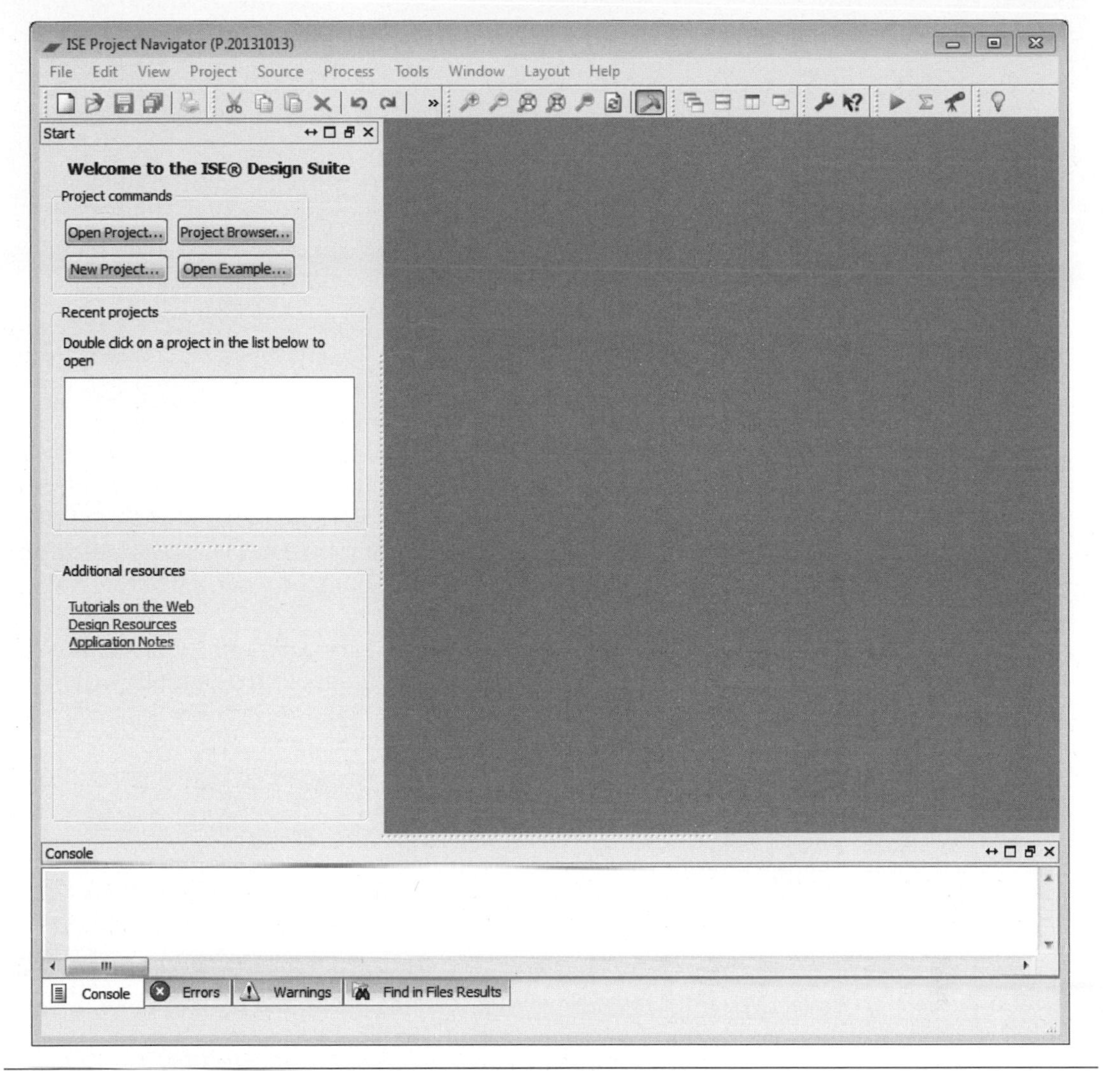

FIGURE A.1 Xilinx ISE Project Navigator main window.

Source: Xilinx

From the ISE menu, select **File > New Project,** or just click on the **New Project** button. You should see the **New Project Wizard: Create New Project** window as shown in Figure A.2.

Type in the name of your project.

- For this lab, type in the project name, `myAndGate`.

Type in the directory location for storing your project. You also can click on the [...] icon next to it to browse to the directory. If the directory does not already exist, then ISE will create it.

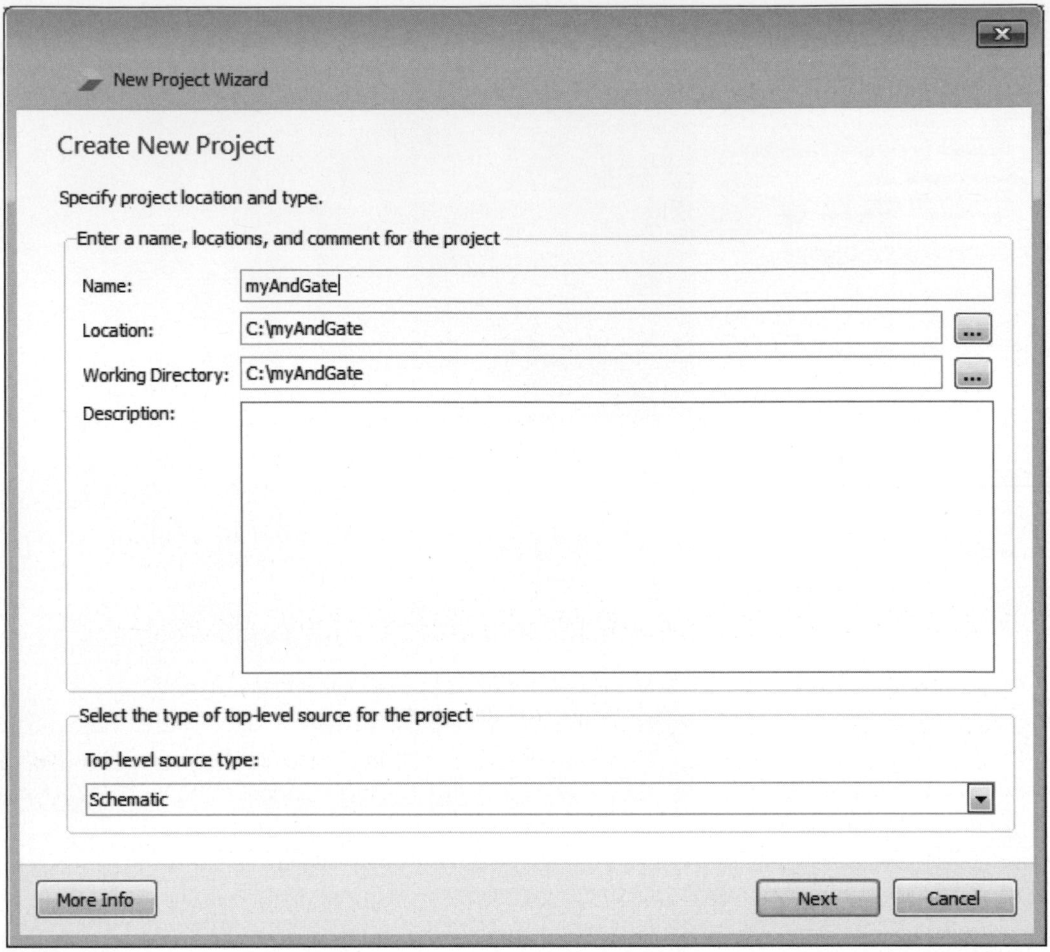

FIGURE A.2 The New Project Wizard: Create New Project window with the project name, location, working directory, and the top-level source type filled in.

Source: Xilinx

- For this lab, type in `c:\myAndGate` to create a folder named `myAndGate` in the root directory of the C drive.

Typically, you want the project location to be the same as the working directory. Your project can contain many source files, and they can be written in different types—Schematic or HDL. You need to specify the source type for your project's top-level file.

- For this lab, select Schematic as the top-level source type.

A.1.2 Specifying the FPGA

Click **Next** to continue to the next window. In the next **New Project Wizard: Project Settings** window, you need to select the correct target FPGA device on which you will implement your circuit. If your development board is listed in the **Evaluation Development Board** drop-down list, then simply select it, and the remaining necessary information about your FPGA will be filled in automatically and disabled. In Figure A.3, the **Spartan-3E Starter Board** is selected, and the **Family**, **Device**, **Package**, and **Speed** for the FPGA on that board are filled in automatically.

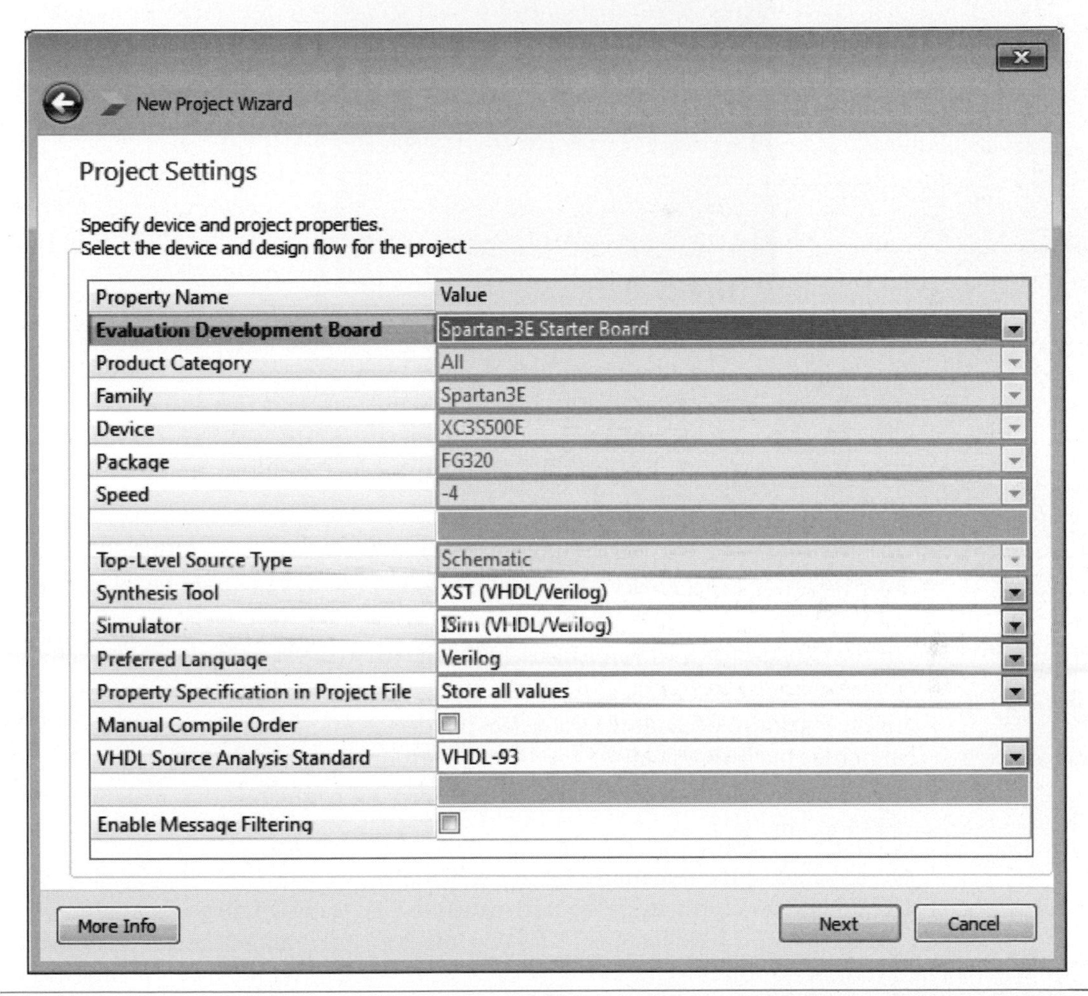

FIGURE A.3 The New Project Wizard: Project Settings window with the Spartan-3E Starter Board selected.

Source: Xilinx

If your development board is not listed in the **Evaluation Development Board** drop-down list, then you will first need to find the device, package and speed of the FPGA chip that is used on your development board. The FPGA chip will look similar to the following.

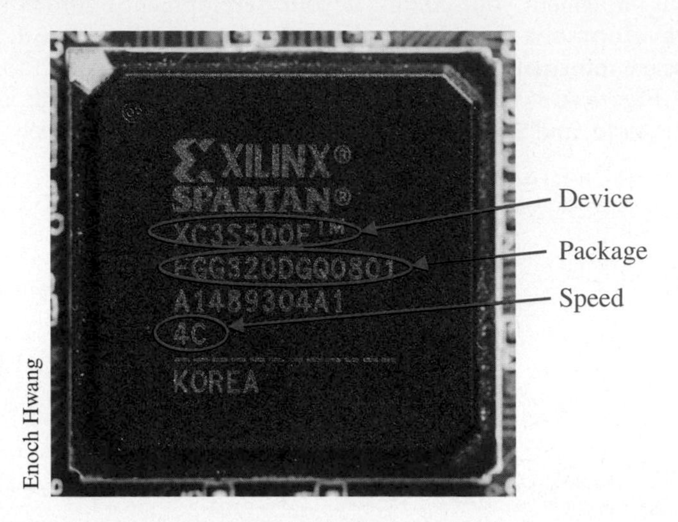

Enoch Hwang

In the picture, the family for this FPGA chip is Spartan3E, the device for this FPGA chip is XC3S500E, the package is FG320, and the speed is −4. This is the FPGA chip used on the Spartan-3E development board. After you have obtained the information for your FPGA chip, you can fill in the information as shown in Figure A.4.

- In the **Evaluation Development Board** drop-down list, select **None Specified**.
- In the **Family** drop-down list, select **Spartan3E** or the one that matches your FPGA chip.
- In the **Devices** drop-down list, select the device **XC3S500E** or the one that matches your FPGA chip.
- In the **Package** drop-down list, select the device **FG320** or the one that matches your FPGA chip.
- In the **Speed** drop-down list, select the speed **−4** or the one that matches your FPGA chip.

Do not change the remaining default selections.

Click **Next** to continue to the next window. The final window is a summary of the choices that you have just made. Review the information and then click **Finish** to create your new project.

You now will see the main **ISE Project Navigator** window as shown in Figure A.5. The window contains three sections. The left section shows the various

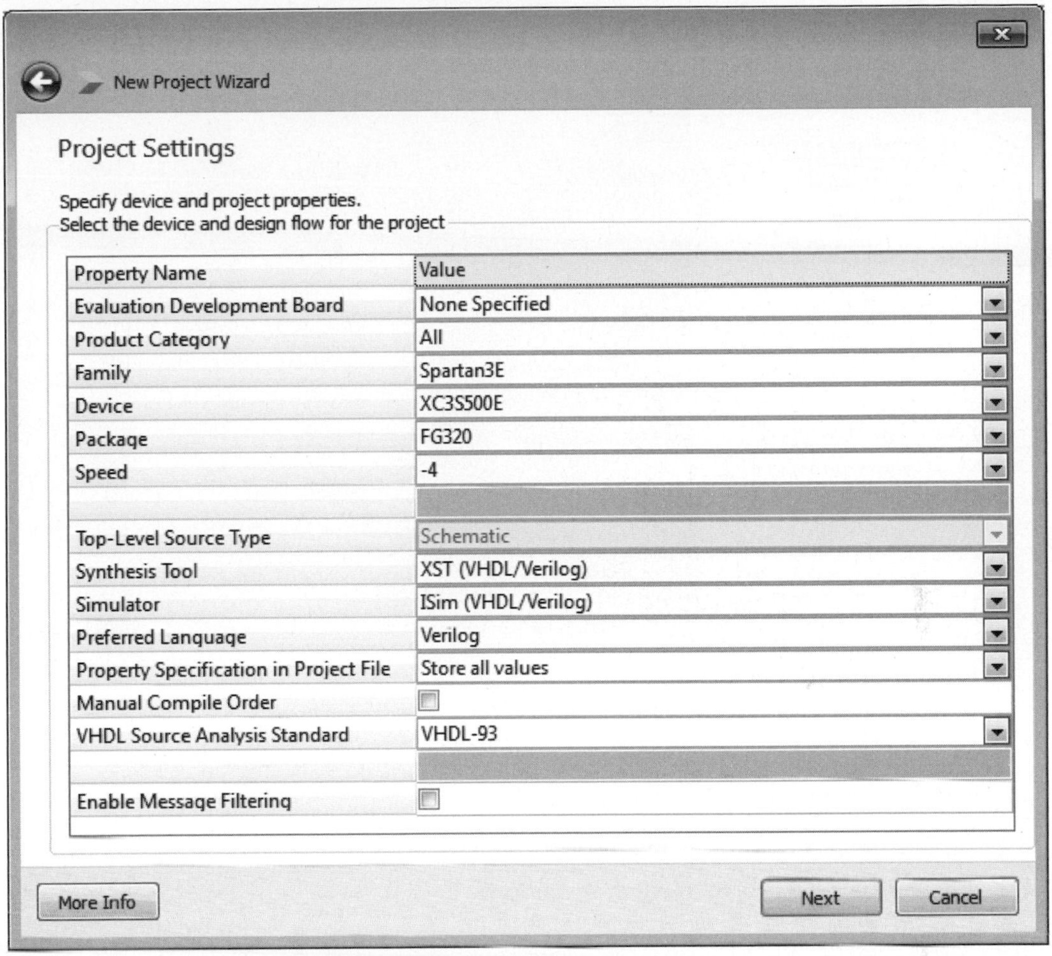

New Project Wizard

Project Settings

Specify device and project properties.
Select the device and design flow for the project

Property Name	Value
Evaluation Development Board	None Specified
Product Category	All
Family	Spartan3E
Device	XC3S500E
Package	FG320
Speed	-4
Top-Level Source Type	Schematic
Synthesis Tool	XST (VHDL/Verilog)
Simulator	ISim (VHDL/Verilog)
Preferred Language	Verilog
Property Specification in Project File	Store all values
Manual Compile Order	☐
VHDL Source Analysis Standard	VHDL-93
Enable Message Filtering	☐

More Info Next Cancel

FIGURE A.4 The New Project Wizard: Project Settings window with the Spartan3E family, XC3S500E device, FG320 package and −4 speed selected.

Source: Xilinx

operations that you can perform on your project and the files in your project. Notice the many tabs at the bottom of this section. Various operations can be performed, depending on which tab you select. The main tab is the **Design** tab, which allows you to see your overall project files. At the top of this section, you can select between one of two views: **Implementation** or **Simulation**. Selecting the **Implementation** view allows you to perform the necessary operations needed to implement your project onto an FPGA.

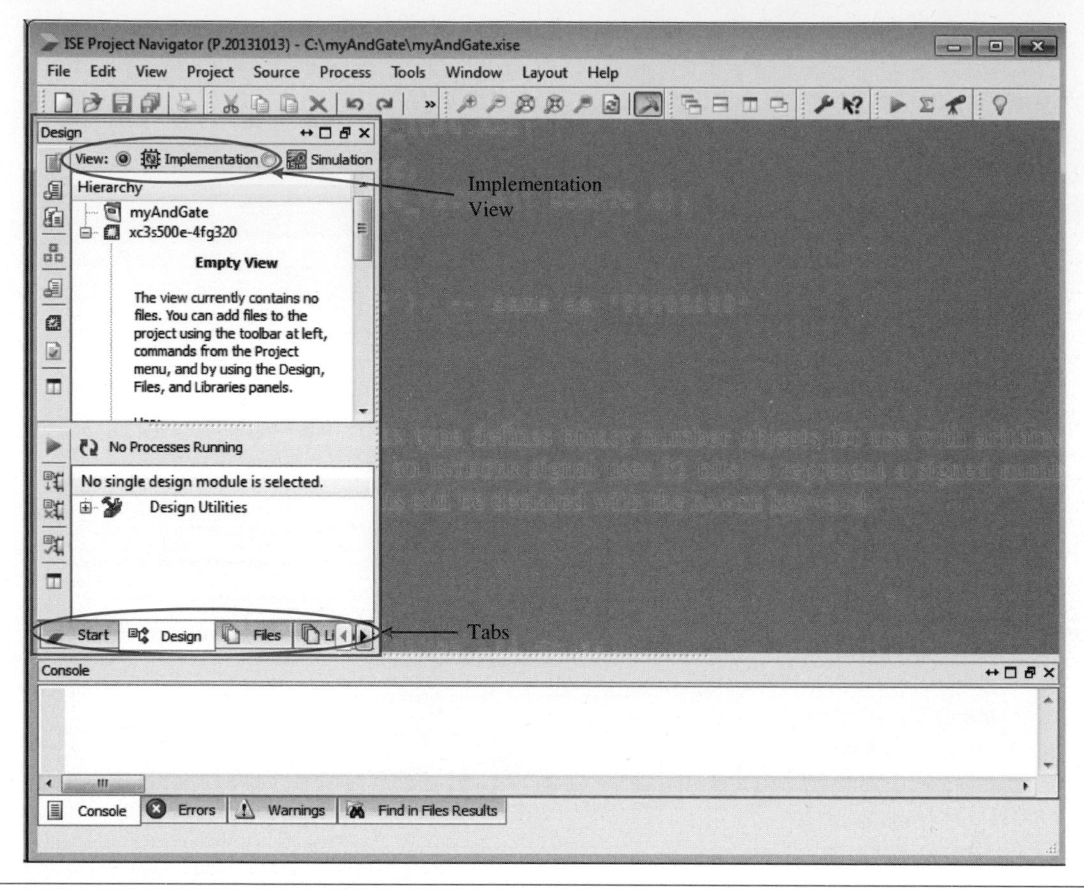

FIGURE A.5 The main ISE Project Navigator window.

Source: Xilinx

A.2 Creating a New Schematic Source File

After creating a new project, we are ready to add a new schematic source file to our project. The schematic editor allows you to manually draw a schematic circuit.

The steps to create a new schematic source file are as follows and also shown in Figure A.6:

1. From the **ISE Project Navigator** menu, select **Project** > **New Source**, or click the **New Source** toolbar button .
2. In the **New Source Wizard** window as shown in Figure A.6, select **Schematic** as the source type.
3. Type in the file name `myAndGate` for this file.
4. Make sure that the **Location** is your project's location.
5. Make sure that the **Add to project** option is checked.
6. Click **Next**.

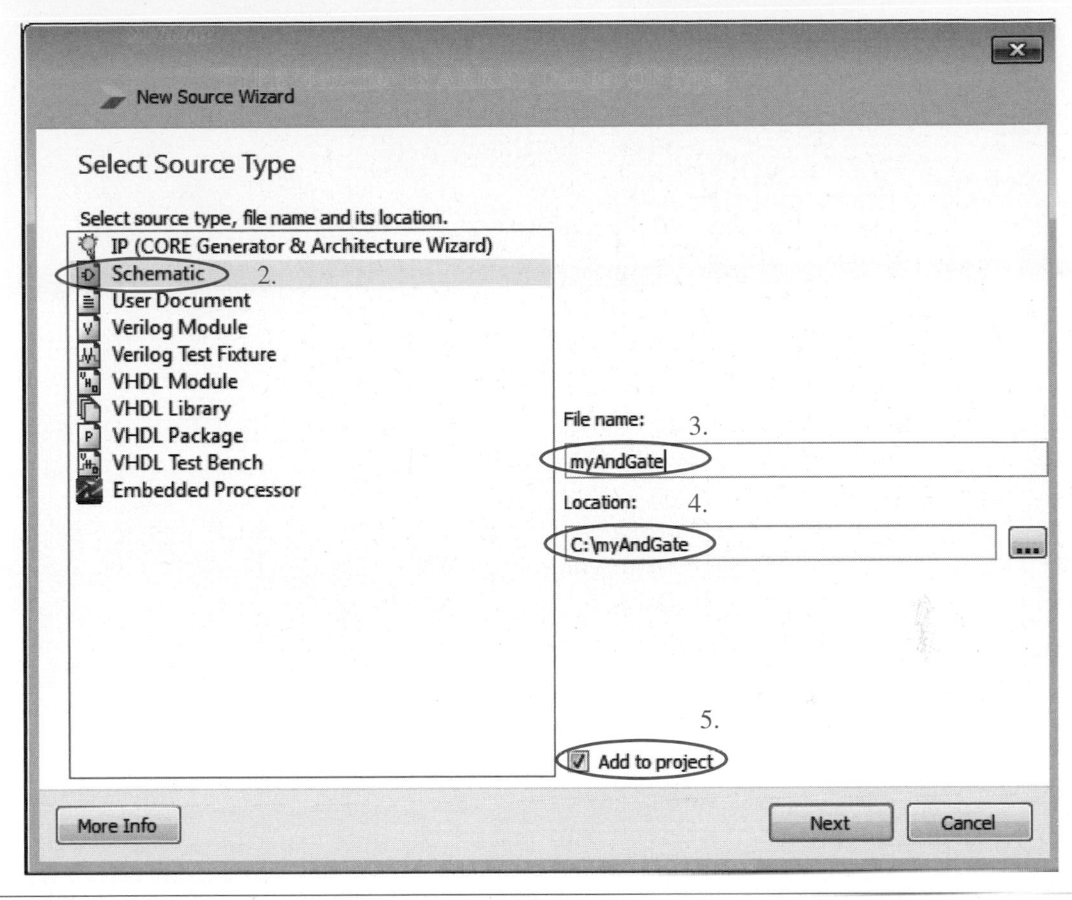

FIGURE A.6 Creating a new schematic source file.

Source: Xilinx

Verify the information in the **Summary** window and then click **Finish**.

A.2.1 Drawing Your Schematic Circuit

You should see the **Schematic Editor** window similar to the one shown in Figure A.7. Any schematic circuit diagram can be drawn in this Schematic Editor window. In the figure, the **Symbols** tab has been selected and a list of all of the available symbols from the library is listed.

Drawing Tools

In Figure A.7, the tools for drawing schematic circuits in the **Schematic Editor** are shown in the toolbar in the middle. The **Selection** tool allows you to select and move objects, such as logic symbols and connection wires. The **Add Symbol** tool allows you to add logic symbols from the library or from your own design files into your schematic drawing. The **Add I/O Marker** tool allows you to add input and output pins to your schematic drawing. The **Add Wire** tool allows you to make connections between logic symbols.

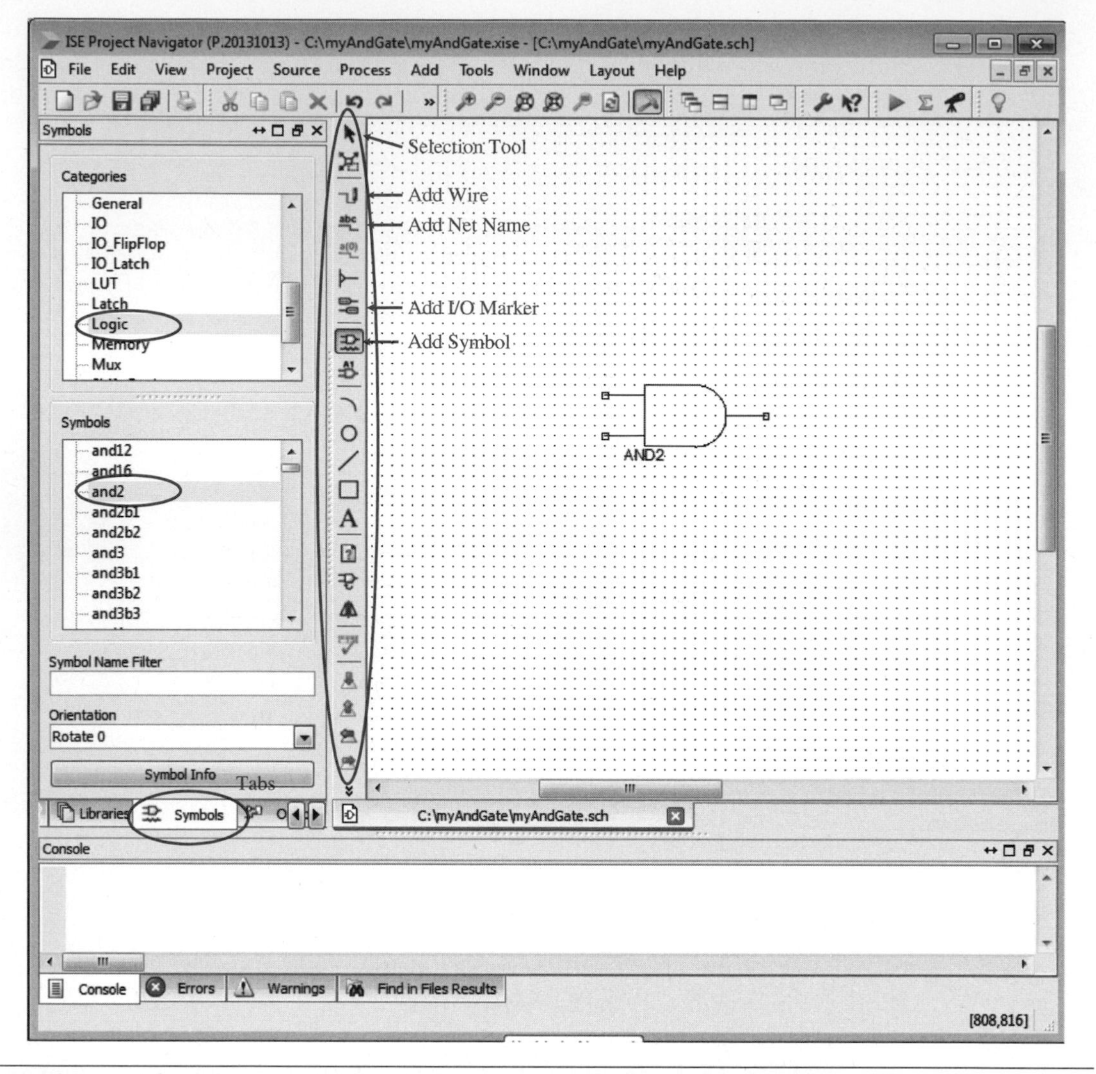

FIGURE A.7 The Schematic Editor window with the Symbols tab selected and the schematic toolbar being shown.

Source: Xilinx

Inserting Logic Symbols

To insert a logic symbol, either click on the **Add Symbol** button in the toolbar or select the **Symbols** tab. All of the available symbols will be listed and grouped by categories. Most of the basic logic symbols that we will use are listed under the **Logic** category. You also can type in the first few letters of the symbol name in the **Symbol Name Filter** box to narrow down the search.

After you have found and selected the symbol that you want, move your mouse to the schematic drawing. Your mouse will change to a crosshair with an outline of the symbol. Click on a spot in the schematic drawing to place the symbol in the schematic. You can add multiple instances of the same symbol by clicking continually on the schematic drawing. Press the **Esc** key when you have finished adding that symbol.

For this lab, insert the following symbol to the schematic drawing:

- A 2-input AND gate (**and2**) found in the **logic** category.

Notice the small square connection points at the end of each line attached to the AND gate symbol in Figure A.7. You will make connections to other symbols and wires by clicking on these connection points.

Inserting I/O Markers

All I/O signals in your circuit must have I/O markers attached. Click on the **Add I/O Marker** button in the toolbar. Now click on the three small square connection points from the AND gate. An I/O marker will attach automatically to that connection point as shown next.

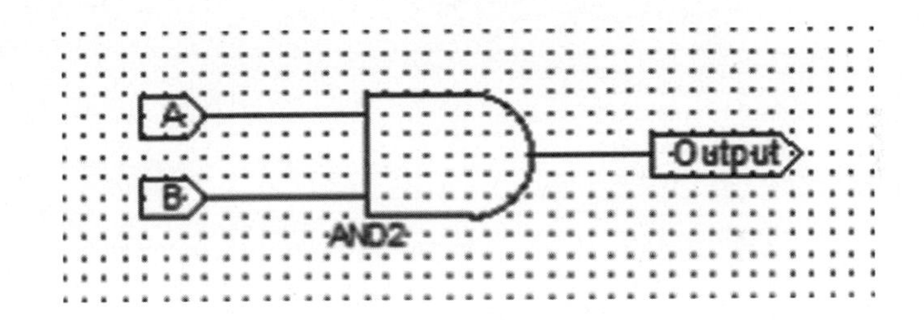

You can rename the I/O markers by right-clicking on the marker and then select **Rename Port**. For this lab, perform the following operations:

- Rename the two input connectors to `A` and `B`.
- Rename the output connector to `Output`.

A.2.2 Creating and Using a Schematic Symbol

Complex circuits usually are designed in a hierarchical fashion. To do this, we first create a schematic symbol for a low-level circuit, and then that symbol is used in the next level circuit above it. Schematic symbols are like black boxes that hide the details of a circuit. Only the input and output signals for the circuit are shown. The input and output signals for the schematic symbol are obtained directly from the input and output signal markers that are connected in the circuit.

The steps to create a schematic symbol for a circuit are as follows and also shown in Figure A.8:

1. In the **View** pane of the **Design** panel, select **Implementation**.
2. In the **Hierarchy** pane, select the source file that contains the design module for which you want to create a logic symbol.
3. In the **Processes** pane, expand **Design Utilities**, and double-click **Create Schematic Symbol**. If a symbol already has been created for this circuit, then

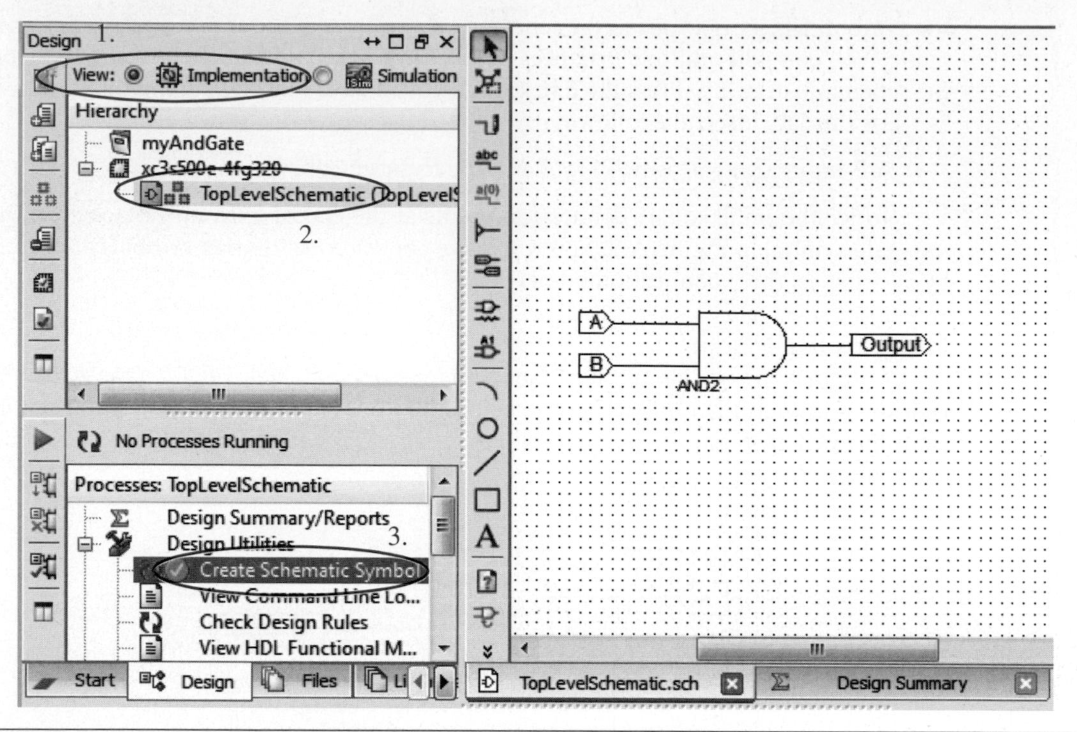

FIGURE A.8 **Steps to create a schematic symbol from a source file.**

Source: Xilinx

you will need to right-click on **Create Schematic Symbol** and select **ReRun** from the pop-up menu. If you get an error message saying that you cannot overwrite the existing symbol file, then you will first need to turn on this property by selecting from the ISE main menu **Process** > **Process Properties**, and check the **Overwrite Existing Symbol** property.

A green check mark will be shown next to the **Create Schematic Symbol** item after the symbol has been created successfully.

To see the detail circuit for your symbol, right-click on the symbol and select **Symbol** > **Push into Symbol** from the pop-up menu.

The steps to use a schematic symbol that you have created for a circuit in the current project are exactly the same as using a symbol from the library. The only difference is that you select your symbol instead. Symbols that you have created will be listed in a separate category in the Symbols pane as shown in Figure A.9.

For this lab, create another new schematic source file and give it the name `Top`. We will use this as our higher-level schematic source file. Insert the **myAndGate** schematic symbol to this schematic drawing. Add I/O markers to the three input and output signals. Rename the three I/O markers to `InputA`, `InputB`, and `Output` as shown in Figure A.9.

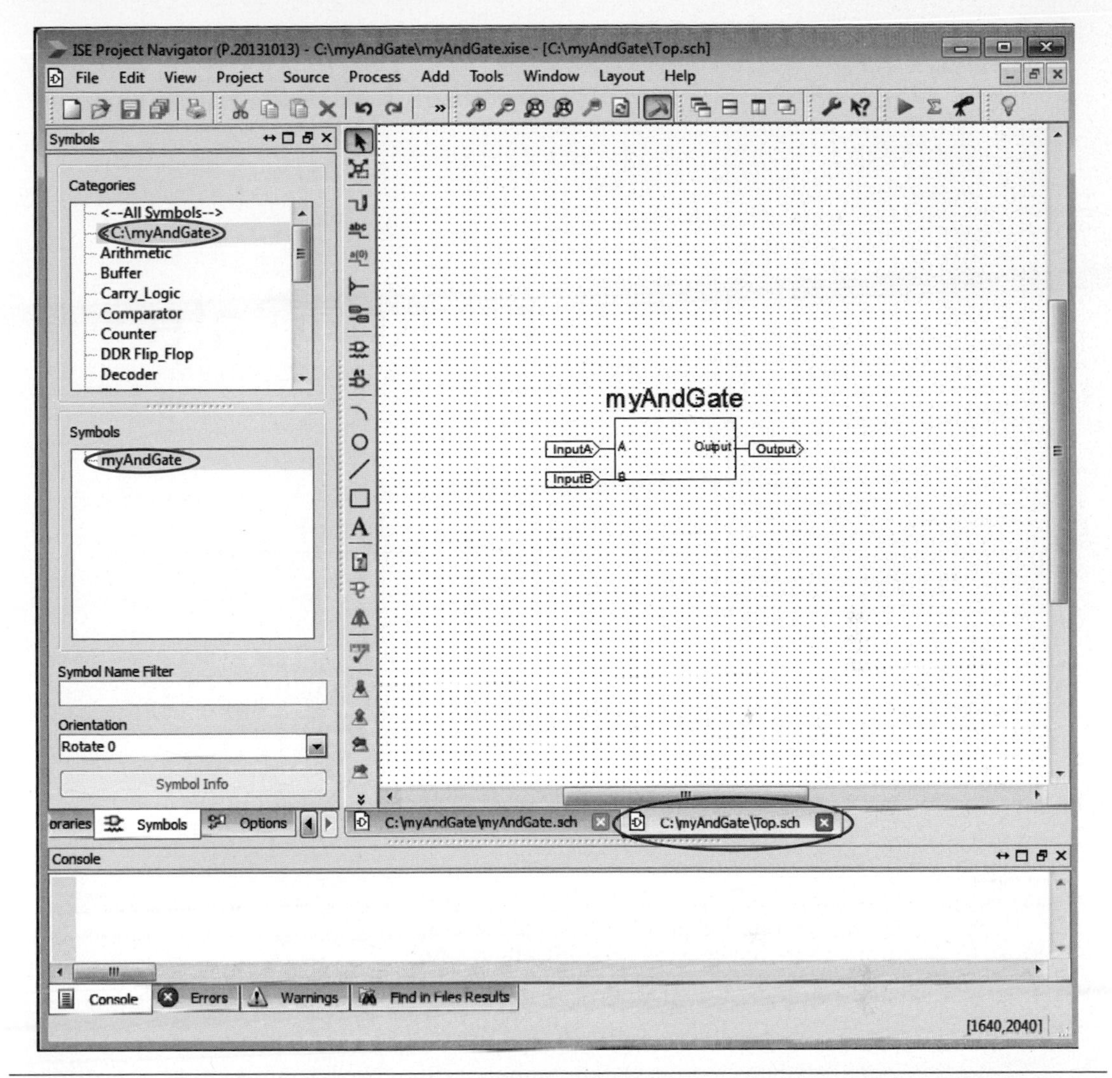

FIGURE A.9 Selecting and placing the myAndGate schematic symbol into another schematic drawing.

Source: Xilinx

A.2.3 Editing a Schematic Symbol

You can edit the schematic symbol by right-clicking on the symbol and then select **Symbol** > **Edit Symbol** from the pop-up menu. The symbol editor appears as shown in Figure A.10.

In the **Symbol Editor**, click the **Add Pin** toolbar button to add a new connection pin. In the **Add Pin Options** that appear in the **Options** panel, set the pin name and direction polarity. For bus names, use the format such as `data(0:7)`. Click in the symbol window to place the pin.

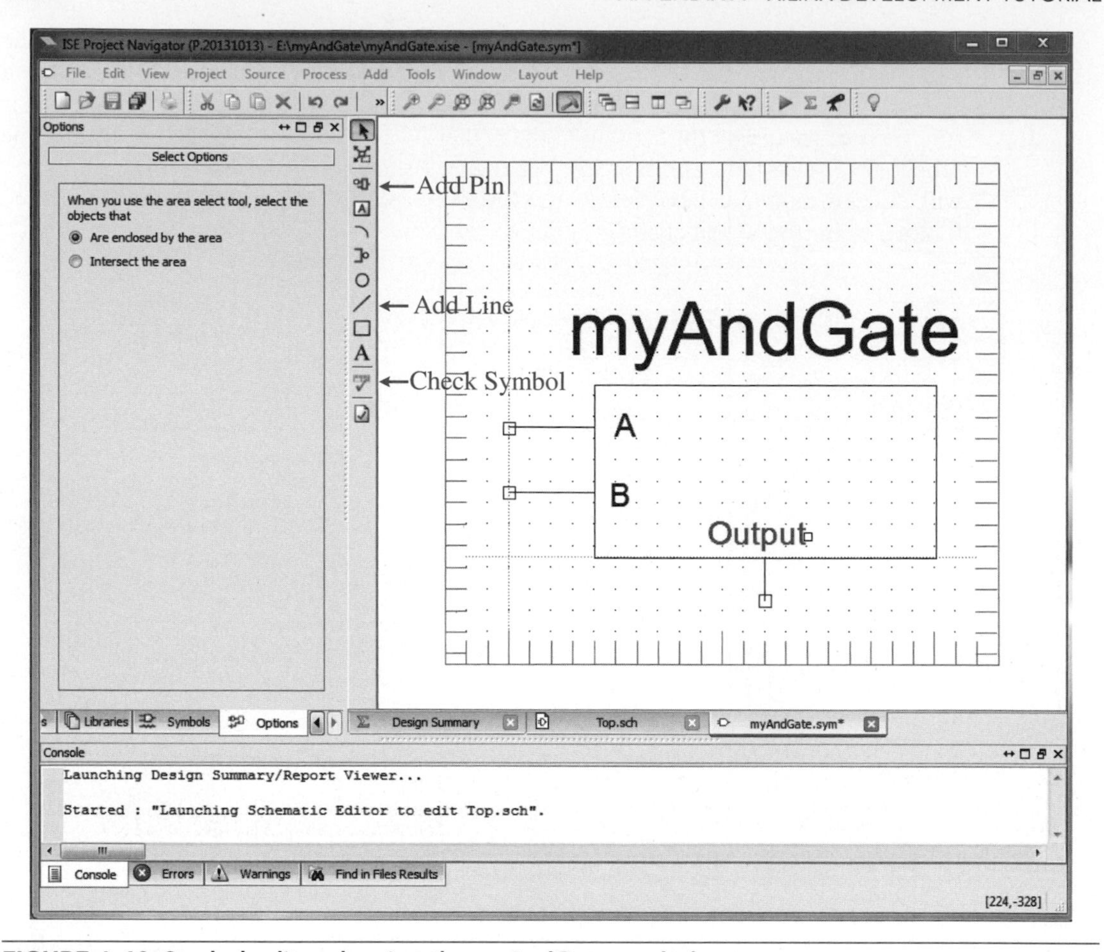

FIGURE A.10 Symbol editor showing the myAndGate symbol.

Source: Xilinx

Click on the **Add Line** toolbar button to draw a connection line from the center of the square connection pin to the symbol outline. The line must be either horizontal or vertical. Click on the **Check Symbol** toolbar button to make sure that there are no errors. You will get an error message if either the connection line is neither horizontal nor vertical, or it does not terminate in the center of the square pin as shown next.

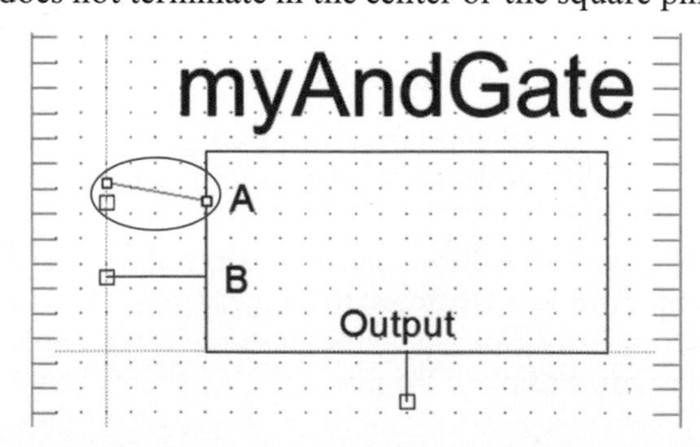

After you have made some changes to the symbol, you will need to update the symbol in the higher-level schematic drawing that uses that symbol. Otherwise, when you synthesize the project you likely will get an error message saying that the symbol is out of date. To update the symbol in the higher-level schematic drawing, first close the schematic drawing file, and then re-open it. The **Open Schematic File Errors** window will come up as shown next. Select the symbol that is out of date and click the **Update Instances** button. Then click **OK** to close the window.

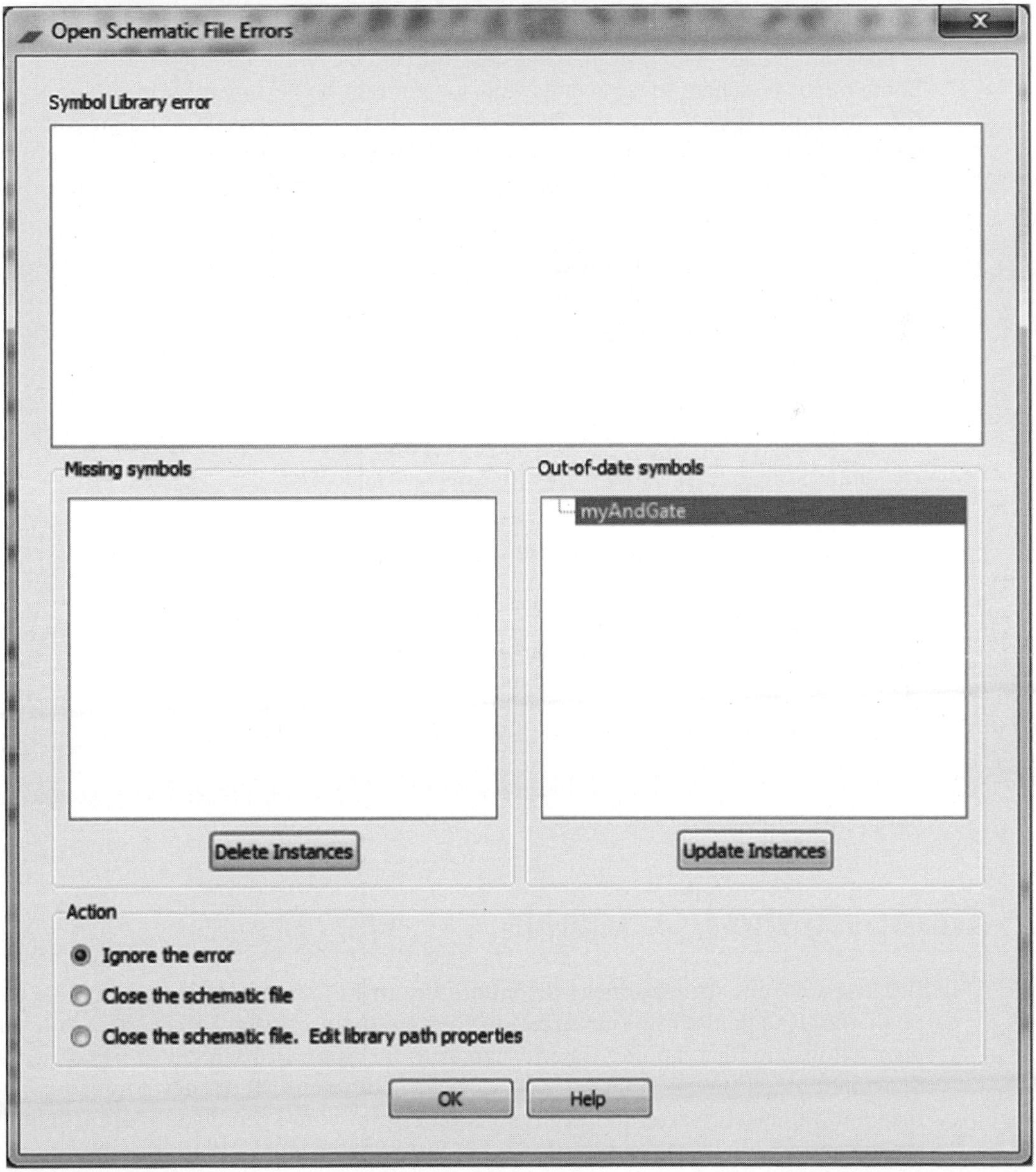

Source: Xilinx

A.2.4 Using a Schematic Symbol in Another Project

To use a circuit and its associated schematic symbol in another project, select **Project > Add Source**, or click the **Add File** toolbar button. In the **Add Source** dialog box, browse to the source file and select it. Click **Open**. In the **Adding Source Files** dialog box, select the **Implementation** Design View.

A.3 Creating a New Verilog or VHDL Source File

A project typically will have many design source files, which can be of different types. Some might be schematic drawings, and some might be Verilog and/or VHDL source files. Adding a new Verilog or VHDL source file to your project is similar to adding a new schematic source file, except in the **New Source Wizard** window, you select **Verilog Module** or **VHDL Module** instead of **Schematic** as the source type.

The steps are as follows:

1. From the **ISE Project Navigator** menu, select **Project > New Source**, or click the **New Source** toolbar button.
2. In the **New Source Wizard** window, select either **Verilog Module** or **VHDL Module** as the source type.
3. Type in the file name.
4. Make sure that the **Location** is your project's location.
5. Make sure that the **Add to project** option is checked.
6. Click **Next**.
7. Verify the information in the **Summary** window and then click **Finish**.

A.4 Setting the Top-Level Module Design File

To make a particular file your top module file for the project, right-click on the file name listed in the **Design** section. In the pop-up list, select **Set as Top Module**. Alternatively, you can select the file and click on the **Set Module as Top** icon. This option is disabled if the file is already set to be the top module file.

A.5 Mapping the I/O Signals

Because we want to implement the circuit on an FPGA, we need to assign or map all of the I/O signals from our circuit to the actual pins on the FPGA. The following instructions are for mapping the I/O signals to the Spartan XC3S500E FPGA on the Spartan3E development board. If you are using a different FPGA development board, then you will need to refer to the documentations for your specific development board for the correct pin assignments.

A.5.1 Using PlanAhead for Mapping the Pins

We will use the I/O Pin Planning (PlanAhead) tool to map each of the I/O signals from our circuit to the pins on the Spartan 3E chip. The steps to start the PlanAhead tool are as follows:

1. From the **Project Navigator** menu, select **Tools** > **PlanAhead** > **I/O Pin Planning (PlanAhead) – Pre-Synthesis** to start the PlanAhead tool. Alternatively, in the **Processes** pane of the **Design** panel, expand **User Constraints**, and double-click on **I/O Pin Planning (PlanAhead) – Pre-Synthesis.**
2. The first time that you do this, the message window shown in Figure A.11 will pop up saying that you need an Implementation Constraint File (UCF) added to the project. Click **Yes** to create this file.

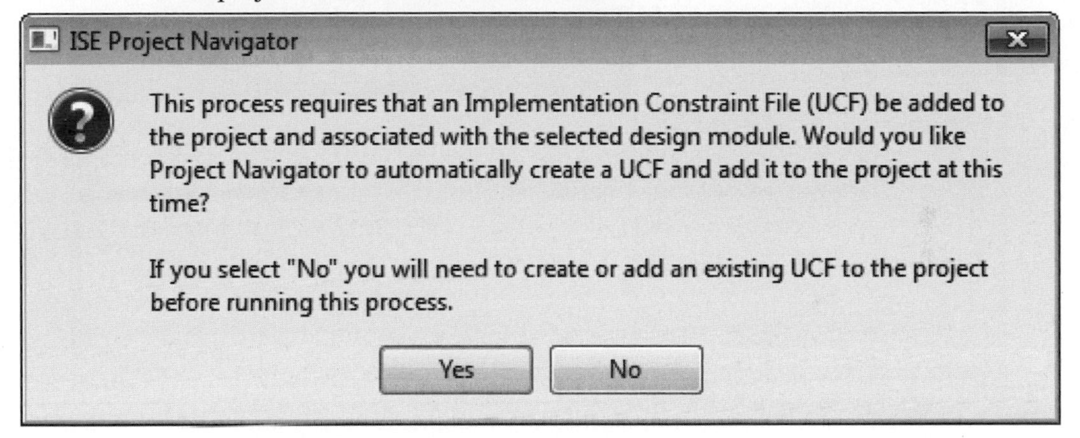

FIGURE A.11 Message window saying that an Implementation Constraint File needs to be added to your project.

Source: Xilinx

3. You also might see the message window shown in Figure A.12 saying that another editor is already editing constraints for this project. If you are sure that you do not have another PlanAhead editor running, then you can click **Yes** to continue opening PlanAhead.

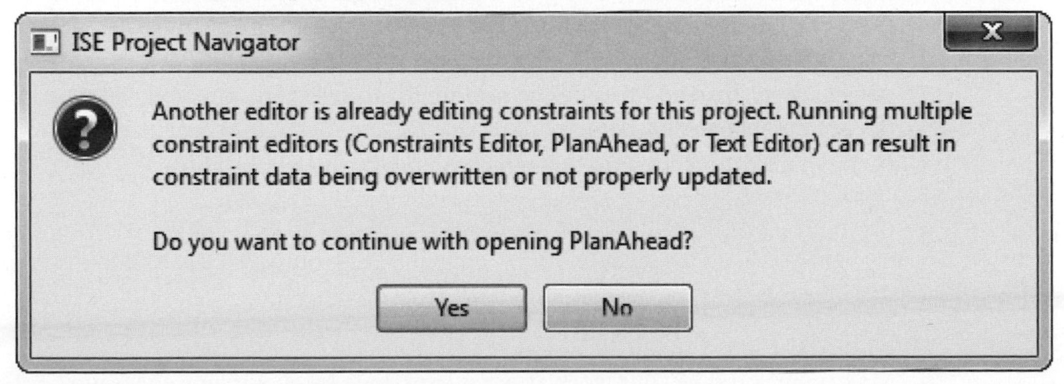

FIGURE A.12 Message window saying that another editor is editing constraints for this project.

Source: Xilinx

4. When the PlanAhead tool opens, you will see the **PlanAhead** window similar to Figure A.13.

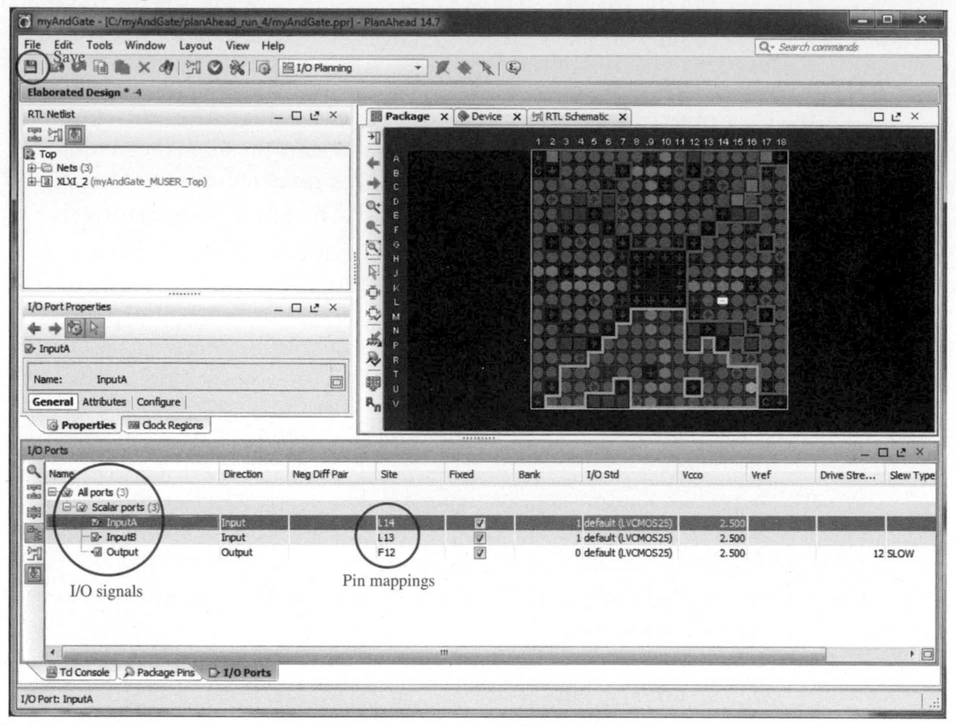

FIGURE A.13 The PlanAhead tool showing the pin assignments for the Spartan 3E development board.

Source: Xilinx

5. All of the available I/O signals from the circuit will be listed in the **I/O Ports** pane. Expand the **Scalar ports**, and you will see the I/O signal names for your circuit.
6. For each I/O signal name, single-click on the cell next to the signal name under the **Site** column to bring up a pop-up list of all the assignable pins from the FPGA. Select the pin number that you want to assign to that I/O signal. Figure A.13 shows that Pin L14 is assigned to the signal InputA.
7. Alternatively, instead of using the pop-up list to select the pin, you can type in the pin number such as L14.
8. Perform the following signal-to-pin mapping for the FPGA chip.

Node Name (I/O Signal)	Location (Pin)
InputA	L14
InputB	L13
Output	F12

9. Make sure you save your work after you are done by clicking on the **Save Constraints** button.

A.5.2 **Faster Alternative Method for Mapping the Pins**

A much faster pin mapping method is to directly edit the pin mapping file. The name of this file is <project name>.ucf located in your project directory. Use any text editor such as Windows Notepad to open and edit this file. There is a NET line for each pin mapping. To change the pin mapping, simply type in a new pin location for that node name, for example, change the Output net from F12 to E11. Do not modify or delete any other lines in the file. For our project's pin mappings, we will have the following three lines in the file.

```
NET "InputA" LOC = L13;
NET "InputB" LOC = L14;
NET "Output" LOC = F12;
```

After you save this file, the new pin mappings will be changed.

A.6 **Synthesis and Implementation**

After drawing your circuit and mapping the pins, the next step is to synthesize it. During this step, ISE collects all of the necessary information about your circuit, and produces a netlist for it. This step is hardware independent. The steps to synthesize a design are as follows:

1. In the **View** pane of the **Design** panel, select **Implementation**.
2. In the **Hierarchy** pane, select the top module.
3. In the **Processes** pane, double-click the **Synthesize** line.

A green check mark will be shown next to the **Synthesize – XST** line after the design has been synthesized successfully as shown in Figure A.14. If there are errors, then you will have to go back and fix them first and then synthesize the design again.

After the successful synthesis of your design, the next step is to implement it. In this step, ISE will map and place the netlist generated from the previous step to fit on your FPGA chip. This and the succeeding steps are hardware dependent. So if you remap the hardware pins, you will have to redo from this step forward. The steps to implement a design are as follows:

4. In the **Processes** pane, double-click the **Implement Design** line.

A green check mark will be shown next to the **Implement Design** line after the design has been implemented successfully as shown in Figure A.14.

Instead of performing the above two steps one at a time, you can click on the **Implement Top Module** button to perform them at once.

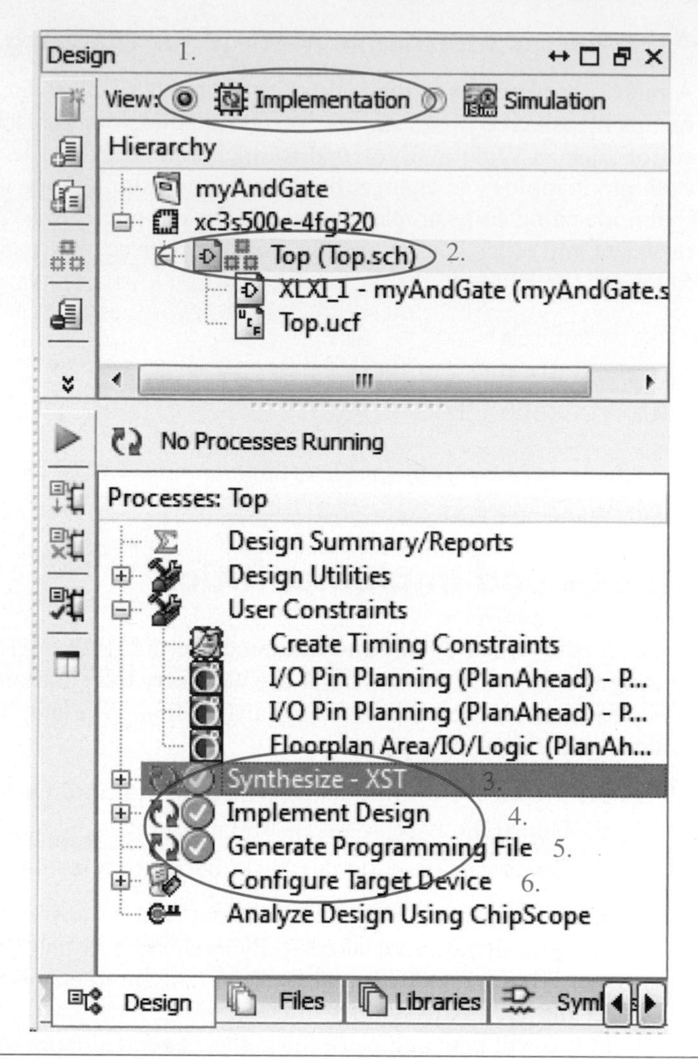

FIGURE A.14 Successful synthesis and implementation of the circuit.

Source: Xilinx

After the successful implementation of your design, the next step is to generate the programming file.

5. In the **Processes** pane, double-click the **Generate Programming File** line.

A green check mark will be shown next to the **Generate Programming File** line after the programming file has been generated successfully as shown in Figure A.14.

A.7 Programming the Circuit to the FPGA

After you have generated the programming file successfully, you are ready to configure the FPGA. The steps are as follows:

1. Plug in the Xilinx Spartan-3E board to the computer using the USB cable, and turn on the power to the board.
2. In the **Processes** pane, double-click the **Configure Target Device** line.

   Click **OK** if you see the following warning message window.

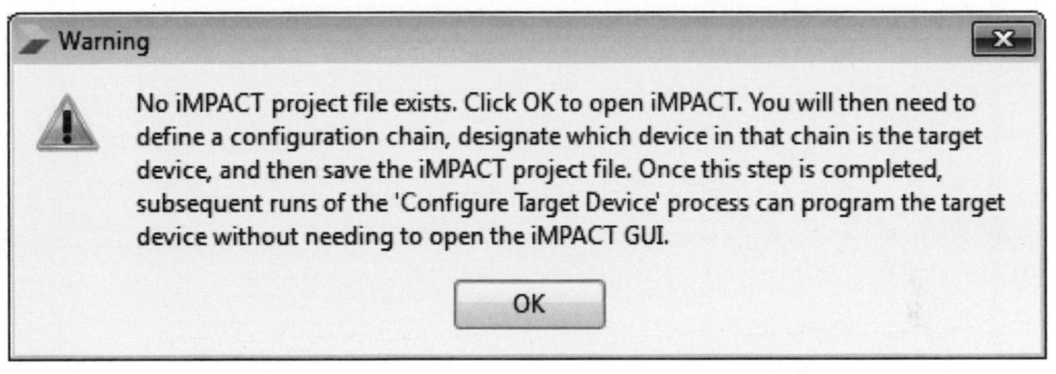

Warning
No iMPACT project file exists. Click OK to open iMPACT. You will then need to define a configuration chain, designate which device in that chain is the target device, and then save the iMPACT project file. Once this step is completed, subsequent runs of the 'Configure Target Device' process can program the target device without needing to open the iMPACT GUI.
OK

Source: Xilinx

   The ISE iMPACT window as shown in Figure A.15 opens up.

3. Under the **iMPACT Flows** pane, double-click on the **Boundary Scan** line.
4. Right-click on the message line **Right click to Add Device or Initialize JTAG chain**.
5. From the pop-up menu, select **Initialize Chain**. If you get a warning message about not finding the cable, then make sure that the board is connected properly using the USB cable and that the power is turned on.
6. The **Assign New Configuration File** window as shown in Figure A.16 is displayed. Select the file to configure the FPGA with. The name of this file is the name of your project and with the extension `bit`. In our case it is `top.bit`. Click **Open** after selecting the file.
7. You will next see the **Attach SPI or BPI PROM** window as shown next. Select **No**.
8. Because there are three configuration devices on the Spartan-3E board, the same **Assign New Configuration File** window as shown in Figure A.16 will come up a second and third time. Click on the **Bypass** button for both the second and third time that this window comes up.
9. Click **OK** in the next **Device Programming Properties** window as shown in Figure A.18.

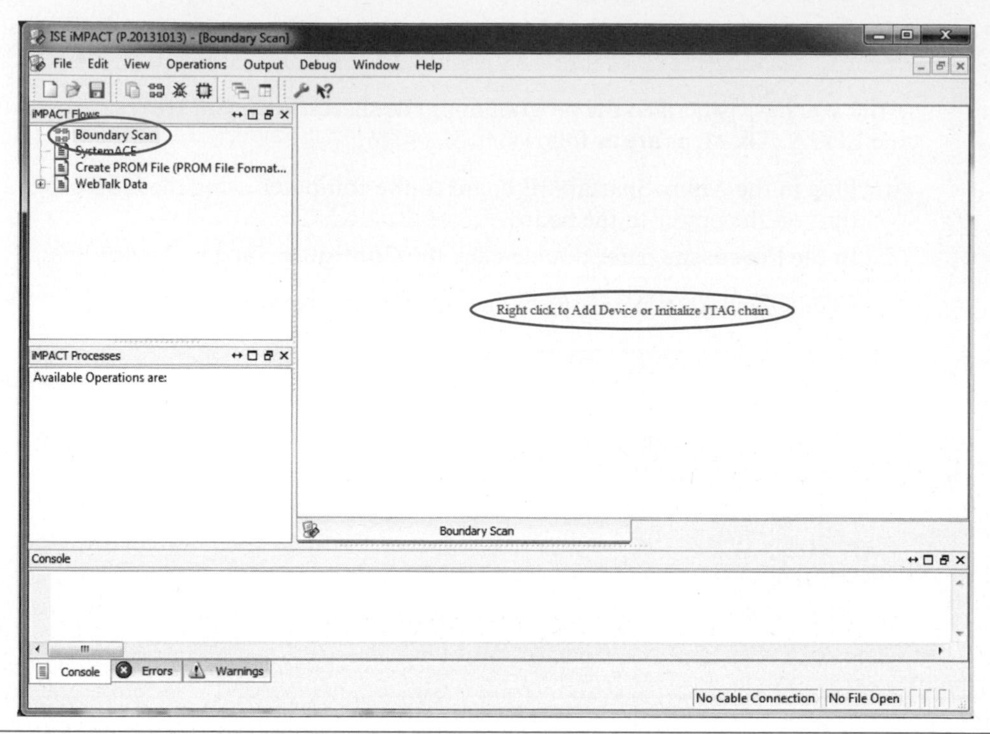

FIGURE A.15 The ISE iMPACT FPGA programming window.

Source: Xilinx

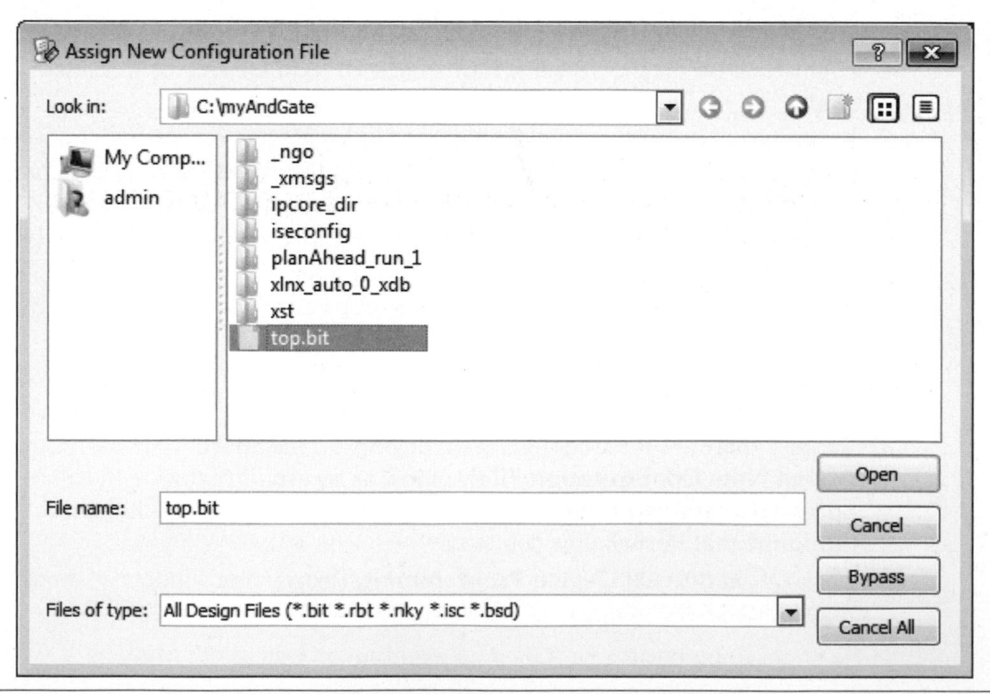

FIGURE A.16 The Assign New Configuration File window with the programming file selected.

Source: Xilinx

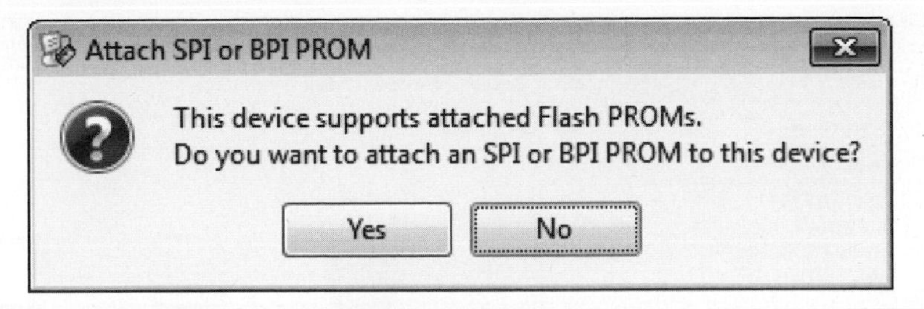

FIGURE A.17 The Attach SPI or BPI PROM window.

Source: Xilinx

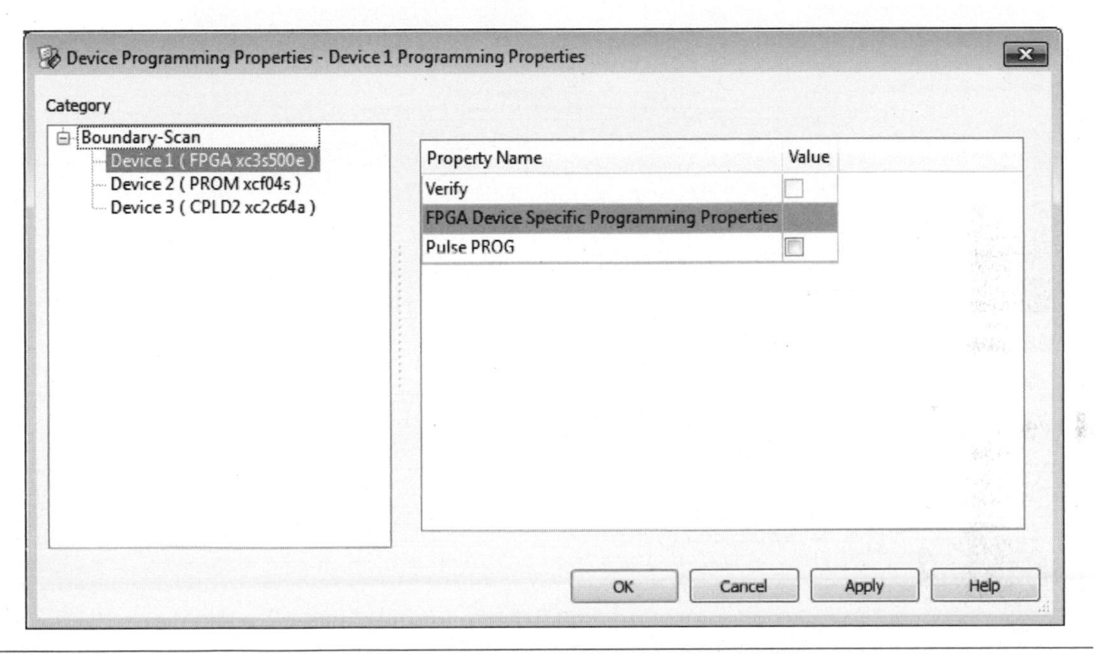

FIGURE A.18 The Device Programming Properties window with the first device selected.

Source: Xilinx

10. You are back at the original **ISE iMPACT** window showing the three FPGA chips with the first one having been assigned the file `top.bit` and the last two bypassed. Make sure that the first chip is selected and is green.
11. In the **iMPACT Processes** pane under the **Available Operations**, double-click on the **Program** line. After a few seconds you will see the **Program Succeeded** message as shown in Figure A.19 if the FPGA chip is programmed successfully.

    Perform Experiment 1 to verify the operation of your circuit running in hardware.

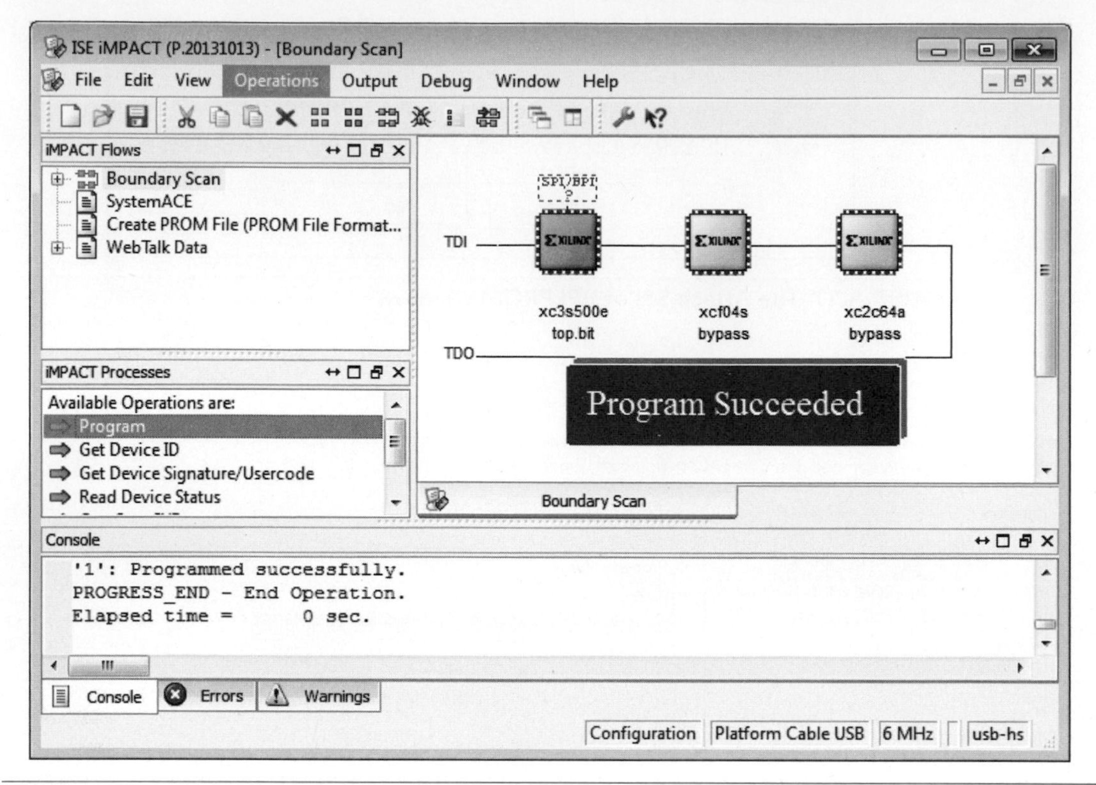

FIGURE A.19 The Programmer window showing that the chip was programmed successfully with your circuit.

Source: Xilinx

A.8 PROBLEMS

A.1. A switch is on when it is in the up position, and off when it is in the down position. Set the two switches (SW0 and SW1) to be either on or off, and observe whether the LED (LD0) output is on or off. Write down your observation in the table below. This table that you have derived is the truth table for the 2-input AND gate.

A	B	Output
0 (off)	0 (off)	
0 (off)	1 (on)	
1 (on)	0 (off)	
1 (on)	1 (on)	

A.2. Replace the 2-input AND gate in your circuit with a 2-input OR gate. Derive the truth table for the 2-input OR gate by filling in the table below.

A	B	Output
0	0	
0	1	
1	0	
1	1	

A.3. Replace your circuit with a NOT (inverter) gate. You'll have to remove one of the input pins. Derive the truth table for the NOT (inverter) gate by filling in the table below.

A	Output
0	
1	

A.4. Replace your circuit with a 3-input AND gate. You will need to have three input pins. Refer to your board's pin mapping to find the pin number for the third switch (SW2). Derive the truth table for the 3-input AND gate by filling in the table below.

A	B	C	Output
0	0	0	
0	0	1	
0	1	0	
0	1	1	
1	0	0	
1	0	1	
1	1	0	
1	1	1	

A.5. Replace your circuit with a 3-input OR gate. Derive the truth table for the 3-input OR gate by filling in the table below.

A	B	C	Output
0	0	0	
0	0	1	
0	1	0	
0	1	1	
1	0	0	
1	0	1	
1	1	0	
1	1	1	

A.6. Draw and implement the following circuit. This circuit is known as the Multiplexor or Mux for short. Derive the truth table for this circuit. Describe the operation of this circuit in as few words as possible. What do you think the letter *s* for one of the inputs stands for?

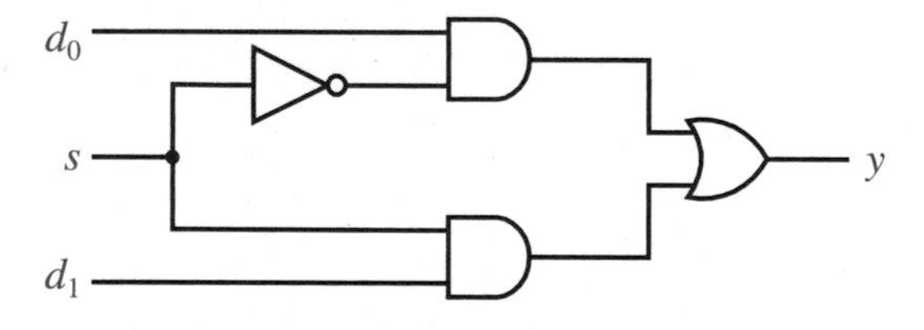

s	d_1	d_0	y (Output)
0	0	0	
0	0	1	
0	1	0	
0	1	1	
1	0	0	
1	0	1	
1	1	0	
1	1	1	

A.7. Repeat Problem A.1 but use two push buttons (BTN West and BTN East) instead of the two switches. What do you notice about the operation of the push buttons?

A.8. Connect a circuit having one switch and one LED. Make the LED turn on when the switch is on, and off otherwise.

A.9. Draw a random circuit having three inputs and one output. Randomly connect several AND gates, OR gates, and NOT gates together between the inputs and output. Connect the three inputs to the switches and the output to a LED. Derive the truth table for it.

APPENDIX B

Altera Development Tutorial

The Altera Quartus II development software and an Altera FPGA (field-programmable gate array) development board, such as the DE1 development board, provide all of the necessary tools to implement and try out all of the circuits, including building the final general-purpose microprocessor. The Quartus II software offers a completely integrated development tool and easy-to-use graphical-user interface for the design, and synthesis of digital logic circuits. Together with the DE1 development board, these circuits actually can be implemented in hardware. The main component on the DE1 development board is an FPGA chip that is capable of implementing very complex digital logic circuits. After synthesizing a circuit and downloading it onto the FPGA, you can see the operation of the circuit in hardware.

The Web Edition version of the Quartus II software can be downloaded for free from the Altera website at www.altera.com. This lab assumes that you are familiar with the Windows environment, and that the Quartus II software already has been installed on your computer. The rest of this lab will provide a step-by-step instruction for the schematic entry of a 2-input AND gate circuit.

B.1 Starting Quartus

After the successful installation of the Quartus II software, a link to the program, named **Quartus II 12.1 Web Edition,** should be under the Windows' **Start** button. Click on this link to start the program. You should see the main **Quartus II** window similar to Figure B.1.

B.1.1 Creating a New Project

Each circuit design in Quartus II is called a project. Each project should be placed in its own folder, because the program creates many associated working files for a project. Perform the following steps to create a new project and a new folder to store the project files.

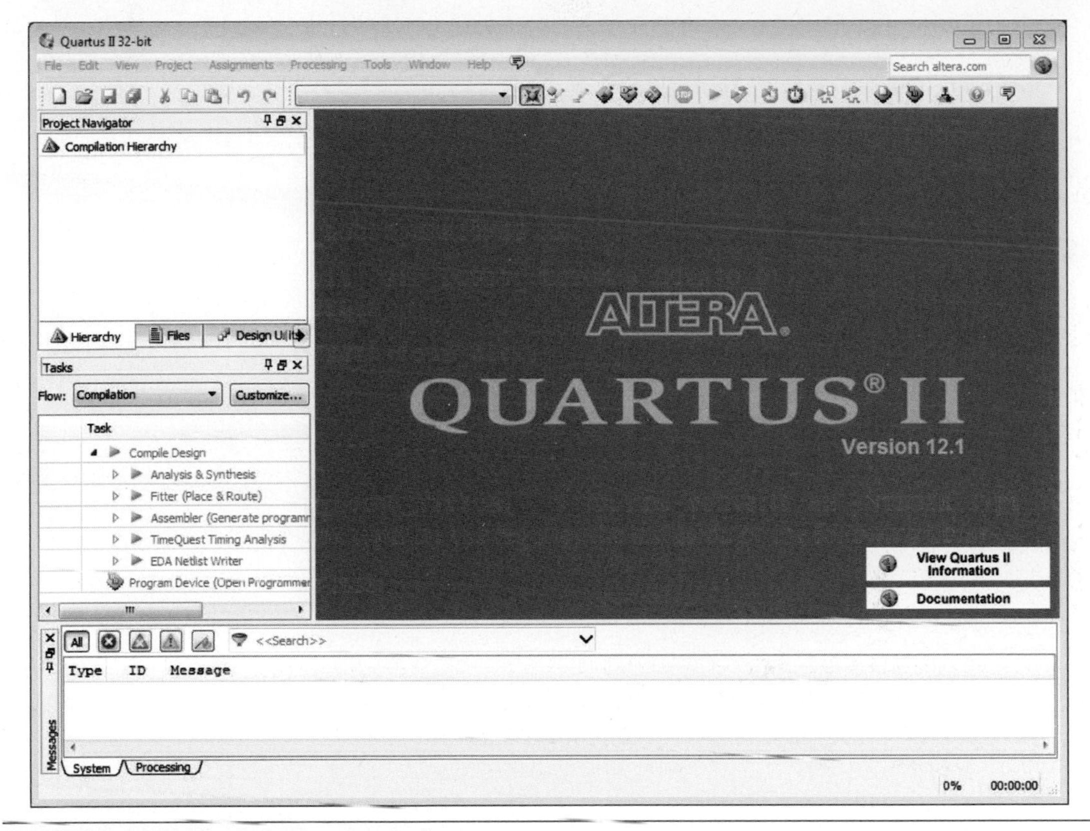

FIGURE B.1 The Quartus II main window.

Source: Altera

From the Quartus II menu, select **File > New Project Wizard**. If the **New Project Wizard Introduction** screen appears and you don't want to see it again the next time you start the new project wizard, you can select the check box that says **Don't show me this introduction again**, and then click **Next** to go to the next screen. You should see the **New Project Wizard: Directory, Name, Top-Level Entity [page 1 of 5]** window as shown in Figure B.2.

Type in the directory for storing your project. You can also click on the [...] icon next to it to browse to the directory.

- For this lab, type in `c:\myAndGate` to create a folder named `myAndGate` in the root directory of the C drive.

You also need to give the project a name.

- For this lab, type in the project name `myAndGate`.

A project might have more than one design file. Whether your project has one or more files, you need to specify which design file is the top-level design entity. The default name given is the same as the project name. However, you can use a different name.

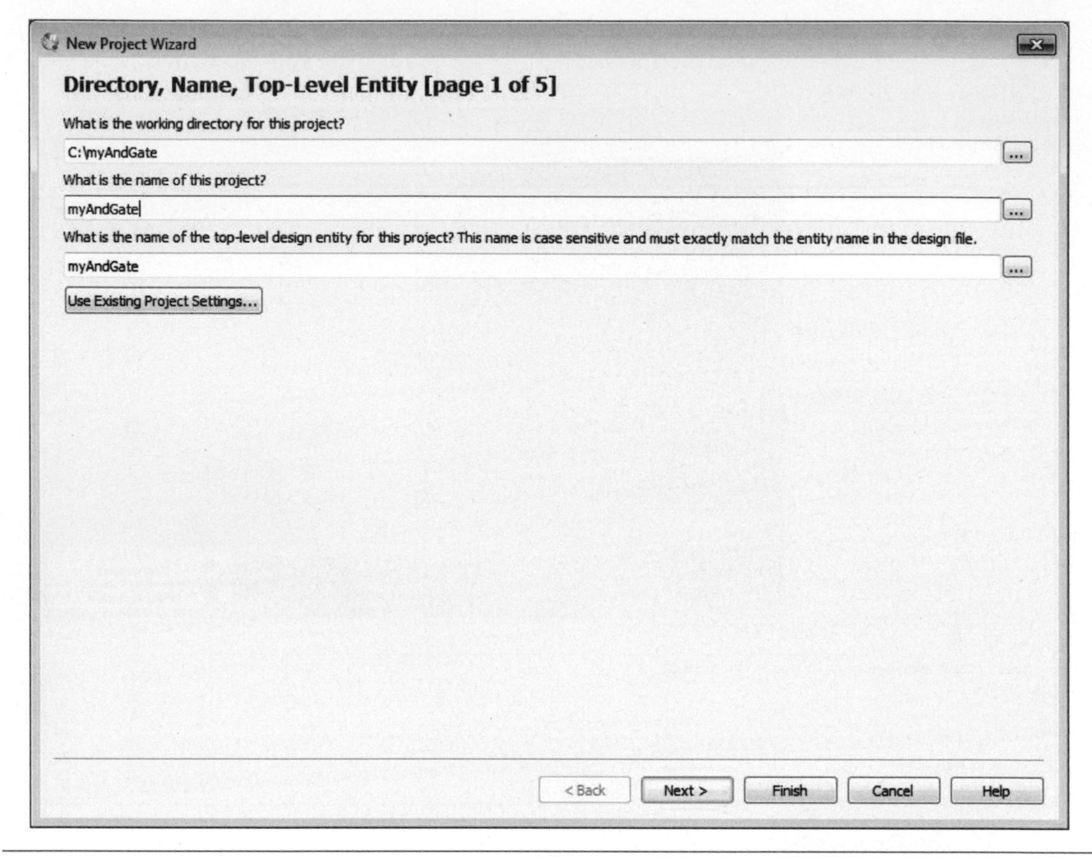

FIGURE B.2 The New Project Wizard: Directory, Name, Top-Level Entity window with the working directory, the project name, and the top-level entity name filled in.

Source: Altera

- For this lab, leave the top-level file name as `myAndGate`, and click **Next** to continue to the next window.

Since the directory `c:\myAndGate` does not yet exist, Quartus II will inform you of that and will ask whether you want to create this new directory. Click **Yes** to create the directory.

In the **New Project Wizard: Add Files [page 2 of 5]** window, you can add existing circuit source files associated with your project. For example, if you have a source file created in another project and want to use it in this project, you can specify that here.

- Click **Next** to continue to the next window because we are starting a new project and do not yet have any source files to add.

B.1.2 **Specifying the FPGA**

In the **New Project Wizard: Family & Device Settings [page 3 of 5]** window as shown in Figure B.3, we select the target FPGA device on which we will implement the circuit. You need to find the device family and name of the FPGA chip that is used on your development board. The FPGA chip will look similar to the following. The device family for the FPGA chip in the picture is Cyclone II and the name is EP2C20F484C7N.

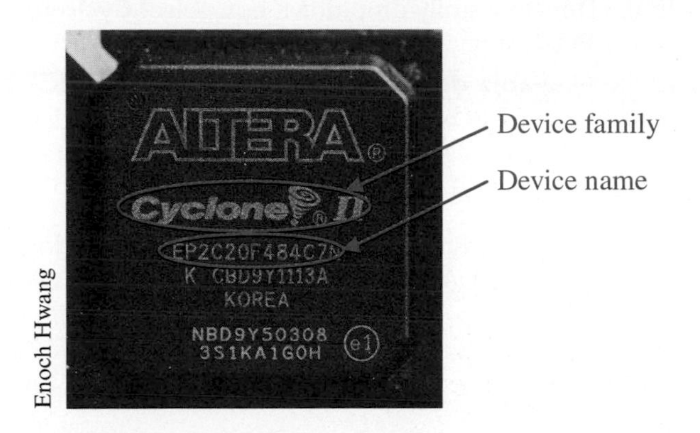

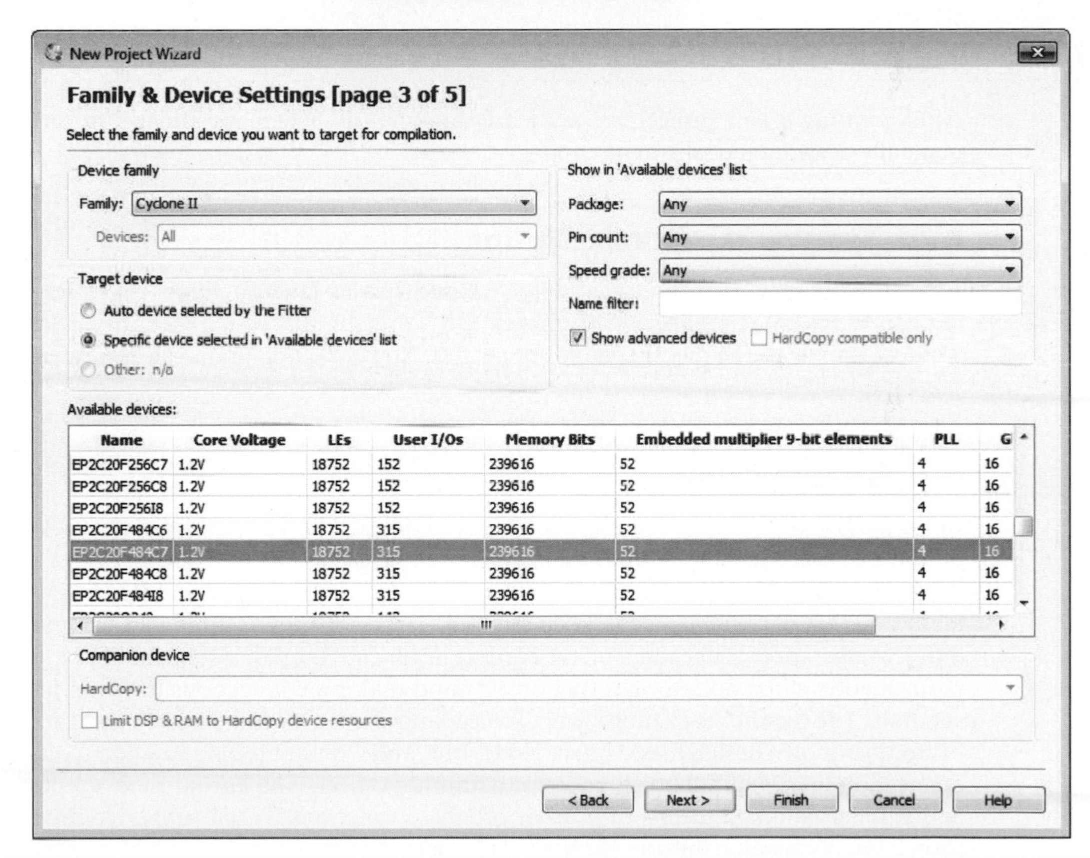

FIGURE B.3 The New Project Wizard: Family & Device Settings window with the device EP2C20F484C7 selected.

Source: Altera

The DE1 development board uses the Cyclone II EP2C20F484C7N FPGA chip. If you are using a different FPGA chip, then you need to make the appropriate changes in the following instructions.

- In the **Device Family** drop-down box, select **Cyclone II** or the one that matches your FPGA chip.
- In the **Available devices** list, select the device **EP2C20F484C7** or the one that matches your FPGA chip. If this device is not listed, then you need to reinstall the Quartus II program with the Cyclone II device family option checked.
- Click **Next** to continue to the next window.

In the next **New Project Wizard: EDA Tool Settings [page 4 of 5]** window, we do not have any EDA tools to use for this project, so click **Next** to continue to the next window.

The final window is a summary of the choices that you have just made. Click **Finish** to create your new project.

B.2 Using the Graphic Editor

After creating a new project, we are ready to start the Schematic Block Editor for manually drawing the schematic circuit.

B.2.1 Starting the Graphic Editor

From the Quartus II menu, select **File** > **New**. Under **Design Files**, select **Block Diagram/Schematic File**, and then click **OK**. You should see the **Graphic Editor** window similar to the one shown in Figure B.4. Any circuit diagram can be drawn in this Graphic Editor window.

B.2.2 Drawing Tools

In Figure B.4, the tools for drawing circuits in the Block Editor are shown in the toolbar on the left side. The default location for this tool bar is at the top. There are the standard tools such as text writing, zoom, flip and rotate, and line and shape drawing. The main tool that you will use is the Selection tool. This tool allows you to perform many different operations, depending on the context in which it is used. Two main operations performed by this tool are selecting objects and making connections between logic symbols. The Symbol tool allows you to select and use logic symbols from the library or from your own design files. The six Node, Bus, and Conduit tools allow you to draw connection lines that are not connected to another object. The Partial Line Selection and Rubberbanding buttons turn on or off these functions. When rubberbanding is turned on, connection lines are adjusted automatically when symbols are moved from one location to another. When rubberbanding is turned off, moving a symbol will not affect the lines connected to it.

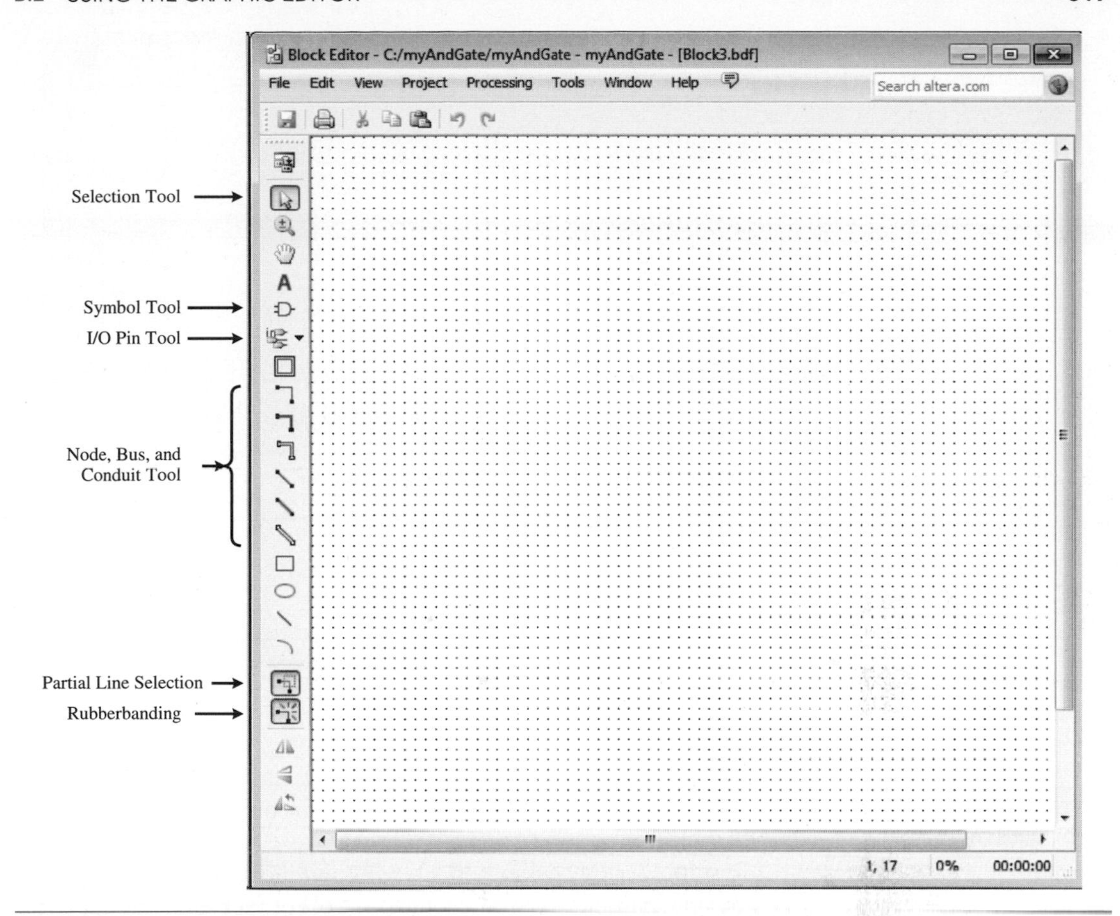

FIGURE B.4 The Graphic Editor window with the graphics toolbar on the left.

Source: Altera

B.2.3 Inserting Logic Symbols

To insert a logic symbol, first select the Selection tool, and then double-click on an empty spot in the **Block Editor** window. You should see the **Symbol** window as shown in Figure B.5.

- Alternatively, you can click on the Symbol tool icon in the toolbar to bring up the **Symbol** window.

Available symbol libraries are listed in the **Libraries** box. These libraries include the standard primitive gates, standard combinational and sequential components, and your own logic symbols located in the current project directory.

All of the basic logic gates, latches, flip-flops, and input and output connectors that you need are located in the **primitives** folder. If this folder is not listed, then click on the plus (+) sign to expand the libraries folder. Within the **primitives** folder are several subfolders. The basic gates are in the **logic** subfolder; the latches and flip-flops are in

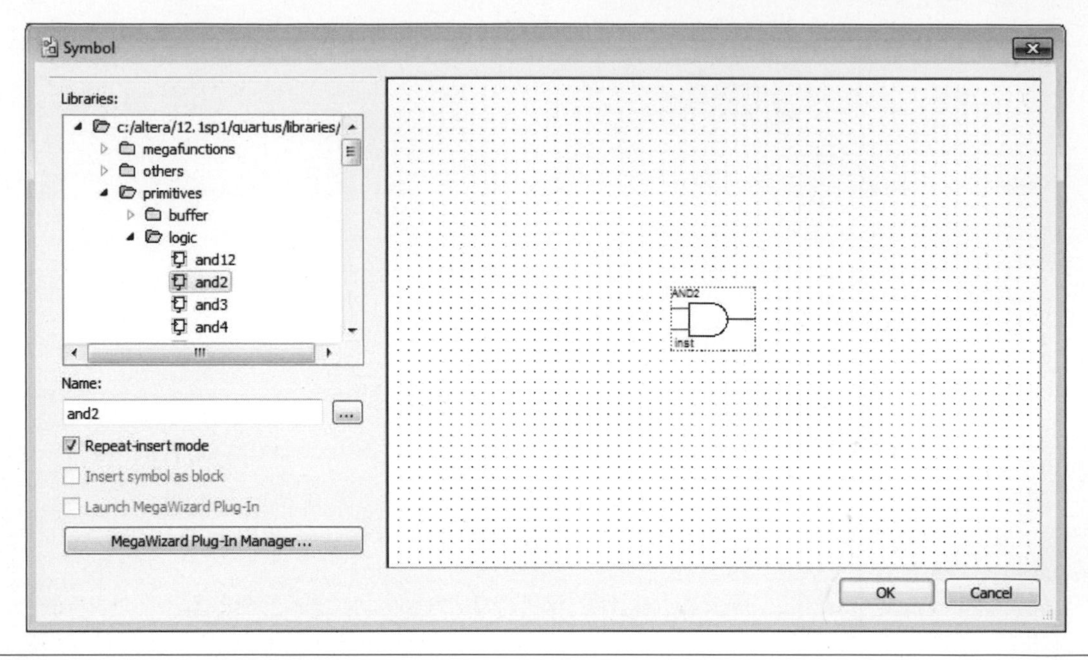

FIGURE B.5 The Symbol selector window.

Source: Altera

the **storage** subfolder; and the input and output connectors are in the **pin** subfolder. Your own circuits, if there are any that you want to reuse in building larger circuits, will be listed in the **Project** folder.

Expand the **logic** subfolder by clicking on the plus sign next to it to see a list of logic gate symbols available in that library. The logic symbols are sorted in alphabetical order. Select the logic symbol name that you want to use, or alternatively, you can type in the name of the logic symbol in the **Name** field. Click on the **OK** button to insert the symbol in the Graphic Editor. If the Repeat-insert mode box is checked, then you can insert several instances of the same symbol until you press the **Esc** key.

For this lab, insert the following symbols into the Graphic Editor:

- A 2-input AND gate (**and2**) found in the **logic** subfolder.
- An input signal connector (**input**) found in the **pin** subfolder.
- An output signal connector (**output**) found in the **pin** subfolder.

A unique number is assigned to each instance of a symbol and is written at the lower-left corner of the symbol. This number is used only as a reference number in the output netlist and report files. The numbers that you see might be different from those in the examples.

B.2.4 Selecting, Moving, Copying, and Deleting Logic Symbols

To select a logic symbol in the Block Editor, click on the symbol using the Selection tool. To select multiple symbols, you can hold down the **Ctrl** key and select each one individually, or you can draw a rectangle around the objects that you want to select. All objects inside the rectangle will be selected.

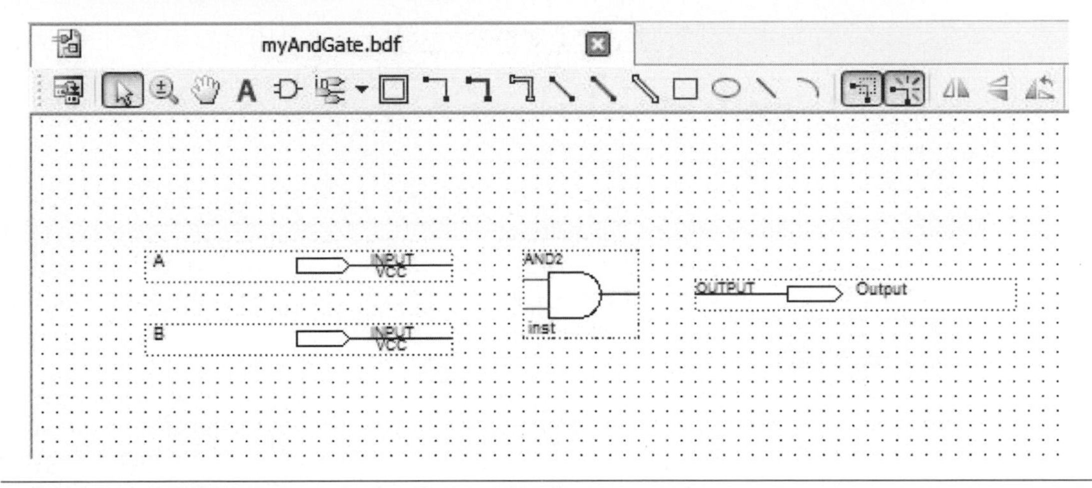

FIGURE B.6 Symbol placements for the AND circuit.

Source: Altera

To de-select a symbol, click on an empty spot in the Block Editor.

To move a symbol, select it and drag it to a different location.

To copy a symbol, select it and then perform the Copy and Paste operations, or hold down the **Ctrl** key while you drag the symbol to a different location.

To delete a symbol, select it and then press the **Delete** key.

To rotate a symbol, right-click on it, select **Rotate by Degrees** from the pop-up menu, and select the angle to rotate the symbol. Alternatively, you can first select the symbol and then click on one of the Flip or Rotate buttons on the tool bar.

Perform the following operations for this lab:

- Make a copy of the 2-input AND gate.
- Make two more copies of the input signal connector.
- Position the symbols similar to Figure B.6.

B.2.5 **Making and Naming Connections**

To make a connection between two connection points, use the Selection tool and drag from one connection point to the other connection point. Note that when you position the pointer on a connection point, the arrow pointer changes to a crosshair.

To change the direction of a connection line while dragging the line, simply release and press the mouse button again, and then continue to drag the connection line.

You also can make a connection between two connection points by moving a symbol so that its connection point touches the connection point of the second symbol. With rubberbanding turned on, you now can move one symbol away from the second symbol, and a connection line is drawn between them automatically.

If you want to make a connection line that does not start from a symbol connection point, you will need to use either the Orthogonal Node tool or the Diagonal Node tool instead of the Selection tool.

FIGURE B.7 **Making or deleting a connection point.**

Do not use the Line tool to make connections; this tool is only for drawing lines and does not actually make a connection.

After a connection is made to a symbol, you can move the symbol to another location, and, if the rubberbanding function is turned on, the connection line will be adjusted automatically. However, if the rubberbanding function is turned off, the connection will be broken when the symbol is moved.

To make a connection between two lines that cross each other as shown in Figure B.7, you need to use the Selection tool. Right-click on the junction point (i.e., the point where the two lines cross) and then select from the pop-up menu **Toggle Connection Dot**. You can repeat the same process to remove the connection point.

To select a line segment, single-click on it. To select the entire line (with several line segments connected in different directions), you double-click on it.

Use the Orthogonal Bus tool to draw a bus connection.

To change a single node line to a bus line, right-click on the line and select **Bus Line** from the pop-up menu. Select **Node Line** from the pop-up menu to change it back to a node line.

A bus must have a name and a width. To name a connection line, right-click on the line that you want to name. In the pop-up menu, select **Properties** and then type in the name and the width for the bus in the Name box. For example, `data[7..0]` is an 8-bit bus with the name `data`, as shown in Figure B.8.

To change the name, double-click on the name and edit it.

To connect one line to a bus, connect a single line to the bus, and then give it the same name as the bus, with the line index appended to it. For example, `data[2]` is bit two of the `data` bus, as shown in Figure B.8.

To check whether a name is attached to a line correctly, select the line, and the name that is attached to the line also will be selected.

All input and output signals in a circuit must be connected to input and output signal connectors, respectively. To name an input or output signal connector, select its name label by single-clicking it, and then double-clicking it. You can now type in the

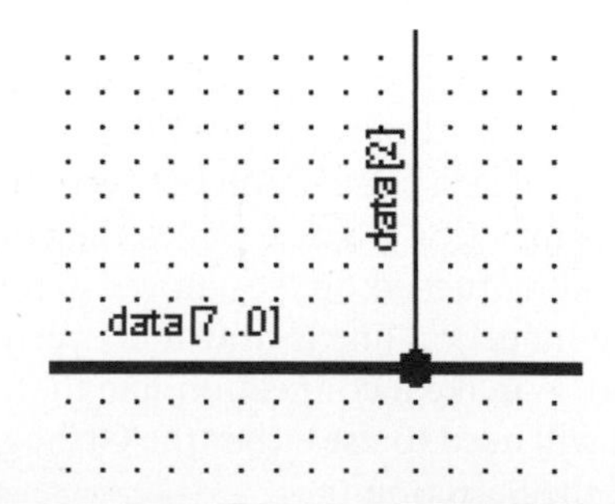

FIGURE B.8 **A single line connected to an 8-bit bus with the name `data`.**

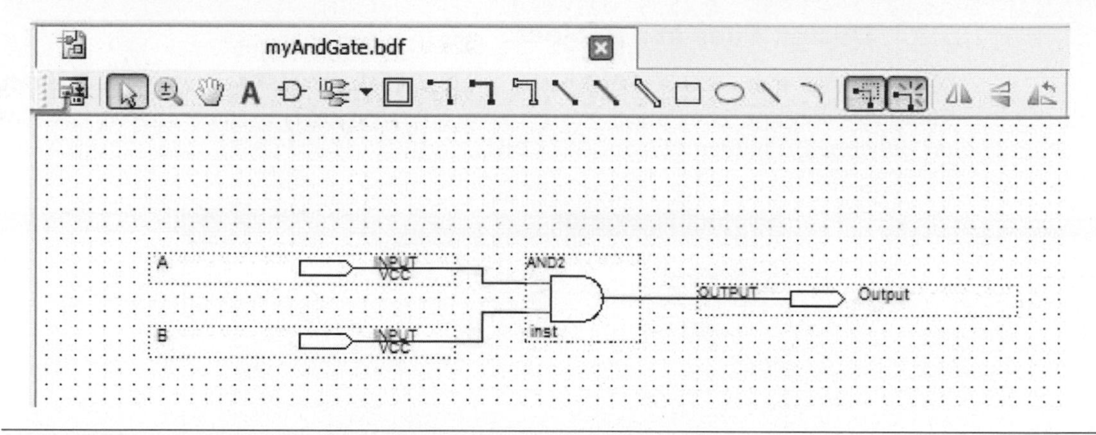

FIGURE B.9 Connections and names for the `myAndGate` circuit.

Source: Altera

new name. Pressing the **Enter** key will move the text entry cursor to the name label for the symbol below the current symbol. Alternatively, you can select the input or output connector and then double-click it. The Properties window for that pin will open up, allowing you to enter the pin name, among other things.

A bus line connected to an input or output connector must have the same bus width as the connector.

For this lab, perform the following operations to look like Figure B.9:

- Name the two input connectors `A` and `B`.
- Name the output connector `Output`.

Select **File** > **Save** to save the design file. Type in `myAndGate` for the file name. The default file extension is `.bdf` (block design file). Recall that when we created the project, we had specified `myAndGate` as the top-level file name. We now will use this file as the top-level source file.

B.2.6 Selecting, Moving, and Deleting Connection Lines

To select a straight connection line segment, single-click on it.

To select an entire connection line with horizontal and vertical segments, double-click on it.

To select a portion of a line segment, turn on the Use Partial Line Selection button, and then drag a rectangle around the line segment. Only the portion of the line segment that is inside the rectangle will be selected.

After a line is selected, it can be moved by dragging.

After a line is selected, it can be deleted by pressing the **Delete** key.

B.3 Managing Files in a Project

A project typically will have many design source files, which can be of different types. Some might be schematic drawings, and some might be Verilog and/or VHDL source files.

B.3.1 **Design Files in a Project**

To see the files that are associated with a project, click on the **Files** tab in the **Project Navigator** window. The Project Navigator shown in Figure B.10 shows that this project has only one file named `myAndGate.bdf`.

FIGURE B.10 Files associated with a project as shown in the Project Navigator window.

Source: Altera

B.3.2 **Creating a New Verilog or VHDL Source File**

The steps to create a new schematic drawing, Verilog or VHDL design file are the same. Select **File** > **New** from the Quartus II menu. In the **New** window under the **Device Design Files** tab, select the type of design file that you want to create: **Block Diagram/ Schematic File**, **Verilog HDL File**, or **VHDL File**. After you save this file, it is added automatically to the project.

B.3.3 **Opening a Design File**

To open a design file, double-click on the file that is listed in the Project Navigator window. Depending on the type of file, the associated editor will be used. The Block Editor is used to edit a Block Diagram/Schematic File, and a text editor is used to edit a VHDL or Verilog text file.

B.3.4 Adding Design Files to a Project

To add an existing design file to the current project, select **Project** > **Add/Remove Files in Project** from the Quartus II menu. Alternatively, you can right-click on the folder icon labeled **Files** in the **Project Navigator** window, and then select **Add/Remove Files in Project** from the pop-up menu.

This will bring up the **Files Category** under the **Settings** window. From this window, you can choose additional files to be added into the project by either manually typing in the file name or browsing to the directory and then selecting it. Click on the **Add** button to add individual files, or click on the **Add All** button to add all of the files in the selected directory.

B.3.5 Deleting Design Files from a Project

To delete a design file from a project, select it in the **Project Navigator** window, and then press the **Delete** key. Alternatively, you can right-click on the file that you want to delete, and then select **Remove File from Project** from the pop-up menu.

B.3.6 Setting the Top-Level Entity Design File

When you created a new project, you had to specify the name of the top-level design file. If you want to change the top-level entity to another design file, you can do so by right-clicking on the file that you want to be the top-level entity in the **Project Navigator** window. From the pop-up menu, select **Set as Top-Level Entity**.

B.3.7 Saving the Project

Select **File** > **Save Project** to save the project and all of its associated files.

B.4 Analysis and Synthesis

After drawing your circuit with the Graphic Editor, the next step is to analyze and synthesize it. During this step, Quartus II collects all of the necessary information about your circuit, and produces a netlist for it.

- From the Quartus II menu, select **Processing** > **Start** > **Start Analysis & Synthesis** to synthesize the circuit. Alternatively, you can click on the icon .
- If there are no errors in your circuit, you should see the message "Quartus II Analysis & Synthesis was successful" in the **Message** window at the bottom.

Errors found in the circuit will be reported in the **Message** window and highlighted in red. You can double-click on the error message to see where the error is in the circuit. Go back and double-check your circuit with the one shown in Figure B.9 to correct all of the errors.

B.5 Creating and Using a Logic Symbol

If you want to use a circuit as part of another circuit in a schematic drawing, you can create a logic symbol for this circuit. Logic symbols are like black boxes that hide the details of a circuit. Only the input and output signals for the circuit are shown. The input and output signals for the logic symbol are obtained directly from the input and output signal connectors that are connected in the circuit.

To create a logic symbol for a circuit, select the **Block Editor** window containing the circuit that you want as the active window. Select **File** > **Create/Update** > **Create Symbol Files for Current File**. The name of this symbol file will be the same as the name of the current active circuit diagram in the Graphic Editor, but with the file extension `.bsf` (block symbol file).

You can view and edit the logic symbol by first opening the `.bsf` file. Select **File** > **Open** and type in the file name. Click on the **Open** button. A window similar to Figure B.11 will open showing the logic symbol.

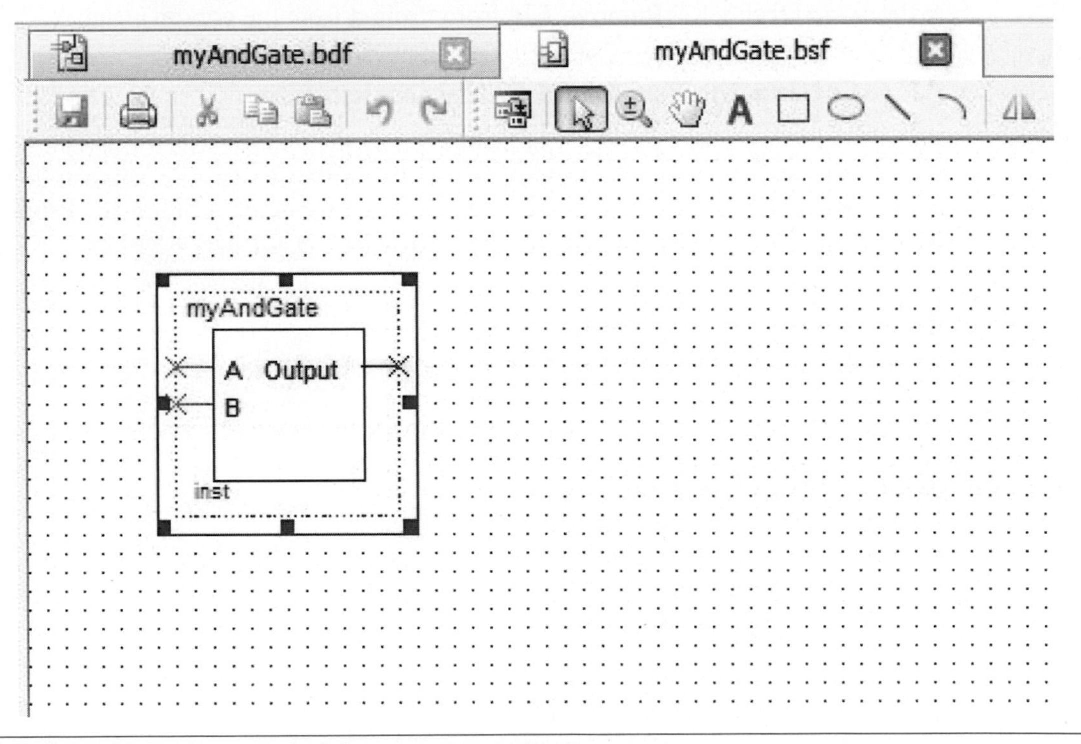

FIGURE B.11 Logic symbol of the `myAndGate` circuit.

Source: Altera

The placements of the input and output signals can be moved to different locations by dragging the signal connection line around the symbol box. The signal label also will be moved. You can then drag the label to another location if you wish. The size of the symbol also can be changed by dragging the edges of the symbol box.

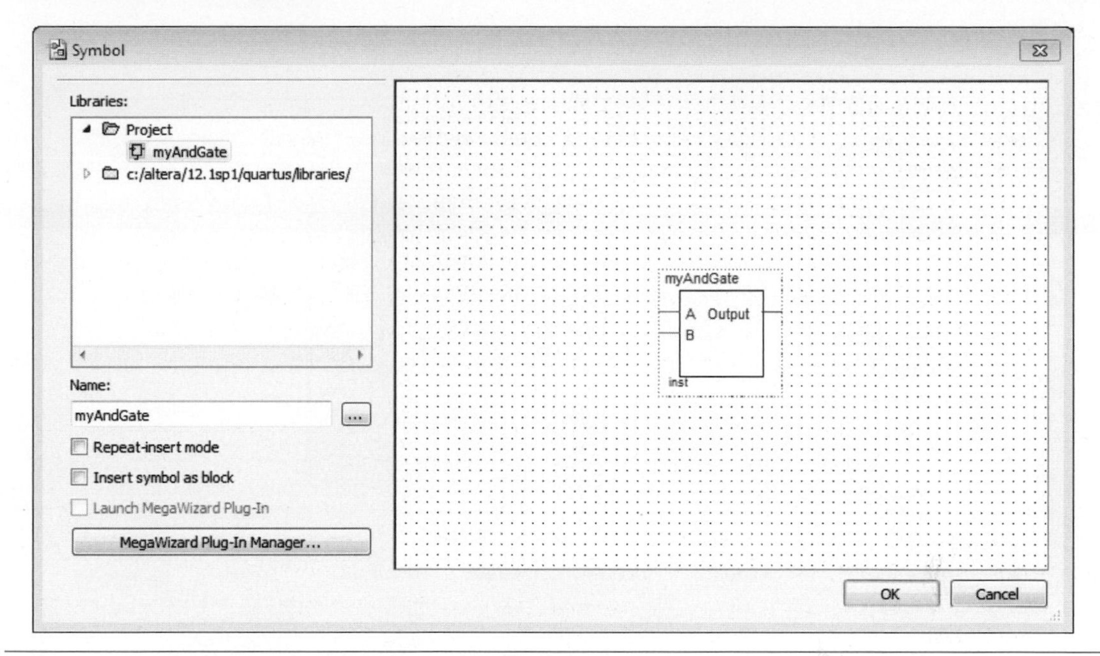

FIGURE B.12 Selecting the `myAndGate` logic symbol to be inserted into another circuit design.

Source: Altera

This new symbol can now be used in the Block Editor. It will show up in the **Symbol** window under the **Project** folder as shown in Figure B.12. You can follow the same steps as discussed earlier for inserting built-in logic symbols to insert this logic symbol into another schematic circuit design.

To use a circuit that is represented by its logic symbol in another project, you need to first copy the `.bsf` symbol file and the corresponding `.bdf` circuit design file to the other project's directory. It then will be available in the **Symbol** window inside the **Project** folder as shown in Figure B.12.

You now can select and use this component just like the standard components from the library.

B.6 Mapping the I/O Signals

Because we want to implement the circuit on an FPGA, we need to assign or map all of the I/O signals from our circuit to the actual pins on the FPGA. The following instructions are for mapping the I/O signals to the Cyclone II EP2C20F484C7 FPGA on the DE1 development board. If you are using a different FPGA development board, then you will need to refer to the documentation for your development board for the correct pin assignments.

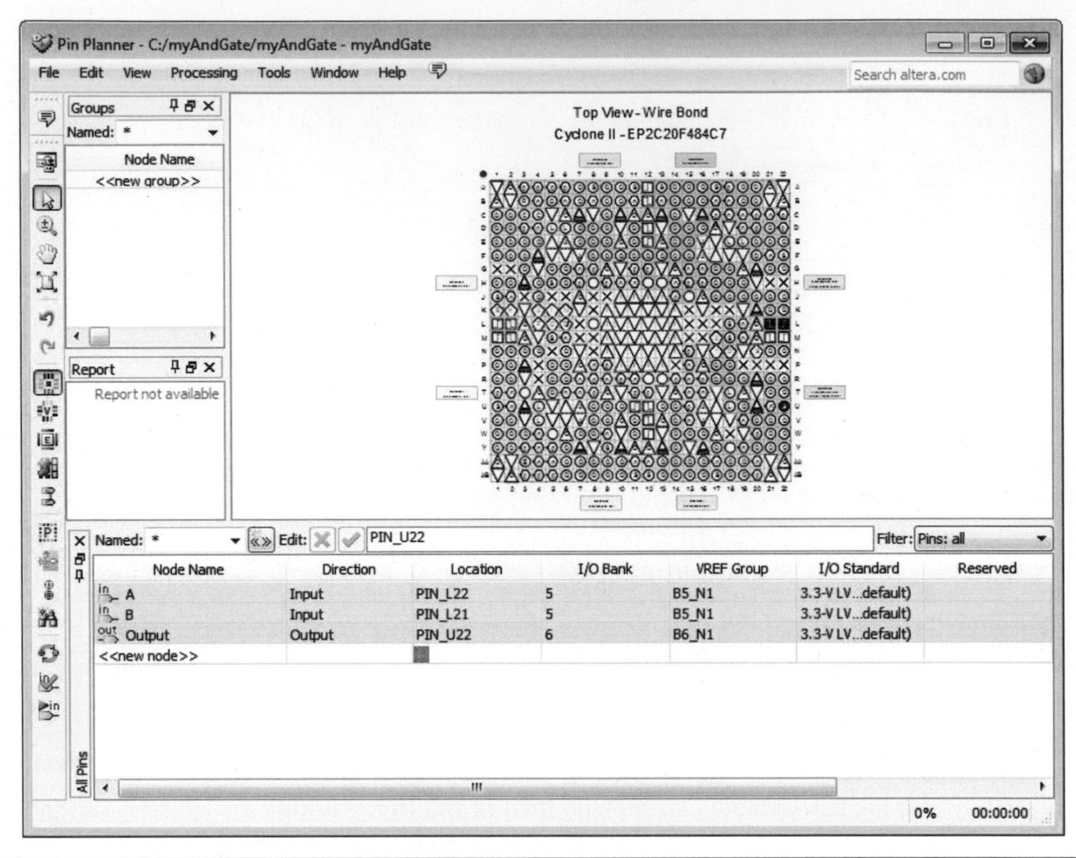

FIGURE B.13 The Pin Planner showing the pin assignments of the EP2C20F484C7 chip.

Source: Altera

We will use the Pin Planner to map each of the I/O signals from our circuit to the pins on the Cyclone II chip. From the Quartus II menu, select **Assignments** > **Pin Planner** to bring up the Pin Planner similar to Figure B.13.

1. Alternatively, you can click on the **Pin Planner** icon.
2. All of the available I/O signals from the circuit will be listed under the **Node Name** column. If the I/O signals are not listed, then you need to go back and do the Analysis and Synthesis step in Section B.4.
3. For each I/O signal name, double-click on the cell next to the signal name under the **Location** column to bring up a pop-up list of all the assignable pins from the FPGA. Select the pin number that you want to assign to that I/O signal.

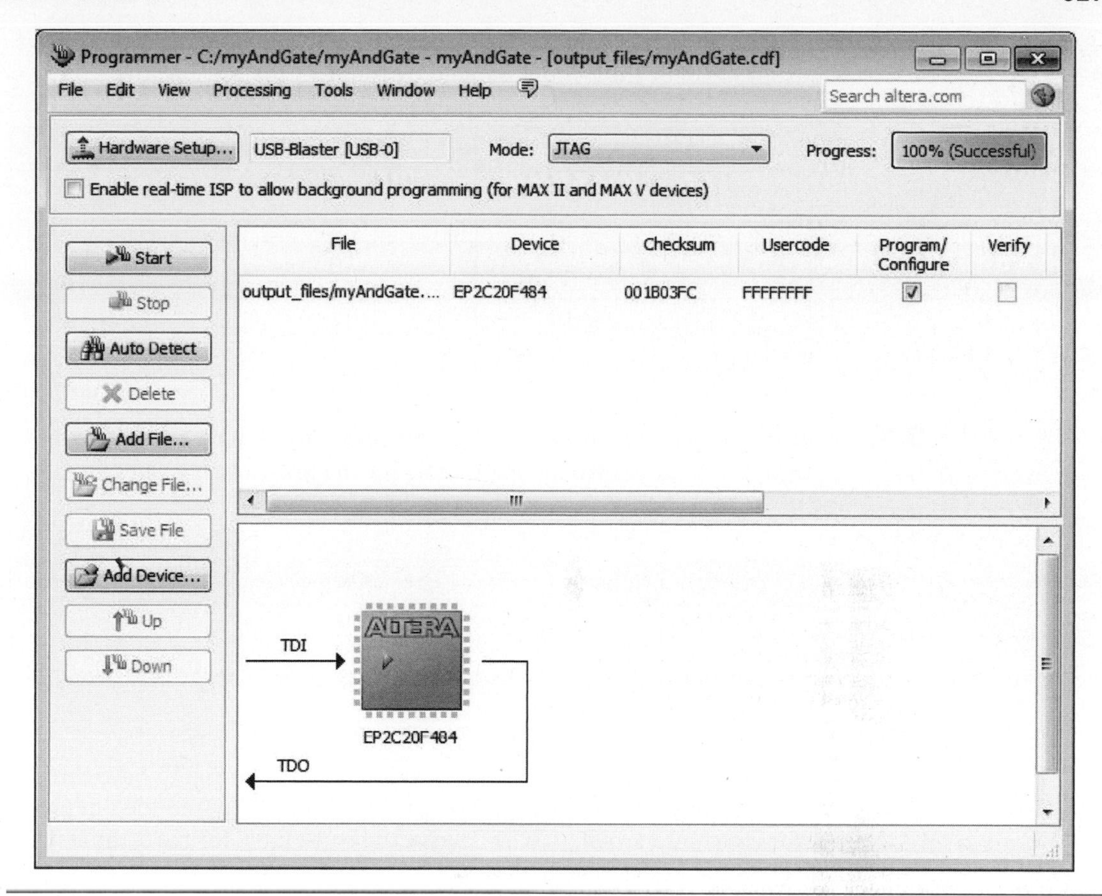

FIGURE B.15 The Programmer window showing that the chip was programmed successfully with your circuit.

Source: Altera

B.9 PROBLEMS

B.1. There are ten switches on the DE1 board. A switch is on when it is in the up position, and off when it is in the down position. Set the two switches (SW0 and SW1) to be either on or off, and note whether the green output LED (LEDG0) is on or off. Write down your observation in the table below. This table that you have derived is the truth table for the 2-input AND gate.

A	B	Output
0 (off)	0 (off)	
0 (off)	1 (on)	
1 (on)	0 (off)	
1 (on)	1 (on)	

B.2. Replace the 2-input AND gate in your circuit with a 2-input OR gate. Derive the truth table for the 2-input OR gate by filling in the table below.

A	B	Output
0	0	
0	1	
1	0	
1	1	

B.3. Replace your circuit with a NOT (inverter) gate. You'll have to remove one of the input pins. Derive the truth table for the NOT (inverter) gate by filling in the table below.

A	Output
0	
1	

B.4. Replace your circuit with a 3-input AND gate. You will need to have three input pins. Refer to the DE1 pin mappings document to find the pin number for your third switch (SW2). Derive the truth table for the 3-input AND gate by filling in the table below.

A	B	C	Output
0	0	0	
0	0	1	
0	1	0	
0	1	1	
1	0	0	
1	0	1	
1	1	0	
1	1	1	

B.5. Replace your circuit with a 3-input OR gate. Derive the truth table for the 3-input OR gate by filling in the table below.

A	B	C	Output
0	0	0	
0	0	1	
0	1	0	
0	1	1	
1	0	0	
1	0	1	
1	1	0	
1	1	1	

B.6. Draw and implement the following circuit. This circuit is known as the Multiplexor or Mux for short. Derive the truth table for this circuit. Describe the operation of this circuit in as few words as possible. What do you think the letter s for one of the inputs stands for?

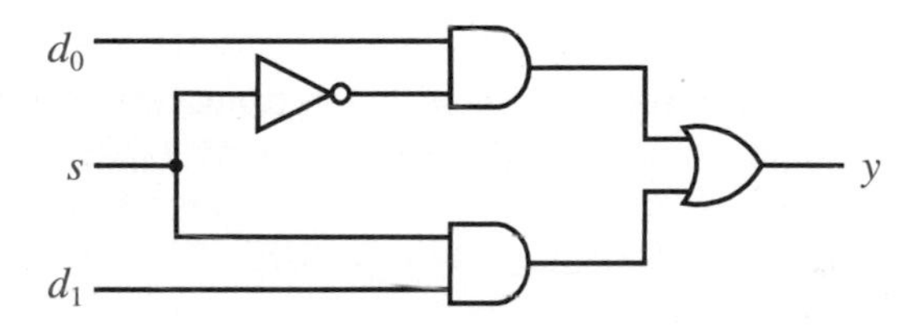

s	d_1	d_0	y (Output)
0	0	0	
0	0	1	
0	1	0	
0	1	1	
1	0	0	
1	0	1	
1	1	0	
1	1	1	

B.7. Repeat Problem B.1 but use two push buttons (PB0 and PB1) instead of the two switches. What do you notice about the operation of the push buttons?

B.8. Connect a circuit having one switch and one LED. Make the LED turn on when the switch is on, and off otherwise.

B.9. Make the seven-segment (HEX) displays to display the number 256 by turning on the appropriate LEDs for each digit. The seven LEDs for each digit is named as follows:

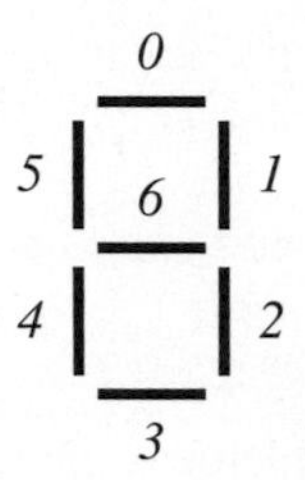

The LEDs in the seven-segment displays are turned on with a 0 rather than a 1 as in the discrete LEDs. The rightmost digit is named HEX0 on the development board.

B.10. Draw some random circuit having three inputs and one output. Randomly connect several AND gates, OR gates, and NOT gates together between the inputs and output. Derive the truth table for it.

APPENDIX C

Verilog Summary

The Verilog language is a hardware description language (HDL) for modeling digital circuits that can range from the simple connection of gates to complex systems. Verilog originally was designed as a proprietary verification and simulation tool. Later, logic and behavioral synthesis tools were added. The language was standardized in 1995 by IEEE, followed by a revision in 2001. This appendix gives a brief summary of the basic Verilog elements and its syntax. Many advanced features of the language are omitted. Interested readers should refer to other references for detailed coverage.

C.1 Basic Language Elements

C.1.1 Keywords

The Verilog language is case sensitive, and all of the keywords are in lower case. Figure C.1 shows a partial list of the Verilog keywords.

always	and	assign	automatic	begin	buf
bufif0	bufif1	case	casex	casez	default
defparam	else	end	endcase	endfunction	endgenerate
endmodule	endtask	event	for	forever	function
generate	genvar	if	include	initial	inout
input	integer	library	module	nand	negedge
nor	not	notif0	notif1	or	output
parameter	posedge	reg	signed	supply0	supply1
task	tri	tri0	tri1	unsigned	wand
while	wire	wor	xnor	xor	@

FIGURE C.1 Partial list of Verilog keywords.

C.1.2 Comments

Single-line comments are preceded by two consecutive slashes (//) and are terminated at the end of the line.

EXAMPLE

```
// This is a single line comment
```

Multiple-line comments begin with the two characters /* and end with the two characters */.

EXAMPLE

```
/* This is a
   multiple line comment
*/
```

C.1.3 Identifiers

Verilog identifiers are user-given names. Verilog identifiers must use the following syntax:

- A sequence of one or more uppercase letters, lowercase letters, digits, and the underscore (_).
- Upper and lowercase letters are treated differently (i.e., case sensitive).
- The first character must be a letter or the underscore.
- The length of the identifier must be 1024 characters or less.

C.1.4 Signals

Signals in Verilog have one of four values. These are:

- 0 for the logic 0.
- 1 for the logic 1.
- ?, X, or x for don't-care or unknown.
- Z or z for high-impedance tri-state.

C.1.5 Numbers and Strings

Numbers

Number constants can be specified in any one of the four bases: decimal, hexadecimal, octal, or binary. An unsized decimal number can also be specified using just the digits from 0 to 9.

Syntax:

a'sfn

where:

a is the number of bits (specified as an unsigned decimal number) of the constant.

s optionally specifies that the value is to be considered as a signed number.

f specifies the base of the number. It is replaced by one of the letters: d (decimal), h (hexadecimal), o (octal), or b (binary).

n is the value of the constant specified in the given base.

EXAMPLE

```
48          // an unsized decimal number 48
4'b 1001    // a 4-bit binary number 1001
8'd 28      // an 8-bit decimal number 28
'o 537      // an unsized octal number 537
12'h 7e9    // a 12-bit hexadecimal number 7e9
```

Strings

String constants are enclosed within double quotes.

EXAMPLE

```
reg [1:8] MyString;
MyString = "This is a string";
```

C.1.6 Constants

Identifiers can be defined with a constant value. After it is defined, the identifier then can be used in place of the constant. The compiler directive starting with the single opening quote ('), followed by the word **define** is used to define the identifier. When using the identifier, the single opening quote must always precede the identifier name.

Syntax: definition:

'define identifier constant

Syntax: usage:

'identifier

EXAMPLE

```
'define buswidth 'd8              // define buswidth to be constant 8
wire ['buswidth-1:0] databus;    // using buswidth
```

C.1.7 **Data Types**

Nets and registers are two main kinds of data types.

- Nets, defined with the **wire** keyword, are used to model electrical connections between components. They are used to connect instances together to transmit logic values between them. Nets do not store values and have to be driven continuously. An optional range [start:end] can be given for the bit width.
- Registers, defined with the **reg** keyword, are used to represent storage elements. Registers can store their values from one assignment to the next. An optional range [start:end] can be given for the bit width. Furthermore, the optional **signed** keyword can be used to denote that the data in the register is to be treated as a signed (two's complement) number.

Syntax:

> **wire** [range] identifier1, identifier2, ...;
> **reg [signed]** [range] identifier1, identifier2, ...;

EXAMPLE

```
wire x, y;          // two 1-bit wire
wire [1:4] bus;     // a 4-bit wire with bit 1 being the most
                    // significant
reg z;              // a 1-bit register
reg [7:0] s;        // an 8-bit register with bit 7 being the most
                    // significant
```

C.1.8 **Data Operators**

Some of the more commonly used Verilog operators are listed in Figure C.2.

Logical Operators	**Operation**	**Example**
&&	Logical AND	if ((a > b) && (c < d))
||	Logical OR	if ((a > b) || (c < d))
!	Logical NOT	if !(a > b)
&	Bitwise AND of individual bits	n = a & b
|	Bitwise OR of individual bits	n = a | b
~	Bitwise NOT of individual bits	n = ~a
^	Bitwise XOR of individual bits	n = a ^ b
Arithmetic Operators	**Operation**	**Example**
+	Addition	n = a + b
−	Subtraction	n = a − b
*	Multiplication (integer or floating point)	n = a * b

FIGURE C.2 Verilog built-in data operators. *(continued on next page)*

Arithmetic Operators	Operation	Example
/	Division (integer or floating point)	n = a / b
%	Modulus; remainder (integer)	n = a % b
**	Power	n = a ** 2
Relational Operators	**Operation**	**Example**
==	Logical equal	if (a == b)
!=	Logical not equal	if (a != b)
<	Less than	if (a < b)
<=	Less than or equal	if (a <= b)
>	Greater than	if (a > b)
>=	Greater than or equal	if (a >= b)
===	Bitwise equal. All bits must match. Can include **x** and **z** values.	if (a === b)
!==	Bitwise not equal. True if only one bit is different. Can include **x** and **z** values.	if (a !== b)
Shift and Other Operators	**Operation**	**Example**
<<	Logical left shift. Pad with zero	n = 7'b1001010 << 2
>>	Logical right shift. Pad with zero	n = a >> 1
<<<	Arithmetic left shift. Pad with zero	n = a <<< 3
>>>	Arithmetic right shift. Pad with sign bit	n = a >>> 2
{, }	String concatenation	n = {a, b, c}
{m{ }}	Repetition where m is repetition number	n = {3{a}}

FIGURE C.2 Verilog built-in data operators.

C.1.9 Module

In Verilog, a **module** represents a logical component in a digital system. Each module has an interface to specify the signals for communication with other modules. These port signals are declared within parenthesis, and can be of types **input**, **output**, or **inout** (for bidirectional communication). A module's body contains statements that describe the actual operation of the logical component.

The operational description of the module can be written using one of three different models: behavioral, dataflow, or structural. Behavioral modeling describes the abstract operation of the circuit using a high-level construct, and does not take into consideration how the circuit is actually implemented. Dataflow modeling specifies the circuit in a form that is closely related to a Boolean equation. Structural modeling describes a circuit in terms of how the primitive gates are interconnected together.

Syntax:

module module_name
(**input** port_name_list,
output port_name_list,
inout port_name_list);

statements;
endmodule

EXAMPLE: BEHAVIORAL MODEL

```
// a 4-bit wide 2-to-1 multiplexer written in behavioral model
module multiplexer (
  input s,
  input [3:0] d0,
  input [3:0] d1,
  output reg [3:0] y
);

  always @(s, d0, d1) begin
    if (~s)
      y = d0;                        // assign d0 to y
    else
      y = d1;
  end

endmodule
```

EXAMPLE: DATAFLOW MODEL

```
// a 1-bit wide 2-to-1 multiplexer written in dataflow model
module multiplexer (
  input s,
  input d0,
  input d1,
  output reg y
);

  assign y = (~s & d0) | (s & d1);

endmodule
```

EXAMPLE: STRUCTURAL MODEL

```
// a 1-bit wide 2-to-1 multiplexer written in structural model
module multiplexer (
  input s, d0, d1,
  output y
);
```

```
  wire sn, snd0, sd1;  // define 3 nets for connecting the components
  not U1(sn,s);        // an instance of the NOT gate. sn is the output
  and U2(snd0,d0,sn); // an AND gate. snd0 is the output
  and U3(sd1,d1,s);
  or U4(y,snd0,sd1);

endmodule
```

C.1.10 Module Parameter

A **module** can have an optional parameter list. This list of parameters, with optional default values, allows us to define generic information about the module. The **parameter** keyword is used to specify identifiers with optional default values assigned to them. The identifier is assigned an external value when the module is instantiated, or is assigned the default value when no external value is given. The identifier can then be used in placed of a constant.

Syntax: Declaration:

> **module** module_name
> **#(parameter** identifier = default_value, identifier = default_value)
> **(input** port_name_list,
> **output** port_name_list,
> **inout** port_name_list);
>
> statements;
> **endmodule**

Syntax: Instantiation:

> module_name #(constant) instance_name (parameter_list);

EXAMPLE: DECLARATION

```
// a default 8-bit 2-to-1 multiplexer written in dataflow model
module multiplexer
  #(parameter width = 8)  // a parameter constant with a default
                             value of 8

  (input [width-1:0] d0, d1,
  input s,
  output [width-1:0] y);

  assign y = (~s) ? d0:d1;     // assigns d0 to y if s is 0,
                                 // otherwise assigns d1 to y
endmodule
```

EXAMPLE: INSTANTIATION

```
// instantiating a 4-bit 2-to-1 multiplexer
multiplexer #(4) U1(input1, input2, select, output);
```

C.2 Behavioral Model

The behavioral model allows statements to be executed sequentially similar to a regular computer program. An **always** block, containing one or more sequential statements, forms the basis of the behavioral model. The **always** block is like a process with its independent thread of control, and continuously executes all the statements that are inside it. All sequential statements, including many of the standard constructs, such as variable assignments, if-then-else, and loops, must be written inside an **always** block.

C.2.1 Assignment

Signal assignments are performed using the symbol = for blocking, or <= for non-blocking. These assignment statements are different from the **assign** keyword used for assignments in the dataflow model.

Syntax:

```
register_identifier = expression;        // blocking (immediate) assignment
register_identifier <= expression;       // non-blocking (concurrent) assignment
```

These assignment statements must be used inside an **always** block or an **initial** block. The identifier on the left side of the equal sign must be of type **reg**, but this does not mean that a memory element will always be used for the identifier.

EXAMPLE: BLOCKING

```
reg a, c;

always begin
  a = b;
  c = a;
end
```

Blocking assignment statements are executed sequentially, so the ordering of the assignment statements does matter. The register variable on the left side is assigned with the value from the right side immediately before continuing to the next statement. In the above example, the result of the two blocking assignment statements is that both variables *a* and *c* will take on the same value of *b*.

EXAMPLE: NON-BLOCKING

```
reg a, c;

always begin
  a <= b;
  c <= a;
end
```

Non-blocking assignment statements are executed in parallel, so the ordering of the assignment statements does not matter. In the above example, the first statement will make *a* take on the value of *b*, but in the second statement, *c* will take on the original value that *a* has as if the first statement never occurred. In other words, all of the right side expressions in the non-blocking assignment statements will be evaluated *before* any of the left side registers are updated. Hence, the ordering of these statements does not affect the resulting output.

Below is a list of general guidelines that you should follow as to when to use the blocking assignment and when to use the non-blocking assignment statements.

- Use blocking assignments (=) when modeling combinational logic inside an **always** block.
- Use non-blocking assignments (<=) when modeling sequential logic inside an **always** block.
- Use non-blocking assignments (<=) when there is a **posedge** or **negedge** clause in the **always** sensitivity list or modeling latches.
- Use non-blocking assignments (<=) when modeling both sequential and combinational logic within the same **always** block.
- Do not mix blocking and non-blocking assignments in the same **always** block.
- Do not make assignments to the same variable from more than one **always** block.

Your code still will synthesize if you do not follow these guidelines, but your simulation might be wrong.

C.2.2 initial

The **initial** block is executed only once at the beginning and terminates after executing all of the statements inside it. The sequential behavioral statements inside the block are executed sequentially. Non-blocking (<=) assignments still will be executed in parallel.

Syntax:

> **initial**
>
> statement;

EXAMPLE

```
reg [7:0] rom[2**4-1:0];

initial begin
  rom[0] <= 8'b01100000;
  rom[1] <= 8'b10000000;
  rom[2] <= 8'b10100000;
  rom[3] <= 8'b11000001;
  rom[4] <= 8'b11111111;
end
```

C.2.3 always

The **always** block is similar to the **initial** block, but repeats continuously, executing the statements that are inside it similar to an endless loop. The **always** block itself is a concurrent statement, so a behavioral module might contain multiple **always** blocks, and they all will be executed concurrently. The sequential behavioral statements inside the block are executed sequentially. Non-blocking (<=) assignments still will be executed in parallel.

Syntax:

always
    statement;

EXAMPLE

```
always begin
   term_1 = D | V;
   S = term_1 & M;
end
```

An **always** construct usually is used in conjunction with an event control (**@**) to create either a combinational or sequential logic.

C.2.4 Event Control

The event control statement, which uses the **@** symbol, waits for the specified event to occur and then executes the statement associated with it. The event is specified in the form of a sensitivity list, which is a comma-separated list of nets. Whenever a signal in the sensitivity list changes value, the associated statement will be executed.

Syntax:

@ (sensitivity_list)
    statement;

EXAMPLE

```
// synthesizes to a combinational logic

module Siren (
  input M, D, V,
  output reg S
);

  reg term_1;

  always @(M, D, V) begin
    term_1 = D | V;
    S = term_1 & M;
  end

endmodule
```

If the sensitivity list contains every variable on the right side of an assignment statement or the condition in an **if** statement, then a combinational logic is created. The asterisk (*) symbol can be used as a shorthand notation to denote all of the variables.

Syntax:

```
@ (*)
    statement;
```

EXAMPLE

```
// synthesizes to a combinational logic
always @(*)        // equivalent to always @(a, b, c)
  if (a == 1)
    x = b;
  else
    x = c;
```

The nets specified in the sensitivity list might be qualified with the keywords **posedge** or **negedge** so that the control statement watches only for the positive or negative transition, respectively, of the given signal before it executes the statement. In this case, non-blocking assignment (<=) statements should be used inside the **always** block and a sequential logic is created.

Syntax:

```
@ (posedge signal)
    statement;

@ (negedge signal)
    statement;
```

Note that the sensitivity list cannot contain both edge triggered signals (with either **posedge** or **negedge** qualifiers) and level sensitive signals (with no **posedge** or **negedge** qualifiers).

EXAMPLE OF A BASIC D FLIP-FLOP

```
module D_flipflop (
  input clock, data,
  output reg q
);

  always @(posedge clock)
    q <= data; // q gets the value of data at next rising clock edge
endmodule;
```

EXAMPLE OF A D FLIP-FLOP WITH SYNCHRONOUS ACTIVE-HIGH *ENABLE* AND ASYNCHRONOUS ACTIVE-HIGH *CLEAR* SIGNALS

```
module D_flipflop (
  input D,
  input Clock,
  input Enable,
  input Clear,
  output reg Q
);

  // execute on rising clock edge or Clear
  always @(posedge Clock or posedge Clear) begin
    if (Clear) begin
     Q <= 1'b0;                    // assign 0 to Q on clear
    end else if (Enable) begin
     Q <= D;  // assign value from D to Q only if Enable is asserted
    end
  end

endmodule
```

C.2.5 begin-end

A block of sequential statements can be grouped together to form a single block with the use of the **begin** and **end** keywords.

Syntax:

```
begin
    statement1;
    statement2;
    ...
end
```

EXAMPLE

```
if (a == 1) begin
   x = 1'b0;
   y = 1'b1;
end
```

C.2.6 if-then-else

Syntax:

```
if (condition)
    statement1;
else
    statement2;

or

if (condition)
    statement1;
else if (condition)
    statement2;
else
    statement3;
```

EXAMPLE

```
if (count != 10)  // not equal
  count = count + 1;
else
  count = 0;
```

C.2.7 case, casex, casez

Syntax:

```
case (expression)
    constant1:  statement1;
    constant2:  statement2;
    ...
    default:  statement3;
endcase
```

The **casex**, and **casez** statements have the same syntax as the **case** statement, except for the replaced keyword. The **casez** statement allows for z values to be treated as don't-care values, while the **casex** statement allows for both x and z values to be treated as don't-cares.

EXAMPLE

```
module mux4
  #(parameter width = 8)
  (input [1:0] s,
  input [width-1:0] d3, d2, d1, d0,
  output reg [width-1:0] y);
```

```
  always @(s, d0, d1, d2, d3) begin
    case (s)
      2'b00: begin
        y = d0;
        end
      2'b01: begin
        y = d1;
        end
      2'b10: begin
        y = d2;
        end
      default: begin
        y = d3;
        end
    endcase
  end
endmodule
```

C.2.8 for

Syntax:

for (id = low_range; id < high_range; id = id + step)
    statement;

EXAMPLE

```
module TestFOR
  (sum);

  inout reg [7:0] sum = 'd0;    // initialize with decimal 0
  reg [3:0] i;

  always
    begin
    for (i = 0; i < 10; i = i + 1)
      begin
      sum = sum + i;
    end
  end

endmodule
```

C.2.9 while

Syntax:

while (condition)
    statement;

EXAMPLE

```
module TestWHILE
  (sum);

  inout reg [7:0] sum = 'd0;
  reg [3:0] i;

  always
   begin
   i = 0;
   while (i < 10)
     begin
     sum = sum + i;
     i = i + 1;
     end
   end

endmodule
```

C.2.10 function

Syntax: Function definition:

function function_name (parameter_list);
    // register declarations
    // wire declarations
    begin
    statement;
    ...
    end
endfunction

Syntax: Function call:

function_name (parameters);

EXAMPLE

```
module TestFunction
  (input [7:0] bitstring,
  output [7:0] result);
```

```
  assign result = Shiftright(bitstring);    // function call

// function to perform a shift right
function [7:0] Shiftright
  (input [7:0] string);

  Shiftright = {1'b0,string[7:1]};
endfunction

endmodule
```

C.2.11 Behavioral Model Example

The following example shows the behavioral code for a binary coded decimal to 7-segment LED decoder.

EXAMPLE: BEHAVIORAL CODE FOR A BCD TO 7-SEGMENT DECODER

```
module decoder
  (input [3:0] I,
  output reg a,b,c,d,e,f,g);

  always @(*) begin
    case(I)
    4'b0000: {a,b,c,d,e,f,g} = 7'b1111110;    // 0
    4'b0001: {a,b,c,d,e,f,g} = 7'b0110000;    // 1
    4'b0010: {a,b,c,d,e,f,g} = 7'b1101101;    // 2
    4'b0011: {a,b,c,d,e,f,g} = 7'b1111001;    // 3
    4'b0100: {a,b,c,d,e,f,g} = 7'b0110011;    // 4
    4'b0101: {a,b,c,d,e,f,g} = 7'b1011011;    // 5
    4'b0110: {a,b,c,d,e,f,g} = 7'b1011111;    // 6
    4'b0111: {a,b,c,d,e,f,g} = 7'b1110000;    // 7
    4'b1000: {a,b,c,d,e,f,g} = 7'b1111111;    // 8
    4'b1001: {a,b,c,d,e,f,g} = 7'b1110011;    // 9
    4'b1010: {a,b,c,d,e,f,g} = 7'b1110111;    // A
    4'b1011: {a,b,c,d,e,f,g} = 7'b0011111;    // b
    4'b1100: {a,b,c,d,e,f,g} = 7'b1001110;    // C
    4'b1101: {a,b,c,d,e,f,g} = 7'b0111101;    // d
    4'b1110: {a,b,c,d,e,f,g} = 7'b1001111;    // E
    4'b1111: {a,b,c,d,e,f,g} = 7'b1000111;    // F
    default: {a,b,c,d,e,f,g} = 7'b0000000;    // all off
    endcase
  end
endmodule
```

C.3 Dataflow Model

The dataflow model specifies the circuit in a form similar to Boolean algebra. Hence, this model is best suited for describing a circuit when given a set of Boolean equations.

C.3.1 Continuous Assignment

The **assign** statement is used to provide continuous assignment of values to nets outside of an **always** block. The **assign** statement is evaluated whenever any of its inputs changes value and the result of the evaluation is assigned to the output net. Continuous assignment statements and conditional assignment statements are executed in parallel, so the ordering of the statements does not matter.

Syntax

assign net_identifier = expression; // assign statement

wire net_identifier = expression; // declaration and initialization of net

The net identifier on the left side of the equal sign must be of type **wire**. The expression on the right side can be either a logical or arithmetic expression.

EXAMPLE

```
module logic (
  input a, b, c,
  output f
);

  wire w = a & b;

  assign f = w | c;

endmodule
```

C.3.2 Conditional Assignment

The conditional signal assignment statement selects one of two different values to assign to a net. This statement is executed whenever an input in any one of the expressions or condition changes. Continuous assignment statements and conditional assignment statements are executed in parallel, so the ordering of the statements does not matter.

Syntax:

assign net_identifier = (condition) ? expression1:expression2;

If the condition is true, then the result of expression1 is assigned to the net, otherwise the result of expression2 is assigned to the net.

EXAMPLE

```
// assigns in to out if enable is true otherwise assigns a z to out
assign out = (enable) ? in:1'bz;
```

C.3.3 Dataflow Model Example

This example describes a full adder (FA) circuit using the dataflow model. The Boolean equations for describing the full adder circuit are:

$$c_{out} = xy + c_{in}(x \oplus y)$$

$$sum = x \oplus y \oplus c_{in}$$

The following example translates the above two equations into the corresponding two **assign** statements.

EXAMPLE: DATAFLOW CODE FOR A FULL ADDER

```
module fa
  (input x, y, cin,
  output cout, sum);

  assign cout = (x & y) | (cin & (x ^ y));
  assign sum = x ^ y ^ cin;
endmodule
```

C.4 Structural Model

The structural model allows the manual connection of primitive gates and module components together using nets to build larger modules. This model is best suited if you already have the schematic drawing for a component and you want to recreate the component based exactly on the schematic diagram.

C.4.1 Built-in Gates

All of the basic gates are built in to the Verilog language. Furthermore, you can use the same syntax for instantiating a user-defined module. This way, you can define a module at a lower level and then use this module at a higher level.

Syntax:

and	instance_name (parameter_list);	// implements the primitive logic function
nand	instance_name (parameter_list);	
nor	instance_name (parameter_list);	
or	instance_name (parameter_list);	
xor	instance_name (parameter_list);	
xnor	instance_name (parameter_list);	
buf	instance_name (output, input);	// implements the noninverting buffer
not	instance_name (output, input);	// implements the inverter

bufif0 instance_name (output, input, enableN); // implements the tri-state buffer with active-low enable
bufif1 instance_name (output, input, enable); // implements the tri-state buffer with active-high enable
notif0 instance_name (output, input, enableN); // implements the tri-state inverter with active-low enable
notif1 instance_name (output, input, enable); // implements the tri-state inverter with active-high enable

Each instance of a gate can have an optional instance name. The parameter list consists of comma-separated input and output signals. All of the input/output signals are nets of type **wire**. For the predefined primitive gates, the first parameter specified is always the output signal, followed by as many input nets as needed.

EXAMPLE

```
and U1(out, in1, in2, in3, in4);    // a 4-input AND gate
or U2(out, in1, in2);               // a 2-input OR gate
```

C.4.2 User-Defined Module

User-defined module can be used to build circuits in a hierarchical fashion.

The syntax for using a user-defined module is the same as for a primitive gate. For the user-defined module, the inputs and outputs in the parameter list depend on how the parameter list in the module definition is specified. The parameter list can be specified using either the positional or named method. The constant number used here will override the number specified in the parameter clause in the module definition.

Syntax:

user_defined_module_name instance_name (parameter_list);

or

user_defined_module_name #(constant) instance_name (parameter_list);

EXAMPLE: POSITIONAL ASSOCIATION

```
fa U2(X[0], Y[0], c[0], c[1], S[0]);
```

EXAMPLE: NAMED ASSOCIATION

```
fa U3(.x(X[0]), .y(Y[0]), .cin(c[0]), .cout(c[1]), .sum(S[0]));
```

EXAMPLE: CONSTANT PARAMETER

```
TriState_Buffer #(8) U4(.E(OE), .D(dp_sum), .Y(Output));
```

C.4.3 Structural Model Example

This structural model example is for the 4-bit ripple-carry adder based on the following circuit. It requires the full adder (fa) module that is defined in Section C.3.3.

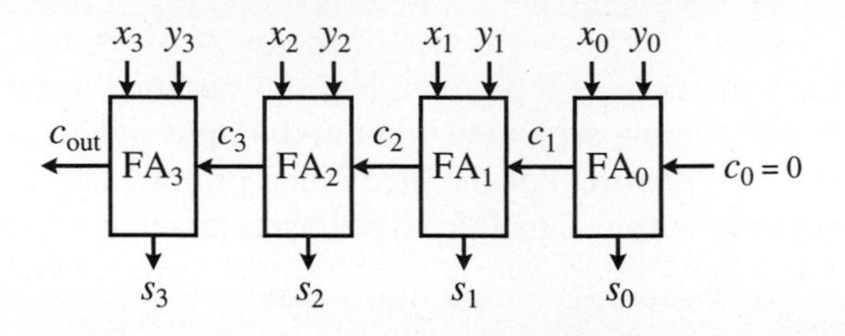

EXAMPLE: STRUCTURAL CODE FOR A 4-BIT ADDER

```
module adder (
  input [3:0] X, Y,
  output [3:0] S,
  output Cout
);
  wire [3:0] c;

  assign c[0] = 1'b0;

  fa U0(.x(X[0]), .y(Y[0]), .cin(c[0]), .cout(c[1]),
        .sum(S[0]));
  fa U1(.x(X[1]), .y(Y[1]), .cin(c[1]), .cout(c[2]),
        .sum(S[1]));
  fa U2(.x(X[2]), .y(Y[2]), .cin(c[2]), .cout(c[3]),
        .sum(S[2]));
  fa U3(.x(X[3]), .y(Y[3]), .cin(c[3]), .cout(Cout),
        .sum(S[3]));

endmodule
```

APPENDIX D

VHDL Summary

VHDL is a hardware description language for modeling digital circuits that can range from the simple connection of gates to complex systems. VHDL is an acronym for VHSIC Hardware Description Language, and VHSIC in turn is an acronym for Very High Speed Integrated Circuits. This appendix gives a brief summary of the basic VHDL elements and their syntax. Many advanced features of the language are omitted. Interested readers should refer to other references for detailed coverage.

D.1 Basic Language Elements

D.1.1 Keywords

The VHDL language is not case sensitive, so the keywords can be in either uppercase or lowercase. In this book, the VHDL keywords are written in uppercase. Figure D.1 shows a partial list of the VHDL keywords.

ABS	AND	ARCHITECTURE	ARRAY	BEGIN	BIT
BIT_VECTOR	BODY	BOOLEAN	CASE	COMPONENT	CONSTANT
DOWNTO	ELSE	ELSIF	END	ENTITY	EXIT
FOR	FUNCTION	GENERIC	IF	IN	INTEGER
IS	LIBRARY	LOOP	MAP	MOD	NAND
NATURAL	NEXT	NOR	NOT	OF	OR
OTHERS	PACKAGE	PORT	POSITIVE	PROCEDURE	PROCESS
RANGE	REM	RETURN	ROL	ROR	SELECT
SIGNAL	SLA	SLL	SRA	SRL	STD_LOGIC
STD_LOGIC_VECTOR	USE				

FIGURE D.1 Partial list of VHDL keywords.

D.1.2 **Comments**

Comments are preceded by two consecutive hyphens (--) and are terminated at the end of the line.

EXAMPLE

```
-- This is a comment
```

D.1.3 **Identifiers**

VHDL identifier syntax:

- A sequence of one or more uppercase letters, lowercase letters, digits, and the underscore (_).
- Upper and lowercase letters are treated the same (i.e., case insensitive).
- The first character must be a letter.
- The last character cannot be an underscore.
- Two underscores cannot be together.

D.1.4 **Data Objects**

There are three kinds of data objects: signals, variables, and constants.

- The data object SIGNAL represents logic signals on a wire in the circuit. A signal does not have memory; thus, if the source of the signal is removed, the signal will not have a value.
- A VARIABLE object remembers its content and is used for computations in a behavioral model.
- A CONSTANT object must be initialized with a value when declared, and this value cannot be changed.

EXAMPLE

```
SIGNAL x: BIT;
VARIABLE y: INTEGER;
CONSTANT one: STD_LOGIC_VECTOR(3 DOWNTO 0) := "0001";
```

D.1.5 **Data Types**

BIT and BIT_VECTOR

The BIT and BIT_VECTOR types are predefined in VHDL. Objects of these types can have the values 0 or 1. The BIT_VECTOR type is simply a vector of type BIT. A vector with all bits having the same value can be obtained using the OTHERS keyword.

EXAMPLE

```
SIGNAL x: BIT;
SIGNAL y: BIT_VECTOR(7 DOWNTO 0);
x <= '1';
y <= "00000010";
y <= (OTHERS => '0'); -- same as "00000000"
```

STD_LOGIC and STD_LOGIC_VECTOR

The STD_LOGIC and STD_LOGIC_VECTOR types provide more values than the BIT type for modeling a real circuit more accurately. Objects of these types can have the following values:

'0' – normal 0
'1' – normal 1
'Z' – high impedance
'-' – don't-care
'L' – weak 0
'H' – weak 1
'U' – uninitialized
'X' – unknown
'W' – weak unknown

The STD_LOGIC and STD_LOGIC_VECTOR types are not predefined, and so the following two library statements must be included in order to use these types.

```
LIBRARY IEEE;
USE IEEE.STD_LOGIC_1164.ALL;
```

If objects of type STD_LOGIC_VECTOR are to be used as binary numbers in arithmetic manipulations, then either one of the following two USE statements also must be included:

```
USE IEEE.STD_LOGIC_SIGNED.ALL;
```

for signed number arithmetic, or

```
USE IEEE.STD_LOGIC_UNSIGNED.ALL;
```

for unsigned number arithmetic. A vector with all bits having the same value can be obtained using the OTHERS keyword, as shown in the next example.

EXAMPLE

```
LIBRARY IEEE;
USE IEEE.STD_LOGIC_1164.ALL;

SIGNAL x: STD_LOGIC;
SIGNAL y: STD_LOGIC_VECTOR(7 DOWNTO 0);

x <= 'Z';
y <= "0000001Z";
y <= (OTHERS => '0'); -- same as "00000000"
```

INTEGER

The predefined INTEGER type defines binary number objects for use with arithmetic operators. By default, an INTEGER signal uses 32 bits to represent a signed number. Integers using fewer bits can be declared with the RANGE keyword.

EXAMPLE

```
SIGNAL x: INTEGER;
SIGNAL y: INTEGER RANGE -64 TO 64;
```

BOOLEAN

The predefined BOOLEAN type defines objects having the two values TRUE and FALSE.

EXAMPLE

```
SIGNAL x: BOOLEAN;
```

Enumeration TYPE

An enumeration type allows the user to specify the values that the data object can have.

Syntax:

TYPE identifier IS (value1, value2, …);

EXAMPLE

```
TYPE state_type IS (S1, S2, S3);
SIGNAL state: state_type;
state <= S1;
```

ARRAY

The ARRAY type groups single data objects of the same type together into a one-dimensional or multi-dimensional array.

Syntax:

TYPE identifier IS ARRAY (range) OF type;

EXAMPLE

```
TYPE byte IS ARRAY(7 DOWNTO 0) OF BIT;
TYPE memory_type IS ARRAY(1 TO 128) OF byte;
SIGNAL memory: memory_type;
memory(3) <= "00101101";
```

SUBTYPE

A SUBTYPE is a subset of a type, that is, a type with a range constraint.

Syntax:

SUBTYPE identifier IS type RANGE range;

EXAMPLE

```
SUBTYPE integer4 IS INTEGER RANGE -8 TO 7;

SUBTYPE cell IS STD_LOGIC_VECTOR(3 DOWNTO 0);
TYPE memArray IS ARRAY(0 TO 15) OF cell;
```

Some standard subtypes include:

- NATURAL—an integer in the range 0 to INTEGER'HIGH
- POSITIVE—an integer in the range 1 to INTEGER'HIGH

D.1.6 Data Operators

The VHDL built-in operators are listed in Figure D.2.

Logical Operators	Operation	Example
AND	AND	n <= a AND b
OR	OR	n <= a OR b
NOT	NOT	n <= NOT a
NAND	NAND	n <= a NAND b
NOR	NOR	n <= a NOR b
XOR	XOR	n <= a XOR b
XNOR	XNOR	n <= a XNOR b

FIGURE D.2 VHDL built-in data operators. *(continued on next page)*

Arithmetic Operators	Operation	Example
+	Addition	n <= a + b
−	Subtraction	n <= a − b
*	Multiplication (integer or floating point)	n <= a * b
/	Division (integer or floating point)	n <= a / b
MOD	Modulus (integer)	n <= a MOD b
REM	Remainder (integer)	n <= a REM b
**	Exponentiation	n <= a ** 2
&	Concatenation	n <= 'a' & 'b'
ABS	Absolute	
Relational Operators	**Operation**	**Example**
=	Equal	IF (n = 10) THEN
/=	Not equal	IF (n /= 10) THEN
<	Less than	IF (n < 10) THEN
<=	Less than or equal	IF (n <= 10) THEN
>	Greater than	IF (n > 10) THEN
>=	Greater than or equal	IF (n >= 10) THEN
Shift Operators	**Operation**	**Example**
SLL	Shift left logical	n <= "1001010" SLL 2
SRL	Shift right logical	n <= "1001010" SRL 1
SLA	Shift left arithmetic	n <= "1001010" SLA 2
SRA	Shift right arithmetic	n <= "1001010" SRA 1
ROL	Rotate left	n <= "1001010" ROL 2
ROR	Rotate right	n <= "1001010" ROR 3

FIGURE D.2 VHDL built-in data operators.

D.1.7 ENTITY

An ENTITY declaration declares the external or user interface of the module similar to the declaration of a function. It specifies the name of the entity and its interface. The interface consists of the signals to be passed into the entity or out from it using the two keywords IN and OUT, respectively.

Syntax:

```
ENTITY entity-name IS
   PORT (list-of-port-names-and-types);
END entity-name;
```

EXAMPLE

```
LIBRARY IEEE;
USE IEEE.STD_LOGIC_1164.ALL;

ENTITY Siren IS PORT (
  M:  IN STD_LOGIC;
  D:  IN STD_LOGIC;
  V:  IN STD_LOGIC;
  S:  OUT STD_LOGIC);
END Siren;
```

D.1.8 ARCHITECTURE

The ARCHITECTURE body defines the actual implementation of the functionality of the entity. This is similar to the definition or implementation of a function. The syntax for the architecture varies, depending on the model (dataflow, behavioral, or structural) you use.

Syntax: Dataflow model:

```
ARCHITECTURE architecture-name OF entity-name IS
   signal-declarations;
BEGIN
   concurrent-statements;
END architecture-name;
```

The concurrent statements are executed concurrently.

EXAMPLE

```
ARCHITECTURE Siren_Dataflow OF Siren IS
  SIGNAL term_1: STD_LOGIC;
BEGIN
  term_1 <= D OR V;
  S <= term_1 AND M;
END Siren_Dataflow;
```

Syntax: Behavioral model:

```
ARCHITECTURE architecture-name OF entity-name IS
   signal-declarations;
   function-definitions;
   procedure-definitions;
BEGIN
   PROCESS-blocks;
   concurrent-statements;
END architecture-name;
```

Statements within the PROCESS block are executed sequentially. However, the PROCESS block itself is a concurrent statement.

EXAMPLE

```
ARCHITECTURE Siren_Behavioral OF Siren IS
  SIGNAL term_1: STD_LOGIC;
BEGIN
  PROCESS (D, V, M)
  BEGIN
    term_1 <= D OR V;
    S <= term_1 AND M;
  END PROCESS;
END Siren_Behavioral;
```

Syntax: Structural model:

```
ARCHITECTURE architecture-name OF entity-name IS
   component-declarations;
   signal-declarations;
BEGIN
   instance-name: PORT MAP-statements;
   concurrent-statements;
END architecture-name;
```

Every component declaration used must have a corresponding entity and architecture. The PORT MAP statements are concurrent statements.

EXAMPLE

```
ARCHITECTURE Siren_Structural OF Siren IS
  COMPONENT myOR PORT (
    in1, in2: IN STD_LOGIC;
    out1: OUT STD_LOGIC);
  END COMPONENT;

  SIGNAL term1: STD_LOGIC;

BEGIN
  U0: myOR PORT MAP (D, V, term1);
  S <= term1 AND M;
END Siren_Structural;
```

D.1.9 **GENERIC**

Generics allow information to be passed into an entity so that, for example, the size of a vector in the PORT list does not have to be known until elaboration time. Generics of an entity are declared with the GENERIC keyword before the PORT list declaration for the entity. An identifier that is declared as GENERIC is a constant that can be only read. The identifier then can be used in the entity declaration and its corresponding architectures wherever a constant is expected.

Syntax: In an ENTITY declaration:

```
ENTITY entity-name IS
GENERIC (identifier: type);          -- with no default value
...
```

or

```
ENTITY entity-name IS
GENERIC (identifier: type := constant);   -- with a default value given by
                                          -- the constant
...
```

EXAMPLE

```
ENTITY Adder IS
-- declares the generic identifier n having a default value 4
GENERIC (n: INTEGER := 4);
PORT (
  -- the vector size is 3 downto 0 since n is 4
  A, B:  IN STD_LOGIC_VECTOR(n-1 DOWNTO 0);
  Cout:  OUT STD_LOGIC;
  SUM:  OUT STD_LOGIC_VECTOR(n-1 DOWNTO 0));
  S:  OUT STD_LOGIC);
END Adder;
```

The value for a generic constant also can be specified in a component declaration or a component instantiation statement.

Syntax: In a component declaration:

```
COMPONENT component-name
   GENERIC (identifier: type := constant);   -- with an optional value given
                                             -- by the constant
   PORT (list-of-port-names-and-types);
END COMPONENT;
```

Syntax: In a component instantiation:

```
label: component-name GENERIC MAP (constant) PORT MAP (association-list);
```

EXAMPLE

```
ARCHITECTURE ...

  COMPONENT mux2 IS
    -- declares the generic identifier n having a default value 4
```

```
    GENERIC (n: INTEGER := 4);
    PORT (
      S: IN STD_LOGIC;              -- select line
      D1, D0: IN STD_LOGIC_VECTOR(n-1 DOWNTO 0);-- data bus input
      Y: OUT STD_LOGIC_VECTOR(n-1 DOWNTO 0));  -- data bus output
  END COMPONENT;

  ...

BEGIN

  U0: mux2 GENERIC MAP (8) PORT MAP (mux_select, A, B, mux_out);
  -- change vector to size 8

  ...
```

D.1.10 **PACKAGE**

A PACKAGE provides a mechanism to group declarations together and share them between several entity units. A package itself includes a declaration and, optionally, a body. The PACKAGE declaration and body usually are stored together in a separate file from the rest of the design units. The file name given for this file must be the same as the package name. In order for the complete design to synthesize correctly using Quartus II, you first must synthesize the package as a separate unit. After that, you can synthesize the unit that uses that package.

PACKAGE Declaration and Body

The PACKAGE declaration contains declarations that may be shared between different entity units. It provides the interface, that is, items that are visible to the other entity units. The optional PACKAGE BODY contains the implementations of the functions and procedures that are declared in the PACKAGE declaration.

Syntax: PACKAGE declaration

```
PACKAGE package-name IS
    type-declarations;
    subtype-declarations;
    signal-declarations;
    variable-declarations;
    constant-declarations;
    component-declarations;
    function-declarations;
    procedure-declarations;
END package-name;
```

Syntax: PACKAGE BODY declaration

```
PACKAGE BODY package-name IS
    function-definitions;       -- for functions declared in the package
                                -- declaration
```

```
        procedure-definitions;    -- for procedures declared in the package
                                  -- declaration
    END package-name;
```

EXAMPLE

```
LIBRARY IEEE;
USE IEEE.STD_LOGIC_1164.ALL;

PACKAGE my_package IS
  SUBTYPE bit4 IS STD_LOGIC_VECTOR(3 DOWNTO 0);
  -- declare a function
  FUNCTION Shiftright (input: IN bit4) RETURN bit4;
  SIGNAL mysignal: bit4;  -- a global signal
END my_package;

PACKAGE BODY my_package IS
  -- implementation of the Shiftright function
  FUNCTION Shiftright (input: IN bit4) RETURN bit4 IS
  BEGIN
    RETURN '0' & input(3 DOWNTO 1);
  END shiftright;
END my_package;
```

Using a PACKAGE

To use a package, you simply include a LIBRARY and USE statement for that package. Before synthesizing the module that uses the package, you need to first synthesize the package by itself as a top-level entity.

Syntax:

```
LIBRARY WORK;
USE WORK.package-name.ALL;
```

EXAMPLE

```
LIBRARY WORK;
USE WORK.my_package.ALL;

ENTITY test_package IS PORT (
  x: IN bit4;
  z: OUT bit4);
END test_package;

ARCHITECTURE Behavioral OF test_package IS
BEGIN
  mysignal <= x;
  z <= Shiftright(mysignal);
END Behavioral;
```

D.2 Behavioral Model—Sequential Statements

The behavioral model allows statements to be executed sequentially as in a regular computer program. Sequential statements include many of the standard constructs, such as variable assignments, if-then-else statements, and loops.

D.2.1 PROCESS

The PROCESS block contains statements that are executed sequentially. However, the PROCESS statement itself is a concurrent statement and so multiple PROCESS blocks in an architecture will be executed simultaneously. These process blocks can be combined together with other concurrent statements.

Syntax:

```
process-name: PROCESS (sensitivity-list)
   variable-declarations;
BEGIN
   sequential-statements;
END PROCESS process-name;
```

The sensitivity list is a comma-separated list of signals, to which the process is sensitive. In other words, whenever a signal in the list changes value, the process will be executed (i.e., all of the statements in the sequential order listed). After the last statement has been executed, the process will be suspended until the next time that a signal in the sensitivity list changes value, when it is executed again.

EXAMPLE

```
Siren: PROCESS (D, V, M)
BEGIN
  term_1 <= D OR V;
  S <= term_1 AND M;
END PROCESS;
```

D.2.2 Sequential Signal Assignment

The sequential signal assignment statement assigns a value to a signal. This statement is just like its concurrent counterpart, except that it is executed sequentially (i.e., only when execution reaches it).

Syntax:

```
signal <= expression;
```

EXAMPLE

```
y <= '1';
z <= y AND (NOT x);
```

D.2.3 **Variable Assignment**

The variable assignment statement assigns a value or the result of evaluating an expression to a variable. The value always is assigned to the variable as soon as this statement is executed.

Variables are declared only within a PROCESS block.

Syntax:

signal := expression;

EXAMPLE

```
y := '1';
yn := NOT y;
```

D.2.4 **WAIT**

When a process has a sensitivity list, the process always suspends after executing the last statement. An alternative to using a sensitivity list to suspend a process is to use a WAIT statement, which also must be the first statement in a process.

Syntax:

WAIT UNTIL condition;

EXAMPLE

```
-- suspend until a rising clock edge
WAIT UNTIL clock'EVENT AND clock = '1';
```

D.2.5 **IF-THEN-ELSE**

Syntax:

```
IF condition THEN
   sequential-statements1;
ELSE
   sequential-statements2;
END IF;
```

or

```
IF condition1 THEN
   sequential-statements1;
ELSIF condition2 THEN
   sequential-statements2;
...
ELSE
   sequential-statements3;
END IF;
```

EXAMPLE

```
IF count /= 10 THEN  -- not equal
  count := count + 1;
ELSE
  count := 0;
END IF;
```

D.2.6 **CASE**

Syntax:

```
CASE expression IS
WHEN choices => sequential-statements;
WHEN choices => sequential-statements;
...
WHEN OTHERS => sequential-statements;
END CASE;
```

EXAMPLE

```
CASE sel IS
WHEN "00" => z <= in0;
WHEN "01" => z <= in1;
WHEN "10" => z <= in2;
WHEN OTHERS => z <= in3;
END CASE;
```

D.2.7 **NULL**

The NULL statement does not perform any actions.

Syntax:

```
NULL;
```

D.2.8 **FOR**

Syntax:

```
FOR identifier IN start [TO | DOWNTO] stop LOOP
   sequential-statements;
END LOOP;
```

Loop statements must have locally static bounds. The identifier is implicitly declared, so no explicit declaration of the variable is needed.

EXAMPLE

```
sum := 0;
FOR count IN 1 TO 10 LOOP
  sum := sum + count;
END LOOP;
```

D.2.9 **WHILE**

Syntax:

```
WHILE condition LOOP
   sequential-statements;
END LOOP;
```

D.2.10 **LOOP**

Syntax:

```
LOOP
   sequential-statements;
   EXIT WHEN condition;
END LOOP;
```

D.2.11 **EXIT**

The EXIT statement can be used only inside a loop. It causes execution to jump out of the innermost loop and usually is used in conjunction with the LOOP statement.

Syntax:

```
EXIT WHEN condition;
```

D.2.12 **NEXT**

The NEXT statement can be used only inside a loop. It causes execution to skip to the end of the current iteration and continue with the beginning of the next iteration. It usually is used in conjunction with the FOR statement.

Syntax:

NEXT WHEN condition;

EXAMPLE

```
sum := 0;
FOR count IN 1 TO 10 LOOP
  NEXT WHEN count = 3;
  sum := sum + count;
END LOOP;
```

D.2.13 **FUNCTION**

Syntax: Function declaration:

FUNCTION function-name (parameter-list) RETURN return-type;

Syntax: Function definition:

FUNCTION function-name (parameter-list) RETURN return-type IS
BEGIN
  sequential-statements;
END function-name;

Syntax: Function call:

function-name (actuals);

Parameters in the parameter list can be either signals or variables of mode IN only.

EXAMPLE

```
LIBRARY IEEE;
USE IEEE.STD_LOGIC_1164.ALL;

ENTITY test_function IS PORT (
  x: IN STD_LOGIC_VECTOR(3 DOWNTO 0);
  z: OUT STD_LOGIC_VECTOR(3 DOWNTO 0));
END test_function;

ARCHITECTURE Behavioral OF test_function IS

  SUBTYPE bit4 IS STD_LOGIC_VECTOR(3 DOWNTO 0);

  FUNCTION Shiftright (input: IN bit4) RETURN bit4 IS
  BEGIN
    RETURN '0' & input(3 DOWNTO 1);
  END shiftright;
```

```
  SIGNAL mysignal: bit4;

BEGIN
  PROCESS
  BEGIN
    mysignal <= x;
    z <= Shiftright(mysignal);
  END PROCESS;
END Behavioral;
```

D.2.14 PROCEDURE

Syntax: Procedure declaration:

PROCEDURE procedure-name (parameter-list);

Syntax: Procedure definition:

PROCEDURE procedure-name (parameter-list) IS
BEGIN
  sequential-statements;
END procedure-name;

Syntax: Procedure call:

procedure-name (actuals);

Parameters in the parameter-list are variables of modes IN, OUT, or INOUT.

EXAMPLE

```
LIBRARY IEEE;
USE IEEE.STD_LOGIC_1164.ALL;

ENTITY test_procedure IS PORT (
  x: IN STD_LOGIC_VECTOR(3 DOWNTO 0);
  z: OUT STD_LOGIC_VECTOR(3 DOWNTO 0));
END test_procedure;

ARCHITECTURE Behavioral OF test_procedure IS

  SUBTYPE bit4 IS STD_LOGIC_VECTOR(3 DOWNTO 0);

  PROCEDURE Shiftright (input: IN bit4; output: OUT bit4) IS
  BEGIN
    output := '0' & input(3 DOWNTO 1);
  END shiftright;

BEGIN
  PROCESS
```

```
    VARIABLE mysignal: bit4;
  BEGIN
    Shiftright(x, mysignal);
    z <= mysignal;
  END PROCESS;
END Behavioral;
```

D.2.15 Behavioral Model Example

EXAMPLE: BEHAVIORAL CODE FOR A BCD TO 7-SEGMENT DECODER

```
LIBRARY IEEE;
USE IEEE.STD_LOGIC_1164.ALL;

ENTITY bcd IS PORT (
  I: IN STD_LOGIC_VECTOR(3 DOWNTO 0);
  Segs: OUT STD_LOGIC_VECTOR(1 TO 7));
END bcd;

ARCHITECTURE Behavioral OF bcd IS
BEGIN
  PROCESS(I)
  BEGIN
    CASE I IS
    WHEN "0000" => Segs <= "1111110";
    WHEN "0001" => Segs <= "0110000";
    WHEN "0010" => Segs <= "1101101";
    WHEN "0011" => Segs <= "1111001";
    WHEN "0100" => Segs <= "0110011";
    WHEN "0101" => Segs <= "1011011";
    WHEN "0110" => Segs <= "1011111";
    WHEN "0111" => Segs <= "1110000";
    WHEN "1000" => Segs <= "1111111";
    WHEN "1001" => Segs <= "1110011";
    WHEN OTHERS => Segs <= "0000000";
    END CASE;
  END PROCESS;
END Behavioral;
```

D.3 Dataflow Model—Concurrent Statements

Concurrent statements used in the dataflow model are executed concurrently, so their order does not affect the resulting output.

D.3.1 Concurrent Signal Assignment

The concurrent signal assignment statement assigns a value or the result of evaluating an expression to a signal. This statement is executed whenever a signal in its expression changes value. The actual assignment of the value to the signal, however, takes place after a certain delay and not instantaneously as for variable assignments. The expression can be any logical or arithmetical expressions.

Syntax:

signal <= expression;

EXAMPLE

```
y <= '1';
z <= y AND (NOT x);
```

A vector with all bits having the same value can be obtained using the OTHERS keyword as shown next.

EXAMPLE

```
SIGNAL x: STD_LOGIC_VECTOR(7 DOWNTO 0);
x <= (OTHERS => '0');   -- 8-bit vector of 0, same as "00000000"
```

D.3.2 Conditional Signal Assignment

The conditional signal assignment statement selects one of several different values to assign to a signal based on different conditions. This statement is executed whenever a signal in any one of the values or conditions changes.

Syntax:

```
signal <= value1 WHEN condition ELSE
          value2 WHEN condition ELSE
          ...
          value3;
```

EXAMPLE

```
z <= in0 WHEN sel = "00" ELSE
     in1 WHEN sel = "01" ELSE
     in2 WHEN sel = "10" ELSE
     in3;
```

D.3.3 Selected Signal Assignment

The selected signal assignment statement selects one of several different values to assign to a signal based on the value of a select expression. All possible choices for the expression must be given. The keyword OTHERS can be used to denote all remaining choices. This statement is executed whenever a signal in the expression or any one of the values changes.

Syntax:

```
WITH expression SELECT
   signal <= value1 WHEN choice1,
             value2 WHEN choice2 | choice3,
             ...
             value4 WHEN OTHERS;
```

In the above syntax, if *expression* is equal to *choice1*, then *value1* is assigned to *signal*. Otherwise, if *expression* is equal to *choice2* or *choice3*, then *value2* is assigned to *signal*. If *expression* does not match any of the above choices, then *value4* in the optional WHEN OTHERS clause is assigned to *signal*.

EXAMPLE

```
WITH sel SELECT
  z <=  in0 WHEN "00",
        in1 WHEN "01",
        in2 WHEN "10",
        in3 WHEN OTHERS;
```

D.3.4 Dataflow Model Example

This example describes a full adder (FA) circuit using the dataflow model. The Boolean equations for describing the full adder circuit are:

$$c_{out} = xy + c_{in}(x \oplus y)$$
$$sum = x \oplus y \oplus c_{in}$$

The following example translates the above two equations into the corresponding two concurrent signal assignment statements.

EXAMPLE: DATAFLOW CODE FOR A FULL ADDER

```
LIBRARY IEEE;
USE IEEE.STD_LOGIC_1164.ALL;

ENTITY fa IS PORT (
  x, y, cin: IN STD_LOGIC;
  cout, sum: OUT STD_LOGIC);
END fa;

ARCHITECTURE Dataflow OF fa IS
BEGIN
  cout <= (x AND y) OR (cin AND (x XOR y));
  sum <= x XOR y XOR cin;
END Dataflow;
```

D.4 Structural Model—Concurrent Statements

The structural model allows the manual connection of several components together using signals. All components used must first be defined with their respective ENTITY and ARCHITECTURE sections, which can be in the same file or can be in separate files.

In the topmost module, each component used in the netlist is declared first using the COMPONENT statement. The declared components then are instantiated with the actual components in the circuit using the PORT MAP statement. SIGNALS are then used to connect the components together based on the schematic diagram.

D.4.1 COMPONENT Declaration

The COMPONENT declaration statement declares the name and the interface of a component that is used in the circuit description. For each COMPONENT declaration used, there must be a corresponding ENTITY and ARCHITECTURE. The declaration name and the interface in the COMPONENT must match exactly the name and interface that is specified in its ENTITY section.

Syntax:

```
COMPONENT component-name IS
   PORT (list-of-port-names-and-types);
END COMPONENT;
```

or

```
COMPONENT component-name IS
   GENERIC (identifier: type := constant);
   PORT (list-of-port-names-and-types);
END COMPONENT;
```

The keyword IS is optional.

EXAMPLE

```
COMPONENT fa IS PORT (
  xi, yi, cin: IN STD_LOGIC;
  cout, si: OUT STD_LOGIC);
END COMPONENT;
```

EXAMPLE: USING GENERIC

```
COMPONENT TriState_Buffer IS
  GENERIC (n: INTEGER := 4);
  PORT (
    E: IN STD_LOGIC;
    D: IN STD_LOGIC_VECTOR(n-1 DOWNTO 0);
    Y: OUT STD_LOGIC_VECTOR(n-1 DOWNTO 0));
END COMPONENT;
```

D.4.2 PORT MAP

The PORT MAP statement instantiates a declared component with an actual component in the circuit by specifying how the connections to this instance of the component are to be made.

Syntax:

label: component-name PORT MAP (association-list);

or

label: component-name GENERIC MAP (constant) PORT MAP (association-list);

The association list can be specified using either the positional or named method. The constant number used here in the GENERIC MAP will override the number specified in the GENERIC clause in the COMPONENT declaration.

EXAMPLE: POSITIONAL ASSOCIATION

```
SIGNAL x0, x1, y0, y1, c0, c1, c2, s0, s1: STD_LOGIC;
U1: fa PORT MAP (x0, y0, c0, c1, s0);
```

EXAMPLE: NAMED ASSOCIATION

```
SIGNAL x0, x1, y0, y1, c0, c1, c2, s0, s1: STD_LOGIC;
U1: fa PORT MAP (cout=>c1, si=>s0, cin=>c0, xi=>x0, yi=>y0);
```

EXAMPLE: USING GENERIC MAP

```
U1: TriState_Buffer GENERIC MAP(8) PORT MAP(E=>OE, D=>dp_sum,
                                            Y=>Output);
```

D.4.3 OPEN

The OPEN keyword is used in the PORT MAP association list to signify that a particular output port is not connected or used. It cannot be used for an input port.

EXAMPLE

```
U1: fa PORT MAP (x0, y0, c0, OPEN, s0);
```

D.4.4 GENERATE

The GENERATE statement works like a macro expansion. It provides a simple way to duplicate similar components.

Syntax:

```
label: FOR identifier IN start [TO | DOWNTO] stop GENERATE
    port-map-statements;
END GENERATE label;
```

EXAMPLE

```
-- using a FOR-GENERATE statement to generate four instances of the
-- full adder component for a 4-bit adder
LIBRARY IEEE;
USE IEEE.STD_LOGIC_1164.ALL;

ENTITY Adder4 IS PORT (
  Cin: IN STD_LOGIC;
  A, B: IN STD_LOGIC_VECTOR(3 DOWNTO 0);
  Cout: OUT STD_LOGIC;
  SUM: OUT STD_LOGIC_VECTOR(3 DOWNTO 0));
END Adder4;

ARCHITECTURE Structural OF Adder4 IS
  COMPONENT fa IS PORT (
    ci, xi, yi: IN STD_LOGIC;
    co, si: OUT STD_LOGIC);
  END COMPONENT;

  SIGNAL Carryv: STD_LOGIC_VECTOR(4 DOWNTO 0);

BEGIN
  Carryv(0) <= Cin;

  Adder: FOR k IN 3 DOWNTO 0 GENERATE
    FullAdder: fa PORT MAP (Carryv(k), A(k), B(k), Carryv(k+1),
                            SUM(k));
  END GENERATE Adder;

  Cout <= Carryv(4);
END Structural;
```

D.4.5 Structural Model Example

This structural model example is for the 4-bit ripple-carry adder based on the following circuit. It requires the full adder (fa) module that is defined in Section D.3.4.

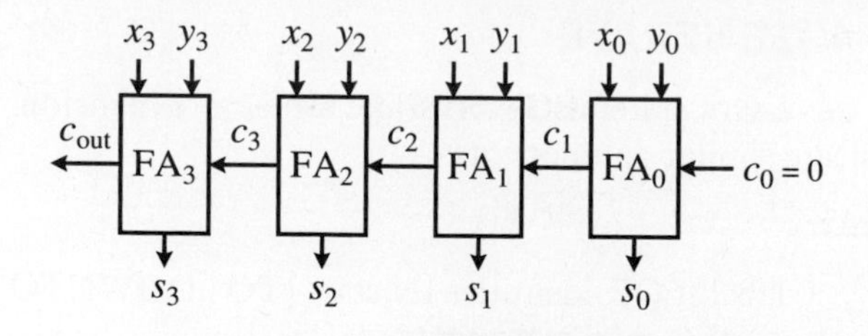

EXAMPLE: STRUCTURAL CODE FOR A 4-BIT ADDER

```
LIBRARY IEEE;
USE IEEE.STD_LOGIC_1164.ALL;

ENTITY adder IS
GENERIC (n: INTEGER := 4);
PORT(
  X, Y: IN STD_LOGIC_VECTOR(n-1 DOWNTO 0);
  S: OUT STD_LOGIC_VECTOR(n-1 DOWNTO 0);
  Cout: OUT STD_LOGIC);
END adder;

ARCHITECTURE Structural OF adder IS
  COMPONENT fa IS PORT (
    x, y, cin: IN STD_LOGIC;
    cout, sum: OUT STD_LOGIC);
  END COMPONENT;

  SIGNAL c: STD_LOGIC_VECTOR(n-1 DOWNTO 0);

BEGIN
    U0: fa PORT MAP (x=>X(0), y=>Y(0), cin=>c(0), cout=>c(1),
                     sum=>S(0));
    U1: fa PORT MAP (x=>X(1), y=>Y(1), cin=>c(1), cout=>c(2),
                     sum=>S(1));
    U2: fa PORT MAP (x=>X(2), y=>Y(2), cin=>c(2), cout=>c(3),
                     sum=>S(2));
    U3: fa PORT MAP (x=>X(3), y=>Y(3), cin=>c(3), cout=>Cout,
                     sum=>S(3));
END Structural;
```

D.5 Conversion Routines

D.5.1 CONV_INTEGER( )

The CONV_INTEGER( ) routine converts a STD_LOGIC_VECTOR type to an INTEGER type. Its use requires the inclusion of the following library.

```
LIBRARY IEEE;
USE IEEE.STD_LOGIC_UNSIGNED.ALL;
```

Syntax:

CONV_INTEGER(std_logic_vector)

EXAMPLE

```
LIBRARY IEEE;
USE IEEE.STD_LOGIC_UNSIGNED.ALL;

SIGNAL four_bit: STD_LOGIC_VECTOR(3 DOWNTO 0);
SIGNAL n: INTEGER;

n := CONV_INTEGER(four_bit);
```

D.5.2 CONV_STD_LOGIC_VECTOR(,)

The CONV_STD_LOGIC_VECTOR(,) routine converts an INTEGER type to a STD_LOGIC_VECTOR type. Its use requires the inclusion of the following library.

LIBRARY IEEE;

USE IEEE.STD_LOGIC_ARITH.ALL;

Syntax:

CONV_STD_LOGIC_VECTOR (integer, number_of_bits)

EXAMPLE

```
LIBRARY IEEE;
USE IEEE.STD_LOGIC_ARITH.ALL;

SIGNAL four_bit: STD_LOGIC_VECTOR(3 DOWNTO 0);
SIGNAL n: INTEGER;

four_bit := CONV_STD_LOGIC_VECTOR(n, 4);
```

INDEX

D

E